Symbolic Dynamics
and its Applications

Recent Titles in This Series

(Continued in the back of this publication)

CONTEMPORARY
MATHEMATICS

135

Symbolic Dynamics
and its Applications

Peter Walters
Editor

American Mathematical Society
Providence, Rhode Island

EDITORIAL BOARD

The AMS Conference on Symbolic Dynamics and its Applications, held in honor of Roy Adler, was held at Yale University, New Haven, Connecticut, July 28–August 2, 1991.

1991 *Mathematics Subject Classification.* Primary 28D99, 54H20; Secondary 60J10, 58F99.

Library of Congress Cataloging-in-Publication Data

Symbolic dynamics and its applications: proceedings of a conference in honor of Roy L. Adler held July 28 to August 2, 1991/Peter Walters, editor.
    p. cm.—(Contemporary mathematics; v. 135)
    Proceedings of a conference held at Yale University.
    ISBN 0-8218-5146-2
    1. Topological dynamics—Congresses. I. Adler, Roy L., 1931–II. Walters, Peter, 1943–. III. Series: Contemporary mathematics (American Mathematical Society); v. 135.
QA611.5.S96 1992   92-20203
514′74—dc20   CIP

Contents

This volume is dedicated to
Roy L. Adler
on the occasion of his 60th birthday
with the affection and respect of his colleagues and friends.

Roy and Audrey Adler

Introduction

This volume contains the proceedings of the conference *Symbolic Dynamics and Its Applications* held at Yale University in New Haven, Connecticut, from July 28 to August 2, 1991. The conference was held in honor of Roy L. Adler. It was organized by Ethan Coven, Shizuo Kakutani (Roy's advisor), Bruce Kitchens, Brian Marcus, Don Ornstein, and Benjamin Weiss.

The conference focused on symbolic dynamics and its applications to other fields. These include ergodic theory, smooth dynamical systems, information theory, automata theory, and statistical mechanics. There were eighteen one-hour talks, including two on Roy Adler's work, and eighteen thirty-minute talks. There were one hundred and thirty-nine participants from thirteen countries, representing mathematics, applied mathematics, electrical engineering, and physics departments at universities as well as industry. The conference was supported by grants from the NSF and IBM.

Roy is honored for both his mathematical and personal contributions to the scientific community. Among his scientific achievements are the invention of topological entropy, the use of Markov partitions for invertible maps, and the application of symbolic dynamics to practical coding problems in information theory. His ideas were crucial in the formation of the subject we call symbolic dynamics. The simple but elegant way he approached difficult problems led many of us into dynamics.

Roy had a wonderful partner. He was married to Audrey Wanner Adler from March, 1953 until she died in June, 1990. Together they were a central part of the mathematical community. They made all of us feel at home. As hosts, they cooked hotdogs in Stanford, hamburgers in Warwick, turkeys in Chapel Hill, and lobsters at home. In each location, they provided the central place to gather: a place to sit, talk, and eat. They helped many of us appreciate beautiful mathematics, good food, and really good friends. Thanks.

Program
One Hour Talks

1. Jonathan Ashley, *Uniqueness of a certain decomposition of conjugacies of shifts of finite type*
2. Mike Boyle, *Quotients of subshifts*
3. Elise Cawley, *Gibbs measures, symbolic dynamics, and toral diffeomorphisms*
4. Leo Flatto, *On a theorem of Mahler, or much ado about nothing*
5. David Handelman, *Shift equivalent matrices that aren't shift equivalent*
6. Mike Keane, *Random walk isomorphisms*
7. Linda Keen, *Pleating coordinates for the Riley slice of Schottky space*
8. Rick Kenyon, *Automatic number systems*
9. Wolfgang Krieger, *Towards a theory of algebraic subshifts*
10. Doug Lind, *The entropies of renewal systems*
11. Brian Marcus, *The Roy Adler story, Part 2*
12. Klaus Schmidt, *Relative entropies for higher dimensional Markov shifts*
13. Caroline Series, *From coding curves to deforming Kleinian groups: An example*
14. Paul Trow, *Degrees of factor maps between sofic systems*
15. Selim Tuncel, *Finite equivalence and positivity of polynomials*
16. Jack Wagoner, *Methods of algebraic topology in symbolic dynamics*
17. Benjamin Weiss, *On Roy Adler's work in ergodic theory*
18. Susan Williams, *Core dimension group constraints for factors of sofic shifts*

Thirty Minute Talks

1. Alexander Blokh, *Abstract sets of periodic points for graph maps*
2. Annie Broglio, *Automaton and prediction*
3. Valerio DeAngelis, *Polynomial beta functions*
4. Doris Fiebig, *Common closing extensions for synchronized systems*
5. Ulf Fiebig, *Covers for coded systems*
6. Roza Galeeva, *Growth rates for the number of periodic points in countable Markov chains*
7. Charlie Jacobson, *Totally amalgamatible one sided sofic systems and the information rate function*
8. Yuval Peres, *Slicing sofic sets for toral automorphisms*
9. Raj Prasad, *Mixing homeomorphisms of prescribed mean rotation vectors*

10. Jim Propp, *The fundamental group of a $\mathbb{Z}^2$ shift*
11. Charles Radin, *2-dimensional uniquely ergodic symbolic dynamics of finite type*
12. Mitsuhiro Shishikura, *The boundary of the Mandelbrot set*
13. Meir Smorodinsky, *Finitary isomorphism of 1-dependent processes*
14. Boris Somolyak, *β expansions and spectral theory of substitutions and adic transformations*
15. Jeff Steif, *An exact critical value for a cellular automaton*
16. Serge Troubetzkoy, *Recurrence properties of Lorentz lattice gas cellular automata*
17. Tom Ward, *Distribution of periodic points for $\mathbb{N}^2$ actions on compact groups*
18. Zhou Zuo-Ling, *The topological Markov chain and the nonnegative matrix*

PARTICIPANTS

Roy Adler, IBM Watson Research Center
Jawad Yusuf Al-Khal, University of Maryland
Paul Algoet, Stanford University
Venkat Anantharam, Cornell University
Jonathan Ashley, IBM Almaden Research Center
Joseph Auslander, University of Maryland
Kirby Baker, University of California, Los Angeles
Marie-Pierre Beal, Université Paris 7
Alexandra Bellow, Northwestern University
Ken Berg, University of Maryland
Daniel Berend, Ben-Gurion University
Vitaly Bergelson, Ohio State University
Chris Bernhardt, Fairfield University
S. I. Bezuglyi, Universitat Heidelberg
Lou Block, University of Florida
Alexander Blokh, Wesleyan University
Mike Boyle, University of Maryland
Rob Bradley, Northwestern University
Jonathan Brezin, IBM Watson Research Center
Annie Broglio, Université de Provence
Karen Brucks, University of Wisconsin-Milwaukee
Bob Burton, Oregon State University
Catherine Carroll, Northwestern University
Elise Cawley, M.S.R.I
Hsin-Ta Frank Cheng, City University of New York
J. Choksi, McGill University
Ethan Coven, Wesleyan University
Larry J. Cummings, University of Waterloo
Karma Dajani, University of South Alabama
Andres del Junco, University of Toronto
Valerio DeAngelis, University of Washington
David Delchamps, Cornell University
Beverly Diamond, College of Charleston
Stanley Eigen, Northeastern University
Jacob Feldman, University of California
Sebastien Ferenczi, CNRS - URA 225

Doris Fiebig, Universitat Heidelberg
Ulf Fiebig, Universitat Heidelberg
Adam Fieldsteel, Wesleyan University
Albert Fisher, Yale University
Leo Flatto, Bell Laboratories, Inc.
Christiane Frougny, LITP, Institut Blaise Pascal
Roza Galeeva, LCTA JINR
Will Geller, University of Maryland
Bob Gilman, Stevens Institute of Technology
Richard Goldberg, IBM Watson Research Center
Geoff Goodson, Towson State University
Matthew Grayson, IBM Watson Research Center
Piotr Grzegorczyk, Stanford University
Kamel Haddad, University of Maryland
Arshag B. Hajian, Northeastern University
Michael Handel, CUNY, Graduate Center
David Handelman, University of Ottawa
Georges Hansel, Université de Rouen
Jane Hawkins, University of North Carolina
Nicolai Haydn, University of Southern California
Chris Heegard, Cornell University
G. A. Hedlund, Yale University
Melissa Hidalgo, University of Hartford
Alan Hoffman, IBM Watson Research Center
Jun Hu, City University of New York
Dan-run Huang, University of Maryland
Kristin Hubner, Wesleyan University
Charlie Jacobson, Elmira College
Aimee Johnson, Tufts University
Natasha Jonoska, SUNY at Binghamton
Shizuo Kakutani, Yale University
Steve Kalikow, Ithaca, NY
Anatole Katok, Pennsylvania State University
Svetlana Katok, Pennsylvania State University
Mike Keane, Delft University of Technology
Linda Keen, Lehman College, CUNY
Rick Kenyon, Institut des Hautes Etudes Scientifiques
Jun Kigami, Osaka University
Bruce Kitchens, IBM Watson Research Center
Wolfgang Krieger, Universitat Heidelberg
Cor Kraaikamp, University of Washington
Takashi Kumagai, Osaka University
Jeff Lagarias, AT&T Bell Labs
Michel L. Lapidus, University of California
Yuri Latushkin, San Francisco, CA
François Ledrappier, Université Paris 6
Shihai Li, University of Florida
Pierre Liardet, Université de Provence

Douglas Lind, University of Washington
Zhou Zuo-Ling, Lingnan College
Brian Marcus, IBM Almaden Research Center
Christian Mauduit, Université Claude Bernard - Lyon 1
Dan Mauldin, University of North Texas
John Milnor, SUNY at Stony Brook
Donna Molinek, University of North Carolina
Brigitte Mosse, Université de Provence
Shahar Mozes, M.S.R.I.
Irene Mulvey, Fairfield University
Masakazu Nasu, Mie University
Kyewon Koh Park, Ajou University
Alan Parks, Lawrence University
Yuval Peres, Yale University
Dominique Perrin, Université Paris 7
Karl Petersen, University of North Carolina
Mark Pollicott, Universidade do Porto
Brenda Praggastis, University of Washington
V. S. Prasad, University of Lowell
Jim Propp, Massachusetts Institute of Technology
Charles Radin, University of Texas
Frank Rhodes, University of Southampton
Ted Rivlin, IBM Watson Research Center
Ronny Roth, IBM Almaden Research Center
Dan Rudolph, University of Maryland
Ibrahim Salama, North Carolina Central University
Klaus Schmidt, University of Warwick
Talal Shamoon, Cornell University
Caroline Series, University of Warwick
Paul Shields, University of Toledo
Sandra Shields, College of William and Mary
Mitsuhiro Shishikura, SUNY at Stony Brook
Michael Shub, IBM Watson Research Center
Fred Shultz, Wellesley College
Cesar Silva, Williams College
N. T. Sindhushayana, Cornell University
Meir Smorodinsky, Tel-Aviv University
Boris Solomyak, University of Washington
Matthew Stafford, University of Texas
Jeff Steif, Chalmers University of Technology
Jean-Paul Thouvenot, Université Paris 6
Charles Tresser, IBM Watson Research Center
Serge Troubetzkoy, Universitaet Bielefeld
Paul Trow, Memphis State University
Selim Tuncel, University of Washington
Jack Wagoner, University of California
Peter Walters, University of Warwick
Tom Ward, Ohio State University

Benjamin Weiss, Hebrew University
Robert F. Williams, University of Texas
Susan Williams, University of South Alabama
Carol Wood, Wesleyan University
Michael Yakobson, University of Maryland
Michiko Yuri, University of Maryland
George Zettler, Columbia University

Contemporary Mathematics
Volume **135**, 1992

THE TORUS AND THE DISK

ROY L. ADLER

ABSTRACT. This paper is a survey of a coherent program of mathematics spanning 28 years. It begins with questions concerning classification and structure in ergodic theory and abstract dynamical systems and describes the author's involvement with toral automorphisms, topological entropy, iteration of maps on the interval, symbolic dynamics, and ultimate engineering applications. It serves as case study of how unplanned-for practical applications can result from the pursuit of mathematics for its own sake.

The first item in the title refers to a mathematical abstraction, while the second a successful product of the computer industry.

The torus is a compact group and its automorphisms preserve Haar measure. These are classic examples of *dynamical systems with invariant probability measures*, the objects of study in ergodic theory. The basic abstract object of this subject is designated by (X, α, μ) where X is a Lebesgue space (that is, a space endowed essentially with the measure-theoretic structure of the unit interval), α is a measurable mapping of X onto itself, and μ is a probability measure with the property $\mu(E) = \mu(\phi^{-1}E)$ for any measurable subset E of X. A principal question of this subject is one of isomorphism: when does there exist a measure preserving change of variables? In ergodic theory two dynamical systems (X, α, μ), (Y, β, ν) are said to be *measure-theoretically isomorphic (metrically isomorphic* for short) if there exists a mapping γ of X onto Y such that $\nu(E) = \mu(\gamma^{-1}E)$ and the *conjugacy relation*

$$\gamma\alpha\gamma^{-1} = \beta(a.e.)$$

holds. For invertibility of γ all that is required here is that γ^{-1} exists almost everywhere.

Dynamical systems can be considered from other points of view. For example, in the subject of *topological dynamics* we designate a dynamical system[1] by (X, α) where X is a compact metric space and α a continuous map of X onto itself. Here two dynamical systems (X, α), (Y, β) are said to be *topologically isomorphic.* (or alternately *homeomorphically conjugate*), if γ in the conjugacy relation is a homeomorphism of X onto Y. This is the strongest sense of equivalence from a

This article is reprinted from the *IBM Journal of Reseach and Development*, Vol. 13, No. 2, March 1987; copyright 1987 by IBM; reprinted with permission.

[1] For a comprehensive treatment of such systems see [DGS]

1991 Mathematics Subject Classification. Primary 54H20, 58FXX; Secondary 28DXX.

purely topological point of view. But later we shall elaborate on another slightly weaker one, more in the spirit of measure theory, in the sense that we do not insist that the conjugacy relation hold everywhere. Similarly in the theory of *smooth dynamical systems*, the spaces in question are manifolds, the mappings diffeomorphisms; and we would call the notion of isomorphism *diffeomorphic conjugacy*.

The other object in the title is the magnetic storage disk. More generally, we are referring to any data storage or transmission system. In information theory these are portrayed within the framework of a *channel*, as in Figure 1.

Figure 1 Information channel

The basic question here concerns the construction of finite state automata which encode and decode data in order to pass it through input restricted channels. Later we discuss some typical channel constraints.

It is not difficult to suspect a vague connection between the isomorphism question of dynamical systems and the coding problem of information theory: after all, in both subjects one set is being transformed into another. The discovery that these are really different interpretations of the same problem is a consequence of what turned out to be a coherent program of research spanning 27 years and serves as a good example of how unplanned-for practical applications can result from the pursuit of mathematics for its own sake.

It began for me as a graduate student in the late fifties. The central problem in ergodic theory was one of metric isomorphism between *Bernoulli shifts*. These are dynamical systems representing stochastic processes like coin tossing experiments. The problem was: when could the sequences of independently identically distributed results from one probabilistic experiment be coded into another in an invertible and measurable way so that corresponding events under the coding have equal probability? A major breakthrough occurred when Kolmogorov [Ko] indicated how Shannon's concept of entropy might be utilized as a metric isomorphism invariant, and Sinai [S1] supplied proofs necessary to calculate the entropy of Bernoulli shifts. This established in an effective way that shifts of different entropy are not metrically isomorphic. A decade later Ornstein [O1] was to prove the converse, that Bernoulli shifts with the same entropy are metrically isomorphic. This led to tremendous progress and a profound understanding of the basic structure of stationary stochastic processes.

As a graduate student I had come across a similar problem concerning automorphisms of the torus. Toral automorphisms are given by members of $GL(n, \mathbb{Z})$, i.e. matrices of integers with determinant ± 1. They preserve Haar measure and, therefore, are metrically isomorphic if they are *algebraically conjugate* - i.e., they are conjugate elements in the group $GL(n, \mathbb{Z})$. Naturally one would be tempted to prove the converse. I managed to prove such a converse if metric conjugacy was replaced by diffeomorphic conjugacy [A]. A few years later Richard Palais showed me how to improve this to homeomorphic conjugacy [AP]. In the meantime this was also proved by Arov [Ar]. But the original metric conjugacy conjecture turned out to be false.

In the early sixties, the notion of topological entropy was suggested to me by Kolmogorov's notion of ϵ-entropy [KT] which measures complexity of function spaces. I realized that a dynamical invariant could by defined for continuous maps by formal analogy with the Kolmogorov-Sinai probabilistic entropy for measure preserving transformations (see [AKMc]). This is done by replacing measurable partitions with open covers and the number $\Sigma P(A_i) \log P(A_i)$ (called *entropy of the partition*) with the log of the cardinality of a minimum sub-cover. The *topological entropy* of a continuous map on a compact space can then be defined as the largest possible growth rate of this number as covers are successively refined by action of the map's inverse. There was no more of an idea to it than that.[2] Originally I thought it a mere curiosity. Its main property is that continuous maps which are homeomorphically conjugate have the same topological entropy. But I knew of no maps that could not be distinguished with other invariants more easily; and the converse, as to whether maps with the same topological entropy are homeomorphically conjugate, could be easily shown to be false for any interesting class of continuous maps one would care to mention. Yet there was a striking fact: namely, the Kolmogorov-Sinai entropy (with respect to Haar measure) and the topological entropy are equal for toral automorphisms. The entropy of a toral automorphism is the logarithm of the product of eigenvalues of modulus ≥ 1 of the associated integer matrix.

The significance of this emerged a few years later. In the fall of 1966, Leopold Flatto told Benjamin Weiss and me of a new problem which has since gained enormous notoriety: what is the dynamical behavior of the map $x \to ax(1-x)$ on the unit interval for choices of the parameter a, $1 \leq a \leq 4$? For instance, when is the orbit of the critical point $1/2$ infinite? This has yet to be answered and perhaps is the type of problem that can never be completely settled. We tried our hand on a simpler version - namely, analyze $x \to a - 2a|x - \frac{1}{2}|$, $1 \leq a \leq 4$. Weiss and I noticed that for certain values of the parameter, a, there exists a partition having Markov behavior under the map. This gives rise to a symbolic expansion for the points on the interval which totally describes the dynamical behavior of the map in much the same way as the binary expansion of numbers describes the dynamical behavior of multiplication by two.

We examine this situation in more detail. Let (X, α) where $\alpha : x \to (2x)$ and X is the unit interval with 0 and 1 identified to make α continuous. Let Σ_2 denote the set of all binary expansions of numbers in the unit interval or equivalently the set of all infinite paths (sequences of edges) on the directed graph G illustrated in figure 2.

---

[2]Some years later Bowen [B1] and Dinaburg [Di] independently showed the equivalence of the above definition with one derived more directly from Kolomogorov's ϵ-entropy :namely, the largest growth rate as, $\epsilon \to \infty$, of the number of ϵ-separated orbits of length n (two orbits of length n are ϵ-separated if the distance between some pair of corresponding members is $\geq \epsilon$. Furthermore, this definition brings into clearer focus the fact that topological entropy is a natural generalization of Shannon's noiseless channel capacity.

Let σ denote the shift transformation which shifts the symbol sequences in Σ_2 by one to the left and drops off the initial digit.

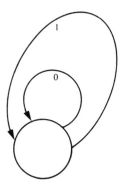

Figure 2 Directed graph for Σ_2

It is easy to define a metric on Σ_2 which makes sequences closer the longer their initial segments agree. This makes Σ_2 a compact metric space, in fact the Cantor discontinuum, and σ a continuous map. We make the elementary observation that the map π of Σ_2 onto X defined by π (binary expansion of x) $= x$ is continuous, onto, and commutes in the sense that $\sigma\pi = \pi\alpha$. Such maps are called, *factor maps* (though "quotient" would be a better term), the system (X, α) a *factor* of (Σ_2, σ), and (Σ_2, σ) an *extension* of (X, α). Furthermore, we call π a *finite* factor map since it is nowhere infinite-to-one; in fact it is at most two-to-one, and we call it *essentially one-to-one* since it is one-to-one except for a certain set of rationals which is "negligible" compared to the totality of all numbers. The existence of an extension by a symbolic system which represents a dynamical system in such a simple fashion arises from certain geometrical properties of the map. For example, consider the partition of X into the intervals $[0, \frac{1}{2}]$ and $[\frac{1}{2}, 1]$ (See Figure 3). If the first interval is labeled 0 and the second 1, then orbits of the system (X, α) have histories through the partition identical with sequences of Σ_2 which are described by the above directed graph.

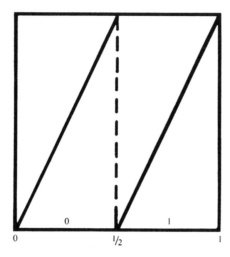

Figure 3 Plot of α

This happens here because the image of each element of the partition is a union of some others. Partitions that behave like this with respect to a map are called *Markov*. Ambiguities occur when the orbit of a point hits a boundary point of one of the intervals in the partition. Such an occurrence is atypical and is a reflection of the same fact that certain rationals have more than one expansion. In order to get a simple description of the set of allowable expansions, one pays the price by having non-uniqueness of symbolic representation. This is a characteristic feature of decimal expansions in arithmetic and symbolic representations of orbits in dynamical systems.

Consider another example (X, α) where β has a plot as in Figure 4.

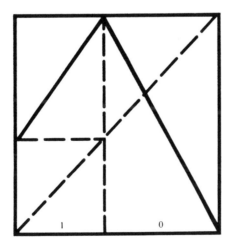

Figure 4 Plot of β

Here the image of the left interval is the right one while the image of the right is the union of both. This Markov partition gives a symbolic extension (Σ_G, σ) where G is the directed graph in Figure 5 and Σ_G is the set of infinite paths (here sequences of nodes) on G. This set can be topologized just as Σ_2 and the defined shift σ is continuous. Furthermore, there is an obvious finite essentially one-to-one factor map π which maps histories to points in the interval.

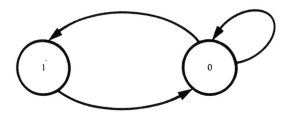

Figure 5 Directed graph G

Unfortunately, in analyzing the dynamical behavior of the maps $x \to ax(1-x)$, or $x \to a - 2a|x - \frac{1}{2}|$ Weiss and I found the above considerations useful only for special values of the parameter a. So we abandoned the problem in favor of trying to understand why the two entropies for toral automorphisms yield identical numbers. To our surprise we discovered (or rather rediscovered what K. Berg [Be] had found shortly before in research for his Ph.D. thesis) that two dimensional hyperbolic toral automorphisms have simple Markov partitions. These give rise to symbolic representations, paths on directed graphs, just like those we had been playing with a short time before.

A brief account of our result is as follows. Let $X = R^2/Z^2$ denote the two-dimensional torus and α an automorphism of X. Here points $(x + m, y + n)$ in the plane for $m, n \in \mathbb{Z}$ are identified; and α is given by

$$\alpha(x, y) = (ax + cy, bx + cy) = (x, y)A$$

where A is a matrix with integer entries and determinant ± 1. Haar measure here is merely the projection of area measure in the plane and area measure is preserved by α because $|\det A| = 1$. The matrix A has two eigenvalues λ and κ with $\lambda\kappa = \pm 1$. This transformation is called hyperbolic if, say, $|\lambda| > 1$ which forces $|\kappa| < 1$. Only the hyperbolic case in dimension 2 is of interest from the dynamical point of view. The geometry of a hyperbolic automorphism is as follows. In the plane there are two distinct directions: one in which distances expand by a factor of $|\lambda|$ under the action of A, and the other in which they contract by $|\kappa|$. Because of this fact one can construct Markov partitions and hence a symbolic extension given by a directed graph. For example, consider the case

$$A = \begin{pmatrix} 2 & 1 \\ 1 & 1 \end{pmatrix}.$$

In the Figure 6 we draw a Markov partition for the automorphism and in Figure 7 the associated directed graph G.

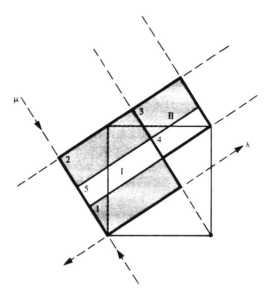

Figure 6 Markov partition

The idea here is the following. Instead of the unit square another fundamental region for the torus is drawn with sides parallel to the expanding and contracting eigenvectors. This region is then partitioned into two parallelograms I and II. Under the action of the automorphism, these get stretched in one direction and shrunk in the other. Weiss and I did a simple minded thing. On one sheet of transparent graph paper we drew the fundamental region and on a second the image of it under the automorphism. We placed one sheet on top of the other and slid them around to see how the two set partition got refined. New lines appeared in the expanding direction; but to our amazement no new ones in the shrinking. This has a simple geometric explanation and profound consequences. As shown in Figure 6, the two basic parallelograms I and II get refined into five smaller ones. The image of rectangle 1 is a collection of rectangles stretching across 1, 2, and 3; similarly for the images of 2 and 5: the image of 3 stretches across 4 and 5; similarly for 4. These facts are summarized by the transitions allowed in the graph depicted in Figure 7.

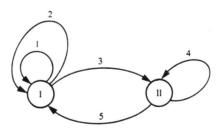

Figure 7 Directed graph G

We now make a slight change in our notion of Σ_G. We assume from now on that it consists of bi-infinite paths in the graph G which makes the shift σ an invertible map on Σ_G. One of the consequences of the fact that there are no new lines in the shrinking direction under repeated applications of the automorphism is the existence of a finite essentially one-to-one factor map π of Σ_G onto X. This map associates a bi-infinite path in G to a unique point of X having that path as a history through the partition under the action of α. We shall not give the proof of this, but suffice it to say that it follows from some elementary plane geometry.

The areas of the parallelograms in the Markov partition are numbers with special meaning. We found that a symbolic system (Σ_G, σ) which is an extension of a toral automorphism satisfies a variational principle: namely, the topological entropy is the same as the maximum probabilistic entropy[3] which in turn is the same as the entropy of the toral automorphism. Also encouraged by Meshalkin's [Me] success in coding between certain Bernoulli shifts, we found that we could construct metric conjugacies (i.e. measure preserving changes of variables) by coding between these symbolic systems representing toral automorphisms whenever they had the same entropy. Here a simplification occurred. Inherent in the power of our method we merely had to construct a measurable change of variables: the measure preserving property was forced to accompany it by virtue of the fact that topological entropy and maximum probabilistic entropy coincide. Answering the question on which I had been stuck as a graduate student, we were able to prove the following.

Theorem. *Two 2-dimensional hyperbolic toral automorphisms are metrically isomorphic if and only if they have the same entropy, i.e., the same corresponding* $|\lambda|$.

This was the first natural class of dynamical systems to be classified by entropy. In the early seventies Ornstein [O2] made a vast generalization.

Our work [AW] combined two new important ideas: finding Markov partitions for smooth dynamical systems and coding between symbolic systems associated with the partitions. Each idea has stimulated mathematical activity.

With respect to the first one, R. Bowen [B2-3] and Ya. Sinai [S3] established the existence of Markov partitions for a more general class of smooth dynamical systems. In particular these results give Markov partitions for hyperbolic toral automorphisms of any dimension. Furthermore, Sinai [S2] made an application of Markov partitions to some basic questions in statistical mechanics. For an entrance into the literature on the use of Markov partitions in smooth systems, one can consult [KSS].

Weiss and I concentrated our further work on the second idea. We observed that the codes which we were constructing between symbolic systems were stronger than metric conjugacies yet weaker than homeomorphic ones. Also we could code between examples of symbolic systems with the same topological entropy. The codes we were constructing were almost but not quite invertible. They failed to be one-to-one on a small set of exceptional symbol sequences.

---

[3] Another rediscovery. The topological entropy of (Σ_G, σ) is the same thing as Shannon's noiseless channel capacity, and the fact that it equals the maximum probabilistic entropy was known to him for Markov measures. Parry [P1] rediscovered this fact and generalized it to arbitrary measures.

This is also the case for metric conjugacies in general, but our set of exceptional points was universally negligible with respect to any regular invariant probability measure rather than a fixed one. Later (see [AM, G]) when their nature was better understood we called these codes "almost homeomorphic conjugacies". It is fashioned from the relationship of a binary expansion and the number it represents to give a relation, between topological systems, which has the appearance of being only slightly weaker than homeomorphic conjugacy. Two topological dynamical systems (X, α) and (Y, β) are said to be *almost homeomorhpically conjugate* if they are factors of a common extension, say (Z, ρ) and the factor maps are finite and essentially one-to-one. Here *essentially one-to-one* means that the factor maps are one-to-one on the doubly transitive points - that is, the points whose future orbits and past orbits are both dense. In systems which satisfy a standard irreducibility condition the nondoubly transitive points comprise a negligible set in the sense of measure and category just like those numbers which have more than one binary expansion. Two basic facts can be proved: *almost homeomorphic conjugacy is an equivalence relation, and topological entropy is an invariant.*

The symbolic systems, with which we were dealing (namely, bi-infinite paths on directed graphs), we called *topological Markov shifts* because they could be specified by non-negative *transition* matrices. The name was chosen because these matrices resemble stochastic ones except the positive transition probabilities have been replaced by non-negative integers. The relevant transition matrix is one with entries 0,1 and is specified as follows: there is a 1 in the i-th row and j-th column if and only if edge i leads next to edge j. We could just as easily label nodes in which case: there is a n in the i-th row and j-th column if and only if there are n paths from node i is to node j. Sometimes it is more convenient to work with nodes, sometimes edges: e.g., if n assumes values other than 0 or 1, then edge labelling avoids an ambiguity which arises using nodes. A zero-one transition matrix specifies a space of admissible sequences of symbols (the labels of nodes or edges), and along with the the shift transformation we get a *symbolic dynamical system*. These systems go under other names: *topological Markov chains* [P1] and *shifts of finite type* [Sm]. We call a directed graph as well as its transition matrix and the dynamical system it specifies *irreducible* if any pair of nodes is connected by a directed path and *aperiodic* if any pair of nodes is connected by a directed path of the same length. The topological entropy of a topological Markov shift is the log of the largest eigenvalue of its transition matrix. This eigenvalue is called the *Perron* value.

We considered first the case where the Perron value was an integer N. A row sum N matrix has Perron value N, but not conversely. We conceived of a proof consisting of two parts. Part 1: to prove that for a system whose matrix has Perron value N there exists a code to a new system where the associated matrix has row sum N. Part 2: to prove there exists a code between a row sum N system and the system given by an $N \times N$ matrix of all ones (such a system is called the *full N-shift* and its space of sequences is denoted by Σ_N). We could prove part 1 by a method which has come to be known as *state splitting*, but the proof of part two was elusive. This problem became known as the "road problem".

Here is the simplest version of the *road problem*. A group of cities is connected by an aperiodic network of one way roads, each city having 2 exit roads (a city

having a road leading to itself is not excluded). The highway department has 2
colors, say red and blue, with which to paint the roads. Each city has one red
exit and one blue. Is it possible to color the roads in such a way that there is a
sequence of colors that leads everyone simultaneously to the same city, say city
1, no matter where he starts?

It is still unsolved; but the most general result to date, I believe, has been
done by O'Brien [Ob]. In 1975 at an ergodic theory symposium at the University
of Warwick, L.W. Goodwyn found a way to bypass the problem by observing
that it was sufficient to solve the road problem for a higher order edge graph
which is easier to do than for the original one. This is described in [AGW].

At that same symposium Furstenburg proved that two irreducible nonnegative
integer matrices A, B (not necessarily of the same dimension) have the same
Perron value if and only if there exists a positive integer matrix F such that
$AF = FB$. Based on this result Brian Marcus and I were able to prove a
topological analogue to Ornstein's isomorphism theorem, to wit:

Theorem. [4] *Two aperiodic topological Markov shifts are almost homeomorphically conjugate if and only if they have the same topological entropy [AM].*

If the transition matrix is irreducible but not aperiodic, another invariant, the
period, must be included with the entropy. Our method is based on a technique
which we call "filling in tableaux" which constructs a new graph from two others
with the same Perron value. In the new graph the outgoing edges at the various
nodes are like those of one of the original graphs, while the incoming edges are
like those of the other.

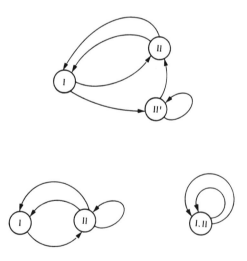

Figure 8 A common extension

This new graph gives a dynamical system which is a common extension of the

---

[4] Likewise Parry [P2] used Furstenburg's result to obtain a version of this theorem which
stops short of getting the one-to-one condition in almost homeomorphic conjugacy. Thus he
obtains a weaker relation, which he calls *finite equivalence*, between two topological Markov
shifts with the same entropy.

original ones and the almost homeomorphism is constructed from factor maps. The factor maps are defined by merging appropriate nodes.

A corollary of this and Bowen's result [B2-3] is the fact that the theorem of Adler and Weiss can be generalized to hyperbolic toral automorphisms of all dimensions:

Theorem. *Two hyperbolic toral automorphism are almost homeomorphically conjugate (hence metrically conjugate) if and only if they have the same entropy.*

An isomorphism theory (at least the type we are discussing) has three elements: an equivalence relation, an invariant, and a special class of systems for which the invariant is a complete one. For ergodic theory they are metric conjugacy, probabilistic entropy, and Bernoulli shifts; while for topological dynamics they are almost homeomorphic conjugacy, topological entropy, and topological Markov shifts. What about an isomorphism theory based on homeomorphic conjugacy? The outstanding problem concerning shifts of finite type is to give an algorithm (or prove its nonexistence) for determining homeomorphic conjugacy. For material concerning this problem see [Ba, KR, PW, Wi]. If two dynamical systems are homeomorphically conjugate, then they have the same topological entropy [AKMc]. However it is not hard to demonstrate that the converse is not true even for shifts of finite type. Williams [Wi] gave an algebraic characterization of topological isomorphism for shifts of finite type. Some algebraic invariants, such as the Jordan form away from 0-eigenvalues, result which are stronger than topological entropy. However, all of the currently known computable invariants are inadequate to completely classify these systems with respect to topological isomorphism. The trouble seems to be that homeomorphic conjugacy is too strong an equivalence relation. It is the weaker one, almost homeomorphic conjugacy, with respect to which an isomorphism theory with a simple description can be established.

The class of dynamical systems to which our isomorphism theorem applies can be enlarged to include sofic systems (the term "sofic", introduced by Weiss [W], is derived from the Hebrew word for finite and is supposed to suggest the finitary character of these systems). It is the set of output sequences from a finite state automata. Shannon [Sh] called them "transducers". A *sofic* system (S, σ) is defined by choosing S to be the space of sequences of symbols gotten from bi-infinite paths on directed graphs just like topological Markov shifts except that perhaps the nodes (edges) are not distinctly labeled. See [CP1-3]. If the nodes (edges) have distinct labels the sofic system is a subshift of finite type: but, if not, then it may or may not be homeomorphically conjugate to a topological Markov shift. Generally it is not. In any case topological entropy is a complete invariant with respect to almost homeomorphic conjugacy for aperiodic members of this larger class of symbolic systems.

Marcus [M1] improved the state splitting method introduced by Weiss and me to show that a topological Markov shift with Perron value N is actually homeomorphically conjugate to the one whose transition matrix has row sum N. The row sum N system is a common extension between the original topological Markov shift and the full N-shift. From this follows a stronger statement than the isomorphism theorem for the special case of rational integer Perron values: namely, a topological Markov shift with Perron value N is almost homeomorphically conjugate to a full N-shift via a factor map. His method has practical implications which we mention later.

The practical applications of the isomorphism theory in topological dynamics was first recognized by Martin Hassner [Ha]. While doing research for his Ph.D. in electrical engineering, he was struck by the fact that our notion of almost topological conjugacy is a coding given by a finite algorithm so that an engineering application ought to be possible. (The metric conjugacies for Bernoulli shifts given by Ornstein are not anything like finite algorithms.) He also was aware that information channels such as ones describing magnetic data storage devices were modeled precisely by topological Markov shifts and sofic systems. As a result of his insight, the attention of mathematicians in ergodic theory and dynamical systems was directed to an area of engineering to which they had previously been oblivious.

Digital information usually takes the form of long binary sequences which we can assume to be arbitrary. When data are to be transmitted or stored, a system doing so may force constraints on the binary sequences. For example, in storing binary data on magnetic surfaces (tapes, disks, drums, etc.) the symbol 1 is ascribed to a transition in the magnetic state of the surface and a 0 to a nontransition. During the read-back process, a transition between magnetic states will cause a voltage pulse while a non-transition will result in an absence of signal in the read head. In such recording systems, the separation of transitions is measured in terms of some basic bit duration unit and constraints are introduced because of the following reasons.

(1) If the transitions are too close, interference between adjacent voltage pulses in the circuits attenuates signals and shifts peaks - trouble for a peak detection scheme. This places a lower limit on the minimum separation between successive transitions.

(2) If the adjacent transitions are too far apart, the absence of a signal may cause a data based clocking scheme, used to correct for drift, to lose synchronization, giving a false measurement of the number of bit duration units. This places an upper limit on the maximum separation between transitions.

This translates into the so-called (d, k) constraints for binary sequences where d is the minimum number of zeros between ones and k the maximum. Consequently, a device is needed to code between arbitrary binary sequences and constrained ones. In order for such a device to be practical it can only process small amounts of data at a time. From its point of view the data that it is processing, though finite, may just as well be infinite both with respect to the past as well as the future. Therefore, symbolic dynamical systems provide a perfect model for such a situation. Arbitrary data are modeled by the full 2-shift (or the full 2^k-shift if one wants to consider blocks of data) and (d, k) constrained data by a topological Markov shift. Furthermore, certain constraints are modelled by sofic systems. For example, in magnetic recording problems may develop due to a d.c. component in the electrical signal especially if the head is coupled by an induction coil to the rest of the system. Besides heating the coil, errors could be introduced in the read back process if the spectrum of the binary sequence contained any power at the frequency zero. Therefore, we may be forced to place spectral constraints on our sequences; and such constraints usually lead to sofic systems [Pe].

The engineer is now faced with two problems.

(1) He must code arbitrary data into constrained, and thereby lose some rate. What coding rates are possible?

(2) Given a possible coding rate, how does one construct practical finite state automata to do the coding and encoding?

The answer to the first question is provided by the notion of topological entropy which is the same as Shannon's noiseless channel capacity.[5] This determines the possible coding rates. The tableau and state-splitting methods construct common extensions and factor maps. This gives a solution to the second problem in the case that topological entropy of input data matches capacity of the channel. To explain this we must discuss the concept of factor map[6] in more detail.

Let L_A and L_B denote an ordered set of symbols for two topological Markov shifts (Σ_A, σ) and (Σ_B, σ), respectively. Also let $\pi : \Sigma_A \to \Sigma_B$ be a factor map. It follows from continuity and the shift commuting property that π is a *k-block* map for some integer k. This means that there is a fixed function $\pi : L_A \times \cdots \times L_A \to L_B$ of k variables (using "π" again by a slight abuse of notation) such that $y_i = \pi(x_{i-j}, x_{i-j+1}, \ldots, x_{i-j+k-1})$ for some fixed $j \in \mathbb{Z}$. (By a suitable change of the symbol set L_A and the transition matrix A we can always arrange $j = 0, k = 1$ so as to obtain a 1-block map). If the mapping π is also invertible, then π^{-1} is also a k-block map, for perhaps a different k. Finite factor maps[7] are not invertible. They do, however, possess a weak type of invertibility. Assuming, without loss of generality, that $\pi(x) = y$ is a 1-block factor map, then it is finite if and only if $(x_{i-p}, \ldots, x_{i+q})$ is uniquely determined from $(y_{i-p}, \ldots, y_{i+q})$, x_{i-p} and x_{i+q} for all i, p, q. Another way of saying this is: If $\pi(u) = \pi(v)$ and the distance between $\sigma^i u$ and $\sigma^i v$ goes to 0 as $i \to \pm\infty$, then $u = v$ ("there are no diamonds in pre-images"). A stronger kind of this weak invertibility is the following. A factor map π is said to be *right closing*[8] if there are fixed integers $p, q \geq 0, r \geq 1$ such that if $\pi(x) = y$ then x_i is uniquely determined from $(y_{i-p}, \ldots, y_{i+q})$ and the coordinates $(x_{i-r}, \ldots, x_{i-1})$ which are to the left of x_i. Another way of saying this is: $\pi(u) = \pi(v)$ and the distance between $\sigma^i u$ and $\sigma^i v$ goes to 0 as $i \to -\infty$ then $u = v$ ("there are no right forks in pre-images"). We call a right closing map *right resolving*, if it is a 1-block map such that $r = 1, p = q = 0$. *Left closing* and *left resolving* are similarly defined.

The concept of a right resolving factor map can be given the following engineering interpretation. $\pi : L_A \to L_B$ determines how to construct a finite state automaton which encodes sequences Σ_B to Σ_A. The present output symbol depends on the present input and the present internal state. The present internal

---

[5]It is instructive at this point to compare Shannon's noiseless coding theorem [Sh] with the results presented here. His theorem states that coding is possible between source and channel data if the source entropy is strictly less than the channel capacity and impossible if the inequality is reversed. He does not treat the case of equality. Furthermore, his theorem says nothing about how coding can be done with automata. On the other hand his source entropy is a probabilistic one, something more general than topological entropy.

[6]For initial literature on factor maps in symbolic dynamics see the paper of Hedlund [H].

[7]Kitchens [Ki] found an important algebraic property of factor maps of topological Markov shifts: *If $\pi : \Sigma_A \to \Sigma_B$ is a finite factor map, then the Jordan form of A, apart from the 0-eigenvalues, contains that of B.*

[8]This concept was known to B. McMillan [Mc] and was called *unifilar* by him.

state depends on the past input and the past internal state. The encoding proceeds from left to right on the sequences, but this specification of the direction of time is merely a convention. The general isomorphism theorem for topological Markov shifts states that given (Σ_A, σ) and (Σ_B, σ) of equal entropy there is a common extension (Σ_C, σ) and two essentially one-to-one 1-block factor maps $\pi_1 : \Sigma_C \to \Sigma_A$ and $\pi_2 : \Sigma_C \to \Sigma_B$. One of these maps is right resolving and one is left. This means that the finite state automaton which encodes sequences from Σ_A via Σ_C to Σ_B proceeds from right to left, whereas the one that decodes goes from left to right.

In the case of Perron value N, an integer, we have a special situation. Encoding from (Σ_N, σ) to (Σ_A, σ) is derived, as in general, from a right resolving factor map $\pi_1 : \Sigma_C \to \Sigma_N$ where Σ_C is a common extension of Σ_A and Σ_N. The graph given by C defines the automaton and the factor map $\pi_2 : \Sigma_C \to \Sigma_A$ specifies its output. However, from Marcus's theorem the factor map $\pi_2 : \Sigma_C \to \Sigma_A$ from the common extension to Σ_A is invertible, which means that the decoding automaton is just the image of a continuous map - namely, $\pi_1 \pi_2^{-1}$ - and does not depend on an internal state. Engineers call this type of decoder a *sliding block decoder*. Thus, while errors in the input of the encoder might cause infinite output errors, error propagation is limited for a sliding block decoder. This is just what is required for encoding and decoding data for the magnetic recording information channel. The channel coder is not responsible for user errors, only channel errors. These he wants to limit in order not to overwhelm an *error correcting code*, another level of coding which has not been part of our discussion.

To sum the last two paragraphs, the engineering significance of almost homeomorphic conjugacy (in fact even the weaker relation, finite equivalence) is that automata can readily be constructed to encode and decode arbitrarily long sequences from one subshift of finite type to another of equal entropy. When the first is the full N-shift, then the decoder can be made sliding block, something not true in general. An added quality of almost homeomorphic conjugacy over finite equivalence is the existence of a finite input which resets the encoding automaton independent of the current internal state (likewise for the decoder). There may be situations where it is important to know the current internal state: for instance, an initial state is an ingredient in specifying an actual device.

To illustrate what we have been saying we return to the previous example in Figure 9 with edges labeled by data symbols.

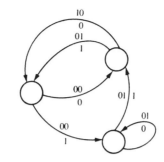

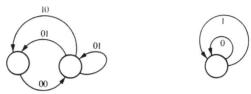

Figure 9 Automata

The graph of the full 2-shift Σ_2 is on the right with its edges labeled by 0 and 1. It represents sequences of arbitrary data. The graph of Σ_A is on the left with its edges labeled by some pairs of zeros and ones. It represents constrained data. The sequences of pairs ostensibly comprise a sofic system. Actually it is topologically isomorphic to the Markov shift Σ_A because there is a one-to-one correspondence between sequences of pairs and sequences of edges. Here sequences of pairs obey constraints which imply run-length ones. One can read from the graph of the common refinement Σ_C on top how to encode arbitrary data to constrained data at a rate 1:2 and also decode. The internal states are 6 edges of graph C which should be viewed as having distinct labels. Encoding proceeds from left to right (reverse for decoding). On the other hand, this is a case of integer Perron value (two). For this example the state-splitting method and the tableau-method construct the same common extension. A consequence of the state-splitting method is that we can decode in the same direction as encoding, namely, from left to right. In this method to determine a present unconstrained symbol, one just looks ahead in the sequence of constrained pairs.

Marcus's result can be used to solve the second engineering problem of achieving error limited decoding with a sliding block decoder for rate $p : q$ in the case where the Perron value $= 2^{p/q}$, the log of which is the entropy. This applies to the case where the input system taken as p-blocks of user data has the same topological entropy as that of q-blocks of channel constrained data. Constraints rarely have such entropies: so his method had to be extended to include Perron values $\neq 2^{p/q}$; and this was done by Adler, Coppersmith, and Hassner [ACH]. One can always get the source entropy less than or equal to the channel capacity by choosing p and q properly.

The theorem of Marcus [M1] for factoring topological Markov shifts onto the full N-shift was generalized by him [M2] to include a special class of sofic systems, he called *almost finite type*. These are described by finite state automata with the property that output sequences lying in some open set determine sequences of internal states - i.e., nodes of the defining graph. This mathematical notion was inspired by engineering applications, which impose constraints on signals representing bit strings, like having no d.c. component. Spectral constraints of this sort lead to this type of sofic system. Actually Marcus's result [M2] for sofic systems was not as strong as the one for topological Markov shifts because it just gives a factoring of (S, σ^k) onto (Σ_N, σ^k) for some k.[9] The following remark will perhaps shed some light on the practical significance of this. Patel [Pa] invented a code for the so-called *zero-modulation channel* used in the IBM 3850, a mass storage magnetic tape system. It satisfies the (1,3) run-length constraint along with a spectral constraint to eliminate the d.c. component in the electrical signals. This makes it a sofic system. Patel discovered a code with a simple sliding block decoder having small error propagation. The encoder, however, is based on a finite factor map which is not right-resolving. In order to design an encoding automaton for this particular system, a small amount of rate had to be sacrificed. As a by-product of Marcus's and Karabed's research on sofic systems it can now be done without this sacrifice. Admittedly, there is a substantial increase in complexity and error propagation in the decoder.

It is appropriate to mention here the interesting work of Franaszek [F] and of Cohn and Lempel [LC]. They developed methods similar to the aforementioned in the sense that all are based on the Perron-Frobenius spectral theory of non-negative matrices. Mathematical clarity which is missing from their work can now be supplied in terms of notions which have become standard in symbolic dynamics.

Finally, patents [AHM, CK, LS] and products (IBM 9332 Hard Disk File) have accrued as dividends of this kind work. Moreover, in the past it might take an engineer several months to design, by ingenious ad hoc methods, the code tables and the logic for an automaton to code data to fit constraints such as (d, k) ones. Now some of the most complicated cases can be routinely handled; but the main point is that these can often be done in less than a day.

REFERENCES

[A]   R.L. Adler, *Diffeomorphic Conjugacy of Automorphisms on the Torus.*, RC 1117, IBM Research Center, Yorktown Heights, NY (1964).

[ACH]   R.L. Adler, D. Coppersmith and Hassner, M., *Algorithms for sliding block codes*, IEEE-IT **29** (1983), 5–22.

[AGW]   R.L. Adler, L.W. Goodwyn and B. Weiss, *Equivalence of topological Markov shifts*, Israel J. of Math. **27** (1977), 49–63.

[AHM]   R.L. Adler, M. Hassner and J. Moussouris, *Method and Apparatus for Generating a Noiseless Sliding Block Code for a (1,7) Channel with rate 2/3*, U.S. Patent 4,413,251 (1983).

[AKMc]   R.L. Adler, A.G. Konheim and M.H. McAndrew, *Topological entropy*, Trans. Amer. Math. Soc. **114** (1965), 309–319.

[AM]   R.L. Adler and B. Marcus, *Topological entropy and equivalence of dynamical systems*, Memoir of AMS **219** (1979).

---

[9] A hot new result at IBM Research San Jose by him and a collaborator, R. Karabed, is the fact that $k = 1$, just as in in the Markovian case. In addition, Marcus was then able to extend this theorem to arbitrary sofic systems by inventing a new notion, he called a *non-catastrophic decoder*, which was also inspired by engineering requirements. It is a generalization of the notion of sliding block decoder and beautifully fits the mathematics of the situation.

[AP] R.L. Adler and R. Palais, *Homeomorphic Conjugacy of Automorphisms on the Torus*, Proc. of Amer. Math. Soc. **16** no. 6 (1965), 1222–1225.

[AW] R.L. Adler and B. Weiss, *Similarity of Automorphisms of the torus*, Memoir Amer. Math. Soc. **98** (1970).

[Ar] D.Z. Arov, *Topological similitude of automorphisms and translations of compact commutative groups*, Uspehi Mat. Nauk **18** no. 5 (113) (1963), 133–138.

[Ba] K. Baker, *Strong shift equivalence of 2×2 matrices of nonnegative integers associated with topological Markov chains*, Ergod. Th. and Dynam. Sys. **3** (1983), 501–508.

[Be] K. Berg, *On the Conjugacy Problem for K-systems*, Ph.D. Thesis, U. of Minn. (1967).

[B1] R. Bowen, *Entropy for group endomorphisms and homogeneous spaces*, Trans. Amer. Math. Soc. **153** (1971), 401–414.

[B2] R. Bowen, *Equilibrium states and the ergodic theory of Anosov diffeomorphisms*, Springer Lecture Notes in Math. (1975).

[B3] R. Bowen, *Markov partitions for axiom A diffeomorphisms*, Amer. J. Math. **91** (1970), 725–747.

[CK] D. Coppersmith and B. Kitchens, *Run-length limited code without dc level*, Patent pending.

[CP1] E.M. Coven and M.E. Paul, *Endomorphisms of irreducible subshifts of finite type*, Math Systems Theory **8** (1974), 167–175.

[CP2] E.M. Coven and M.E. Paul, *Sofic Systems*, Israel J. of Math **20** (1975), 165–177.

[CP3] E.M. Coven and M.E. Paul, *Finite Procedures for Sofic Systems*, Monatsh. Math. **83** (1977), 265–278.

[DGS] M. Denker, C. Grillenberger and K. Sigmund, *Ergodic theory on compact spaces*, Springer Lecture notes in Math. **527** (1976).

[Di] E.I. Dinaburg, *The relation between topological entropy and metric entropy*, Dokl. Akad. Nauk SSSR **190** no. 1 (1970), 13–16.

[F] P. Franaszek, *Construction of bounded delay codes for discrete channels*, IBM J. Res. Dev. **26** (1982), 506–514.

[G] R.L. Gray, *Generalizing Period and Topological Entropy To Transitive Non-wandering Systems*, Masters thesis, U. of North Carolina, Chapel Hill (1978).

[Ha] M. Hassner, *A Non-probabilistic Source and Channel Coding Theory*, Ph.D. dissertation, UCLA (1980).

[H] G.A. Hedlund, *Endomorphisms and automorphisms of the shift dynamical system*, Math System Theory **3** (1969), 320–375.

[KSS] A. Katok, Ya. G. Sinai and A.M. Stepin, *Theory of dynamical systems and general transformation groups with an invariant measure*, J. of Soviet Math (1977), 974–1064.

[Ki] B. Kitchens, *An invariant for continuous factors of Markov shifts*, Proc. Amer. Math. Soc. **83** (1981), 825–828.

[KR] K.H. Kim and F.W. Roush, *Some results on decidability of shift equivalence*, J. of Combinatorics, Information and System Sciences 4 (1979), 123–146.

[Ko] A.N. Kolmogorov, *A new metric invariant of transitive automorphisms of Lebesgue spaces*, Dokl. Akad. Nauk SSSR **119** no. 5 (1958), 861–864.

[KT] A.N. Kolmogorov and V.M. Tihomirov, *ϵ-entropy and ϵ-capacity of sets in functional spaces*, Amer. Math. Soc. Transl.(2), **17** no. 2 (1961), 277–364.

[LS] G. Langdon and P. Siegel, *Direction constrained ternary codes using peak and polarity detection*, U.S. Patent 4,566,044 (1986).

[LC] A. Lempel and M. Cohn, *Look ahead coding for input restricted channels*, IEEE Trans Inform. Theory **IT-28** (1982), 933–937.

[Mc] B. McMillan, *The basic theorems of information theory*, Ann. Math. Stat. **24** (1953), 196–219.

[M1] B. Marcus, *Factors and extensions of full shifts*, Monatsh. für Math. **88** (1979), 239–247.

[M2] B. Marcus, *Sofic systems and encoding data*, IEEE-IT **31** (1985), 366–377.

[Me] L.D. Meshalkin, *A case of isomorphism of Bernoulli schemes*, Dokl. Akad. Nauk SSSR **128** (1959), 41–44.

[Ob] G.L. O'Brien, *The road colouring problem*, Israel J. Math. **39** (1981), 145–154.

[O1] D.S. Ornstein, *Bernoulli shifts with the same entropy are isomorphic*, Adv. in Math. **5** (1970), 337–352.

[O2]   D.S. Ornstein, *Yale Math. Monographs 5*, Yale Univ. Press, 1974.

[P1]   W. Parry, *Intrinsic Markov chains*, Trans. Amer. Math. Soc. 112 (1964), 55–66.

[P2]   W. Parry, *A finitary classification of topogical Markov chains and sofic systems*, Bull. London Math. Soc. 9 (1977), 86–92.

[PW]   W. Parry and R.F. Williams, *Block coding and a zeta function for finite Markov chains*, Proc. London Math. Soc. 35 (1977), 483–495.

[Pa]   A. Patel, *Zero modulation encoding in magnetic recording*, IBM J. Res. Dev. 19 no. 4 (1975), 366–378.

[Pe]   K. Petersen, *Chains, entropy, and coding*, Ergod. Th. and Dynam. Sys. (to appear)..

[Sh]   C. Shannon, *A mathematical theory of communication*, Bell Sys. Tech J. (1948), 623–656.

[S1]   Ya. G. Sinai, *On the concept of entropy of a dynamical system*, Dokl. Akad. Nauk SSSR 124 (1959), 768–771.

[S2]   Ya. G. Sinai, *Ergodic and kinetic properties of the Lorentz gas*, Anal. of N.Y. Acad. of Sci. 357 (1980), 143–149.

[S3]   Ya. G. Sinai, *Markov partitions and C-diffeomorphisms*, Func. Anal. and its Appl. 2 (1968), 64–89.

[Sm]   S. Smale, *Differentiable dynamical systems*, Bul. Amer. Math. Soc. (1967), 747–813.

[W]   B. Weiss, *Subshifts of finite type and sofic systems*, Monatsh. Math 77 (1973), 462–474.

[Wi]   R.F. Williams, *Classification of shifts of finite type*, Ann. of Math. 98 (1973), 120–153.

MATHEMATICAL SCIENCES DEPARTMENT, IBM T. J. WATSON RESEARCH CENTER, YORKTOWN HEIGHTS, NY 10598, USA

Contemporary Mathematics
Volume **135**, 1992

On the Work of Roy Adler in Ergodic Theory and Dynamical Systems[1]

Benjamin Weiss[2]

In order to put Roy Adler's work in its proper perspective I shall begin by drawing a sketch of what ergodic theory was like before Roy started to work in the field. After the first ergodic theorems of J. von Neumann and G.D. Birkhoff a number of outstanding mathematicians such as E. Hopf, N. Wiener and S. Kakutani continued to study specific examples and to extend the scope of theory. The 1956 lectures of P. Halmos gave a good summary of the state of the field at that time and also highlighted a more structural approach featuring the isomorphism problem of deciding when two abstract systems were isomorphic. The related field of topological dynamics that had been founded by H. Poincare and G.D. Birkhoff also had its representative at Yale, where Roy studied, namely G. Hedlund, one of the founders of symbolic dynamics. Ergodic theory was a quiet area, with some fascinating problems but not highly competitive and high-pressured. Those who know Roy will easily understand why he was attracted to it and chose S. Kakutani to be his doctoral advisor.

Most of Roy's work has been connected with specific examples rather than with general theories. A striking exception is his invention of topological entropy and the first section will be devoted to this basic concept. The second section will describe his result about topological conjugacy of toral automorphisms, a precursor of many "rigidity" results of algebraic systems. In the third section I'll tell the story of our work on the metric conjugacy of toral automorphisms. The fourth section will be devoted to his work on one dimensional maps while the fifth will tell of his more recent work with L. Flatto on symbolic dynamics for the geodesic flows.

Roy's work on shifts of finite type and its connections with practical encoding

---

[1]This paper is in final form.
[2]Primary: 28D, Ergodic theory, Secondary: 54H, Toplogical dynamics.
1991 Mathematics Subject Classification. Primary 28D; Secondary 54H.

problems will be described in a separate paper by B. Marcus. As a glance at Roy's list of publications will show I have been far from exhaustive in describing all of his contributions to ergodic theory. My purpose is rather to give some sense of the kind of work he has done and of his impact on this burgeoning field.

Finally I must acknowledge a very specific debt to Roy for his help in preparing this paper. Some years ago he wrote a wonderful survey of his work at IBM entitled <u>The Torus and the Disk</u>. I have cribbed extensively from it and can find no better way to end this introduction then by quoting from his abstract:

"This paper is a survey of a coherent program of mathematics spanning 28 years. It begins with questions concerning classification and structure in ergodic theory and abstract dynamical systems and describes the author's involvement with toral automorphisms, topological entropy, iteration of maps on the interval, symbolic dynamics and ultimate engineering applications". (The disk in the title is not $\{|z| \leq 1\}$ but rather the magnetic disk of modern computer memories.)

1 Topological Entropy

The first to bring entropy into mathematics was C. Shannon in his pioneering work on information theory. He used it in order to quantify the intuitive concept of how much information is revealed when we learn the outcome of a random experiment. He went on to define the entropy of a stationary process $\{x_n\}$ as the time average of the information contained in the random variables x_n. It was not unnatural to try to apply this concept to the study of measure preserving systems and indeed von Neumann suggested that it might be possible to use this concept to distinguish between the two shift and the three shift. Although his suggestion didn't directly bear fruit, soon afterwards A.N. Kolmogorov did define the entropy of a transformation and showed with Ya. Sinai, that indeed the two shift was not isomorphic to the three shift. At about the same time, he also introduced the notion of ϵ-entropy as a measure of the complexity of function spaces.

This ϵ-entropy can be applied to any abstract metric space and has had many applications. It was while he was studying this latter work on ϵ-entropy that Roy realized that one could define a numerical invariant of an arbitrary continuous mapping T of a compact space X. For this he followed the pattern of Shannon but in a topological setting. You begin by defining the covering number $c(U)$ of an arbitrary open covering of X as the minimal number of sets of the cover whose union is X. Since X is compact this number is always finite. Next one defines the join of open covers $U_1, \ldots, U_k$ as the open cover whose elements are intersections of the form $u_1 \cap u_2 \cap \cdots u_k$ where the $u_i \in U_i$ range over all elements of the covers U_i. Denote this join by $\overset{k}{\underset{i=1}{\wedge}} U_i$.

The dynamics of T is introduced by setting $U_i = T^{-i}U = \{T^{-i}u : u \in U\}$, and then defining

$$H(T, U) = \lim_{n \to \infty} \frac{1}{n} \log c(\overset{k}{\underset{i=1}{\wedge}} T^{-i}U).$$

As usual in these formulas the existence of the limit follows from an easy argument based on the submultiplicative property of covering number:

$$c(u_1 \wedge u_2) \leq c(u_1) \cdot c(u_2).$$

The topological entropy of T, $h(T)$ is then defined as the supremum of $H(T, U)$ as U ranges over all open covers of X. The definition of $H(T, U)$ is analogous to Shannon's definition of entropy of a fixed stochastic process while to obtain an invariant of T one maximizes over all covers just like Kolmogorov did to define the entropy of a measure preserving transformation. Following the definition, the basic properties of the entropy were obtained such as the facts that

a) $h(T^n) = |n|h(T)$

b) $h(T \times S) = h(T) + h(S)$

where $T \times S$ represents the transformation $(x, y) \to (Tx, Sy)$, the Cartesian product of (X, T) with (Y, S). In addition it was shown that the supremum over all covers could be replaced by a limit over a sequence of covers U_i with the property that $\sup\{diam(u) : u \in U_i\} \to 0$ as $i \to \infty$.

In their first groundbreaking paper it was conjectured that the topological entropy is the supremum of the measure theoretic entropy $h_\mu(T)$ as one ranges over all the T-invariant measures μ. This conjecture was based mainly on what emerged in the examples like the n-shift, and toral automorphisms. Some years later W. Goodwynn and T. Goodman proved this conjecture which was put into its proper perspective by the definition that R. Bowen and E.I. Dinaburg gave for the topological entropy. This formulation puts it closer to the framework of the general ϵ-entropy but requires the introduction of an explicit metric d on X. The ϵ-spanning number of the metric d is the minimum cardinality of a set of balls of radius ϵ (in the d-metric) whose union is all of X. Defining

$$d_n(x, y) = \max_{0 \leq i < n} d(T^i x, T^i y)$$

one forms

$$\overline{\lim_{n \to \infty}} \frac{1}{n} \log[\epsilon - \text{spanning number of } d_n]$$

and then the limit of this quantity on $\epsilon \to 0$ is the topological entropy.

This version of the definition makes more evident the connection between topological entropy and an exponential rate of divergence of the orbits of nearby points. This latter property was called sensitive dependence on initial conditions by D. Ruelle [R 1979] and forms one of the characteristic features of chaotic systems. There is by now a large literature connected with this fundamental concept. It plays a decisive role in the classification of shifts of finite types and other symbolic systems and I am sure will continue to occupy a central position in the theory of dynamical systems for many years to come. For the full story of Roy's program of classifying shifts of finite type up to topological conjugacy see the next article by B. Marcus in this volume.

Let me conclude this section with a few words on some recent work of F. Blanchard. He has been trying to get some kind of topological analogue of the K-automorphisms of ergodic theory that are characterized as those transformations having completely positive entropy. To do this he has returned to Roy's original definition and defines a class of transformations with the property that every cover of X by two open sets, $U \cup V = X$, nontrivial in the sense that $X \neq \overline{U}, X \neq \overline{V}$, satisfies $H(T, \{U, V\}) > 0$. He has recently been able to prove the beautiful result that such transformation are disjoint from all zero entropy minimal transformations. This is the analogue of Furstenberg's basic result that K-automorphisms and zero entropy processes are disjoint-but in a topological setting. We see that there is still much to be learned from the older way of calculating topological entropy.

2 Topological Conjugacy of Toral Automorphisms

As I indicated in the introduction, one of Roy's particular strengths is his ability to see whole theories in specific examples. While still a graduate student he was attracted to the study of algebraic automorphisms of the n-torus which represent the basic linear transformations of a compact space. He set for himself the problem of determining when two such automorphisms are topologically conjugate. Let us make this problem explicit: given two automorphisms A, B of the n-torus $\mathbb{T}^n$, which may be thought of as integer matrices with determinant $+1$, we ask when is there a homeomorphism $\varphi : \mathbb{T}^n \to \mathbb{T}^n$ such that $B = \varphi^{-1} A \varphi$.

It is clear that the algebraic conjugacy of A and B by another automorphism C is a sufficient condition for one may take simply $\varphi = C$.

To get somewhere Roy began by assuming that φ was a diffeomorphism. In that case, rewriting the basic relation as $\varphi B = A \varphi$ and differentiating both sides leads to some algebraic conjugacy between A and B which Roy was able to show leads to the existence of such a C.R. Palais saw how to eliminate the differentiability assumption and together they published one of the first theorems of this type: homeomorphic conjugacy of algebraic automorphisms implies algebraic conjugacy.

The new insight involves looking at the action induced by φ on the homology of the torus. Since the universal covering of $\mathbb{T}^n$ is $\mathbb{R}^n$ and $\mathbb{Z}^n$ is the first homology group, φ induces an automorphism of $\mathbb{Z}^n$ which we may denote by φ^*. It is this algebraic automorphism of $\mathbb{Z}^n$ that quickly leads to the desired result. Once again the mathematical reasoning is not very complicated but a wonderful new paradigm emerges that has been amplified many times since then.

P. Walters extended these results to algebraically defined maps of nil manifolds which behave like toral automorphisms in many ways. In spite of these and many other generalizations, and much more work on these automorphisms some basic question still remain open. To illustrate how much we still have to learn, I will briefly discuss the following open problem: Is there an ergodic automorphism of the infinite dimensional torus $\mathbb{T}^\infty$? with finite entropy. While this question seems to be related to some infinite dimensional object, a related question is: what is the $inf h(A)$ as A ranges over all ergodic automorphisms of finite dimensional tori. If this infinium is zero, then we could find ergodic automorphisms A_k of $\mathbb{T}^{n_k}$ for integers n_k, so that

$$\sum_1^\infty h(A_k) < +\infty$$

and then $\overset{\infty}{\underset{1}{\Pi}} A_k$ is an ergodic automorphism of $\mathbb{T}^\infty$ with finite entropy. It turns out that one can also establish the converse - namely the existence of an ergodic automorphism of $\mathbb{T}^\infty$ with finite entropy implies that the infinium above is zero

The remarkable thing is that this question was raised, in a purely number theoretic setting by D. Lehmer more than fifty years ago! In fact there is a simple formula for the entropy of a toral automorphism:

$$h(A) = \Sigma \log |\lambda_j|$$
$$\{\lambda_j \in sp(A), |\lambda_j| > 1\}.$$

Since the eigenvalues of a toral automorphism are algebraic integers this is equivalent to the logarithm of the product of those conjugates of a certain algebraic integer that lie outside the unit circle. Lehmer found a positive lower bound for this quantity as one ranges over all integers of a fixed degree and conjectured that in fact there was a lower bound independent of n. In spite of the work already devoted to this question and several partial results - the conjecture remains open - as does its dynamical analogue.

There are many more results of this general flavor, namely that algebraically defined actions are classified up to topological conjugacy by their algebraic classification. M. Ratner has many wonderful results of this type for rather general actions of connected Lie groups on homogenous spares and results of that type have had important consequences in seemingly unrelated questions in number theory. The new insights that Roy's study of this seemingly innocuous question led him to have had far reaching implications indeed.

3 Metric conjugacy of toral automorphisms

My mentor in ergodic theory was Roy and it was he who suggested that we study the metric properties of toral automorphisms. We wanted to understand what were the connections between two toral automorphisms with the same entropy and whether or not the fact that they were not necessarily topologically conjugate influenced their measure theoretic properties. The power of symbolic dynamics was already clear to us from another study we had made of the tent like mappings (see §4) and we were delighted to find in one example after another that we analyzed Markov partitions. It is easier to say what a Markov partition is for a non invertible mapping so let me begin by explaining that.

A finite partition $(A_1, \cdots A_k)$ of X is said to be Markov for a transformation T of X if TA_i is a union of other elements of the partition for each i. That is to say, there are sets $J(i)c\{1, \cdots k\}$ such that $TA_i = \underset{j \in J(i)}{\cup} A_j$. If X is a connected space then one usually allows some flexibility in that one demands only that the interiors of the A_i be disjoint etc. Usually the A_i's are closed and one further assumes that the diameter of $\overset{n}{\underset{1}{\cap}} T^{-i} A_{j(i)}$ tends to zero as $n \to \infty$. This latter condition means that any abstract symbolic sequence $u(1), u(2), \cdots$ compatible with the transitions $u(n + 1) \in J(u(n))$ corresponds to the history of a point $x \in X, T^n x \in A_{u(n)}$ for all n.

If a transformation (X, T) possesses such a Markov partition then its dynamics, and especially its measure theory is essentially the same as that of the symbolic shift of finite type defined by the transitions $i \to j, j \in J(i)$. For invertible transformations, which are modelled by the two sided shift, the definition becomes more complicated since TA_i now doesn't cover A_j's completely - but only covers them in one "direction". For two dimensional ergodic automorphism the picture was particularly simple since both eigenvalues have modulus different from one. Thus the two torus is fibered by two families of lines, one of which the automorphism T expands and the other it contracts. Here an element of a Markov partition is easy to describe - it is just a parallelogram with sides parallel to the two foliations just described and a partition into parallelograms is called

Markov if for each i, j such that $TA_i \cap A_j$ has non empty interior the intersection consists of a finite number of parallelograms each of them going all the way across A_j in the expanding direction, and similarly for $T^{-i}A_i$ in the other direction.

Such a Markov partition enables one to identify toral automorphism with a shift of finite type. For the measure theory we had to know what did Lebesgue measure correspond to. Here topological entropy played a key role. We saw that Lebesgue measure was characterized as the unique measure whose metric entropy equalled the topological entropy and that meant that under the correspondence just described Lebesgue measure was carried over to the unique measure on the shift of finite type that maximized its metric entropy. These measures had been previously identified by W. Parry and were easy to write down explicitly.

After experimenting with many special cases we were finally able to establish the existence of these Markov partitions for an arbitrary ergodic automorphism of the two torus. At this point we had also begun to suspect that entropy completely classified the metric conjugacy. Indeed in one special case after another we were able to find codings between the symbolic shifts of finite type that we were getting for rather different automorphisms with the same entropy - or even for different partitions for the automorphism. When we found the general Markov partitions, they had a sufficiently simple structure that the coding methods we had developed could give a proof of the general case and we were able to complete the project that Roy had embarked on while still in graduate school. The following theorem was proved:

THEOREM:. *Two 2-dimensional ergodic toral automorphisms are metrically isomorphic if and only if they have the same entropy.*

All of the ingredients that went into the proof of this rather special theorem have been expanded enormously. This was the first example of a class of transformations being classified by metric entropy. Shortly afterwards D. Ornstein showed that Bernoulli shifts were classified by entropy and in a remarkable series of papers proved that this was essentially the complete story and explanation of our result.

Ya. G. Sinai and R. Bowen proved the existence of Markov partitions for much more general classes of smooth dynamical systems and used them to establish many properties of these systems. One example of the kind of information that Markov partitions provide is the following. In topological dynamics the basic kind of "irreducible" system is a minimal set - which is a closed invariant set for T that has no proper closed invariant subset. It is natural to ask how many properties of the transformation are reflected in its minimal sets. It particular one can ask - is the topological entropy of T equal to the supremum of the topological entropy of T restricted to its minimal sets. It is not hard to see that in general the answer is no. One can construct compact spaces and homeomorphisms (X, T) such that $h(T) > 0$ but the only minimal sets that T has are fixed points. On the other hand for hyperbolic toral automorphisms, or more generally any Anosov system the answer is yes!

The easiest way to see this is to use Markov partitions to obtain a continuous map π form a shift of finite type onto the hyperbolic system which is essentially one to one. It is straightforward to verify that the shift has minimal sets whose topological entropy comes arbitrarily close to the topological entropy of the shift of finite type and then π maps these onto minimal subsets of the hyperbolic system with the same property.

The coding arguments that went into showing that the shifts of finite type that we obtained from the Markov partitions were isomorphic led to a simple combinatorial question that we called the road coloring problem - that surprisingly enough is still open! In the hope that some reader of this paper will be tempted to attack this question and settle it I give it in detail.

Let (V, E) be a directed graph. Assume that for any two vertices v_1, v_2 in V there is a path from v_1 to v_2. Assume also that the graph is <u>aperiodic</u>, i.e. for some n, and any $v_1, v_2 \in V$ there is a path of length n from v_1 to v_2. Finally suppose that the outdegree at each vertex is constant, say d. The edges are thought of as roads, and a coloring of the road is assignment at every vertex of the colors $\{1, 2, \cdots d\}$ a different one to each of the edges. A set of directions is a sequence $D_1 D_2 \cdots D_m$, each $D_i \in \{1, 2, \cdots d\}$ and gives a mapping from V to V where for each v one follows in succession the paths labelled $D_1, D_2, \cdots$ starting out at v. The coloring together with the directions <u>collapses</u> the graph if the range of this function is a single vertex.

Problem: Show that every aperiodic, d-regular directed graph (V, E) has a coloring and direction set that collapses the graph.

It is easy to see that if the graph has a loop such a road coloring always exists. O'Brien showed that the same holds if there is a simple cycle of prime length [OB-1981]. A further extension was given by J. Friedman [Fr-   ] but in general the problem is still open. Together with Goodwynn we found a way, without solving the problem in general, to prove what we wanted to prove with it - namely that shifts of finite type of entropy $\log d$ were almost-homeomorphically conjugate (see the paper by B. Marcus loc.cit. for more details on this). All in all the further developments by Roy, and Brian Marcus of these coding ideas led to a rich theory that is still expanding vigorously today.

Finally I would like to come back to the point that I mentioned earlier - namely the uniqueness o the measure of maximal entropy. W. Parry had proved this for topological Markov shifts by first showing that such a measure must be Markovian and then analyzing the Markovian measure. We gave another proof in which we showed that the measure was unique by seeing that it gave a uniform distribution to the n-blocks, or equivalently to the periodic points. This argument was later exploited by R. Bowen in his studies on axiom A flows and led to his results on the distribution of periodic points - a theme that is still rather active today.

4 One dimensional maps

In the fall of 1966 Leo Flatto, who was a former colleague of Roy's at IBM, and a professor at Yeshiva University at that time, told me about the problem- which has since spawned an entire field- what is the dynamical behavior of the map $x \rightarrow ax(1-x)$ on the unit interval for various choice of the parameter a. We took a piecewise linear version: $x \rightarrow a - 2a|x - \frac{1}{2}|$ and noticed that for certain values of the parameter a there exists a Markov partition for the mapping. When this happens one can give a complete description of the dynamics by passing to the symbolic system defined by the transitions of the Markov partition. It was this study that led us to the Markov partitions of the toral automorphisms that I described above. For the one dimensional maps themselves, the important lesson that we learned was that it is much easier to get metric properties for piecewise linear maps. This led Roy to study the degree to which the iterates of a one dimensional map can be approximated by piecewise linear maps and to a series of studies concerned with one dimensional maps.

Long before the iteration of one dimensional maps became such a popular study he was led to formulate some "folklore theorems" giving criteria which were easy to check for very strong mixing properties of these maps. Here is one version of his theorem which he presented at the NYU seminar in ergodic theory in the early seventies. Suppose that $\{I_j\}_1^\infty$ is partition of $[0, 1]$ into intervals on each of which ϕ is monotone and $\phi(I_j) = [0, 1]$. Denote by $\mathcal{I}$ the partition and by $\mathcal{I}_n = \bigvee_0^{-n+1} \phi^{-i}\mathcal{I}$. Assume further that these partitons become finer and finer and ultimately become points. These are the geometric conditions. As far as the analytic properties of ϕ the assumptions are:

(1) ϕ is of class C^2 on each I_j

(2) for some s, $\inf_j \{\inf_{x \in I_j} |(\phi^s)'(x)|\} > 1$ where ϕ^s denotes the s-iterate of ϕ.

(3) $\sup_j \sup_{x,y,z \in I_j} \left| \frac{\phi''(x)}{\phi'(y)\phi'(z)} \right| < +\infty$

Under these hypotheses Roy goes on to show that there is a unique ϕ-invariant measure that is absolutely continuous with respect to Lebesque measure and with respect to this measure λ, $\mathcal{I}$ is weakly Bernoulli. This latter condition means a strong kind of approximate independence between $\mathcal{I}_n$ and $\phi^{-n-M}\mathcal{I}_m$ that is uniform in n and m and tends to actual independence as M tends to infinity. It is a consequence of Ornstein's theory that this implies that ϕ is then isomorphic to a Bernoulli shift. (More precisely the invertible extension of ϕ is isomorphic to a two sided Bernoulli shift).

This criterion was applied to the classic continued fraction transformation and culminated, in a sense, a series of investigations on this fundamental algorithm that began with Gauss. Another version of this theorem applies to the case where the transformation ϕ when restricted to the intervals of monotonicity is not onto all of $[0,1]$. Here some other geometric conditions must be imposed. As I've already indicated this area has expanded greatly, but these results of Roy still serve very well indeed.

Another pioneering work of his concerns one dimensional maps preserving an infinite measure space. Twenty years ago there were very few results on infinite measure preserving transformations. Even fewer natural examples had been studied. A well known formula of G. Boole

$$\int_{-\infty}^{\infty} f(x)dx = \int_{-\infty}^{\infty} f(x - \frac{1}{x})dx$$

points out that the map $x \to x - \frac{1}{x}$ of $\mathbb{R}$ onto itself preserves Lebesgue measure. Roy and I showed that this transformation was ergodic by considering the induced transformation on a finite interval. This led us to a transformation of the type described above but now the existence of a neutral fixed point precluded the possibility that a condition like the uniform expansiveness of (2) above would be valid. Nonetheless we were able to establish the ergodicity of the transformation. Here too the results have been extended in many directions. A fine recent survey has been written by C.I. Doering and R. Mané where more recent work can be found.

5 Symbolic dynamics for geodesic flows

At the turn of the century symbolic dynamics for geodesic flows was introduced by J. Hadamard. Later M. Morse made essential use of this tool in establishing the existence of geodesics with various properties. This paradigmatic example of a real dynamical system continued to play a central role in ergodic theory down to the present day. Indeed one can write a history of dynamical systems in this century by following the various studies that have enriched our understanding of the behavior of the geodesic flow on a surface of negative curvature. Roy's most recent magnum opus written with Leo Flatto is a beautiful detailed exposition of the symbolic dynamics for geodesic flows. I shall try in this section to describe their work briefly and put it in its proper perspective.

Let us recall the definition of the geodesic flow for a Riemannian manifold M. Let SM be the unit tangent bundle to M and let $g_t : SM \to SM$ be defined by $g_t(x, v) = (\bar{x}, \bar{v})$ where $v \in SM_x$, and to obtain $(\bar{x}, \bar{v})$ we form the unique geodesic $\gamma(t)$ satisfying $\gamma(0) = x, \gamma'(0) = v$ and set $\bar{x} = \gamma(t), \bar{v} = \bar{\gamma}'(t)$. If M is compact, two dimensional with constant negative curvature then its universal covering space is the two dimensional hyperbolic space and M can be visualized as region in the Poincaré disk with certain identifications. This is the classical geodesic flow and all that follows refers to it. The first step in obtaining a symbolic representation lies in choosing a cross section for the flow and defining the appropriate cross section map.

Both Hedlund [H 1934] and Morse ([Mo] these lecture notes date from 1937 but were only circulated in mimeographed form from 1966 onwards, they were not published) treated only a special class of surfaces, with genus $g \geq 2$ namely those with a fundamental domain consisting of a regular $4g$-sided fundamental polygon with interior angles $\pi/2g$. Their symbolic dynamics differed. Hedlund, using an idea of Nielsen [N 1927], represented orbits by a subshift of finite type. Morse began with the natural idea of looking at the succession of fundamental domains encountered by a geodesic in the disk. If the sides of a uniquely chosen fundamental domain are labelled this gives rise to a symbolic sequence. Unfortunately it is difficult to describe the totality of all such sequences and so he passed over to a related sequence by a geometric construction. He then gave rules describing what sequences were obtained. This turns out to be something called a sofic system which is more general than a shift of finite type.

In [AF-1991] all compact surfaces of genus $g \geq 2$ are considered. They give a careful proof of an old result going back to the work of Dehn, Fenchel and Nielsen which gives for any compact Riemann surface a fundamental region whose boundary consists of $8g-4$ geodesic arcs satisfying some special properties needed for their construction of the cross section. Having obtained a two dimensional cross section there is still a difficulty in that the dynamics of that map are not easy to analyze. They proceed to find a different, rectilinear version of this mapping that has the advantage that one of the coordinates now maps as a function of one variable. This is used to define a one-dimensional factor map which was essentially introduced by R. Bowen and C. Series [Bo-S 1979] in their study of orbits of Fuchsian groups.

At this point some new results connected with the folklore theorem that I described above are used to obtain ergodic theoretic properties of this mapping. In particular an analytic formula is obtained for the unique invariant measure equivalent to Lebesque measure. After this discussion the symbolic dynamics is introduced. This is done in a very careful way especially in all that concerns the geodesics that pass through vertices of the fundamental polygon. The discussion in their paper is self contained and shows how all the theories we've been discussing can be combined into one beautiful parade example. Comparing their work with the papers of C. Series [eg. S-1986] one sees that the essential mathematics is the same but the approach and methods are quite different.

Roy's insistence on carefully written complete proofs is nowhere more evident than in this work which will serve as a basic reference for many years to come to all those interested in the dynamics of geodesic flows.

References

[A1] R. Adler *The torus and the disk,* IBM Journal of Res. and Dev. 31 (1987) 224–233

[A2] R. Adler *f-expansions revisited,* Recent advances in topological dynamics, Springer LNM. 318 (1972) 1–5

[AF] R. Adler and L. Flatto *Geodesic flows, interval maps and symbolic dynamics,* Bulletin of the AMS 25 (1991) 229-334

[AGW] R. Adler, W. Goodwynn and B. Weiss *Equivalence of topological Markov shifts,* Israel J. of Math. 27 (1977) 49–63

[AKM] R. Adler, A. Konhein and H. McAndrew *Topological entropy,* Trans. Amer. M. Soc. 114 (1965) 309–319

[AP] R. Adler and R. Palais *Homeomorphic conjugacy of automorphisms on the torus,* Proc. Amer. M. Soc. 16 (1965) 1222–1225

[AW1] R. Adler and B. Weiss *Entropy, a complete metric invariant for automorphisms of the torus,* Proc. of the Nat. Acad. 57 (1967) 1573–1576

[AW2] R. Adler and B. Weiss *Similarity of automorphisms of the torus,* Memoirs of the A.M.S. 98 (1970)

[AW3] R. Adler and B. Weiss *The ergodic infinte measure preserving transformation of Boole,* Israel J. of Math. 16 (1973) 263–277

[Bl] F. Blanchard *A disjointness theorem involving topological entropy,* preprint Dec. 1991

[Bo1] R. Bowen *Entropy for group endomorphisms and homogeneous spaces,* Trans Amer. M. Soc 153 (1971) 401–414

[Bo2] R. Bowen *Markov partitions for axiom A- diffeomorphisms,* Amer. J. of Math. 91 (1970) 725–747

[BoS] R. Bowen and C. Series *Markov maps associated with Fuchsian groups,* Inst. Hautes Etudes Sci. Publ. Math. 50 (1979) 153–170

[De] M. Dehn *Papers on Group Theory and Topology,* edited and translated by John Still well, Springer-verlag, New York, Berlin, Heidelberg (1987)

[Di] E.I. Dinaburg *The ralation between topological entropy and metric entropy,* Dokl. Akad. Nauk SSSR 190 (1970) 13–16

[DM] C.I. Doering and R. Mane *The Dynamics of Inner Functions,* Ensaios Matematicos-Soc. Brasiliera de Mat. vol. 3 (1991)

[F] J. Friedman *On the road coloring problem,* Proc. Amer. M. Soc. 110(1990) 1133-1135

[G] T.N.T. Goodman *Relating topological entropy and measure entropy,* Bull. Lond. M. Soc. 3 (1971) 176–180

[GL] L.W. Goodwynn *Topological entropy bounds measure theoretic entropy*, Proc. AMS 23 (1969) 679–688

[H] G.A. Hedlund *On the metrical transitivity of the geodesics on closed surfaces of constant negative curvature*, Ann. of Math. 35 (1934) 787–808

[K] A.N. Kolmogorov *Entropy per unit time as a metric invariant of automorphisms*, Dokl. Akad. Nauk SSSR 124 (1959) 754–755

[Le] D. Lehmer *Factorization of cyclotomic polynomials*, Ann. of Math. 34 (1933) 461–479

[Li] D. Lind *Ergodic automorphisms of the infinte torus are Bernoulli*, Israel J. of Math. 17 (1974) 162–168

[Mo] M. Morse <u>*Symbolic Dynamics,*</u> Institute for Advanced Study Notes, Notes by Rufus Oldenburger, Princeton (1966) (unpublished)

[N] J. Nielsen *Untersuchungen zur Topologie der geschlossen zweiseitigen Flächen*, Acta Math. 50 (1927) 189–358

[O'B] G.L. O'Brein *The road colouring problem*, Israel. J. of Math. 39(1981) 145–154

[Or] D. Ornstein *Bernoulli shifts with the same entropy are isomorphic*, Adv. in Math. 48 (1970) 337–352

[Ra] M. Ratner *On Raghunathan's measure conjecture*, Ann. of Math. 134 (1991) 545–607

[Ru] D. Ruelle *Sensitive dependence on initial conditions and turbulent behavior of dynamical systems in Bifurcation theory and its applications*, New York Acad. of Sciences 316 (1979)

[Se] C. Series *Geometrical Markov coding of geodesics on surface of constant negative curvature*, Ergod. Th. & Dynam. Sys. 6 (1986) 601–625

[Sh] C. Shannon *A mathematical theory of communication*, Bell System Tech. J. 27 (1948) 379–423, 623–656

[Si1] Ya. Sinai *On the concept of entropy for a dynamical system*, Dokl. Akad Nauk SSSR 124 (1959) 768–771

[Si2] Ya. Sinai *Markov partiotions and C- diffeomorphisms*, Funct. Anal. and Appl. 2 (1968) 64–89

[W] P. Walters *Topological conjugacy of affine transformations of nilmanifolds*, Math Systems Theory 4, (1970) 327–333

DEPARTMENT: The Hebrew University, Institute of Mathematics, Jerusalem, Israel 91904

E-mail address: weiss@sunrise.huji.ac.il

Contemporary Mathematics
Volume **135**, 1992

THE IMPACT OF ROY ADLER'S WORK ON SYMBOLIC DYNAMICS AND APPLICATIONS TO DATA STORAGE [1] [2]

(or "THE ROY ADLER STORY, PART II")

Brian Marcus

Roy Adler's paper, "The Torus and the Disk"([Ad]), which is reprinted in this volume, is an excellent summary of Roy's work in ergodic theory, symbolic dynamics and applications. While I will repeat and elaborate on some of the material from that paper, I will focus mostly on the impact that Roy's work has had on the development of symbolic dynamics and applications to data storage. Of course, I have concentrated on those things that I know best.

Before that, however, I would like to share a few things that I have learned from Roy. These have helped me a great deal to keep my sanity while doing research.

First, Roy showed me how to do mathematics in an experimental way – to work with a concrete example until insights come rising to the surface. Second, he showed me how to play with mathematics in an innocent, intuitive, and light-hearted manner. Finally, he taught me not to be afraid to consider variations on central problems – you never know where they will lead. In particular, his work on classification of shifts of finite type focused on a new notion of equivalence that lies in between measure isomorphism and topological conjugacy; this ultimately led to a major breakthrough in the theory of data recording.

On the following page, I have highlighted some of Roy's seminal work and the flow from this work into other areas. The parenthetical acronyms refer to some of Roy's seminal papers (see the bibliography). Some of these items (including topological entropy, measure isomorphism for toral automorphisms, and the road problem) are discussed in the article by B. Weiss [W1] in this volume.

---

[1]1991 Mathematics Subject Classification: Primary 54H20, 58F11

[2]This paper is in final form and no version of it will be submitted for publication elsewhere.

Figure 1

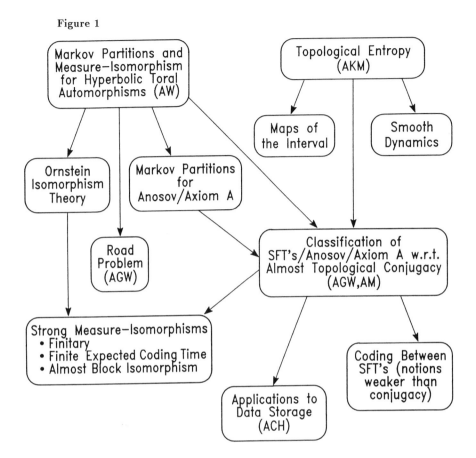

The Flow of Roy Adler's Work

1. ALMOST TOPOLOGICAL CONJUGACY AND ITS IMPACT ON SYMBOLIC DYNAMICS

1.1 Topological Conjugacy and Measure Isomorphism

The following table (which I have plagiarized from one of Roy's talks) is a perspective which compares topological dynamics with measure-preserving dynamics:

Topological Dynamics	*Measure − Preserving Dynamics*
Topological Dynamical System (X, ϕ)	*Measure − Preserving System* (X, ϕ, μ)
Topological Conjugacy	*Measure Isomorphism*
Topological Entropy $h(X, \phi)$	*Measure Entropy* $h_\mu(X, \phi)$
Shifts of Finite Type (X_A, σ)	*Markov Chains* (X_A, σ, μ)

Here, X is a compact metric space, ϕ is a homeomorphism of X, and μ is a ϕ-invariant (Borel probability) measure on X. Topological conjugacy (resp., measure isomorphism) are, of course, the natural equivalence relations used to classify topological dynamical systems (resp., measure-preserving dynamical systems). The entropies, topological (resp., measure), are crude, yet very meaningful and computable invariants. And shifts of finite type, abbreviated SFT, (resp., Markov chains) are classes of simple and concrete, yet rich and chaotic, dynamical systems. An SFT X_A is the set of bi-infinite trips on the finite directed graph whose adjacency matrix is A, and σ is the shift map. We often abbreviate (X_A, σ) as simply X_A, the shift map being understood. And we often refer to topological entropy as simply entropy. For simplicity, we will focus on mixing SFT's (i.e., for some m all entries of A^m are strictly positive) – although most of what I will say can be extended to irreducible SFT's. We refer the reader to [AM], [PT1], or [Wa] for background in symbolic dynamics and ergodic theory.

Any mixing SFT has a unique shift-invariant measure μ_{max} of maximal entropy ([P1], [Sh]). And it is not hard to see, by virtue of the uniqueness, that any topological conjugacy between mixing SFT's must be measure-preserving with respect to μ_{max}; on the other hand, it is quite possible (even typical) for two mixing SFT's to be measure isomorphic w.r.t. μ_{max} without being topologically conjugate. Thus, for mixing SFT's, we may regard topological conjugacy as a stronger (finer) equivalence relation than measure isomorphism.

Partly motivated by the Adler-Weiss classification of hyperbolic automorphisms of the two-dimensional torus [AW], Friedman and Ornstein [FO] proved that measure entropy is a complete measure isomorphism invariant for mixing Markov chains (of course, Ornstein's isomorphism theory did quite a bit more). As explained in [Ad], Roy thought that there ought to be an analogue in topological dynamics. He also knew that the obvious analogue ("topological entropy is a complete topological conjugacy invariant for mixing SFT's") was not true: topological entropy does not even capture the fixed point information. But Roy knew that the isomorphisms that he and Weiss had constructed, while not continuous, had much stronger properties than the run-of-the-mill measure isomorphism.

1.2 Almost Topological Conjugacy and Finitary Isomorphism

Roy introduced almost topological conjugacy (ATC) as a meaningful equivalence relation, with respect to which mixing SFT's could be completely classified by the simple and computable invariant: topological entropy. ATC lies in between measure isomorphism and topological conjugacy.

We define ATC as follows.

Definition: Two mixing SFT's X_A, X_B are *almost topologically conjugate* (ATC) if there exists a third mixing SFT X_C and finite-to-one, 1-1 a.e. factor maps $X_C \to X_A$ and $X_C \to X_B$. The two maps are called *legs*.

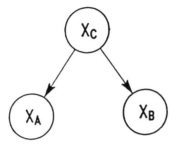

Figure 2

By a factor map, we mean a continuous, shift-commuting, onto map – in the language of information theory a factor map is an onto sliding block code.

Here, 1-1 a.e. can be taken in a topological sense (1-1 on the doubly transitive points) or in a measure sense (1-1 a.e with respect to any ergodic, fully-supported measure – in particular, μ_{max}); these senses turn out to be equivalent in this context. An ATC yields a measure isomorphism between the SFT's w.r.t μ_{max}.

Topological entropy is an invariant of ATC since entropy is preserved by finite-to-one factor maps. Roy and I showed that it is a complete invariant.

Theorem 1. *([AM]) Topological entropy is a complete almost topological conjugacy invariant for mixing SFT's.*

A central and motivating special case, SFT's with entropy $= \log(n), n \in \mathcal{Z}^+$, was handled earlier by Adler, Goodwyn and Weiss [AGW]. The essential ideas are, to a large extent, already in place in that beautiful paper. Also, with the exception of the 1-1 a.e. condition, the result was obtained independently by Parry [P2].

In fact, since Axiom A basic sets can be modeled by SFT's via Markov partitions (see [Bo],[Si]), Theorem 1 can be extended to the more general setting of mixing Axiom A basic sets: topological entropy is a complete ATC invariant for mixing Axiom A basic sets ([AM]).

For mixing SFT's, one way in which the strength of ATC (as opposed to mere measure isomorphism) can be seen is that the measure isomorphism $\phi : (X_A, \sigma, \mu_{max}) \to (X_B, \sigma, \mu_{max})$ generated by an ATC and its inverse, ϕ^{-1}, are both *finitary* with *finite expected coding time*. That is, for a.e. $x \in X_A$ $\phi(x)_0$, the zeroth coordinate of the image, depends only on $x_{-n(x)} \ldots x_{n(x)}$ where $n(x)$ satisfies $\int n(x) d\mu_{max} < \infty$, and likewise for ϕ^{-1}. In fact, in the construction used to prove Theorem 1, ϕ looks only to the future (i.e., $\phi(x)_0$ depends only on $x_0 \ldots x_{n(x)}$), and ϕ^{-1} looks only to the past.

Theorem 1 gives in particular a finitary isomorphism theorem for a significant class of mixing Markov chains (namely, the Markov chains of maximal entropy for mixing SFT's). This was extended to all mixing Markov chains by Keane and Smorodinsky [KeS]. It turns out that, in general, equality of measure entropy is far from sufficient to guarantee isomorphisms with finite expected coding time between mixing Markov chains. See [Kr1],[P3],[PS] [Sc]. The analogue, called almost block isomorphism, of ATC for mixing Markov chains has been studied in [PS], [MT], and [PT2].

1.3 Other Weakenings of conjugacy

The success of the classification up to ATC led to the hope that there might be other interesting weakenings of topological conjugacy with respect to which SFT's could be understood. Here are some examples.

1.3.a Imbeddings

Krieger's Generator Theorem [Kr2] implies that every mixing (actually ergodic will do) measure-preserving transformation (X, ϕ, μ) is measure isomorphic to a stationary stochastic process based on an alphabet of at most $\lfloor exp(h_\mu(X, \phi)) \rfloor + 1$ symbols. Roy thought that there ought to be an analogue of such a theorem for mixing SFT's with respect to some notion of "imbedding". One analogue would be that any mixing SFT X_A is almost topologically conjugate to a subsystem of the full shift on $\lfloor exp(h(X_A)) \rfloor + 1$

symbols. Roy showed that this is indeed true; in fact, only $\lceil exp(h(X_A)) \rceil$ symbols are needed (because if $h(X_A) = \log(n)$, then, by Theorem 1, only n symbols are needed). Krieger proved the following generalization which can be viewed as an unequal entropy version of Theorem 1.

Theorem 2. ([Kr3]): *Let X_A, X_B be mixing SFT's. A necessary and sufficient condition for the existence of an SFT $X_D \subseteq X_B$ such that X_A, X_D are almost topologically conjugate is that $h(X_A) \leq h(X_B)$.*

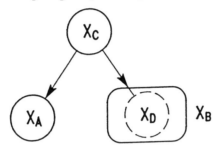

Figure 3

Note that in the particular case that X_B is a full shift, this is Roy's analogue.

Krieger pushed this line of research further. Rather than "lifting" X_A up to an extension and then "pushing" the extension down into X_B, one might hope to directly imbed X_A into X_B. Krieger showed that two simple and checkable necessary conditions are actually sufficient for the existence of a *proper imbedding* $X_A \to X_B$ (i.e., a 1-1 continuous shift-commuting map of X_A onto a proper subshift of X_B).

Theorem 3. ([Kr3]) *(Imbedding Theorem) Let X_A, X_B be mixing SFT's. There is a proper imbedding $X_A \to X_B$ if and only if*
    1. $h(X_A) < h(X_B)$
    and
    2. *For all n, $|Per_n(X_A)| \leq |Per_n(X_B)|$*
    (Per_n denotes the number of periodic points of least period n).

This result has, in turn, had an enormous impact on symbolic dynamics and its applications (for one striking application, see [BH]).

1.3.b Common Factors

Since finite-to-one factor maps preserve entropy, we know from Theorem 1 that if mixing SFT's X_A and X_B have the same entropy, then they have a common extension with the same entropy. When the entropy is the log of a positive integer n, then they have a common factor with the same entropy because every SFT of entropy $\log(n)$ factors onto the full n-shift ([M1]). Kitchens [Ki] gave an example to show that this is not true in general. Faced with this evidence, Roy asked the following question.

Q: Given a number h which occurs as the entropy of a mixing SFT, are there finitely many mixing SFT's $X_{A_1}, \ldots, X_{A_k}$ of entropy h, such that every mixing SFT X_A of entropy h factors onto one of the X_{A_i}.

The answer turns out to be no and yes. The answer to the problem as stated is no; Lind ([L2]) gave a broad class of counterexamples. However, Trow (see [T], [BMT]) showed that the answer is yes if we replace the term 'factors' by the term 'eventually factors' (X_A *eventually factors* onto X_B if and only if for all sufficiently large m, X_{A^m} factors onto X_{B^m}) and we allow negative entries in the matrices A_i (but, of course, $(A_i)^m$ must be nonnegative for large m). In fact, the minimal number k of SFT's needed is the number of ideal classes in a particular ring associated to the given entropy.

1.3.c Finite-to-One Factors

Suppose that the mixing SFT's X_A and X_B have the same entropy. It is natural to wonder under what conditions X_A directly factors onto X_B – without having to go through a common extension X_C. Here, we would like the factor map to have the additional properties, finite-to-one and 1-1 a.e., that the legs of an ATC have. For the moment, I will focus only on the finite-to-one aspect.

A result of Coven and Paul ([CP1]) says that for mixing SFT's, finite-to-one factor maps turn out to be the same as (topological) entropy-preserving factor maps. So, the finite-to-one factor problem for mixing SFT's is really the problem of when one given mixing SFT factors onto another given mixing SFT of the same entropy.

There are serious obstructions involving the zeta function and the Jordan form (see [Ki], [N]), and the exact conditions are not yet known. On the other hand, there is a good deal that is known. For example, as mentioned earlier, every SFT with entropy $= \log(n)$ factors (finite-to-one) onto the full n-shift. This result has been extended in a series of papers by Ashley, Boyle, Kitchens, Trow and myself ([T], [BMT], [KMT], [As1]). At this point, both a weakened and a strengthened version of the problem have been solved: necessary and sufficient algebraic conditions are known for (1) [KMT] the existence of an *eventual finite-to-one factor map* – that is, finite-to-one factor maps $X_{A^m} \to X_{B^m}$ for all sufficiently large m; and (2) [As1] the existence of a special, but important kind of finite-to-one factor map – *right closing*, which we will define in section 1.3.d.

Now, we return to the finite-to-one, 1-1 a.e. factor problem. First, let's look at the case where the mixing SFT's have entropy $= \log(n), n \in \mathcal{Z}^+$. Suppose, in addition, that one of the SFT's X_A has rowsum $= n$ (i.e., the row sums of the matrix A are all equal to n), and the other SFT X_B is the full n-shift. In this case, it is very easy to construct a finite-to-one factor map: at each state of the graph corresponding to A, label the n outgoing edges by the distinct labels $\{0, \dots, n-1\}$; the factor map simply reads off the labels of the edges. Such maps, called *road colorings*, were introduced in [AGW].

This takes care of the finite-to-one requirement. What about the 1-1 a.e. requirement? For road colorings, this is equivalent to the condition that there is an n-ary word w (called a *magic word*) so that all paths that produce w end at the same state: this guarantees that every doubly transitive point has exactly one preimage. The problem of constructing a road coloring with a magic word has come to be known as the Road Problem [AGW]. As mentioned in [Ad], this problem is still unsolved.

In [AGW], the Road Problem was bypassed: instead of solving it for the original SFT X_A, it was solved for an SFT conjugate to X_A. This was good enough to get a finite-to-one, 1-1 a.e. factor map $X_A \to X_B$.

So, here is an instance where the existence of a particular kind of finite-to-one factor map (a road coloring) implies the existence of a finite-to-one, 1-1 a.e. factor map (a road coloring with a magic word). Roy asked if something like this were true in greater generality, and Ashley established the following far-reaching generalization.

Theorem 4. *([As2]) Let X_A, X_B be mixing SFT's. Suppose that there exists a finite-to-one factor map from X_A onto X_B (in particular, $h(X_A) = h(X_B)$). Then there exists a finite-to-one, 1-1 a.e. factor map from X_A onto X_B.*

So, for mixing SFT's, the finite-to-one factor problem is the same as the finite-to-one, 1-1 a.e. factor problem.

1.3.d "One-sided" isomorphisms

A factor map is said to be *right resolving* if it is generated by a graph homomorphism (i.e., a pair of maps π_S on states and π_E on edges that respect initial/terminal states) with the following unique lifting property: for each state I in the domain and each outgoing edge f from state $\pi_S(I)$, there is a unique edge e outgoing from I such that $\pi_E(e) = f$:

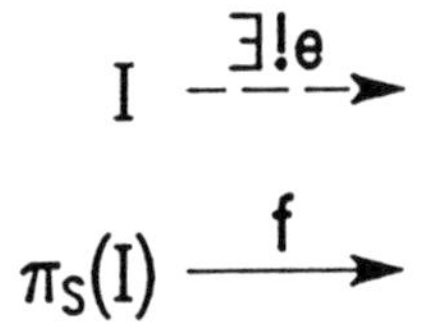

$$I \xrightarrow{\quad \exists! e \quad}$$

$$\pi_S(I) \xrightarrow{\quad f \quad}$$

Figure 4

Road colorings are examples (in fact, the motivating examples) of right resolving maps.

The definition of *left resolving* factor map is obtained by replacing each appearance of the term 'outgoing' by the term 'incoming' in the definition of right resolving. It is not hard to see that right and left resolving factor maps are finite-to-one.

Recall that the almost topological conjugacies constructed in the proof of Theorem 1 yield measure isomorphisms with the following property: ϕ looks only to the future and ϕ^{-1} looks only to the past. This is a consequence of the nature of the legs of the ATC: one is left resolving, and the other is right resolving.

This naturally led to the question of whether it is possible to construct the almost topological conjugacy so that both ϕ and ϕ^{-1} look to the same direction. This would result from an almost topological conjugacy in which both legs are resolving in the same direction - say right resolving. While this relation would be a weakening of the one-sided version of conjugacy, it is not a weakening of conjugacy; for instance, it is not hard to see that the set of row sums is an invariant of this relation, but not an invariant of conjugacy.

However, if we allow a "bounded delay" in the resolvingness of our maps, then we do get a weakening of conjugacy. This concept is embodied in the following notion: a factor map is *right closing* iff it is the composition of a right resolving map and a conjugacy. If we now require both legs of an ATC to be right closing, then the induced measure isomorphism and its inverse will both look only to the past except for a mere bounded look to the future. Such an isomorphism is called a *regular isomorphism*. This gives a relation which lies in between ATC and conjugacy. It refines each ATC class into finitely many subclasses. The complete set of invariants for this equivalence relation adds the ideal class $I(X_A)$, an equivalence class of ideals based on the eigenvector corresponding to the maximal eigenvalue of A, to the complete ATC invariant, topological entropy:

Theorem 5.(*[BMT]*) *Let X_A, X_B be mixing SFT's. Then the following are equivalent.*

(1) X_A, X_B are almost topologically conjugate by an ATC in which both legs are right closing.

(2) (X_A, σ, μ_{max}) and (X_B, σ, μ_{max}) are regularly isomorphic.

(3) $h(X_A) = h(X_B)$ and $I(X_A) = I(X_B)$.

1.4 Entropies of SFT's

Once mixing SFT's were classified (up to ATC), Roy thought it natural to ask:

Q: Which numbers appear as entropies of mixing SFT's?

This was answered by the beautiful theorem, due to Lind, which characterizes the set of entropies.

Theorem 6. *([L1]) The entropies of mixing SFT's are the numbers* $\log(\lambda)$ *such that*

1. λ *is an algebraic integer, and*
2. λ *strictly dominates (in modulus) its algebraic conjugates.*

A number which satisfies conditions 1 and 2 is called a *Perron number*. The set of Perron numbers has several nice algebraic properties, which suggest that it is a natural generalization of the set of positive integers. The set of entropies of all (not necessarily mixing) SFT's is the set of all $\frac{\log(\lambda)}{n}$ where λ is a Perron number and n is a positive integer.

Roy has also introduced the following fascinating problem. A *renewal system* is the set of all sequences obtained by freely concatenating a finite list of words (the words may be of varying lengths). Roy asked

Q: Is every mixing SFT topologically conjugate to a renewal system?

This problem is still unsolved. However, Goldberger, Lind and Smorodinsky were able to show that for mixing SFT's the answer to Roy's question is 'Yes' up to almost topological conjugacy:

Theorem 7. *([GLS]) The entropies of the SFT's are the same as the entropies of the renewal systems. In particular, every mixing SFT is almost topologically conjugate to a renewal system.*

The entropies of the uniquely decipherable systems (a special class of renewal systems) have recently been characterized by Handelman (see [Ha] in this volume).

Renewal systems are not always SFT's. However, they always do belong to the more general class of sofic systems, introduced by B. Weiss [W2]. A *sofic system* is the set of sequences obtained by reading off the labels of a finite directed labeled graph. The previous theorem extends to the class of sofic systems since sofic systems and SFT's have the same sets of entropies (see [CP2]) and since Theorem 1 generalizes to sofic systems. However, S. Williams showed that not every mixing sofic system is conjugate to a renewal system ([Ws]).

2. APPLICATIONS TO DATA RECORDING

2.1 Constraints, Encoders and Decoders

In recording data on storage devices, one is faced with the problem of encoding sequences from a full shift to sequences from a sofic system (sofic systems were defined at the end of the last section).

The sofic system is supposed to represent a constraint on the set of sequences that are allowed to be transmitted over a noisy channel or stored on a (noisy) disk drive. For this reason, we sometimes refer to a sofic system as a *constraint*. The idea is that some sequences (the "good" sequences) are less likely to be corrupted by noise. The engineer designs a sofic system which represents the "good" sequences. On the other hand, one would like to transmit any possible message – hence we have the problem of encoding the full shift into a sofic system.

Figure 5

One class of sofic systems, that arises frequently in practice, is the class of (d, k) run-length limited systems: for nonnegative integers $d \leq k$, the (d, k) *run-length limited system* is the set of all binary sequences where the lengths of runs of 0's are constrained to lie in between d and k.

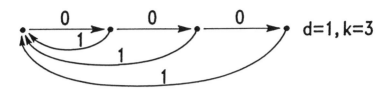

Figure 6

See [Ad] for a discussion of why these systems are useful in practice.

A run-length limited system is actually conjugate to the SFT X_A defined by the directed graph above – roughly because all sufficiently long words are magic words. In the remainder of the paper, we take SFT to mean a sofic system that is conjugate to an SFT (this is consistent with the ambiguity in the literature of the use of the term SFT). So, run-length limited systems may be regarded as SFT's.

What is an acceptable encoding/decoding device? A *finite-state (S, n)-encoder at rate $p:q$* is a finite state machine that accepts as input a p-block $\bar{x}_i = x_{[ip,(i+1)p)}$ in the full n-shift and produces as output a q-block $\bar{y}_i = y_{[iq,(i+1)q)}$ in the sofic system S; the output block depends on the input block and the internal state (one of finitely many); so, the encoder can be written as a function

$$\bar{y}_i = E(\bar{x}_i, z_i).$$

where z_i is the internal state. The ratio p/q is known as the *coding ratio* of the code; it measures how many unconstrained symbols can be represented by one constrained symbol. Transmission is most efficient when the coding ratio is as high as possible.

Such an encoder has a *sliding block decoder* if we can recover the input block $\bar{x}_i$ from the output block $\bar{y}_i$ together with a bounded amount of memory (m) and a bounded amount of anticipation (a) in the output stream; so the decoder can be written

$$\bar{x}_i = D(\bar{y}_{i-m}, \ldots, \bar{y}_{i+a})$$

Notice the lack of symmetry: the decoder has a different form than the encoder. The reason for this is as follows. Since the decoder does not involve any internal state information, it is clear that if an error is caused by noise in the channel, then it will give rise to a bounded number $(\leq m + a + 1)$ of decoded errors, and these errors will be confined to a window of bounded length $(\leq m + a + 1)$. Thus, errors in the channel can propagate by only a bounded distance.

A finite state (S, n)-encoder at rate $p : q$ with a sliding block decoder can be represented by the following commutative diagram:

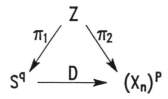

Figure 7

Here, S^q denotes the sofic system S taken in non-overlapping blocks of length q, $(X_n)^p$ is the full n-shift taken in non-overlapping blocks of length p, and Z is the SFT defined by the states of the encoder and transitions from one state to another. The map π_2 is a road coloring, and D (the sliding block decoder) is a factor map from the image of π_1 onto $(X_n)^p$. Both maps π_1, π_2 are finite-to-one (in fact, π_2 is right resolving and π_1 is right closing), and π_2 is onto. Thus, one necessary condition for the existence of such codes is

$$h(S^q) \geq h(Z) = h((X_n)^p),$$

and thus

$$h(S) \geq \frac{p}{q} \log(n).$$

This scenario is clearly quite closely related to Theorems 1 and 2 – though these results do not quite apply. Nevertheless, using ideas developed in the

ATC classification, as well as state splitting ideas ([Wr] and [M1]), Adler, Coppersmith and Hassner were able to prove the following result.

Theorem 8. *([ACH]) Let S be an SFT. Then there is a finite-state (S, n)-encoder at rate $p : q$ with sliding block decoder if and only if*

$$h(S) \geq \frac{p}{q} \log(n).$$

More importantly, the construction of the encoder and decoder is given by an effective algorithm, now known as the state splitting algorithm. While considered a major breakthrough in data storage applications, the algorithm is related to heuristic techniques known earlier e.g., [Fr] , [Pa]. Descriptions of the algorithm are now available in several places e.g., [Bl], [Im], [KN], [MSW], [SC].

The paper [ACH] was immediately recognized as a significant contribution to information theory; it won the best paper award of the IEEE-Information Theory Society for the year 1983. The algorithm is now widely used in data recording applications. The following is a list of some of the papers and patents in which the algorithm has been used to construct codes: [AHM], [FWD], [HI], [Ho], [KS], [MPS], [MP], [WW], [We].

2.2 Further work and variations on the state splitting algorithm

The state splitting algorithm has motivated a good deal of further theoretical work in the subject; this work has the potential for significant influence on practical applications as well. I would like to describe a few of these items.

2.2.a Generalization to Sofic Systems

As stated above, the state splitting algorithm applies to SFT constraints. What about the more general sofic constraints? These can be practical too. One example that comes up is the class of charge constrained sequences. This class of systems is parameterized by a parameter c, and each system is defined as the set of all ±1-valued sequences such that the sum of the symbols in any block is bounded in absolute value by c.

Figure 8

The state splitting algorithm was generalized to the sofic setting in the papers [M2] and [KM]. In the case of excess entropy (i.e., when $h(S) > \frac{p}{q}\log(n)$), the same result holds as for SFT's: there is a finite state (S, n)-encoder at rate $p : q$ with sliding block decoder.

When there is no excess in entropy (i.e., when $h(S) = \frac{p}{q}\log(n)$), there is always a finite state (S, n)- encoder at rate $p : q$ which can be decoded, but it may be impossible to decode it with a sliding block decoder. However, one can always construct the encoder so that the number of decoded errors produced by a channel error is guaranteed to be bounded – although these decoded errors may occur arbitrarily far apart. But the algorithm for constructing such codes in general is enormously complex.

2.2.b Variable Length State Splitting

The state splitting algorithm constructs encoders by an iterative state splitting procedure; the procedure is applied to the graph which corresponds to the matrix that defines the SFT. The edges of such graphs all have the same fixed length. Some constraints are more naturally and more compactly described by graphs whose edges have variable lengths. For example, the (d, k) run-length limited system can be described by the ordinary fixed length graph with $k + 1$ states (shown earlier), but it can also be described by a variable length graph with only one state:

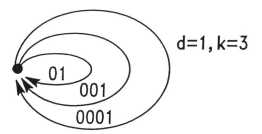

Figure 9

In [AFKM] and [HMS], a variable length version of the state splitting algorithm was developed. This variant is very useful for constraints that would ordinarily be presented by fixed length graphs with enormous numbers of states. However, because the lengths of the edges in the variable length presentation may be large, there can be significant problems with control of error propagation in the decoder. Nevertheless, Belongie and Heegard [BeH] have used this method to construct very nice codes which satisfy some useful constraints and have significant error-correcting power as well.

2.2.c Bounds on encoder size and decoder error propagation

In practice, one wants the encoder to be relatively small and the extent of decoder error propagation to also be relatively small. There has been some work which establishes bounds on the sizes of these two very important parameters.

In [As3], Ashley gave a construction that shows one can always obtain for SFT's, using a variant of the state splitting algorithm, finite state encoders with decoder error propagation that is at most quadratic as a function of the number of states of the original presentation of the SFT. In [Ka], Kamabe found the smallest decoder windows for certain run length limited constraints at certain rates.

In [MS], Siegel and I showed how to use state merging, in conjunction with the state splitting algorithm, to reduce the size of the encoders. In [MR], Roth and I gave lower bounds on the size of an encoder that can realize any given $(S, n), p$ and q. These bounds are quite general, fairly computable and, in some special classes, even definitive. For some natural classes of systems, this shows that the state splitting algorithm yields encoders with the smallest possible number of states.

ACKNOWLEDGEMENTS: I would like to thank Paul Algoet for reading an earlier draft of this paper, Selim Tuncel for helpful discussions, and Roy Adler for being Roy Adler.

BIBLIOGRAPHY

[Ad] R. Adler, "The Torus and the Disk" *IBM J. Res. & Dev.*, 31 (1987), 224-234.

[ACH] R. Adler, D. Coppersmith, and M. Hassner, "Algorithms for sliding-block codes," *IEEE Trans. Info. Th.*, vol. IT-29, No. 1, pp. 5-22, January 1983.

[AFKM] R. Adler, J. Friedman, B. Kitchens, and B. Marcus, "State splitting for variable-length graphs," *IEEE Trans. Info. Th.*, vol. IT-32, no. 1, pp. 108-113, January 1986.

[AGW] R. Adler, L. Goodwyn, and B. Weiss, "Equivalence of topological Markov shifts," *Israel J. Math.*, vol 27, pp. 49-63, 1977.

[AHM] R. Adler, M. Hassner, and J. Moussouris, "Method and apparatus for generating a noiseless sliding block code for a (1,7) channel with rate 2/3," U.S. Patent 4,413,251, 1982.

[AKM] R. Adler, A. Konheim, and M. McAndrew, "Topological Entropy" *Trans. Amer. Math. Soc.*, 114 (1965), 309-319.

[AM] R. Adler and B. Marcus, "Topological entropy and equivalence of dynamical systems", *Mem. AMS* 219 (1979).

[AW] R. Adler and B. Weiss, "Similarity of automorphisms of the torus" *Mem. Amer. Math. Soc.*, 98 (1970).

[As1] J. Ashley, "Resolving factor maps for shifts of finite type with equal entropy", *Ergod. Th. & Dynam. Sys.*, 11 (1991), 219-240.

[As2] J. Ashley, " Bounded-to-1 factors of an aperiodic shift of finite type are 1-to-1 almost everywhere factors also" *Ergod. Th. and Dynam. Sys.* 10 (1990), 615-626.

[As3] J. Ashley, "A linear bound for sliding-block decoder window size," *IEEE Trans. Info. Th.*, vol. IT-34, no. 3, pp. 389-399, May 1988.

[BeH] M. Belongie and C. Heegard, "Run length limited codes with coding gains from variable length graphs", preprint.

[Bl] R. E. Blahut, Digital Transmission of Information Reading, Massachusetts: Addison-Wesley, 1990.

[Bo] R. Bowen, "Markov partitions for Axiom A diffeomorphisms" *Amer. J. Math.* 91 (1970) 725-747.

[BH] M. Boyle and D. Handelman, "The spectra of nonnegative matrices via symbolic dynamics", *Annals of Math.*, 133 (1991), 249-316.

[BMT] M. Boyle, B. Marcus and P. Trow, "Resolving maps and the dimension group for shifts of finite type", *Mem. AMS* 377 (1987).

[CP1] E. Coven and M. Paul, "Endomorphisms of irreducible shifts of finite type", *Math. Sys. Theory* 8 (1974), 167-175.

[CP2] E. Coven and M. Paul, "Sofic systems", *Israel J. Math.* 20 (1975), 165-177.

[Fr] P. Franaszek, "Construction of bounded delay codes for discrete noiseless channels," *IBM J. Res. Dev.*, vol. 26, pp. 506-514, 1982.

[FWD] C. French, J. Wolf and G. Dixon, "Signaling with special run length constraints for a digital recording channel" *IEEE-Magnetics*, 24 (1988), 2092-2097.

[FO] N. Friedman and D. Ornstein, "An isomorphism of weak Bernoulli tranformations" *Advances in Math*, 5 (1971), 365-394.

[GLS] J. Goldberger, D. Lind, and M. Smorodinsky, "The entropies of renewal systems" *Israel J. Math.*, 33 (1991), 1-23.

[Ha] D, Handleman, "Spectral radii of primitive integral companion matrices and log concave polynomials", this volume.

[HMS] C. Heegard, B. Marcus, and P. Siegel, "Variable length state splitting with applications to average runlength constrained (ARC) codes," *IEEE Trans. Info. Th.*, vol. 37, no. 3, pt. II, pp. 759-777, May 1991.

[HI] H. Hollmann and K. Immink, "Prefix-synchronized run-length codes", *IEEE-Journal on Selected Areas in Communication*, 10(1992), 214-222.

[Ho] T. Howell, "Statistical properties of selected recording codes," *IBM J. Res. Dev.*, vol. 33, no. 1, pp. 60-73, January 1989.

[Im] K. A. Schouhamer Immink, Coding Techniques for Digital Recorders, Englewood Cliffs, New Jersey: Prentice-Hall, 1991.

[Ka] H. Kamabe, "Minimum scope for sliding-block decoder mappings," *IEEE Trans. Info. Th.*, vol. 35, no. 6, pp. 1335-1340, November 1989.

[KM] R. Karabed and B. Marcus, "Sliding-block coding for input-restricted channels," *IEEE Trans. Info. Th.*, vol. IT-34, No. 1, pp. 2-26, January 1988.

[KS] R. Karabed and P. Siegel, "Matched spectral-null codes for partial-response channels," *IEEE Trans. Info. Th.*, vol. 37, no. 3, pt. II, pp. 818-855, May 1991.

[KeS] M. Keane and M. Smorodinsky, "Finitary isomorphisms of irreducible Markov shifts, *Israel J. Math* 34 (1979), 281-286.

[KN] A. Khayrallah and D. Neuhoff, "Subshift models and finite-state codes for input-constrained noiseless channels: a tutorial," preprint (based upon Ph.D. dissertation by A. Khayrallah, U. Mich., 1989).

[Ki] B. Kitchens, "An invariant for continuous factors of Markov shifts", *Proc. AMS* 83 (1981), 825-828.

[KMT] B. Kitchens, B. Marcus and P. Trow, "Eventual factor maps and compositions of closing maps", *Ergod. Th. & Dynam. Sys.* 11 (1991) 85-113.

[Kr1] W. Krieger, "On the finitary isomorphisms of Markov shifts that have finite expected coding time", *Z. Wahr.* 65 (1983), 323-328.

[Kr2] W. Krieger, "On entropy and generators of measure preserving transformations" *Trans. Amer. Math. Soc.*, 149 (1970) 453-464; erratum: 168 (1972), 519.

[Kr3] W. Krieger, "On the subsystems of topological Markov chains", *Ergod. Th. & Dynam. Sys.*, 2 (1982) 195-202.

[L1] D. Lind, "The entropies of toplogical Markov shifts and a related class of algebraic integers" *Ergod. Th. & Dynam. Sys.*, 4 (1984) 283-300.

[L2] D. Lind, "The spectra of topological Markov shifts", *Ergod. Th. & Dynam. Sys.*, 6 (1986), 571-582.

[M1] B. Marcus, "Factors and extensions of full shifts," *Monatshefte fur Math.*, vol. 88, pp. 239-247, 1979.

[M2] B. Marcus, "Sofic systems and encoding data," *IEEE Trans. Info. Th.*, vol. IT-31, No. 3, pp. 366-377, May 1985.

[MPS] B. Marcus, A. Patel, and P. Siegel, "Method and apparatus for implementing a PRML code," U.S. Patent 4,786,890, November 1988.

[MR] B. Marcus and R. Roth, "Bounds on the number of states in encoder graphs for input-constrained channels," *IEEE Trans. Info. Th.*, vol. 37, no. 3, pt. II, pp. 742-758, May 1991.

[MS] B. Marcus and P. Siegel, "Constrained codes for partial response channels," *Beijing Int. Workshop Info. Th.*, DI1.1-DI1.4, July 1988.

[MSW] B. Marcus, P. Siegel, and J. Wolf "A tutorial on finite-state modulation codes" *IEEE-Journal on Selected Areas in Communication*, 10(1992) 5-37.

[MT] B. Marcus and S. Tuncel "The weight-per-symbol polytope and scaffolds of invariants associated with Markov chains", *Ergod. Th. & Dynam. Sys.*, 11 (1991) 129-180.

[MP] M. Monti and Pierobon, "Codes with multiple spectral null at zero frequency", *IEEE-Information Theory*, 35 (1989), 463-472.

[N] M. Nasu, "Uniformly finite-to-one and onto extensions of homomorphisms between strongly connected graphs", *Discrete Math*, 39 (1982) 171-197.

[P1] W. Parry, "Intrinsic Markov chains" *Trans. Amer. Math. Soc.* 112 (1964), 55-66.

[P2] W. Parry, "A finitary classification of topological Markov chains and sofic systems", *Bull. Lond. Math. Soc.* 9 (1977), 86-92.

[P3] W. Parry, "Finitary isomorphisms with finite expected code-lengths" *Bull. London Math. Soc*, 11 (1979) 170-176.

[PS] W. Parry and K. Schmidt, "Natural coefficients and invariants for Markov shifts", *Invent. Math.* 76 (1984), 15-32.

[PT1] W. Parry and S. Tuncel, Classification Problems in Ergodic Theory (LMS Lecture Notes, v. 67), Cambridge Univ. Press, 1982.

[PT2] W. Parry and S. Tuncel, "On the classification of Markov chains by finite equivalence", *Ergod. Th. and Dynam. Sys.* 1 (1981), 303-335.

[Pa] A. Patel, "Zero modulation encoding in magnetic recording," *IBM J. Res. Dev.*, vol. 19, no. 4, pp. 366-378, July 1975.

[Sc] K. Schmidt, "Invariants for finitary isomorphisms with finite expected code lengths", *Invent. Math.* 76 (1984), 33-40.

[Sh] C. Shannon, "A mathematical theory of communication", *Bell. Sys. Tech. J.* 27 (1948), 379-423; 623-656.

[Si] Ya. G. Sinai, "Markov partitions and C-diffeomorphisms" *Funct. Anal. & Appl.* 2 (1968) 64-89.

[SC] N. Swenson and J. Cioffi, "A simplified design approach for run-length limited sliding block codes," preprint (based upon Ph.D. Dissertation by N. Swenson, Stanford Univ., 1991).

[T] P. Trow, "Resolving maps which commute with a power of the shift" *Ergod. Th. & Dynam. Sys.* 6 (1986) 281-293.

[Wa] P. Walters, An introduction to Ergodic theory, Graduate Texts in Math 79, Springer, New York, 1982.

[WW] A. Weathers and J. Wolf, "A new rate 2/3 sliding-block code for the (1,7) runlength constraint with the minimal number of encoder states,"

IEEE Trans. Info. Th., vol. 37, no. 3, pt. II, pp. 908-913, May 1991.

[We] T. Weigandt, "Magneto-optic recording using a (2,18,2) run-length-limited code," SM Thesis, MIT, 1991.

[W1] B. Weiss, "On the work of Roy Adler in ergodic theory and dynamical systems", this volume.

[W2] B. Weiss, "Subshifts of finite type and sofic systems", *Monats. fur Math.*, 77 (1973), 462-474.

[Wr] R. Williams, "Classification of shifts of finite type", *Ann. Math.* 98 (1973), 120-153; Errata, Ann. Math. 99 (1974), 380-381.

[Ws] S. Williams, "Notes on renewal systems" *Proc. Amer. Math. Soc.*, 110 (1990) 851-853.

IBM Almaden Research Center, San Jose, CA, 95120

Contemporary Mathematics
Volume **135**, 1992

LR Conjugacies of Shifts of Finite Type are Uniquely So

JONATHAN ASHLEY*

Abstract

According to a construction originated by Williams and modified by Kitchens, any conjugacy between shifts of finite type, when composed with a sufficiently high power of the shift, can be decomposed into the product of the inverse of a left-resolving conjugacy and a right-resolving conjugacy. We show that this decomposition is essentially unique.

Key words: shift of finite type, conjugacy, automorphism.

Subject index primary classifications: 58F11.

*IBM Research Division, Almaden Research Center, 650 Harry Road, San Jose, CA 95120.

This paper is in final form and no version of it will be published elsewhere.
1991 Mathematics Subject Classification. Primary 58F11.

is the 1-block conjugacy induced by the graph isomorphism from $\mathcal{G}_C$ to $\mathcal{G}_{C'}$.

This result is already proved in [Nas]. We hope our proof technique is of interest.

To *any* conjugacy $\phi : \Sigma_A \to \Sigma_B$ we can associate the least integer n such that $\phi \circ \sigma^n$ is an LR conjugacy. For each $m \geq n$, there is a unique decomposition of $\phi \circ \sigma^m$ as in theorem 1. For $m > n$, the decompositions are simply derived from that for n by replacing Σ_C by the higher block system $\Sigma_C^{[m-n]}$, and replacing λ and ρ by the natural compositions of maps.

2. Background

Starting with any finite set F of symbols we define the *full shift* to be the product space $F^{\mathbb{Z}}$ with topology given by the product of the discrete topologies on F. The shift map $\sigma : F^{\mathbb{Z}} \to F^{\mathbb{Z}}$ defined by

$$\sigma(x)_i = x_{i+1}$$

is a continuous bijection.

A closed, shift-invariant subset of $F^{\mathbb{Z}}$ is called a *shift space*.

A continuous mapping $\phi : S \to T$ between shift spaces that preserves the shift structure ($\sigma_T \circ \phi = \phi \circ \sigma_S$) is called a *blockmap*. For any block map ϕ there are non-negative integers m and a (called the *memory* and *anticipation* of ϕ) such that $\phi(x)_i$ depends only on the finite sequence $x_{i-m} \ldots x_i \ldots x_{i+a}$. To emphasize the length of this window, ϕ is called a $(m + a + 1)$-block map.

A continuous bijection $\phi : S \to T$ between shift spaces that preserves the shift structure is called a *conjugacy*. One would like to classify shift spaces into conjugacy classes. The greatest progress has been made in classifying a particular class of shift spaces, the *shifts of finite type.*

A *shift of finite type* is the subset of $F^{\mathbb{Z}}$ remaining after discarding all points that contain any occurance of any word from a finite list W of (finite length) forbidden words. To be a bit more formal, the shift of finite type Σ defined by the list W is

$$\Sigma = \left\{ x \in F^{\mathbb{Z}} : \text{no word } w \in W \text{ occurs in the sequence } x \right\}.$$

We actually use a more special definition of SFT in terms of directed graphs. Given a finite directed graph $\mathcal{G}$ with edges E and states V, we define the SFT $\Sigma_{\mathcal{G}}$ as

$$\Sigma_{\mathcal{G}} = \left\{ x \in E^{\mathbb{Z}} : \forall i \text{ edge } x_{i+1} \text{ follows edge } x_i \text{ in } \mathcal{G} \right\},$$

which is the set of bi-infinite paths in $\mathcal{G}$. This is a special case of the forbidden word definition, but not really so special: any SFT is conjugate to some SFT $\Sigma_{\mathcal{G}}$ defined in this way by a graph $\mathcal{G}$.

A 1-block map $\phi : S \to T$ is *left resolving* if for any two distinct 2-blocks $x'x$ and $x''x$ occuring in S, the symbols x' and x'' have distinct images under ϕ. Similarly, a 1-block map $\phi : S \to T$ is *right resolving* if for any two distinct 2-blocks xx' and xx'' occuring in S, the symbols x' and x'' have distinct images under ϕ. If a 1-block map is both left and right resolving, we say it is *bi-resolving.*

A 1-block map ϕ whose domain is an SFT $\Sigma_{\mathcal{G}}$ can be regarded as being given by an *edge labeling* ϕ_* of the graph $\mathcal{G}$: each edge e

1. Introduction

In [Wil73, Wil74], Williams proved that two shifts of finite type Σ_A and Σ_B are conjugate if and only if the square non-negative integer matrices A and B are strong shift equivalent. Kitchens [Kit87] later modified the construction Williams used in the course of proving that conjugacy implies strong shift equivalence. This gave the following theorem.

Theorem 1. *If $\phi : \Sigma_A \to \Sigma_B$ is a conjugacy between shifts of finite type such that $\phi(x)_0$ depends only on $x_0 x_1 \ldots x_n$ and $\phi^{-1}(y)_0$ depends only on $y_{-n} y_{-n+1} \ldots y_0$, then the conjugacy $\phi \circ \sigma^n$ can be decomposed as $\phi = \rho \circ \lambda^{-1}$, where $\lambda : \Sigma_C \to \Sigma_A$ is a left-resolving 1-block conjugacy and $\rho : \Sigma_C \to \Sigma_B$ is a right-resolving 1-block conjugacy from a shift of finite type Σ_C.*

We give Kitchens' construction in section 3. For any conjugacy ψ there is an integer N such that $\phi = \psi \circ \sigma^n$ satisfies the hypothesis of the theorem for any $n \geq N$. We give the simple upper bound on N in section 3.

Nasu [Nas] calls an automorphism $\psi : \Sigma_A \to \Sigma_A$ that can be so decomposed an *LR automorphism*. Likewise, we call a conjugacy that can be decomposed as in theorem 1 an *LR conjugacy*. Our main result, theorem 2, is that the LR decomposition $\phi = \rho \circ \lambda^{-1}$ (where $\lambda : \Sigma_C \to \Sigma_A$ and $\rho : \Sigma_C \to \Sigma_B$) is unique in the following sense. If $\phi = \rho' \circ (\lambda')^{-1}$ (where $\lambda' : \Sigma_{C'} \to \Sigma_A$ is a left-resolving conjugacy and $\rho' : \Sigma_{C'} \to \Sigma_B$ is a right-resolving conjugacy) is another such decomposition of ϕ, then the graphs $\mathcal{G}_C$ and $\mathcal{G}_{C'}$ are graph isomorphic. Furthermore, $\lambda = \lambda' \circ \iota$ and $\rho = \rho' \circ \iota$, where $\iota : \Sigma_C \to \Sigma_{C'}$

of $\mathcal{G}$ is labeled by $\phi_*(e) = \phi(x)_0$, where $x \in \Sigma_{\mathcal{G}}$ is any point with $x_0 = e$. In terms of graph labelings, a 1-block map is right resolving if and only if its corresponding labeling is (forwardly) deterministic. Similarly, ϕ is left resolving if and only if its labeling is backwardly deterministic.

Finally we need some notation for *cylinder sets* contained in a SFT $\Sigma_{\mathcal{G}}$. For any edge e of $\mathcal{G}$, denote the initial state of e by $is(e)$ and the terminal state by $ts(e)$. A point x of $\Sigma_{\mathcal{G}}$ is a bi-infinite sequence of edges $\ldots x_{-1}x_0x_1\ldots$ and therefore determines a bi-infinite sequence of states. Just to fix notation, we call $is(x_i)$ the ith state along the bi-infinite sequence x. Denote

$$_j[e_0e_1\ldots e_n]_{j+n+1} = \left\{x \in \Sigma_{\mathcal{G}} : x_{j+i} = e_i,\ 0 \le i \le n\right\},$$

where the subscripts indicate that the cylinder specification begins at time j (with the j-th state being $is(e_0)$) and ends at time $j+n+1$ (with the $(j+n+1)$-th state being $ts(e_n)$). If s is a state of $\mathcal{G}$, we denote

$$_j[s]_j = \left\{x \in \Sigma_{\mathcal{G}} : is(x_j) = s\right\}.$$

3. The Proofs

We begin this section with Kitchens' proof of theorem 1.

Proof of theorem 1. We are given a conjugacy $\phi : \Sigma_{\mathcal{G}} \to \Sigma_{\mathcal{H}}$ such that $\phi(x)_0$ depends only on $x_0x_1\ldots x_n$ and $\phi^{-1}(y)_0$ depends only on $y_{-n}y_{-n+1}\ldots y_0$. Our aim is to construct a graph $\mathcal{G}'$; a left resolving conjugacy $\lambda : \Sigma_{\mathcal{G}'} \to \Sigma_{\mathcal{G}}$; and a right resolving conjugacy $\rho : \Sigma_{\mathcal{G}'} \to \Sigma_{\mathcal{H}}$ such that $\phi \circ \sigma^n = \rho \circ \lambda^{-1}$. We first specify the graph

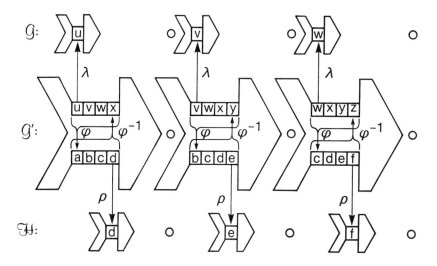

Figure 1: An LR decompostion of $\phi \circ \sigma^n$

$\mathcal{G}'$. Figure 1 illustrates the construction.

The edges of $\mathcal{G}'$ are the pairs $(x_0 \ldots x_n, \; y_0 \ldots y_n)$ obtained by reading off coordinates 0 through n of a pair of points $(x, y) \in \Sigma_\mathcal{G} \times \Sigma_\mathcal{H}$ having $y = \phi(x)$. If $n > 0$, the initial and terminal states of edge $(x_0 \ldots x_n, \; y_0 \ldots y_n)$ are defined to be the pairs $(x_0 \ldots x_{n-1}, \; y_0 \ldots y_{n-1})$ and $(x_1 \ldots x_n, \; y_1 \ldots y_n)$ respectively. Otherwise (if $n = 0$), the initial and terminal states of edge $(x_0, \; y_0)$ are defined to be $(is(x_0), \; is(y_0))$ and $(ts(x_0), \; ts(y_0))$ respectively.

The 1-block map $\lambda : \Sigma_{\mathcal{G}'} \to \Sigma_\mathcal{G}$ is defined by mapping the edge $(x_0 \ldots x_n, \; y_0 \ldots y_n)$ of $\mathcal{G}'$ to the edge x_0 of $\mathcal{G}$; and the 1-block map

$\rho : \Sigma_{\mathcal{G}'} \to \Sigma_{\mathcal{H}}$ is defined by mapping the edge $(x_0 \ldots x_n, \; y_0 \ldots y_n)$ of $\mathcal{G}'$ to the edge y_n of $\mathcal{H}$.

We first verify that λ and ρ are conjugacies and that $\phi \circ \sigma^n = \rho \circ \lambda^{-1}$. It is easy to show that if $z \in \Sigma_{\mathcal{G}'}$, then $z_i = (x_i \ldots x_{i+n}, \; y_i \ldots y_{i+n})$ for some $x \in \Sigma_{\mathcal{G}}$ and $y \in \Sigma_{\mathcal{H}}$ with $y = \phi(x)$. In fact the mapping $z \mapsto (x, \phi(x))$ is a 1-block conjugacy $\psi : \Sigma_{\mathcal{G}'} \to \Sigma$, where $\Sigma = \{(x, y) \in \Sigma_{\mathcal{G}} \times \Sigma_{\mathcal{H}} : y = \phi(x)\}$. The projections $\pi_1 : \Sigma \to \Sigma_{\mathcal{G}}$ and $\pi_2 : \Sigma \to \Sigma_{\mathcal{H}}$ are also 1-block conjugacies. Now observe that $\lambda = \pi_1 \circ \psi$ and $\rho = \sigma^n \circ \pi_2 \circ \psi$, so

$$
\begin{aligned}
\rho \circ \lambda^{-1}(x) &= \sigma^n \circ \pi_2 \circ \psi \circ \psi^{-1} \circ \pi_1^{-1}(x) \\
&= \sigma^n \circ \pi_2 \circ \pi_1^{-1}(x) \\
&= \sigma^n \circ \pi_2(x, \phi(x)) \\
&= \sigma^n \circ \phi(x).
\end{aligned}
$$

We now show that λ is left resolving. The proof that ρ is right resolving is similar. We give the proof for the case $n > 0$; the case $n = 0$ is even simpler. Let $z_0 = (x_0 \ldots x_n, \; y_0 \ldots y_n)$ be an edge of $\mathcal{G}'$. The 1-block map λ maps z_0 to the edge x_0 in $\mathcal{G}$. We must show that for each edge x_{-1} preceding edge x_0 in $\mathcal{G}$, there is at most one edge z_{-1} preceding edge z_0 in $\mathcal{G}'$ such that $\lambda(z_{-1}) = x_{-1}$. If $z_{-1} = (u_0 \ldots u_n, \; v_0 \ldots v_n)$ is any edge preceding the edge z_0 then $u_1 \ldots u_n = x_0 \ldots x_{n-1}$ and $v_1 \ldots v_n = y_0 \ldots y_{n-1}$. If in addition, we assume $\lambda(z_{-1}) = x_{-1}$, then $u_0 = x_{-1}$. The only data defining the edge z_{-1} that we have not yet shown to be determined is the symbol v_0. But $v_0 = \phi(u)_0$ depends only on the $(n+1)$-block $u_0 u_1 \ldots u_n$, and is therefore determined also. Thus, λ is left resolving. $\qquad \square$

We remark that if $\psi : \Sigma_{\mathcal{G}} \to \Sigma_{\mathcal{H}}$ is *any* conjugacy, then for all sufficiently large k, $\psi \circ \sigma^k$ is an LR conjugacy. We outline the proof. If ψ has memory m and anticipation a, and ψ^{-1} has memory $m_\star$ and anticipation $a_\star$, then $\phi = \psi \circ \sigma^{\max(m, a_\star)}$ and $n = 1 + \max(m, a_\star) + \max(m_\star, a)$ satisfy the hypothesis of theorem 1, so

$$\psi \circ \sigma^{1+2\max(m,a_\star)+\max(m_\star,a)}$$

is an LR conjugacy.

Our main objective is to show LR deompositions are unique.

Theorem 2. *The LR decomposition of an LR conjugacy ϕ : $\Sigma_\mathcal{G} \to \Sigma_\mathcal{H}$ is unique in the following sense. If $\phi = \rho_1 \circ \lambda_1^{-1}$ and $\phi = \rho_2 \circ \lambda_2^{-1}$ are two LR decompositions of ϕ, where the common domain of λ_1 and ρ_1 is $\Sigma_{\mathcal{G}_1}$ and the common domain of λ_2 and ρ_2 is $\Sigma_{\mathcal{G}_2}$, then there is a 1-block conjugacy $\iota : \Sigma_{\mathcal{G}_1} \to \Sigma_{\mathcal{G}_2}$ induced by a graph isomorphism $\iota_\star : \mathcal{G}_1 \to \mathcal{G}_2$ satisfying $\lambda_1 = \lambda_2 \circ \iota$ and $\rho_1 = \rho_2 \circ \iota$.*

Figure 2 gives a commutative diagram illustrating the situation.

We remark that the theorem holds even when ϕ is only a *right closing* block map. The proof is essentially the same.

We shall state four lemmas, lemmas 1, 2, 3, and 4, and then prove theorem 2 using these lemmas. We shall defer the proofs of these lemmas until after the proof of theorem 2 because the lemmas' proofs are more technical.

Lemma 1. *The inverse λ^{-1} of a left resolving conjugacy λ : $\Sigma_{\mathcal{G}'} \to \Sigma_\mathcal{G}$ has memory 0.*

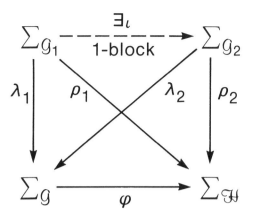

Figure 2: A commutative diagram with two LR decompositions of ϕ.

The next lemma is key. Its statement is illustrated in figure 3.

Lemma 2. *Let $\phi : \Sigma_{\mathcal{G}} \to \Sigma_{\mathcal{H}}$ be an n-block map with memory 0. There is a left resolving conjugacy $\lambda_0 : \Sigma_{\mathcal{G}_0} \to \Sigma_{\mathcal{G}}$ such that:*

(1) the composition $\rho_0 = \phi \circ \lambda_0$ is a 1-block map from $\Sigma_{\mathcal{G}_0}$ to $\Sigma_{\mathcal{H}}$,

(2) if $\lambda_1 : \Sigma_{\mathcal{G}_1} \to \Sigma_{\mathcal{G}}$ is any left resolving conjugacy such that the composition $\rho_1 = \phi \circ \lambda_1$ is a 1-block map, then there is a left resolving conjugacy $\iota : \Sigma_{\mathcal{G}_1} \to \Sigma_{\mathcal{G}_0}$ with $\lambda_1 = \lambda_0 \circ \iota$.

Thus there is a 'left resolving least upper bound' λ_0 for ϕ. We remark without proof that $\lambda_0 : \Sigma_{\mathcal{G}_0} \to \Sigma_{\mathcal{G}}$ is unique up to 1-block conjugacies on $\Sigma_{\mathcal{G}_0}$ induced by graph isomorphisms on $\mathcal{G}_0$.

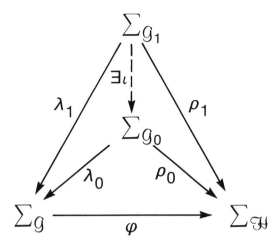

Figure 3: λ_0 is a left resolving least upper bound for ϕ.

Lemma 3. [BMT87, (4.11)]. *If $\iota : \Sigma_{\mathcal{G}_1} \to \Sigma_{\mathcal{G}_0}$ and $\rho_0 : \Sigma_{\mathcal{G}_0} \to \Sigma_{\mathcal{H}}$ are any 1-block maps whose composition $\rho_0 \circ \iota$ is a right resolving 1-block map, then ι is itself right resolving.*

Lemma 4. *If $\iota : \Sigma_{\mathcal{G}_1} \to \Sigma_{\mathcal{G}_0}$ is a bi-resolving conjugacy, then it is induced by a graph isomorphism $\iota_* : \mathcal{G}_1 \to \mathcal{G}_0$.*

Proof of theorem 2. Suppose $\phi : \Sigma_{\mathcal{G}} \to \Sigma_{\mathcal{H}}$ has the LR decompostion $\phi = \rho_1 \circ \lambda_1^{-1}$, where $\lambda_1 : \Sigma_{\mathcal{G}_1} \to \Sigma_{\mathcal{G}}$ is a left resolving conjugacy and $\rho_1 : \Sigma_{\mathcal{G}_1} \to \Sigma_{\mathcal{H}}$ is a right resolving map. By lemma 1, λ_1^{-1} has memory 0. Thus ϕ, being the composition of a memory 0 map λ_1^{-1} with a 1-block map ρ_1, has memory 0 as well. Now we can apply lemma 2 to ϕ to construct a left resolving conjugacy $\lambda_0 : \Sigma_{\mathcal{G}_0} \to \Sigma_{\mathcal{G}}$ satisfying:

(1) the composition $\rho_0 = \phi \circ \lambda_0$ is a 1-block map from $\Sigma_{\mathcal{G}_0}$ to $\Sigma_{\mathcal{H}}$,

(2) if $\lambda_1 : \Sigma_{\mathcal{G}_1} \to \Sigma_{\mathcal{G}}$ is any left resolving conjugacy such that the composition $\rho_1 = \phi \circ \lambda_1$ is a 1-block map, then there is a left resolving conjugacy $\iota : \Sigma_{\mathcal{G}_1} \to \Sigma_{\mathcal{G}_0}$ with $\lambda_1 = \lambda_0 \circ \iota$.

In the present situation, $\rho_1 = \rho_0 \circ \iota$ is right resolving, so we may apply lemma 3 to conclude that ι is right resolving as well. Hence, $\iota : \Sigma_{\mathcal{G}_1} \to \Sigma_{\mathcal{G}_0}$ is bi-resolving. We apply lemma 4 to conclude that ι is induced by a graph isomorphism $\iota_* : \mathcal{G}_1 \to \mathcal{G}_0$. Thus, the given LR decomposition $\phi = \rho_1 \circ \lambda_1^{-1}$ is the same (up to graph isomorphism) as the particular decomposition $\phi = \rho_0 \circ \lambda_0^{-1}$ provided by lemma 2.

$\square$

The remainder of this section provides proofs of lemmas 1, 2, 3, and 4.

Proof of lemma 1. Let $\lambda : \Sigma_{\mathcal{G}'} \to \Sigma_{\mathcal{G}}$ be a left resolving conjugacy. We show that λ^{-1} has memory 0.

Suppose that λ^{-1} has memory m and anticipation a. Then for all $(m+a+1)$-paths $y_{-m} \ldots y_0 \ldots y_a$ in the graph $\mathcal{G}$, there is an edge e_0 in $\mathcal{G}'$ such that

$$\lambda^{-1}\big({}_{-m}[y_{-m} \ldots y_0 \ldots y_a]_{a+1}\big) \subseteq {}_0[e_0]_1.$$

We must show that in fact

$$\lambda^{-1}\big({}_0[y_0 \ldots y_a]_{a+1}\big) \subseteq {}_0[e_0]_1.$$

Since λ is left resolving and since

$$_{-m}[y_{-m} \ldots y_0 \ldots y_a]_{a+1} \subseteq \lambda({}_0[e_0]_1),$$

there is a unique path $e_{-m} \ldots e_{-1}e_0$ in $\mathcal{G}'$ that is λ-labelled by the symbol sequence $y_{-m} \ldots y_{-1}y_0$. So, in fact

$$_{-m}[y_{-m} \ldots y_0 \ldots y_a]_{a+1} \subseteq \lambda({}_{-m}[e_{-m} \ldots e_{-1}e_0]_1).$$

Applying σ^{-m} to both sides, we can assert that for any $(m+a+1)$-path $z_0 \ldots z_{m+a+1}$ in $\mathcal{G}$, there is a unique edge f_0 in $\mathcal{G}'$ with

$$_0[z_0 \ldots z_{m+a}]_{m+a+1} \subseteq \lambda({}_0[f_0]_1).$$

Now if $y_{a+1} \ldots y_{a+m}$ is any continuation of the path $y_{-m} \ldots y_a$ in $\mathcal{G}$, we have

$$_{-m}[y_{-m} \ldots y_a \ldots y_{a+m}]_{m+a+1} \subseteq \lambda({}_0[e_0]_1),$$

so, in fact

$$_0[y_0 \ldots y_{m+a}]_{m+a+1} \subseteq \lambda({}_0[e_0]_1).$$

But the path $y_{a+1} \ldots y_{a+m}$ is an arbitrary continuation of $y_{-m} \ldots y_a$, so finally,

$$_0[y_0 \ldots y_a]_{a+1} \subseteq \lambda(_0[e_0]_1),$$

as was to be shown.   □

Before proving lemma 2, we must state and prove four preparatory lemmas that relate left resolving conjugacies having range $\Sigma_{\mathcal{G}}$ with partitions of the range $\Sigma_{\mathcal{G}}$. These lemmas might be of independent interest.

Lemma 5. *Fix a graph $\mathcal{G}$. Left resolving conjugacies $\lambda : \Sigma_{\mathcal{G}'} \to \Sigma_{\mathcal{G}}$ are in one-to-one correspondence with partitions S of $\Sigma_{\mathcal{G}}$ satisfying:*

(1) The partition S depends on a finite number of coordinates; and these are non-negative.

(2) $S \geq \{_0[s]_0 : s$ is a state of $\mathcal{G}\}$.

(3) For all edges e of $\mathcal{G}$,

$$S|_{_0[e]_1} \leq (\sigma^{-1}S)|_{_0[e]_1},$$

(where $S|_U$ is the partition on $U \subseteq \Sigma_{\mathcal{G}}$ induced by S),

via the correspondence

$$\lambda \leftrightarrow S_\lambda = \{\lambda(_0[s]_0) : \ s \text{ is a state of } \mathcal{G}'\}.$$

Proof. We first show that if $\lambda : \Sigma_{\mathcal{G}'} \to \Sigma_{\mathcal{G}}$ is a left resolving conjugacy, then the partition S_λ satisfies conditions (1), (2), and (3).

The atoms of S_λ can be expressed as

$$\lambda({}_0[s]_0) = \bigcup_{is(e)=s} \lambda({}_0[e]_1),$$

so S_λ is coarser than the partition of Σ_G into the sets $\lambda({}_0[e]_1)$, which by lemma 1 is in turn coarser than the partition of Σ_G into all the $(a+1)$-blocks ${}_0[y_0 \ldots y_a]_{a+1}$ where a is the anticipation of λ^{-1}. Thus S_λ depends on at most the coordinates 0 through a, so S_λ satisfies (1).

Since λ is a 1-block map, if e and e' have the same initial state s in G, then $\lambda_* e$ and $\lambda_* e'$ must have the same initial state $\lambda_* s$ in G' (which is determined as the terminal state of $\lambda_* e''$ where e'' is any edge of G terminating at state s). Also, for each edge e of G', $\lambda({}_0[e]_1) \subseteq {}_0[\lambda_* e]_1$. Thus, $\lambda({}_0[s]_0) \subseteq {}_0[\lambda_* s]_0$, giving (2).

Fix an edge e in the graph G and let t be a state of G' with ${}_0[e]_1 \cap \lambda({}_1[t]_1) \neq \emptyset$. To show (3) we must show that there is a state s of G' with

$$ {}_0[e]_1 \cap \lambda({}_1[t]_1) \subseteq {}_0[e]_1 \cap \lambda({}_0[s]_0).$$

Since λ is left resolving there is a *unique* edge e' in the graph G' with $ts(e') = t$ and $\lambda_* e' = e$. Set $s = is(e')$ and observe

$$
\begin{aligned}
{}_0[e]_1 \cap \lambda({}_1[t]_1) &= \lambda({}_0[e']_1) \cap \lambda({}_1[t]_1) \\
&= \lambda({}_0[e']_1) \\
&\subseteq \lambda({}_0[s]_0),
\end{aligned}
$$

so

$$ {}_0[e]_1 \cap \lambda({}_1[t]_1) \subseteq {}_0[e]_1 \cap \lambda({}_0[s]_0),$$

giving (3).

We now start with a partition S of Σ_G satisfying (1), (2), and (3) and construct a graph $\mathcal{G}'$ and a left resolving conjugacy $\lambda : \Sigma_{\mathcal{G}'} \to \Sigma_G$ with $S_\lambda = S$.

The states of $\mathcal{G}'$ are defined to be the elements of the partition S. For each edge e of $\mathcal{G}$, and for each pair of states $s, t \in S$ of $\mathcal{G}'$ such that

$$s \cap {}_0[e]_1 \cap \sigma^{-1}(t) \neq \emptyset,$$

there is defined to be an edge e' in $\mathcal{G}'$ from state s to state t with label $\lambda_* e' = e$.

We first show that $s = \lambda({}_0[s]_0)$ for each $s \in S$, establishing that $S = S_\lambda$.

$\lambda({}_0[s]_0) \subseteq s$: Set $s = s_0$, and suppose $y \in \lambda({}_0[s_0]_0)$. There is a sequence $(\ldots s_{-1}, s_0, s_1, \ldots)$ of states of $\mathcal{G}'$ such that

$$s_i \cap {}_0[y_i]_1 \cap \sigma^{-1}(s_{i+1}) \neq \emptyset, \quad i \in \mathbb{Z}.$$

By (2), for each s_i, there is a state t_i in the graph $\mathcal{G}$ with $s_i \subseteq {}_0[t_i]_0$. Thus y_i is an edge in $\mathcal{G}$ from state t_i to state t_{i+1}, and therefore $y \in \Sigma_G$. Because

$$S|_{{}_0[y_i]_1} \leq (\sigma^{-1} S)|_{{}_0[y_i]_1},$$

we have

$$\emptyset \neq {}_0[y_i]_1 \cap \sigma^{-1} s_{i+1} \subseteq s_i$$

and so, by induction on k,

$$\emptyset \neq {}_0[y_0 \ldots y_k]_{k+1} \cap \sigma^{-k-1} s_{k+1} \subseteq s_0, \quad k \geq 0.$$

Since membership in s_0 depends on a finite number of coordinates, and these are non-negative coordinates, we have $y \in s_0$. Thus $\lambda({}_0[s_0]_0) \subseteq s_0$.

$s_0 \subseteq \lambda(_0[s_0]_0)$: Assume $y \in s_0$ and define $s_i \in \mathcal{S}$ to be that partition element satisfying $y \in \sigma^{-i}s_i$. Then

$$s_i \cap {}_0[y_i]_1 \cap \sigma^{-1}(s_{i+1}) \neq \emptyset, \quad i \in \mathbb{Z},$$

so there is an edge x_i from s_i to s_{i+1} in $\mathcal{G}'$ that has label $\lambda_*(x_i) = y_i$. Thus $y \in \lambda(_0[s_0]_0)$, so $s_0 \subseteq \lambda(_0[s_0]_0)$.

Using $s = \lambda(_0[s]_0)$, we immediately see that $\lambda : \Sigma_{\mathcal{G}'} \to \Sigma_{\mathcal{G}}$ is onto. We can also see that λ is one-to-one: Suppose $\lambda(x) = y$. Let $s_i \in \mathcal{S}$ be such that $y \in \sigma^{-i}s_i$. Then $x \in \sigma^{-i}(_0[s_i]_0) = {}_i[s_i]_i$ and $\lambda_*x_i = y_i$, so x_i must be the unique edge in $\mathcal{G}'$ from state s_i to state s_{i+1} that is labeled y_i.

Finally we show that λ is left resolving. Fix a state t in the graph $\mathcal{G}'$ and an edge e preceding the state λ_*t in the graph $\mathcal{G}$. We must show that there is at most one edge e' preceding the state t with $\lambda_*e' = e$. Now $_0[e]_1 \cap \sigma^{-1}(t) \neq \emptyset$, and $\mathcal{S}|_{0[e]_1} \leq (\sigma^{-1}\mathcal{S})|_{0[e]_1}$, so there is a unique $s \in \mathcal{S}$ with $s \cap {}_0[e]_1 \supseteq {}_0[e]_1 \cap \sigma^{-1}t$. So there is a unique $s \in \mathcal{S}$ with $s \cap {}_0[e]_1 \cap \sigma^{-1}t \neq \emptyset$. Hence, by the definition of $\mathcal{G}'$, there is a unique edge e' preceding state t of $\mathcal{G}'$ that has $\lambda_*e' = e$: it is the edge e' corresponding to the non-empty set $s \cap {}_0[e]_1 \cap \sigma^{-1}t$. $\qquad\square$

Lemma 6. *Let $\lambda_0 : \Sigma_{\mathcal{G}_0} \to \Sigma_{\mathcal{G}}$ and $\lambda_1 : \Sigma_{\mathcal{G}_1} \to \Sigma_{\mathcal{G}}$ be left resolving conjugacies. Then $\mathcal{S}_{\lambda_0} \leq \mathcal{S}_{\lambda_1}$ if and only if there exists a left resolving conjugacy $\lambda : \Sigma_{\mathcal{G}_1} \to \Sigma_{\mathcal{G}_0}$ with $\lambda_0 \circ \lambda = \lambda_1$.*

Proof. ($\Leftarrow$). Because $\lambda : \Sigma_{\mathcal{G}_1} \to \Sigma_{\mathcal{G}_0}$ is a 1-block map, for each state s of $\mathcal{G}_1$, there is a state t of $\mathcal{G}_0$ with $\lambda(_0[s]_0) \subseteq {}_0[t]_0$. Hence $\lambda_1(_0[s]_0) = \lambda_0 \circ \lambda(_0[s]_0) \subseteq \lambda_0(_0[t]_0)$, showing $\mathcal{S}_{\lambda_0} \leq \mathcal{S}_{\lambda_1}$.

($\Rightarrow$). We first show that $\lambda = \lambda_0^{-1} \circ \lambda_1$ is a 1-block map. This equivalent to showing that for each edge e of $\mathcal{G}_1$, there is an edge f of $\mathcal{G}_0$ such that $\lambda_0^{-1} \circ \lambda_1(_0[e]_1) \subseteq {}_0[f]_1$, or equivalently, $\lambda_1(_0[e]_1) \subseteq \lambda_0(_0[f]_1)$. But

$$\lambda_1(_0[e]_1) = \lambda_1(_0[is(e)]_0) \cap {}_0[(\lambda_1)_*e]_1 \cap \lambda_1(_1[ts(e)]_1).$$

Because $\mathcal{S}_{\lambda_0} \leq \mathcal{S}_{\lambda_1}$, there are unique states s and t in $\mathcal{G}_0$ with $\lambda_1(_0[is(e)]_0) \subseteq \lambda_0(_0[s]_0)$ and $\lambda_1(_1[ts(e)]_1) \subseteq \lambda_0(_1[t]_1)$. Thus there must be a unique edge f from state s to state t in $\mathcal{G}_0$ with $(\lambda_0)_*f = (\lambda_1)_*e$. Now

$$\begin{aligned}
\lambda_1(_0[e]_1) &= \lambda_1(_0[is(e)]_0) \cap {}_0[(\lambda_1)_*e]_1 \cap \lambda_1(_1[ts(e)]_1) \\
&\subseteq \lambda_0(_0[s]_0) \cap {}_0[(\lambda_0)_*f]_1 \cap \lambda_0(_1[t]_1) \\
&= \lambda_0(_0[f]_1),
\end{aligned}$$

so λ is a 1-block map. Finally we show λ is left resolving. If e and e' are distinct edges preceding a state s of $\mathcal{G}_1$, then $(\lambda_1)_*e = (\lambda_0)_* \circ \lambda_* e$ and $(\lambda_1)_*e' = (\lambda_0)_* \circ \lambda_* e'$ are distinct edges in $\mathcal{G}$ because λ_1 is left resolving. So $\lambda_* e$ and $\lambda_* e'$ must be distinct edges of $\mathcal{G}_0$, giving that λ is left resolving. $\qquad\square$

Lemma 7. *Given a partition $\mathcal{R}$ of $\Sigma_\mathcal{G}$ depending on a finite number of coordinates, and these are non-negative, there is a partition $\mathcal{S}_{\lambda_0}$ corresponding to a left resolving conjugacy $\lambda_0 : \Sigma_{\mathcal{G}_0} \to \Sigma_\mathcal{G}$ such that:*

(1) $\mathcal{S}_{\lambda_0} \geq \mathcal{R}$,

(2) $\mathcal{S}_{\lambda_1} \geq \mathcal{S}_{\lambda_0}$ for any partition $\mathcal{S}_{\lambda_1} \geq \mathcal{R}$ that corresponds to a left resolving conjugacy $\lambda_1 : \Sigma_{\mathcal{G}_1} \to \Sigma_\mathcal{G}$.

Before proving the lemma, we establish some notation that will use in the proofs of this lemma and the next. For $x \in \Sigma_G$, let $x^+ = (x_0 x_1 \ldots)$. For a subset $U \subseteq \Sigma_G$, denote

$$U^+ = \{x^+ : x \in U\}$$

A partition $\mathcal{T}$ of a cylinder set $_{-k}[b]_0 \subseteq \Sigma_G$ that depends only on coordinates $n \geq -k$ induces a partition $\mathcal{T}^+$ of $_0[ts(b)]_0^+$:

$$\mathcal{T}^+ = \{S^+ : S \in \mathcal{T}\}.$$

Proof of lemma 7. Let $\mathcal{S}$ be the partition of Σ_G satisfying:

(i) The partition $\mathcal{S}$ depends only on non-negative coordinates,

(ii) $\mathcal{S} \geq \{_0[s]_0 : s$ is a state of $\mathcal{G}\}$,

(iii)

$$(\mathcal{S}|_{_0[s]_0})^+ = \bigvee_{k \geq 0} \; \bigvee_{|b|=k \; \& \; ts(b)=s} ((\sigma^k \mathcal{R})|_{_{-k}[b]_0})^+.$$

We verify that $\mathcal{S}$ corresponds to a left resolving conjugacy λ_0 (i.e., that $\mathcal{S} = \mathcal{S}_{\lambda_0}$) by verifying that $\mathcal{S}$ satisfies properties *(1)*, *(2)* and *(3)* of lemma 5.

The partition $\mathcal{S}$ satisfies condition *(1)* of lemma 5 by *(i)* and the fact that the wedge product in *(iii)* really has only a finite number of non-trivial terms. Condition *(2)* of lemma 5 is *(ii)*. We show that $\mathcal{S}$ satisfies *(3)* of lemma 5:

$$\mathcal{S}|_{_0[e]_1} \leq (\sigma^{-1}\mathcal{S})|_{_0[e]_1}.$$

We have:

$$((\sigma \mathcal{S})|_{-1[e]_0})^+ = \bigvee_{k \geq 0} \bigvee_{|b|=k \ \& \ ts(b)=s} ((\sigma \sigma^k \mathcal{R})|_{-k-1[be]_0})^+$$

$$\leq \bigvee_{k \geq 0} \bigvee_{|c|=k \ \& \ ts(c)=ts(e)} ((\sigma^k \mathcal{R})|_{-k[c]_0})^+$$

$$= (\mathcal{S}|_{0[ts(e)]_0})^+$$

$$= (\mathcal{S}|_{-1[e]_0})^+.$$

So

$$(\sigma \mathcal{S})|_{-1[e]_0} \leq \mathcal{S}|_{-1[e]_0},$$

or equivalently,

$$\mathcal{S}|_{0[e]_1} \leq (\sigma^{-1}\mathcal{S})|_{0[e]_1}.$$

Thus $\mathcal{S}$ corresponds to a left resolving conjugacy.

Condition (1) of the present lemma holds because, by (iii),

$$(\mathcal{S}|_{0[s]_0})^+ \geq (\mathcal{R}|_{0[s]_0})^+.$$

We show condition (2) of the present lemma. Let $\mathcal{T}$ be a partition corresponding to a left resolving conjugacy onto $\Sigma_{\mathcal{G}}$ and suppose $\mathcal{T} \geq \mathcal{R}$. We must show $\mathcal{T} \geq \mathcal{S}$. By (3) of lemma 5,

$$\mathcal{T}|_{-1[e]_0} \geq (\sigma \mathcal{T})|_{-1[e]_0}$$

for each edge e of $\mathcal{G}$. It follows by induction on k that

$$\mathcal{T}|_{-k[b]_0} \geq (\sigma^k \mathcal{T})|_{-k[b]_0}$$

for each k-path b of $\mathcal{G}$. Since $\mathcal{T} \geq \mathcal{R}$,

$$(\sigma^k \mathcal{T})|_{-k[b]_0} \geq (\sigma^k \mathcal{R})|_{-k[b]_0}.$$

Thus,

$$\mathcal{T}|_{-k[b]_0} \geq (\sigma^k \mathcal{R})|_{-k[b]_0}.$$

By *(1)* of lemma 5, $\mathcal{T}$ depends only on non-negative coordinates, so

$$(\mathcal{T}|_{0[ts(b)]_0})^+ \geq ((\sigma^k \mathcal{R})|_{-k[b]_0})^+.$$

Thus $(\mathcal{T}|_{0[s]_0})^+$ refines the wedge product of the terms $((\sigma^k \mathcal{R})|_{-k[b]_0})^+$ over all $k \geq 0$ and all k-paths terminating at state s, but this wedge product is defined to be $(\mathcal{S}|_{0[s]_0})^+$. Hence, $\mathcal{T} \geq \mathcal{S}$. □

Lemma 8. *Let* $\lambda : \Sigma_{\mathcal{G}'} \to \Sigma_{\mathcal{G}}$ *be a left resolving conjugacy and let* $\phi : \Sigma_{\mathcal{G}} \to \Sigma_{\mathcal{H}}$ *be a block map with memory* 0. *Then* $\phi \circ \lambda$ *is a* 1-*block map if and only if* $\mathcal{S}_\lambda \geq \mathcal{R}$ *where* $\mathcal{R}$ *is the coarsest partition of* $\Sigma_{\mathcal{G}}$ *depending only on non-negative coordinates and satisfying*

$$(\mathcal{R}|_{-1[e]_0})^+ \geq ((\phi^{-1}\sigma\mathcal{F})|_{-1[e]_0})^+, \quad e \text{ is an edge of } \mathcal{G},$$

where $\mathcal{F}$ *is the partition of* $\Sigma_{\mathcal{H}}$ *defined by*

$$\mathcal{F} = \{ {}_0[f]_1 : f \text{ is an edge of } \mathcal{H} \}.$$

Proof. We prove the lemma by a chain of equivalences:

$$\phi \circ \lambda \text{ is a 1-block map}$$

if and only if

$$\forall \text{ edges } e' \text{ of } \mathcal{G}', \exists \text{ edge } f \text{ of } \mathcal{H}, \; \phi \circ \lambda(_{-1}[e']_0) \subseteq {}_{-1}[f]_0$$

if and only if

$$\forall \text{ edges } e' \text{ of } \mathcal{G}', \exists \text{ edge } f \text{ of } \mathcal{H},$$
$$_{-1}[\lambda_* e']_0 \cap \lambda(_0[ts(e')]_0) = \lambda(_{-1}[e']_0) \subseteq \phi^{-1}(_{-1}[f]_0)$$

if and only if

$$\forall \text{ edges } e' \text{ of } \mathcal{G}', \exists \text{ edge } f \text{ of } \mathcal{H},$$
$$_{-1}[\lambda_* e']_0 \cap \lambda(_0[ts(e')]_0) \subseteq {}_{-1}[\lambda_* e']_0 \cap \phi^{-1}(_{-1}[f]_0)$$

if and only if

$$\forall \text{ edges } e' \text{ of } \mathcal{G}', \ (\mathcal{S}_\lambda|_{-1[\lambda \ast e']_0})^+ \geq ((\phi^{-1}\sigma\mathcal{F})|_{-1[\lambda \ast e']_0})^+$$

if and only if

$$\mathcal{S}_\lambda \geq \mathcal{R}.$$

$\square$

Having proved lemmas 5, 6, 7, and 8, we are ready to prove lemma 2.

Proof of lemma 2. We are given an n-block map $\phi : \Sigma_\mathcal{G} \to \Sigma_\mathcal{H}$ having memory 0. For this ϕ, let $\mathcal{R}$ be the partition of $\Sigma_\mathcal{G}$ defined in lemma 8. The partition $\mathcal{R}$ depends only on a finite number of coordinates and these are non-negative. By lemma 8, $\mathcal{R}$ has the property that for all left resolving conjugacies $\lambda : \Sigma_{\mathcal{G}'} \to \Sigma_\mathcal{G}$, $\phi \circ \lambda$ is a 1-block map if and only if $\mathcal{S}_\lambda \geq \mathcal{R}$.

For this partition $\mathcal{R}$, let $\lambda_0 : \Sigma_{\mathcal{G}_0} \to \Sigma_\mathcal{G}$ be the left resolving conjugacy provided by lemma 7. By lemma 7, λ_0 satisfies

(i) $\mathcal{S}_{\lambda_0} \geq \mathcal{R}$,

(ii) $\mathcal{S}_{\lambda_1} \geq \mathcal{S}_{\lambda_0}$ for any partition $\mathcal{S}_{\lambda_1} \geq \mathcal{R}$ that corresponds to a left resolving conjugacy $\lambda_1 : \Sigma_{\mathcal{G}_1} \to \Sigma_\mathcal{G}$.

Property (i) (with lemma 8) gives that the composition $\rho_0 = \phi \circ \lambda_0$ is a 1-block map from $\Sigma_{\mathcal{G}_0}$ to $\Sigma_\mathcal{H}$. This is condition (1) of lemma 2.

We establish condition (2) of lemma 2 as follows. Let $\lambda_1 : \Sigma_{\mathcal{G}_1} \to \Sigma_\mathcal{G}$ be any left resolving conjugacy such that the composition $\rho_1 = \phi \circ \lambda_1$ is a 1-block map. By lemma 8, $\mathcal{S}_{\lambda_1} \geq \mathcal{R}$. By property (ii) of ϕ_0, $\mathcal{S}_{\lambda_1} \geq \mathcal{S}_{\lambda_0}$. By lemma 6, there is a left resolving conjugacy $\iota : \Sigma_{\mathcal{G}_1} \to \Sigma_{\mathcal{G}_0}$ with $\lambda_1 = \lambda_0 \circ \iota$. This establishes condition (2). $\square$

We end this section with proofs of lemmas 3 and 4.

Proof of lemma 3. Suppose $\iota : \Sigma_{\mathcal{G}_1} \to \Sigma_{\mathcal{G}_0}$ and $\rho_0 : \Sigma_{\mathcal{G}_0} \to \Sigma_{\mathcal{H}}$ are 1-block maps whose composition $\rho_0 \circ \iota$ is a right resolving 1-block map. If e and e' are distinct edges following a state s of $\mathcal{G}_1$, then the edges $(\rho_0 \circ \iota)_* e = (\rho_0)_* \circ \iota_* e$ and $(\rho_0 \circ \iota)_* e' = (\rho_0)_* \circ \iota_* e'$ are distinct edges in $\mathcal{H}$ because $\rho_0 \circ \iota$ is right resolving. So $\iota_* e$ and $\iota_* e'$ must be distinct edges of $\mathcal{G}_0$, giving that ι is right resolving. $\square$

Proof of lemma 4. Suppose $\iota : \Sigma_{\mathcal{G}_1} \to \Sigma_{\mathcal{G}_0}$ is a biresolving conjugacy. Because ι is left resolving, we can apply lemma 1 to conclude that ι^{-1} has memory 0. Reverse the sense of time and note that ι is right resolving to conclude that ι^{-1} also has anticipation 0. Thus ι^{-1} is a 1-block map, so the graph homomorphism $\iota_* : \mathcal{G}_1 \to \mathcal{G}_0$ has the inverse $(\iota^{-1})_* : \mathcal{G}_0 \to \mathcal{G}_1$, and is therefore a graph isomorphism. $\square$

4. Examples

Example 1. Define a 'marker' automorphism $\psi : \Sigma_2 \to \Sigma_2$ to act by switching the markers 0100 and 0110: wherever 0100 occurs in x, 0110 occurs in ψx, and wherever 0110 occurs in x, 0100 occurs in ψx; elsewhere ψ acts as the identity map. That ψ is well-defined follows from the fact that the bits of x that are changed by ψ are framed by a single marker. The map ψ is one-to-one because it is an involution.

The composition $\phi = \psi \circ \sigma^2$ has memory 0, so we can apply lemma 2 to construct a left resolving conjugacy $\lambda_0 : \Sigma_{\mathcal{G}_0} \to \Sigma_2$ so that the composition $\rho_0 = \lambda_0 \circ \phi$ is a 1-block map. We will carry out the construction of λ_0 outlined in the proofs of lemmas 7 and 8.

In this example, we have $\phi : \Sigma_{\mathcal{G}} \to \Sigma_{\mathcal{H}}$ where $\Sigma_{\mathcal{G}} = \Sigma_{\mathcal{H}} = \Sigma_2$. The graph $\mathcal{H} \ (= \mathcal{G})$ has a single state s and two edges, 0 and 1, from state s to itself. The partition $\mathcal{F}$ of $\mathcal{H}$ given in lemma 8 is

$$\mathcal{F} = \{{}_0[0]_1, \ {}_0[1]_1\},$$

so $\phi^{-1}\sigma\mathcal{F}$ is the partition of $\Sigma_{\mathcal{G}}$ into the two sets $\phi^{-1}({}_{-1}[0]_0)$ and $\phi^{-1}({}_{-1}[1]_0)$. Now

$$\phi^{-1}({}_{-1}[0]_0) = \bigcup_{w \in W_0} {}_{-1}[w]_3,$$

where

$$W_0 = \{0000, 0001, 0101, 0110, 1000, 1001, 1100, 1101\},$$

and

$$\phi^{-1}({}_{-1}[1]_0) = \bigcup_{w \in W_1} {}_{-1}[w]_3,$$

where

$$W_1 = \{0010, 0011, 0100, 0111, 1010, 1011, 1110, 1111\}.$$

The partition $\mathcal{R}$ of $\Sigma_{\mathcal{G}}$ given in lemma 8 is the partition depending only on non-negative coordinates such that

$$\mathcal{R}^+ = ((\phi^{-1}\sigma\mathcal{F})|_{-1[0]_0})^+ \vee ((\phi^{-1}\sigma\mathcal{F})|_{-1[1]_0})^+.$$

The partitions $((\phi^{-1}\sigma\mathcal{F})|_{-1[0]_0})^+$, $((\phi^{-1}\sigma\mathcal{F})|_{-1[1]_0})^+$, and $\mathcal{R}^+$ depend only on coordinates 0, 1, and 2. They are illustrated schematically

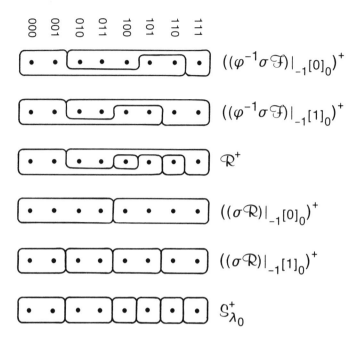

Figure 4: Example 1.

in figure 4. The partition S_{λ_0} of Σ_G given in lemma 7 is the partition depending only on non-negative coordinates and with

$$(S|_0[s]_0)^+ = \bigvee_{k \geq 0} \quad \bigvee_{|b|=k \ \& \ ts(b)=s} ((\sigma^k \mathcal{R})|_{-k}[b]_0)^+.$$

This wedge product really has only three distinct non-trivial terms: $\mathcal{R}^+$, $((\sigma \mathcal{R})|_{-1}[0]_0)^+$, and $((\sigma \mathcal{R})|_{-1}[1]_0)^+$. These and $S_{\lambda_0}^+$ are shown in figure 4.

By construction, for each edge e' of the graph $\mathcal{G}_0$, there is an edge f (either 0 or 1) of the graph $\mathcal{H}$ such that

$$-_1[(\lambda_0)_* e']_0 \cap \lambda_0(_0[ts(e')]_0) \subseteq \phi^{-1}(_{-1}[f]_0).$$

The composition $\rho_0 = \phi \circ \lambda_0$ is a 1-block map with $(\rho_0)_* e' = f$.

We show that ρ_0 is *not* right resolving, and therefore that ϕ is not an LR automorphism of Σ_2. In fact, the state $_0[01]_2$ of the graph $\mathcal{G}_0$ has four follower states: $_0[100]_3$, $_0[101]_3$, $_0[110]_3$, and $_0[111]_3$. This already shows that ρ_0 is not right resolving because these four edges must have $(\rho_0)_*$-images in $\{0, 1\}$. In fact, the sets $\lambda_0(_{-1}[e']_0)$, where e' is an edge following state $_0[01]_2$, are:

$$-_1[0100]_3 \subseteq \phi^{-1}(_{-1}[1]_0)$$
$$-_1[0101]_3 \subseteq \phi^{-1}(_{-1}[0]_0)$$
$$-_1[0110]_3 \subseteq \phi^{-1}(_{-1}[0]_0)$$
$$-_1[0111]_3 \subseteq \phi^{-1}(_{-1}[1]_0).$$

In order for an isomorphism ϕ to be an LR isomorphism it is necessary that ϕ have memory 0 and that ϕ^{-1} have anticipation 0. This example shows that this condition is not sufficient.

Example 2. We redo example 1 with a change: instead of setting $\phi = \psi \circ \sigma^2$, we set $\phi = \psi \circ \sigma^3$. As in example 1, $\phi^{-1}\sigma\mathcal{F}$ is the partition of $\Sigma_\mathcal{G}$ into the two sets $\phi^{-1}(_{-1}[0]_0)$ and $\phi^{-1}(_{-1}[1]_0)$. But now these two sets are

$$\phi^{-1}(_{-1}[0]_0) = \bigcup_{w \in W_0} {}_{-1}[w]_4,$$

where W_0 is the set of binary words of the form $abc0d$, except that 00100 and 10100 are excluded and 00110 and 10110 are included; and

$$\phi^{-1}(_{-1}[1]_0) = \bigcup_{w \in W_1} {}_{-1}[w]_4,$$

where W_1 is the set of binary words of the form $abc1d$, except that 00110 and 10110 are excluded and 00100 and 10100 are included.

Figure 5 gives the distinct non-trivial partitions that are terms in the wedge product that computes $\mathcal{S}_{\lambda_0}$. Table 1 lists the states of $\mathcal{G}_0$ in the first column, their followers via edges that are $(\rho_0)_*$-labeled 0 in the second column, and their followers via edges that are $(\rho_0)_*$-labeled 1 in the third column. By examining the table, one confirms that ρ_0 is right resolving, and hence that $\phi = \psi \circ \sigma^3$ is an LR automorphism.

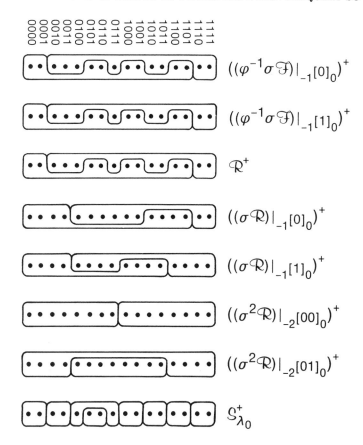

Figure 5: Example 2.

state	0	1
$_0[000]_3$	$_0[000]_3$	$_0[001]_3$
$_0[001]_3$	$_0[0101]_4 \cup {}_0[0110]_4$	$_0[0100]_4 \cup {}_0[0111]_4$
$_0[0100]_4 \cup {}_0[0111]_4$	$_0[100]_3$	$_0[111]_3$
$_0[0101]_4 \cup {}_0[0110]_4$	$_0[110]_3$	$_0[101]_3$
$_0[100]_3$	$_0[000]_3$	$_0[001]_3$
$_0[101]_3$	$_0[0101]_4 \cup {}_0[0110]_4$	$_0[0100]_4 \cup {}_0[0111]_4$
$_0[110]_3$	$_0[100]_3$	$_0[101]_3$
$_0[111]_3$	$_0[110]_3$	$_0[111]_3$

Table 1: Example 2.

References

[BMT87] M. Boyle, B. Marcus, and P. Trow. Resolving maps and the dimension group for shifts of finite type. *Memoirs A.M.S.*, (377), 1987.

[Kit87] B. Kitchens, 1987. personal communication.

[Nas] M. Nasu. Textile systems for endomorphisms and automorphisms of the shift. preprint.

[Wil73] R. Williams. Classification of shifts of finite type. *Anals of Math.*, 98:120–153, 1973.

[Wil74] R. Williams. Classification of shifts of finite type, errata. *Anals of Math.*, 99:380–381, 1974.

Contemporary Mathematics
Volume **135**, 1992

A POLYNOMIAL–TIME ALGORITHM FOR DECIDING THE FORCING RELATION ON CYCLIC PERMUTATIONS

CHRIS BERNHARDT AND ETHAN M. COVEN

ABSTRACT. Implicit in the proof of Sharkovsky's Theorem is the *forcing relation* induced on cyclic permutations of finite ordered sets by extending them to continuous maps of the interval. We give a polynomial-time algorithm for deciding this relation.

INTRODUCTION

This paper is concerned with the *forcing relation* induced on cyclic permutations of finite ordered sets by extending them to continuous maps of the interval. Given two such permutations, θ of $\{1, \ldots, m\}$ and π of $\{1, \ldots, n\}$, we say that θ *forces* π if every continuous map of the interval which has a representative of θ also has a representative of π. A *representative* of π in f is a set $P = \{p_1, \ldots, p_n\}$ such that if P is labelled so that $p_1 < \cdots < p_n$, then each $f(p_i) = p_{\pi(i)}$.

The forcing relation is implicit in the Block-Guckenheimer-Misiurewicz-Young proof of Sharkovsky's Theorem [BGMY]. It was formalized by S. Baldwin [Ba], who showed that forcing is a partial order. He also gave an algorithm for deciding whether θ forces π. Thinking of θ as fixed and π as varying, Baldwin's algorithm is inefficient in that that number of cases to be checked is exponential in n. (The number of steps required to check each case is polynomial in n.) Subsequently, I. Jungreis [J] gave an algorithm which requires checking only one case, but which works only under restrictive hypotheses on θ and π. A geometric version of Jungreis' algorithm appears in [BCMM]. In this paper, we extend the geometric version of Jungreis' algorithm to arbitrary θ and π. The number of cases to be checked is polynomial in n.

Throughout this paper, f will denote a continuous map of a nondegenerate compact interval to itself. θ and π will denote cyclic permutations of $\{1, \ldots, m\}$ and $\{1, \ldots, n\}$ respectively, with $m, n \geq 3$. The restriction that $m, n \geq 3$ is harmless, as will be shown in Section 1. Furthermore, we will assume that θ is not a double. This technical term will be defined in Section 1 and we will show that this assumption too is harmless.

Let $L_\theta : [1, m] \to [1, m]$ denote the *canonical θ-linear map*, defined by $L_\theta = \theta$ on $\{1, \ldots, m\}$ and L_θ is linear on each $[i, i+1]$. The nondegenerate intervals I maximal with respect to "L_θ is strictly monotone on I" are called the *laps* of L_θ. Let M denote the number of laps of L_θ, and define $H : [0, 1] \to [0, 1]$ as follows.

1980 *Mathematics Subject Classification* (1985 *Revision*). Primary 58F20, 58F08, 54H20.

This paper is in final form and no version will be submitted for publication elsewhere.

$H(0) = 0$ or 1 according to whether $\theta(1) < \theta(2)$ or $\theta(1) > \theta(2)$, and H is maps each $[(i-1)/M, i/M]$ linearly onto $[0, 1]$. It follows from [ALM, Theorem 2.7.7] and [BCMM, Lemma 3] that θ forces π if and only if H has representatives P of π and Q of θ such that

$$(*) \qquad \min_{q \in Q} \left| q - \frac{i}{M} \right| \leq \min_{p \in P} \left| p - \frac{i}{M} \right|, \qquad i = 1, \ldots, M - 1.$$

In [BCMM] it was shown that θ has a unique representative Q in H such that for every cyclic permutation π, if $(*)$ holds for P and Q, then for all representatives Q' of θ in H, it holds for P and Q' as well.

It is easy to recover P and Q, and hence the numbers appearing in $(*)$, from the H-itineraries of any of their points. This requires solving just one linear equation. If there were some simple method of finding the H-itineraries of all the representatives of π in H, then we would already have a simple algorithm for deciding whether θ forces π. However, no such simple method is known to the authors. Instead, we introduce a class of representatives of π in H, which we call *fundamental* (in H). The representative Q of θ referred to above is the unique fundamental representative of θ in H. We show that θ forces π if and only if $(*)$ holds for some fundamental representative P of π in H and Q. We provide a simple method of finding the H-itineraries of all the fundamental representatives of π in H.

1. PRELIMINARIES

The standing assumption that $m, n \geq 3$ is harmless. The cycle (1) is the unique permutation of $\{1\}$. It forces only itself, and is forced by every cyclic permutation. The cycle (12) is the only cyclic permutation of $\{1, 2\}$. It forces only itself and (1), and is forced by every cyclic permutation other than (1).

π is a *double* if n is even and π cyclically permutes the $n/2$ sets $\{1, 2\}, \ldots, \{n - 1, n\}$. In that case, the cyclic permutation of $\{1, \ldots, n/2\}$ defined by $i \mapsto j$ if $\pi\{2i - 1, 2i\} = \{2j - 1, 2j\}$ is denoted $\pi/2$. If θ is a double, then θ forces π if and only if $\pi = \theta$ or $\theta/2$ forces π [Be, Theorem 1.12]. Following [BCMM], we inductively define a finite sequence of cyclic permutations as follows. Let $\theta_0 = \theta$; if θ_k is a double, let $\theta_{k+1} = \theta_k/2$. This process terminates at some θ_K which is not a double. Then θ forces π if and only if $\pi = \theta_k$ for some k or θ_K forces π. Thus the assumption that θ is not a double is harmless.

We cannot assume that π is not a double. This introduces certain technical difficulties, but no conceptual difficulties.

Just as we have defined laps for L_θ, if f is piecewise monotone, the finitely many nondegenerate intervals I maximal with respect to "f is strictly monotone on I" are called the *laps* of f. If f has N laps, $[F_1] \leq \cdots \leq [F_N]$ (this notation means that $\max[F_i] \leq \min[F_{i+1}]$), then an *$f$-itinerary* of x is a sequence $(t_j)_{j=0}^{\infty}$ such that $f^j(x) \in [F_{t_j}]$ for all j. All points except those whose orbits pass through a turning point of f have unique f-itineraries. If f is expanding on each lap, a point can be recovered from its f-itinerary, at least in principle. However, in practice, this is difficult unless the point is periodic and the map is sufficiently simple. We introduce a class of such maps below.

A *horseshoe map* with M laps is one of the two maps $H: [0,1] \to [0,1]$, with laps $[0, 1/M] \le \cdots \le [(M-1)/M, 1]$, such that H maps each lap linearly onto $[0,1]$. The horseshoe map of the *same type* as θ is the horseshoe map with the same number of laps as the canonical θ-linear map L_θ, such that $H(0) = 0$ or 1 according to whether $\theta(1) < \theta(2)$ or $\theta(1) > \theta(2)$. (This is the map which appears in (*).)

Suppose L_π has N laps, $[L_1] \le \cdots \le [L_N]$. For $\epsilon > 0$ small enough and for $j = 0, \ldots, n-1$, $L_\pi^j[1, 1+\epsilon]$ is contained in a unique lap $[L_{\bar{\iota}_j}]$ of L_π. We call the finite sequence $\bar{\iota} = (\bar{\iota}_0, \ldots, \bar{\iota}_{n-1})$ the *fundamental (temporal) sequence* of π. (It may be thought of as the "lap version" of the fundamental loop of π. See [Be] or Section 2 below.)

Lemma 1.1. *$\bar{\iota}$ has period $n/2$ or n according to whether π is a double or not.*

Proof. Notice that $\pi^n(1) = 1$ and $\pi^j(1) \in [L_{\bar{\iota}_j}]$ for all j. The period of $\bar{\iota}$ is n/k where k is a divisor of n. In this case, there are k integers, $i_1 < \cdots < i_k$, lying in $[L_1]$, such that for all j, $\pi^j(i_1), \ldots, \pi^j(i_k)$ all lie in the same lap of L_π, and either $\pi^j(i_1) < \cdots < \pi^j(i_k)$ or $\pi^j(i_k) < \cdots < \pi^j(i_1)$. This is imposssible unless $k = 1$ or 2, for otherwise $\pi^j(i_1) \ne i_2$ for all j.

If $k = 2$, then $\pi^{n/2}(1) \in [L_1]$ and for all j, $\pi^j(1)$ and $\pi^{j+n/2}(1)$ lie in the same lap of L_π. As in the preceding paragraph, no point of $\{1, \ldots, n\}$ lies between them. Therefore π is a double.

Conversely, suppose that π is a double. Then $\pi^{n/2}(1) = 2$, and for all j, no point of $\{1, \ldots, n\}$ lies between $\pi^j(1)$ and $\pi^j(2)$. Thus $L_\pi^j[1, 2]$ is contained in a single lap of L_π. That lap must be $[L_{\bar{\iota}_j}]$. But $L_\pi^j[1, 2] = L_\pi^{j+n/2}[1, 2]$, so $\bar{\iota}_j = \bar{\iota}_{j+n/2}$. Thus $k \ge 2$ and hence $k = 2$. $\square$

If π is not a double and $P = \{x, f(x), \ldots, f^{n-1}(x)\}$ is a representative of π in f with $x = \min P$, we say that P is *fundamental* if there exists $\delta > 0$ such that for $j = 0, \ldots, n-1$, f is strictly monotone on $f^j[x, x+\delta]$, increasing or decreasing according to whether L_π is increasing or decreasing on $[L_{\bar{\iota}_j}]$. If π is a double and $P = \{x, f(x), \ldots, f^{n/2-1}(x)\}$ is a representative of $\pi/2$ in f with $x = \min P$, we say that P is a *fundamental representative of* π if there exists $\delta > 0$ such that for $j = 0, \ldots, n/2-1$, f is strictly monotone on $f^j[x, x+\delta]$, increasing or decreasing according to whether L_π is increasing or decreasing on $[L_{\bar{\iota}_j}]$. Note that if π is a double, then the fundamental representatives of π are actually representatives of $\pi/2$. In particular, P may be a fundamental representative of $\pi/2$ but a non-fundamental representative of π. For example, let $\pi = (135246)$ and $f = L_\pi$. For ease of exposition, we will postpone dealing with this somewhat confusing terminology until Section 4, and *assume until then that π is not a double.*

2. WHY THE ALGORITHM WORKS

Let $Q = \{q_1, \ldots, q_m\}$ with $q_1 < \cdots < q_m$. We say that f is *Q-monotone* if $f(x) = f(q_1)$ for $x \le q_1$, $f(x) = f(q_m)$ for $x \ge q_m$, and f is *weakly monotone*, i.e., either non-increasing or non-decreasing, on each $[q_i, q_{i+1}]$. The quintessential Q-monotone map is the canonical θ-linear map L_θ. Here $Q = \{1, \ldots, m\}$.

Theorem 2.1. [ALM, Theorem 2.7.7] *Let Q be a representative of θ in f and suppose that f is Q-monotone. Then θ forces π if and only if π has a representative in f.*

The "art" in this subject comes in choosing the model map f. The problem with using L_θ as the model map is that, although the set of representatives of π in L_θ is in one-to-one correspondence with a subset of the closed paths of length n in a certain directed graph (see [Ba]), there is no efficient way known of determining whether π has a representative in L_θ. Baldwin solved this problem by ignoring it. In his algorithm, one examines all closed paths of length n in the graph to see if any of them correspond to a representative of π. The number of such paths is exponential in n.

A better choice of model map is a "truncated horseshoe map." Truncated maps are formed as follows. Let $R = \{r_1, \ldots, r_K\}$ with $r_1 < \cdots < r_K$. Define $f_{R,i}$ on $[r_i, r_{i+1}]$ by

$$f_{R,i}(x) = \begin{cases} \min\big(f(r_i), f(r_{i+1})\big) & f(x) < \min\big(f(r_i), f(r_{i+1})\big) \\ \max\big(f(r_i), f(r_{i+1})\big) & f(x) > \max\big(f(r_i), f(r_{i+1})\big) \\ f(x) & \text{otherwise.} \end{cases}$$

The R-truncation f_R of f is defined by

$$f_R(x) = \begin{cases} f(r_1) & x \le r_1 \\ f_{R,i}(x) & r_i \le x \le r_{i+1} \\ f(r_K) & x \ge r_K. \end{cases}$$

Theorem 2.2. [BCMM, Lemma 3] *Let H be the horseshoe map of the same type as θ and let Q be the unique fundamental representative of θ in H. If P is a representative of π in H, then P is a representative of π in H_Q if and only if $(*)$ holds.*

Theorems 2.1 and 2.2 together yield the following. With H and Q as above: θ forces π if and only if π has a representative P in H such that $(*)$ holds. As mentioned in the Introduction, there remains the difficult problem of finding all the representatives of π in H. This is where fundamental representatives come in. However, first we introduce the notions of *oriented Markov graph* and *pouring water*.

Let $I_1 \le \cdots \le I_K$ be nondegenerate closed intervals such that on each I_k, f is either non-increasing or non-decreasing, but not constant. The *oriented Markov graph* of f with respect to $(I_1, \ldots, I_K)$ is the graph with vertices $I_1, \ldots, I_K$, and directed, labelled edges

$$I_k \xrightarrow{+} I_\ell \quad \text{if } f(I_k) \supseteq I_\ell \text{ and } f \text{ is non-decreasing on } I_k,$$

$$I_k \xrightarrow{-} I_\ell \quad \text{if } f(I_k) \supseteq I_\ell \text{ and } f \text{ is non-increasing on } I_k.$$

The oriented Markov graph of L_π with respect to $\big([1,2], \ldots, [n-1, n]\big)$ is called the oriented Markov graph of π.

Following [Be], we have for $\epsilon > 0$ small enough and for $j = 0, \ldots, n$, there is a unique λ_j such that $L_\pi^j[1, 1+\epsilon] \subseteq [\lambda_j, \lambda_j + 1]$. The loop

$$[\lambda_0, \lambda_0 + 1] \to \cdots \to [\lambda_{n-1}, \lambda_{n-1} + 1] \to [\lambda_n, \lambda_n + 1]$$

in the oriented Markov graph of π is called the *fundamental loop* of π. (Since $\lambda_0 = \lambda_n = 1$, this really is a loop and not just a path.)

Lemma 2.3. *Suppose that the map* $[k, k+1] \mapsto I_k$ *induces a label-preserving injection of the oriented Markov graph of* π *into the oriented Markov graph of* f *with respect to* $(I_1, \ldots, I_{n-1})$. *Then* f *has a representative* P *of* π *such that* $f^j(\min P) \in I_{\lambda_j}$ *for* $j = 0, \ldots, n-1$.

Proof. The loop in the Markov graph of f with respect to $(I_1, \ldots, I_{n-1})$ corresponding to the fundamental loop is

$$I_{\lambda_0} \to \cdots \to I_{\lambda_{n-1}} \to I_{\lambda_0}.$$

Since π is not a double, it follows from [Be, Lemma 3.6] that the fundamental loop of π is *simple*, i.e., does not consist of multiple repetitions of shorter loops. The result then follows from [Ba, Theorem 3.3]. $\square$

As in [ALM], if $r < s$ and $f[r, s]$ is the closed interval with endpoints $f(r)$ and $f(s)$, the map g obtained by "pouring water" into the graph of f on $[r, s]$ is defined by

$$g(x) = \begin{cases} \min\left(\sup_{r \leq y \leq x} f(y), \ \sup_{x \leq y \leq s} f(y)\right) & r \leq x \leq s \\ f(x) & \text{otherwise.} \end{cases}$$

The *flat pieces* of f are the nondegenerate intervals I maximal with respect to "f is constant on I."

Theorem 2.4. *Let* H *be the horseshoe map of the same type as* θ *and let* Q *be the unique fundamental representative of* θ *in* H. *If* π *has a representative in* H_Q, *then it also has a fundamental representative in* H_Q.

Proof. Without loss of generality, $\pi \neq \theta$ and π has a non-fundamental representative P in H_Q. Write $P = \{p_1, \ldots, p_n\}$ with $p_1 < \cdots < p_n$. Form the P-monotone map G by first truncating H_Q to form $(H_Q)_P$ and then pouring water into the graph of $(H_Q)_P$ on the intervals $[p_1, p_2], \ldots, [p_{n-1}, p_n]$.

Let

$$p_i' = \sup\{x : p_i \leq x \leq p_{i+1}, G(x) = G(p_i)\},$$
$$p_i'' = \inf\{x : p_i \leq x \leq p_{i+1}, G(x) = G(p_{i+1})\}.$$

Then $p_i' < p_i''$. Consider the oriented Markov graph of G with respect to

$$([p_1', p_1''], \ldots, [p_{n-1}', p_{n-1}'']).$$

It is canonically isomorphic, via the label-preserving map $[p_i', p_i''] \mapsto [i, i+1]$, to the oriented Markov graph of π. Then by Lemma 2.3, G has a representative P' of π such that $G^j(\min P') \in [p_{\lambda_j}', p_{\lambda_j}'']$ for $j = 0, \ldots, n-1$. Since P is not fundamental, some $[p_{\lambda_j}', p_{\lambda_j}'']$ is a proper subset of $[p_i, p_{i+1}]$, and hence $P' \neq P$. $P' \neq Q$ because $\pi \neq \theta$. But G maps its flat pieces into $P \cup Q$, and therefore P' does not meet any flat piece of G. It follows that P' is fundamental in G and hence in H_Q. $\square$

This theorem holds under more general hypotheses. It can be shown that if f is piecewise weakly monotone with finitely many flat pieces, π is not a double, and f has a representative of π, then it also has a fundamental representative of π. However, if π is a double, the statement need not hold.

Finally, we observe

Theorem 2.5. *Let H be the horseshoe map of the same type as θ and let Q be the unique fundamental representative of θ in H. If P is a representative of π in H_Q, then P is fundamental in H_Q if and only if it is fundamental in H.*

The theorems of this section reduce the problem of deciding whether θ forces π to the problem of finding all the fundamental representatives of π in H. In [BCMM], this problem was "solved" by considering only those cases where there is exactly one such representative. In the next section, we will solve the problems of finding and counting the number of all fundamental representatives of π in an arbitrary horseshoe map.

3. FINDING FUNDAMENTAL REPRESENTATIVES IN HORSESHOE MAPS

Suppose that H is a horseshoe map with M laps, labelled $[H_1] \leq \cdots \leq [H_M]$ or $[H_0] \leq \cdots \leq [H_{M-1}]$, so that the orientations of H on $[H_i]$ and L_π on $[L_i]$ are the same (and depend only on whether i is odd or even) for all i for which the statement makes sense.

The *fundamental spatial sequence* $\bar{s} = (\bar{s}_1, \ldots, \bar{s}_n)$ of π is defined as follows. For $i = 1, \ldots, n$, let $j(i)$ be the unique integer, $0 \leq j(i) \leq n-1$, such that $\pi^{j(i)}(1) = i$. Let $\bar{s}_i = \bar{t}_{j(i)}$. Thus $\bar{s}$ is the fundamental sequence of π written in spatial order.

Lemma 3.1.

  (1) $\bar{s}_1 \leq \cdots \leq \bar{s}_n$,
  (2) $\bar{s}_{i+1} - \bar{s}_i = 0$, 1 or 2,
  (3) if $\bar{s}_{i+1} = \bar{s}_i + 2$, then $\pi(i) > \pi(i+1)$ or $\pi(i) < \pi(i+1)$ according to whether L_π is increasing or decreasing on $[L_{\bar{s}_i}]$.

We omit the straightforward proof.

In the same way, we can form the *spatial sequence* $s = (s_1, \ldots, s_n)$ of any representative P of π in H. If $x = \min P$, then the H-itinerary of x is of the form t^∞, where $t = (t_0, \ldots, t_{n-1})$. Define s in the obvious way: $s_i = t_{j(i)}$.

Theorem 3.2. *Let P be a fundamental representative of π in H, where the laps of H are labelled as above. Then the spatial sequence s of P satisfies*

  (1) $s_1 \leq \cdots \leq s_n$,
  (2) s_i and $\bar{s}_i$ are both even or both odd,
  (3) if $s_i = s_{i+1}$, then $\bar{s}_i = \bar{s}_{i+1}$.

Conversely, if s satisfies (1)–(3), then s is the spatial sequence of some fundamental representative of π in H.

Proof. (1) follows from the fact that P is a representative of π in H. (2) is the fact that it is fundamental. We show (3).

Suppose $s_i = s_{i+1}$, but $\bar{s}_i \neq \bar{s}_{i+1}$. Then by Lemma 3.1(2) and Theorem 3.2(2), $\bar{s}_{i+1} = \bar{s}_i + 2$. Write $P = \{p_1, \ldots, p_n\}$ with $p_1 < \cdots < p_n$. Then $p_i, p_{i+1} \in [H_{s_i}]$. Suppose that $\pi(i) > \pi(i+1)$. Then by Lemma 3.1(3), L_π is increasing on $[L_{\bar{s}_i}]$ and so by (2), H is increasing on $[H_{s_i}]$. Therefore $H(p_i) < H(p_{i+1})$ and hence P isn't a representative of π. Similarly, $\pi(i) < \pi(i+1)$ leads to a contradiction.

Conversely, suppose s satisfies (1)–(3). Let $\bar{H}$ be the horseshoe map of the same type as π, with laps labelled as above, and let $\bar{P} = \{\bar{x}_1, \ldots, \bar{x}_n\}$, with $\bar{x}_1 < \cdots <$

$\bar{x}_n$, be the unique fundamental representative of π in $\bar{H}$. Let Y and Z be finite subsets of $[0, 1]$ such that

(i) $0, 1 \in Y \cup Z$.

(ii)   The continuous, piecewise linear map $G : [0, 1] \to [0, 1]$ with $G(0) = H(0)$, obtained by "connecting the dots" of
$$\{(\bar{x}, \bar{H}(\bar{x})) : \bar{x} \in \bar{P}\} \cup \{(y, 0) : y \in Y\} \cup \{(z, 1) : z \in Z\},$$
has the same number of laps as H.

(iii) G maps each lap onto $[0, 1]$.

(iv)   The laps $[G_j]$ of G are labelled so that they have the same set of labels and the same order as the laps $[H_j]$ of H, and so that $\bar{x}_i \in [G_{s_i}]$ for $i = 1, \ldots, n$.

By [BC, Theorem 2.7], G is topologically conjugate, via an order-preserving homeomorphism, to H. Since this conjugacy takes lap $[G_j]$ of G to lap $[H_j]$ of H, H has a representative of π with spatial sequence s. By (2) and the way the laps of H are labelled, that representative must be fundamental. $\square$

Theorem 3.3. *Let L_π have N laps, and let H be a horseshoe map with M laps, labelled as above. Then the number of fundamental representatives of π in H is*

$$\binom{n + \left[\frac{M-N}{2}\right]}{\left[\frac{M-N}{2}\right]} \quad or \quad \binom{n + \left[\frac{M-N-1}{2}\right]}{\left[\frac{M-N-1}{2}\right]}$$

according to whether the orientations of L_π and H on their first laps agree or disagree.

(Here $[\cdot]$ denotes "integer part of.")

Proof. Suppose that the orientations of L_π and H on their first laps agree. We first show that the set of sequences $s = (s_1, \ldots, s_n)$ on $\{1, \ldots, M\}$ satisfying (1)–(3) of Theorem 3.2 is in one-to-one correspondence with the set of sequences $u = (u_1, \ldots, u_n)$ on $\{1, \ldots, M - N + 2n - 1\}$ satisfying

(1') $u_1 < \cdots < u_n$,

(2')   each u_i is odd.

Let $u_i = s_i - \bar{s}_i + 2i - 1$. The map $s \mapsto u$ is clearly one-to-one. Since $\bar{s}_1 = 1$ and $\bar{s}_n = N$, u_i takes on values in $\{1, \ldots, M - N + 2n - 1\}$. Let s satisfy (1)–(3) of Theorem 3.2 and let $s \mapsto u$. That (2') holds is clear from (2). To show (1'), note that $u_{i+1} - u_i = (s_{i+1} - s_i) - (\bar{s}_{i+1} - \bar{s}_i) + 2$. Now $\bar{s}_{i+1} - \bar{s}_i = 0, 1$ or 2 and $s_{i+1} \geq s_i$. Therefore $u_{i+1} > u_i$ unless $\bar{s}_{i+1} - \bar{s}_i = 2$ and $s_{i+1} = s_i$. This is impossible by (3).

Conversely, suppose $u = (u_1, \ldots, u_n)$ satisfies (1') and (2') and let $s \mapsto u$. Then $s_i = u_i + \bar{s}_i - (2i - 1)$. It is immediate that s satisfies (1)–(3).

There are clearly

$$\binom{n + \left[\frac{M-N}{2}\right]}{n} = \binom{n + \left[\frac{M-N}{2}\right]}{\left[\frac{M-N}{2}\right]}$$

such sequences.

If the orientations of L_π and H on their first laps disagree, replace M by $M - 1$ in the arguments above. $\square$

Corollary 3.4. [BCMM, Proposition 12] *θ has a unique fundamental representative in the horseshoe map of the same type as θ.*

4. When π is a Double

In this section, we state, without proof, "double" versions of the results in Sections 2 and 3, *all with the (unstated) assumption that π is a double*, using the formal notation below. Together they show how and why the algorithm works when π is a double.

Recall that if π is a double, then the fundamental representatives of π are actually representatives of $\pi/2$. To keep this abused terminology from becoming too confusing, we introduce the formal notations $\text{REP}(\pi, f)$ and $\text{FUND}(\pi, f)$. Thus if π is a double, then $\text{FUND}(\pi, f) \subseteq \text{REP}(\pi/2, f)$.

Theorem 2.1$'$. *Let $Q \in \text{REP}(\theta, f)$ and suppose that f is Q-monotone. Then θ forces π if and only if $\text{REP}(\pi, f) \neq \varnothing$.*

Theorem 2.2$'$. *Let H be the horseshoe map of the same type as θ and let Q be the unique fundamental representative of θ in H. If $P \in \text{REP}(\pi, H)$, then $P \in \text{REP}(\pi, H_Q)$ if and only if $(*)$ holds.*

Lemma 2.3$'$. *Suppose that the map $[k, k+1] \mapsto I_k$ induces a label-preserving injection of the oriented Markov graph of π into the oriented Markov graph of f with respect to $(I_1, \ldots, I_{n-1})$. Then there exists $P \in \text{REP}(\pi/2, f)$ such that $f^j(\min P) \in I_{\lambda_j}$ for $j = 0, \ldots, n/2 - 1$.*

Remark. The relevant fact for the proof is that the fundamental loop of π consists of two repetitions of a simple loop [Be, Lemma 3.6].

Theorem 2.4$'$. *Let H be the horseshoe map of the same type as θ and let Q be the unique fundamental representative of θ in H. If $\text{REP}(\pi, H_Q) \neq \varnothing$, then $\text{FUND}(\pi, H_Q) \neq \varnothing$.*

Theorem 2.5$'$. *Let H be the horseshoe map of the same type as θ and let Q be the unique fundamental representative of θ in H. If $P \in \text{REP}(\pi/2, H_Q)$, then $P \in \text{FUND}(\pi, H_Q)$ if and only if $P \in \text{FUND}(\pi, H)$.*

If π is a double, then the *fundamental spatial sequence* $\bar{s} = (\bar{s}_1, \ldots, \bar{s}_{n/2})$ of π is defined by $\bar{s}_i = \bar{t}_{j(i)}$.

Lemma 3.1$'$.

(1) $\bar{s}_1 \leq \cdots \leq \bar{s}_{n/2}$.
(2) $\bar{s}_{i+1} - \bar{s}_i = 0, 1$ or 2.
(3) if $\bar{s}_{i+1} = \bar{s}_i + 2$, then $\pi(i) > \pi(i+1)$ or $\pi(i) < \pi(i+1)$ according to whether L_π is increasing or decreasing on $[L_{\bar{s}_i}]$.

Theorem 3.2$'$. *Let P be a fundamental representative of π in H, where the laps of H are labelled as at the beginning of Section 3. Then the spatial sequence s of P satisfies*

(1) $s_1 \leq \cdots \leq s_{n/2}$,
(2) s_i *and* $\bar{s}_i$ *are both even or both odd,*
(3) *if* $s_i = s_{i+1}$, *then* $\bar{s}_i = \bar{s}_{i+1}$.

Conversely, if s satisfies (1)–(3), then s is the spatial sequence of some fundamental representative of π in H.

Theorem 3.3′. *Let L_π have N laps and let H be a horseshoe map with M laps, labelled as at the beginning of Section* 3. *Then the number of fundamental representatives of π in H is*

$$\binom{\frac{n}{2} + \left[\frac{M-N}{2}\right]}{\left[\frac{M-N}{2}\right]} \quad \text{or} \quad \binom{\frac{n}{2} + \left[\frac{M-N-1}{2}\right]}{\left[\frac{M-N-1}{2}\right]}$$

according to whether the orientations of L_π and H on their first laps agree or disagree.

Acknowledgment. The authors thank Karen Collins for teaching them how to count.

REFERENCES

[ALM] Ll. Alsedà, J. Llibre, and M. Misiurewicz, *Combinatorial dynamics and entropy in dimension one*, Barcelona, 1990.

[Ba] S. Baldwin, *Generalizations of a theorem of Sarkovskii on orbits of continuous real-valued functions*, Discrete Math. **67** (1987), 111–127.

[Be] C. Bernhardt, *The ordering of permutations induced by continuous maps of the real line*, Ergodic Theory Dynamical Systems **7** (1987), 155–160.

[BCMM] C. Bernhardt, E. Coven, M. Misiurewicz, and I. Mulvey, *Comparing periodic orbits of maps of the interval*, Trans. Amer. Math. Soc. (to appear).

[BC] L. Block and E. M. Coven, *Topological conjugacy and transitivity for a class of piecewise monotone maps of the interval*, Trans. Amer. Math. Soc. **300** (1987), 297–306.

[BGMY] L. Block, J. Guckenheimer, M. Misiurewicz, and L.-S. Young, *Periodic points and topological entropy of one-dimensional maps*, Springer Lecture Notes in Mathematics **819** (1980), 18–34.

[J] I. Jungreis, *Some results on the Sarkovskii partial ordering of permutations*, Trans. Amer. Math. Soc. **325** (1991), 319–344.

DEPARTMENT OF MATHEMATICS AND COMPUTER SCIENCE, FAIRFIELD UNIVERSITY, FAIRFIELD, CONNECTICUT 06430

E-mail: cbernhardt@fair1.bitnet

DEPARMENT OF MATHEMATICS, WESLEYAN UNIVERSITY, MIDDLETOWN, CONNECTICUT 06459

E-mail: coven@jordan.math.wesleyan.edu

Fairfield University and Wesleyan University

Contemporary Mathematics
Volume **135**, 1992

FULLY POSITIVE TOPOLOGICAL ENTROPY
AND TOPOLOGICAL MIXING

F. BLANCHARD

ABSTRACT. A flow (X, T) is said to have uniform positive entropy (u.p.e.) if any cover of X by two non dense open sets has positive topological entropy ; the weaker property of completely positive entropy (c.p.e.) means by definition that any topological factor has positive entropy. It is shown that u.p.e. flows are topologically weakly but not always strongly mixing ; c.p.e. implies existence of an invariant measure with support X, but no degree of topological mixing, not even transitivity.

1 - Introduction:

Let us call *flow* a compact metric set X endowed with a homeomorphism T. One of the various reasons why flows have been investigated is they are frequently met with as supports of particular dynamical systems; quite naturally, many notions in Topological Dynamics are closely related to, or even derived from, notions in metric Ergodic Theory. Topological entropy was introduced in the Theory of Flows [A-K-M] by analogy with the Kolmogorov entropy of measurable dynamical systems, and the Variational Principle of Goodwyn-Dinaburg-Goodman (see [D-G-S]) describes the relationship between the two concepts. Topological mixing properties (among which one may include transitivity) were introduced in the first place as necessary topological conditions for existence of an invariant measure on (X, T) with support X (i.e. such that any nonempty open set of X has positive measure) and having the corresponding mixing properties: for instance ergodicity of μ implies transitivity of the support of μ [G-H], [F].

It is well-known that a K-system, i.e. a dynamical system with completely positive metric entropy, is strongly mixing of all orders [P]. In the topological setting, does there exist some (preferably not too far-fetched) property of fully positive entropy, implying topological mixing of some or all kinds ? The aim of this paper is to give some partial answer to this question, by examining two such properties. The property of specification [B] implies strong mixing, but it is not characterized as an entropy property, though implying u.p.e.; maybe it is also too strong to cover all relevant cases.

The stronger requirement for (X, T) we study in this article is that all covers of X by two non dense open sets have positive entropy: this property we call *uniform positive entropy* (u.p.e.). In par. 3 it is shown u.p.e. implies weak mixing, hence

1980 *Mathematics Subject Classification* (1985 *Revision*). 54H20.
Supported by C.N.R.S. (U.R.A. 225) .

This paper is in final form and no version will be submitted for publication elsewhere.

transitivity (Proposition 2): on this account the answer to our question is partly
yes. But no such implication holds for strong mixing, as shown by another symbolic
example. The proof that u.p.e. holds in this case relies on the system's possessing
a combinatorial property, weaker than specification though it is in fact the one
classically used for proving that flows with specification have positive topological
entropy [B], [D-G-S] (and in this paper, also u.p.e. in Proposition 3).

In view of the Variational Principle, an obvious necessary condition for existence
of a K-measure with support X is *completely positive entropy* (c.p.e.): all topological
factors of (X, T) must have positive topological entropy. With this definition of fully
positive entropy the answer to our question is no; c.p.e. does not imply any kind of
mixing (par. 4). Examples are given of weakly but not strongly mixing, transitive
but not weakly mixing, and even non transitive, flows, all having c.p.e. This clearly
means that the analogy between the topological and metric situations cannot be
carried out too far. Most examples are constructed by submitting subshifts to a
certain kind of flip-off which preserves nonmixing properties while injecting entropy
into all factors. By the way some of these examples establish that c.p.e. is strictly
weaker than u.p.e.: because it does not imply weak mixing, of course, but there are
also weakly mixing flows with c.p.e. and without u.p.e.. The only other positive
result about c.p.e. is that any non empty open set can have positive probability
for some invariant measure, so that there is an invariant measure with support X
(Proposition 6, Corollary 7).

Naturally none of these two properties implies existence of a K-measure with
support X (Example 5); nor does c.p.e. imply existence of an ergodic measure with
support X (Example 8). This was to be expected, since topological mixing of some
kind does not imply existence of measures with support X having the corresponding
measure-theoretic mixing property.

Eli Glasner provided me with some very precious information. It was Benjy Weiss
who suggested the statement and proof of Proposition 6 in its present form, and
whose comments caused me to start anew, eventually finding Example 5. Georges
Hansel took an important part in this research and was originally to cosign this
paper but, for extremely respectable reasons of his own, finally decided not to. The
referee pointed out several mistakes.

2 - Definitions:

A *flow* is a compact metric space X, together with a homeomorphism $T : X \to$
X. An important class of flows is that of subshifts, or symbolic dynamical systems.
Let A be a finite set of symbols endowed with the discrete topology and $A^{\mathbf{Z}}$ the
set of all bi-infinite sequences $(x_n)_{n \in \mathbf{Z}}$ of elements of A endowed with the product
topology; let $\sigma : A^{\mathbf{Z}} \to A^{\mathbf{Z}}$ be the *shift* defined by $\sigma((x_n)_{n \in \mathbf{Z}}) = ((x_{n+1})_{n \in \mathbf{Z}})$; then
σ is a homeomorphism of the compact metrizable space $A^{\mathbf{Z}}$. A *subshift* or *symbolic
dynamical system on A* is a couple (X, T) where X is any closed subset of $A^{\mathbf{Z}}$ which
is invariant under σ and T is the homeomorphism of X induced by σ. We denote
by A^* the set of finite sequences of elements of A, called *words* on A. Equipped
with the concatenation product, A^* is a free monoid. Let $s = (a_n) \in A^{\mathbf{Z}}$ and let
$i \leq j$ be two integers; we denote by $s[i, j]$ the word $a_i a_{i+1} \cdots a_j$. Conversely, let
$u = a_0 a_1 \cdots a_n \in A^*$; we denote by $[u]$ the *cylinder* consisting of all the $s \in A^{\mathbf{Z}}$
such that $s[0, n] = u$ and by $| u |$ the length $n + 1$ of the word u.

Let X be a subshift on A. The set

$$L(X) = s[i,j] \mid s \in X, \ i, \ j \in \mathbf{Z}, \ i \leq j$$

is the *language* of X; it uniquely defines X.

Let (X, T), (X', T') be two flows. A *factor map* is a continuous, onto map $\phi : X \to X'$ changing T to T' (i.e. $T' \circ \phi = \phi \circ T$). The flow (X', T') is called a *factor* of (X, T).

All subsequent definitions for flows may be checked on $L(X)$ whenever X is a subshift.

Transitivity : for any two non empty open sets U, $V \subset X$, there is $n \in \mathbb{N}$ such that $T^n(U) \cap V \neq \emptyset$. In the symbolic case this becomes: for any two words u, $v \in L(X)$, there exists $w \in L(X)$ such that $uwv \in L(X)$.

Weak mixing : the cartesian square $(X \times X, \ T \times T)$ is transitive.

Strong mixing : for any two non empty open sets U, $V \subset X$, there is $n_0 \in \mathbb{N}$ such that for any $n \geq n_0$, $T^n(U) \cap V \neq \emptyset$. In the symbolic case: for any two words u, $v \in L(X)$, there is $n_0 \in \mathbb{N}$ such that for any $n \geq n_0$, there exists $w \in L(X)$ such that $\mid w \mid = n$ and $uwv \in L(X)$.

Property of specification : given $\varepsilon > 0$, there is an integer $N = N(\varepsilon)$ such that for any $k \geq 2$, for any k points $x_1, \ldots, x_k \in X$, for any integers

$$a_1 \leq b_1 < a_2 \leq b_2 < \ldots < a_k \leq b_k \quad \text{with} \quad a_i - b_{i-1} \geq N(\varepsilon), \ 2 \leq i \leq k,$$

and for any integer p with $p \geq N(\varepsilon) + b_k - a_1$, there exists $x \in X$ with $T^p x = x$ such that $d(T^n x, T^n x_i) \leq \varepsilon$ for $a_i \leq n \leq b_i$, $1 \leq i \leq k$.

It is easy to show that specification implies strong mixing, which implies weak mixing, which in its turn implies transitivity. Those four properties are preserved under factor maps; the next one also is.

Minimality : the system (X, T) is said to be *minimal* if there exists no nonempty closed T-invariant set strictly contained in X.

Let us now give all the required definitions of entropy. The proofs of all stated results may be found in [D-G-S].

Let X be a flow. Suppose $\mathcal{R}$ is an open cover of X (i.e. a set of non empty open subsets of X whose union is X). A *subcover* is any subset $\mathcal{S} \subset \mathcal{R}$ also covering X. Given two covers $\mathcal{R}$ and $\mathcal{S}$, $\mathcal{S}$ is said to be *finer* than $\mathcal{R}$ if any element of $\mathcal{S}$ is contained in an element of $\mathcal{R}$; then we write $\mathcal{R} \leq \mathcal{S}$. Denote by $\mathcal{R} \vee \mathcal{S}$ the cover of X made up by all the sets $R \cap S$, $R \in \mathcal{R}$, $S \in \mathcal{S}$. If $\mathcal{R}$ is a cover and $n \in \mathbb{N}$, we write $\mathcal{R}_n = \bigvee_{i=0}^{n} T^i \mathcal{R}$. For an open cover $\mathcal{R}$ of X, let

$$H(\mathcal{R}) = \inf(\log \# \mathcal{R}')$$

where $\mathcal{R}'$ spans the set of all finite subcovers of $\mathcal{R}$ ($\#$ denotes the cardinality).

The *topological entropy of* $\mathcal{R}$ is the number

$$h(\mathcal{R}, T) = \lim_{n \to \infty} \frac{1}{n} H(\mathcal{R}_n).$$

The limit is known to exist and if $\mathcal{R} \leq \mathcal{S}$, then $h(\mathcal{R}, T) \leq h(\mathcal{S}, T)$.

The *topological entropy of the flow* (X, T) is the real number

$$h(X, T) = sup\ h(\mathcal{R}, T),$$

where the sup is taken over all finite open covers $\mathcal{R}$ of X. One proves without major difficulties that in the symbolic case, the topological entropy of X is given by the following formula:

$$h(X, T) = \lim_{n \to \infty} \frac{1}{n} \log \# (L(X) \cap A^n).$$

If μ is a T-invariant probability measure on X, the usual metric entropy of T is denoted by $h_\mu(X, T)$. The *support* of an invariant probability measure μ on X is the intersection of all closed invariant subsets of X with measure 1. One of the most significant facts about topological entropy is the Variational Principle (Goodwyn-Dinaburg-Goodman): for any flow (X, T)

$$h(X, T) = sup\ h(\mu)$$

where the sup is taken over all T-invariant probability measures, and $h(\mu)$ is the usual entropy of the system (X, T, μ).

3. Flows with uniform positive entropy:

Definition 1. A flow (X, T) is said to have *uniform positive entropy* (u.p.e.) if any cover by two non-dense open sets has positive entropy.

Definition 2. A flow (X, T) is said to have *completely positive entropy* (c.p.e.) if any non trivial factor of (X, T) has positive entropy.

Both properties are stable under factor maps. From the definition of entropy it is obvious that u.p.e. flows have c.p.e.; in par. 4 we show that the converse is false.

Remark. Why not assume positive entropy for *all* nontrivial open covers in the definition of u.p.e.? Of course one must exclude trivial covers, i.e. those containing X itself, since they all have entropy zero. But this is not sufficient: covers by two open sets distinct from X, having entropy 0, can be found in any flow. Let $x \in X$, $\mathcal{R} = (\{x\}^c, V)$ with $x \in V$ and V strictly less than X. $\mathcal{R}_n$ contains the open set $\{x, Tx, \cdots, T^{n-1}x\}^c$, so there is a subcover of $\mathcal{R}_n$ by at most $n + 1$ open sets. Thus $H(\mathcal{R}_n) = \log(n + 1)$ and $h(\mathcal{R}) = 0$.

Let us first prove that u.p.e. implies topological weak mixing. The next Lemma will be useful for this.

Lemma 1 (Petersen [Pe]). *A flow (X, T) is not weakly mixing if and only if there exist two nonempty open sets U and V such that for any $n \in \mathbb{N}$, one cannot have simultaneously $U \cap T^n U \neq \emptyset$ and $U \cap T^n V \neq \emptyset$.*

Remark. Note that the condition of Lemma 1 implies that U and V have empty intersection.

Proposition 2. *A flow having u.p.e. is weakly mixing.*

Proof. Let (X, T) be a non weakly mixing flow. We build up a nontrivial two-set open cover with zero entropy. Suppose U and V are open sets as given by Lemma 1. Let $\mathcal{R}' = \{U', V'\}$ be a nontrivial open cover of X such that $V^c \subset U'$, and $U^c \subset V'$. Let $n \in \mathbb{N}$ and $i \in \{0, 1, \cdots, n\}$. There are two disjoint possibilities: either $U \cap T^i U \neq \emptyset$, in which case $U \subset (T^i V)^c \subset T^i U'$; or $U \cap T^i U = \emptyset$, in which case $U \subset T^i V'$. Thus U is contained in one set $W_0 \cap T W_1 \cap \cdots T^n W_n$, $W_i = U'$ or V', of the cover $\mathcal{R}'_n$. Moreover note that if $x \notin U$, then $x \in V'$. Considering for each $x \in X$ the first $i \in \{0, 1 \cdots, nJ\}$ such that $T^{-i} x \in U$, we get that R'_n admits as a subcover the set

$$\mathcal{R}''_n = \{V' \cap T V' \cdots \cap T^{i-1} V' \cap T^i W_0 \cap T^{i+1} W_1 \cap \cdots \cap T^n W_{n-i} \mid i = 0, 1, \cdots, n\}.$$

Hence for all $n \in \mathbb{N}$, $H(\mathcal{R}'_n) \leq \log(n+1)$ and therefore $h(\mathcal{R}', T) = 0$.

The following property implies u.p.e.; this is used for proving specification implies u.p.e, and also in example 5 below.

Definition. A flow (X, T) is said to have *Property P* if for any two nonempty open sets U_0 and U_1 in X there is an integer N such that whatever $k \geq 2$, whatever $s = (s(1), s(2), \cdots, s(k)) \in \{0, 1\}^k$ there exists $x \in X$ with $x \in U_{s(1)}, \cdots, T^{(k-1)N} x \in U_{s(k)}$.

This property is obviously not stronger than the specification property [B]; that it is strictly weaker will be shown further on. The only meaningful case in the present setting is when $U_0 \cap U_1 = \emptyset$.

The following statement is hardly more than a remark: the argument is the one proving that flows with specification have positive topological entropy in [D-G-S].

Proposition 3. *A flow X having property P has u.p.e..*

Proof. Let $\mathcal{R} = (U, V)$ be a cover of X by two nondense open sets; there exist two disjoint nonempty open sets $U_0 \subset U^c$ and $U_1 \subset V^c$. As X possesses property P, there exists an integer N such that for any integer k, any sequence $s = (s(1), s(2), \cdots, s(k)) \in \{0, 1\}^k$, the open set

$$U_{s(1)} \cap T^{-N} U_{s(2)} \cap \cdots \cap T^{-(k-1)N} U_{s(k)}$$

is nonempty, so one may choose a point $z(s)$ belonging to it.

Thus, if s and s' are two different elements in $\{0, 1\}^k$, since $U_0 \subset U^c$ and $U_1 \subset V^c$, the two points $z(s)$ and $z(s')$ cannot be in the same element of the cover $\mathcal{R}_{nN}$. Hence $\# \mathcal{R}_{nN} \geq 2^n$, and consequently $h(\mathcal{R}, T) > \log \frac{2}{N}$.

Thus all flows with the specification property, for instance Anosov flows and aperiodic sofic systems, have u.p.e.. But they are not the only ones.

Before describing an example, in order to simplify the notations, we introduce the idea of pattern, suggested by the construction of Toeplitz sequences.

Given a finite alphabet A and a symbol $\wedge \notin A$ a *pattern on A* is a word on $A' = A \cup \{\wedge J\}$; one may consider $\wedge$ as representing a blank, i.e. a position for which no letter has been assigned. A pattern $\pi = \pi_1 \cdots \pi_n$, $\pi_i \in A'$ may be identified to a subset of A^n : "$u = u_1 \cdots u_n \in \pi$" means "$u_i = \pi_i$ wherever $\pi_i \neq \Lambda$, and u_i is arbitrary elsewhere". A pattern π is said to be non empty in language L if $L \cap \pi \neq \emptyset$. A (doubly) infinite pattern is an element of $A'^{\mathbf{Z}}$, identified to a closed subset of $A^{\mathbf{Z}}$. Here is a useful characterization of Property P in the symbolic case.

Proposition 4. *A subshift X on alphabet A satisfies Property P if for any integer p belonging to some infinite strictly increasing sequence of integers,*
(A)   There exists an integer $N(p)$ such that no pattern $u_1 \wedge^{N(p)}$ $u_2 \wedge^{N(p)}$ $\cdots \wedge^{N(p)} u_k$ is empty, in $L(X)$ for arbitrary k and $u_i \in L(X) \cap A^p$ for $0 < i \leq k$.

The proof is elementary and left to the reader.

Example 5. Here is an uncountable family of flows, having Property P and therefore with u.p.e. (and weakly mixing), but not strongly mixing (thus not having the specification property).

Let us first introduce an arbitrary, strictly increasing sequence of positive integers $(h(n), n \geq 1)$: one associates to it a decreasing sequence $(X_n, n \geq 1)$ of subshifts of finite type on alphabet $A = \{a, b\}$, having some suitable properties and converging to a non strongly mixing subshift X. Then we show $(h(n),\ n \in \mathbb{N})$ may be chosen so that X possesses Property P.

1– Definition of a subshift given an increasing sequence of integers:
To each n associate the language $E_n \in A^{h(n)+2}$: $E_n = \{a A^{h(n)} a,\ n \in \mathbb{N}^*\}$; call X_n the subshift of finite type defined by exclusion of language $F_n = \cup_{0 < i \leq n} E_i$: in X_n an a must never occur in position $h(i) + 1$ after another a $(1 \leq i \leq n)$. Put $X = \cap_{n > 0} X_n$, $[a] = \{x \in X \mid x_0 = a\}$.
Remark that:
(α) X is non empty, as a decreasing intersection of subshifts of finite type.
(β) if $(h(n)) \neq (h'(n))$, then the two corresponding subshifts X and X' are distinct;
(γ) as $(h(n))_{n>0}$ is an infinite sequence and all sets of the form $[a] \cap T^{h(n)+1}[a]$ are empty in the limit subshift X, X cannot be strongly mixing;
(δ) for all n any two words u, v in $L(X_n)$ may be connected by a string of b's with any length greater than $h(n) + 1$. This is a straightforward consequence of the next two facts; if u belongs to $L(X_n)$ then ub also does; and X_n is defined by forbidding words with length less than or equal to $h(n) + 2$.

2 - Choosing a suitable sequence of integers:
We claim that the sequence $(h(n))$ may be chosen so that X possesses Property P. To prove this we construct $(h(n))$ and two auxiliary strictly increasing sequences of integers $f(n)$ and $g(n)$ by an induction.
Initial step. Fix integers $f(1) \geq 1$, $g(1) \geq f(1)$, and $h(1) \in \{f(1)+1, \cdots, f(1)+ g(1)\}$ (mod $f(1) + g(1)$), and define X_1 by forbidding $E_1 = \{a A^{h(n)} a\}$. The main feature of X_1 is that it satisfies assumption (A) in Proposition 4 for $p = f(1)$, $N(p) = g(1)$; to check this it is enough to replace symbols $\wedge$ by b's in patterns described in (A): as $g(1) \geq f(1)$, and because of the condition on $h(1)$, all constraints due to occurrences of letter a in words u_i fall into the $\wedge$ strings, or after the end of the pattern, and they are satisfied by writing b's in place of $\wedge's$.
Furthers steps of the induction. They are hardly more complicated than the first, and the idea is the same: the new set of constraints introduced at step n must preserve assumption (A) for $p = f(i)$, $N(p) = g(i)$ $(0 < i \leq n)$ in X_{n-1}.

Suppose $f(i)$, $g(i)$, $h(i)$, and thus X_i are defined up to $n-1$ and assumption (A) holds in X_{n-1} for any $i \leq n$ for $p = f(i)$, $N(p) = g(i)$.
First choose $f(n) > f(n-1)$, then

$$g(n) \geq sup\{(f(n), h(n-1) + 1, 2(f(n-1) + g(n-1))\},$$

and finally $h(n) > h(n-1)$ such that

(1) $$h(n) \in \{f(n) + 1, ..., f(n) + g(n)\} \quad (\text{mod } f(n) + g(n)).$$

(2) $$h(n) \in \{f(n-1) + 1, ..., f(n-1) + g(n-1)\} \quad (\text{mod } f(n-1) + g(n-1)).$$

...

(n) $$h(n) \in \{f(1) + 1, ..., f(1) + g(1)\} \quad (\text{mod } f(1) + g(1)).$$

Conditions (1) to (n) can be fulfilled simultaneously: since $g(i) \geq 2(f(i-1) + g(i-1))$ for all $i \leq n$, any interval of the integer line of the form

$$[f(i) + 1, \cdots, f(i) + g(i)] + k(f(i) + g(i))$$

corresponding to condition $(n - i + 1)$ contains at least one complete interval of the form
$$[f(i-1) + 1, \cdots, f(i-1) + g(i-1)] + k'(f(i-1) + g(i-1)),$$

and a backward induction leads to existence of a suitable value of $h(n)$; in fact, there is a denumerable set of possible values.
The set of conditions (2) to (n) precisely means this: suppose one of the patterns corresponding to step i which must be nonempty according to the induction hypothesis, call it π, contains an a in the kth position; then the new set of constraints E_n defining subshift X_n in addition to F_{n-1} forbids occurrence of an a in position $k + h(n)$, which necessarily belongs to a string of $\wedge's$ in the same pattern owing to condition $(n - i + 1)$, so it is always possible to satisfy it by writing a b in this position; therefore pattern π is still nonempty in $L(X_n)$. Thus assumption (A) remains valid in $L(X_n)$ for $p = f(i)$, $N(p) = g(i)$, $i = 1, \cdots, n - 1$. Condition (1) ensures the same is true for $i = n$: given some corresponding pattern π, words u_j in it have length $f(n)$, which is less than $h(n)$ by condition (1), so by fact (δ) π is not empty in $L(X_{n-1})$; but as $g(n) \geq f(n)$ and by condition (1) again, all constraints of E_n due to occurrences of letter a in words u_j fall into the $\wedge$ strings, or after the end of the pattern, and they are satisfied by writing b's in place of $\wedge's$; thus π is non empty in $L(X_n)$ and assumption (A) holds in X_n for $p = f(n)$, $N(p) = g(n)$. Therefore the induction hypotheses for next step are fulfilled.
One already knows the limit X of the just constructed sequence (X_n) cannot be strongly mixing (β). We need only check it possesses Property P. Given some integer i, for $n > i$, $L(X_n) \cap A^{f(i)} = L(X_{i-1}) \cap A^{f(i)}$ so one also has $L(X) \cap A^{f(i)} = L(X_{i-1}) \cap A^{f(i)}$. Moreover, given i and $n \geq i$, from assumption (A) applied to $p = f(i)$, $N(p) = g(i)$ in X_n, one deduces by compactness that any infinite pattern

of the form $\cdots u_{-1} \wedge^{g(i)} u_0 \wedge^{g(i)} u_1 \wedge^{g(i)} \cdots$ is included in X_n, therefore in X; thus for any i assumption (A) is also true in X for $p = f(i)$, $N(p) = g(i)$; as $f(n)$ tends to infinity with n, by Proposition 4 Property P is established. By fact (α) and since there is a denumerable set of possible choices for $h(n)$ at step n, there exists an uncountable family of distincts subshifts satisfying all the requirements.

Remarks.

(1) No nontrivial invariant measure on X is K or even strongly mixing, since such a measure gives positive probability to cylinder $[a]$!

(2) These examples also provide weakly but not strongly mixing flows of a new type, since examples in [D-K] and [Pe-S] have topological entropy O.

4 - Flows with completely positive entropy:

The assumption of next statement, existence of a non empty universally null open set, appears in [S-W], where it is related with other properties.

Proposition 6 (B. Weiss). *Let (X, T) be a flow with a non empty open set U such that $\mu(U) = 0$ for any invariant measure μ. Then (X, T) has a non trivial topological factor with topological entropy zero.*

L. et $x_0 \in U$; using Urysohn's lemma construct a map $f : X \to [0, 1]$, continuous with $f(x_0) = 0$ and $f(x) = 1$ on U^c, a closed set disjoint from x_0. Define $F : X \to [0, 1]^{\mathbf{Z}}$ by

$$(F(x))_n = f(T^n x)$$

Let $Y = F(X)$ be endowed with the shift σ. Then (Y, σ) is a non trivial topological factor of (X, T) and the only invariant measure on Y is concentrated on the fixed point $1^{\mathbf{Z}}$. Thus (Y, σ) has topological entropy zero by the variational principle. $\square$

Corollary 7. *If (X, T) has c.p.e. there is an invariant measure μ with support X.*

A. a compact metric space X has a countable base of open sets $(U_n, n \geq 0)$. For each U_i choose a measure μ_i such that $\mu_i(U_i) > 0$. Then any linear combination of the $\mu_i's$ with positive coefficients fulfills the requirement. $\square$

Remark. Example 8 below shows that one cannot replace "invariant" by "ergodic" in Corollary 7.

Typical examples of flows with positive entropy but without c.p.e. are some Toeplitz flows: their continuous eigenfunctions generate zero-entropy factors that are rotations of the 1-torus [M-P], [W].

When one wants to prove a flow has c.p.e., there are two obvious methods:

1) Proving it has u.p.e.

2) Proving there exists a $K-$measure on (X, T) with support X.

Thus any strictly ergodic model of a $K-$system, as given by [H-R], [K], has c.p.e.; so have the topological supports of the Ornstein-Shields $K-$systems [O-S].

Some c.p.e. flows do not have any of these properties. Here is an example:

Example 8. Let $X = \{a, b\}^{\mathbf{Z}} \cup \{a, c\}^{\mathbf{Z}}$. X is not shift-transitive, because it is impossible to have occurrences of b and c in the same sequence. But it has c.p.e.: any factor Y of X is the union of a factor of $\{a, b\}^{\mathbf{Z}}$ and a factor of $\{a, c\}^{\mathbf{Z}}$; if Y has entropy 0, so have those two. But any zero-entropy factor of the full shift is trivial, i.e. reduced to a fixed point. As $\{a,b\}^{\mathbf{Z}}$ and $\{a,c\}^{\mathbf{Z}}$ have non empty intersection, the two fixed points are identical and Y is trivial. In this case one easily sees no ergodic measure can have support X.

The rest of this section is devoted to constructing more sophisticated counterexamples. All the flows we build up have c.p.e. but some of them do not have u.p.e. and/or weak mixing.

Example 9. Let Y be a transitive subshift on an alphabet $A = \{a,b,c\}$ and let X denote the full shift on $\{0,1\}$.

Construct the cartesian product $Y \times X$, and let Z be its image under the $1-$block map $\phi : Y \times X \to (A \cup 0)^{\mathbf{Z}}$ defined by $\phi((y_n, x_n)_{n \in \mathbf{Z}} = (z_n)_{n \in \mathbf{Z}}$ with

$$z_n = \begin{cases} y_n & \text{if } x_n = 1 \\ 0 & \text{if } x_n = 0 \end{cases}$$

Note that $Y \subset Z$, and Z is transitive when Y is.

Proposition 10. *If Y is minimal, Z has c.p.e..*

S. uppose ψ is a factor map from (Z, σ) onto a nontrivial flow (W, T) with metric d. Let $\alpha \in Z$ be the fixed point on the letter 0. If $\varepsilon > 0$ is small enough, there exists $\beta \in Z$ such that

$$d(\psi(\beta), \psi(\alpha)) > 3\varepsilon.$$

Since ψ is continuous, there exists $m \in \mathbb{N}$ such that if z and z' are two point in Z with

$$z[-m, m] = \alpha[-m, m] = 0^{2m+1} \text{ and } z'[-m, m] = \beta[-m, m]$$

then $d(\psi(z), \psi(z')) > \varepsilon$.

Note that this implies $\beta[-m, m] \neq 0^{2m+1}$. Let $v = \beta[-m, m]$. Choose $u \in L(Y)$ and $u' \in L(X)$ such that $\phi([u] \times [u']) = v$. Now take any $y \in Y$. Since Y is minimal, there exists a constant $c > 0$ such that for all $n \in \mathbb{N}$ the word $y[0, n]$ has at least $[cn]$ nonoverlapping occurrences of u. Now one can choose $2^{[cn]}$ distinct points $x_i \in X$, $i = 1, \cdots, 2^{[cn]}$, with occurrences of either u' or 0^{2m+1} in $x_i[0, n]$ corresponding to the occurrences of u in $y[0, n]$ (two different points in this set being distinct in at least one of these occurrences). For each $i = 1, \cdots, 2^{[cn]}$, let $z_i = \psi(\phi(y, x_i))$. By its construction, the set $E = \{z_i \mid i = 1, \cdots 2^{[cn]}\}$ is (n, ε) - separated and we get that $h(W, T) \geq c$. Therefore Z has c.p.e. $\square$

Remark. In fact, as is readily seen from the proof of the proposition, it is sufficient to assume that Y has a dense set of minimal points, i.e. points the orbits of which generate minimal sets. For instance the conclusion is true when Y is a full shift.

The next proposition shows that Z inherits mixing and nonmixing properties of Y.

Proposition 11. *Let Y be a subshift. Then the following conditions are equivalent:*

(1) Y is topologically weakly (strongly) mixing.
(2) Z is topologically weakly (strongly) mixing.

(. 1) implies (2): suppose Y is strongly mixing. The same is clearly true of the subshift $Y \times X$ (recall that X is a full shift). Therefore $Z = \phi(Y \times X)$ is also strongly mixing. The same proof works for weak mixing.

(2) implies (1): let u and v be two words in $L(Y)$. They also belong to $L(Z)$. Since Z is strongly mixing, there is $n_0 \in \mathbb{N}$ such that for all $n \geq n_0$ there exist $w \in L(Z)$ such that $uwv \in L(Z)$ and $\mid w \mid = n$. Let $z \in [uwv]$ and let (y, x) be a preimage of z. By definition of the factor map ϕ, there exists $w' \in L(Y)$ such that $\mid w' \mid = n$ and $y \in [uw'v]$. Hence $uw'v \in L(Y)$ and Y is strongly mixing. The proof for weak mixing is quite similar. $\square$

Finally Z also inherits existence of zero-entropy covers by two non dense open sets.

Proposition 12. *Let Y be a subshift not having u.p.e.; then Z has not u.p.e..*

L. et $\mathcal{R} = (U, V)$ be a zero-entropy cover of Y, U and V being open and non dense. One can find two clopen neighbourhoods $U'^c = \{y \in Y \mid y(-n, n) = x(-n, n)\} \subset U^c$ and $V'^c = \{y \in Y \mid y(-n, n) = x'(-n, n)\} \subset V^c$; of course $U'^c \cap V'^c = \emptyset$, and since (U', V') is coarser than (U, V) it has entropy zero. In Z define $U''^c = \{z \in Z \mid z(-n, n) = x(-n, n)\}$, $V''^c = \{z \in Z \mid z(-n, n) = y(-n, n)\}$: U'' and V'' are non dense and (U'', V'') is a cover of Z. Put $E = \{x \in X \mid x(-n, n) = 1^{2n+1} J\}$: the preimages of U''^c and V''^c by ϕ in $Y \times X$ are $U'^c \times E$ and $V'^c \times E$; those of U'' and V'' are $(U' \times X) \cup (U'^c \times E)$ and $(V' \times X) \cup (V'^c \times E)$; they constitute an open cover of $Y \times X$, coarser than $(U' \times X, V' \times X)$; this last cover has entropy zero just like (U', V'). Therefore $h(U'', V'') = 0$. $\square$

Consequences. It is known there exist non weakly mixing minimal subshifts. Therefore by Propositions 10 and 11 there exist non weakly mixing subshifts with c.p.e.. These, according to Proposition 2, have not got u.p.e. Existence of a fixed point in Z shows such non weakly mixing flows cannot have a non trivial continuous eigenfunction, thus providing a new example alongside those in [K-R].

Also, by Proposition 12 examples of weakly mixing minimal flows with entropy 0 [Pe-S], [D-K] generate examples of weakly mixing flows with c.p.e. but without u.p.e..

Questions.

1– Is there some relationship between strong topological mixing and entropy ? Or else, can one find some entropy property, either stronger than u.p.e. or more to the point, also implying strong mixing ?

2– When (X, T) has u.p.e., does there always exist an ergodic measure with support X?

REFERENCES

[A-K-M] R.L. Adler, A.G. Konheim, M.H. McAndrew, *Topological entropy*, Trans. Amer. Math. Soc. 114 (1965), 309–319.

[B] R. Bowen, *Periodic points and measures for Anosov diffeomorphisms*, Trans. Amer. Math. Soc. 154 (1971), 377-397.

[D-G-S] M. Denker, C. Grillenberger, K. Sigmund, *Ergodic theory on compact spaces*, Lecture notes in Math., vol. 527, Springer, Berlin, 1976.

[D-K] M. Dekking, M. Keane, *Mixing properties of substitutions*, Z. Wahrscheinlichkeist. verw. Gebiete **42** (1978), 22–33.

[F] H. Furstenberg, *Disjointness in ergodic theory, minimal sets, and a problem in diophantine approximation*, Math. Systems Th. 1 (1967), 1–55.

[G-H] W.H. Gottschlak , G.A. Hedlund, *Topological Dynamics.*, Amer. Math. Soc. Colloq. Pub., vol. 36, Amer. Math. Soc. Providence, R.I., 1955.

[H-R] G. Hansel, J.P. Raoult, *Ergodicité, uniformité et unique ergodicité*, Indiana Univ. Math. J. **23** (1973), 221–237.

[K] W. Krieger, *On unique ergodicity.*, Proc. 6 Berkeley Symposium on Math. Stat. and Proba., Univ. of California Press.

[K-R] H. Keynes , J.B. Robertson, *Eigenvalues in topological transformation groups*, Trans. Amer. Math. Soc. **139** (1969), 359–369.

[M-P] N.G. Markley, M.E. Paul, *Almost automorphic minimal sets without unique ergodicity*, Israel J. Math. **34** (1979), 259–272.

[O-S] D. Ornstein, P. Shields, *An uncountable family of K-automorphisms*, Advance in Math. **10** (1973).

[P] W. Parry, *Topics in ergodic theory*, Cambridge Tracts in Mathematics, Cambridge University Press, Cambridge, 1981.

[Pe] K. Petersen, *Disjointness and weak mixing of minimal sets*, Proc. Amer. Math. Soc. **24** (1970), 278–280.

[Pe-S] K. Petersen, L. Shapiro, *Induced flows*, Trans. Amer. Math. Soc. **177** (1973), 375–390.

[S-W] M. Shub, B. Weiss, *Can one always lower topological entropy*, To appear in Ergodic Th. Dyn. Sys..

[Si] K. Sigmund, *On dynamical systems with the specification property*, Trans. Amer. Math. Soc. **190** (1974), 251–255.

[W] S. Williams, *Toeplitz minimal flows which are not uniquely ergodic.*, Z. Wahrscheinlichkeitst. verw. Gebiete **57** (1984), 95–107.

UNIVERSITÉ DE PROVENCE, UFR DE MATHÉMATIQUES, CASE X, 3 PLACE V. HUGO, 13331 MARSEILLE CEDEX, FRANCE

Contemporary Mathematics
Volume **135**, 1992

THE STOCHASTIC SHIFT EQUIVALENCE
CONJECTURE IS FALSE

MIKE BOYLE

INTRODUCTION

Two morphisms A, B in a category are shift equivalent if there are morphisms R,S and a positive integer l (the lag) such that $AR = RB, SA = BS, RS = A^l$ and $SR = B^l$. Two morphisms are strong shift equivalent if they are related by a finite string of shift equivalences each of lag one. These ideas were introduced by R.F. Williams in his study of one-dimensional attractors [W1] and shifts of finite type [W2]. At this point, the classification problems for three classes of symbolic dynamical systems have been addressed with these ideas.

1. SHIFTS OF FINITE TYPE (SFT's).

This is the seminal contribution of Williams [W2]. Here the morphisms are nonnegative integral matrices and composition is matrix multiplication. Two SFT's are isomorphic (i.e. topologically conjugate) if and only if their defining matrices are strong shift equivalent [W2]. They are eventually isomorphic (i.e., all large powers are isomorphic) if and only if those matrices are shift equivalent [W2,KR1].

2. IRREDUCIBLE SFT's WITH MARKOV MEASURE.

Following work of Parry and Williams [PW], this theory was introduced by Parry and Tuncel [PT]. A fine current reference is [MT]. An irreducible SFT with a (supported) Markov measure can be presented by an irreducible stochastic matrix P. Regard the nonzero entries of P as lying not in the reals, but in the integral group ring of the group of positive real numbers under multiplication. The morphisms

Primary Classification #54H20, 58F03, Secondary 60J10.

Partially Supported by NSF.

This paper is in final form and no version of it will be submitted for publication elsewhere.

1991 Mathematics Subject Classification. Primary 54H20, 58F03; Secondary 60J10.

of the category now are the matrices whose entries are elements in this integral
group ring, with nonnegative coefficients, with composition given by matrix multiplication. Isomorphism of two irreducible SFT's with Markov measure is equivalent
to strong shift equivalence of their defining matrices, and eventual isomorphism is
equivalent to shift equivalence [MT].

3. Sofic systems.

A sofic system can be defined by way of a "representation matrix" in a certain
integral semigroup semiring. Again, two sofic systems are isomorphic/eventually
isomorphic if they are defined by strong shift equivalent/shift equivalent elements
of this semiring [BK], where composition of morphisms is given by multiplication
in the semiring. Working in this setting, Kim and Roush proved that eventual
isomorphism of sofic systems is decidable [KR2].

In each of the categories above (especially the first) shift equivalence is a more
algebraic, manageable and well understood relation than strong shift equivalence.
Thus a key step in each classification problem is to understand shift equivalence,
and a natural question arises: does shift equivalence imply strong shift equivalence?
Williams' shift equivalence conjecture [W2] is that the answer is yes for nonnegative
integral matrices. Later, Parry and Tuncel proposed the "possible conjecture" that
shift equivalence implies strong shift equivalence for irreducible SFT's with Markov
measure [PT]. We dub this the "stochastic shift equivalence conjecture". Kim and
Roush showed shift equivalence does not imply strong shift equivalence in the sofic
category [KR3].

Recently Kim and Roush [KR4] refuted Williams' original conjecture, in the reducible case: they exhibited two reducible nonnegative integral matrices which were
shift equivalent but not strong shift equivalent. (In the most important case, the irreducible case, Williams' conjecture remains open.) Their argument makes essential
use of the fundamental results on the dimension representation in [KRW]. Below,
we use the Kim-Roush example to produce a counterexample to the stochastic conjecture. (In particular, it becomes clear that the classification of reducible SFT's
is an essential ingredient for the classification of irreducible SFT's with Markov
measure.) Thus of the shift equivalence conjectures for the three categories above,
only Williams' original conjecture remains standing, in the irreducible case. Understanding the refinement of shift equivalence by strong shift equivalence remains
a fundamental part of the classification problem in all three categories.

The counterexample

As in [MT], it suffices to show shift equivalence does not imply strong shift
equivalence for matrices whose entries are Laurent polynomials with nonnegative
integer coefficients.

Let U be the unimodular matrix used in the Kim-Roush counterexample,

$$U = \begin{bmatrix} 0 & 0 & 1 & 1 \\ 1 & 0 & 0 & 0 \\ 0 & 1 & 0 & 0 \\ 0 & 0 & 1 & 0 \end{bmatrix}$$

Let D be the nonnegative unimodular matrix $(U - I)U^{(20)}$ of that example. We use the letters a,b,...i as variables. Define the matrices

$$A = \begin{bmatrix} aU & bI & cI \\ dD & eU & fI \\ gD & hI & iU \end{bmatrix}$$

$$B = \begin{bmatrix} aU & bD & cD \\ dI & eU & fI \\ gI & hI & iU \end{bmatrix}$$

Now A and B are similar via the integral matrix $R = diag(D, I, I)$. Let $S = A^{(k)}R^{(-1)}$. We claim that for large k, S is nonnegative (in the sense that every entry is a Laurent polynomial with nonnegative coefficients). This holds because

- if C is nonnegative, then AC is nonnegative
- $U^{(k)}D^{(-1)}$ is nonnegative for large k (because U and D commute, U is primitive and D multiplies the principal eigenvector of U by a positive number)
- for $k \geqslant 0$, given that $A^{(k)}R^{(-1)}$ can be written (in block form) as $diag(U^{(k)}D^{(-1)}, 0, 0)$ plus a nonnegative matrix, we multiply by A and compute to see $A^{(k+1)}R^{(-1)}$ is $diag(U^{(k+1)}D^{(-1)}, 0, 0)$ plus a nonnegative matrix.

Therefore for the large enough k, the pair S,R give a shift equivalence of the matrices A and B.

Now suppose A and B are strong shift equivalent. Then [MT] there is a topological conjugacy f from the SFT defined by A to that defined by B which respects the weights on periodic orbits induced by the labelling variables. In particular, if m is one of the letters a,e or i, then f restricts to a conjugacy from the A-subsystem given by mU to the B-subsystem given by mU. We claim additionally that f must restrict to a topological conjugacy between the subSFT's corresponding to the matrices

$$\begin{bmatrix} aU & 0 \\ dD & eU \end{bmatrix} \text{ and } \begin{bmatrix} aU & 0 \\ dI & eU \end{bmatrix}.$$

Such a conjugacy requires a strong shift equivalence between the matrices

$$\begin{bmatrix} U & 0 \\ D & U \end{bmatrix} \text{ and } \begin{bmatrix} U & 0 \\ I & U \end{bmatrix}$$

which by Kim and Roush [KR4] does not exist. So it remains to prove the claim.

We show that f maps a point x weighted like ....eeeedaaaa.... (all e, then one d, then all a) to a point with the same weight pattern (we do not require that the symbols weighted by d in x and f(x) occur in the same coordinate). Suppose not. Then f(x) must have weight of the form ...eeeeWaaaa... (all e, some word W, then all a) where the word W has at least two symbols outside a,e. Using the density of the periodic points, we can construct a periodic point y with a defining block weighted like e...eda...ab, where the lengths of the strings weighted by e and a are longer than the coding length of f, and the symbols within a coding length of the symbol weighted by d match the corresponding symbols of x. Then f(y) must have a defining block weighted like e...eWa...aW', where W' contains at least one symbol outside a,e. But f must respect weights on defining blocks for periodic points [MT]. This is a contradiction. The same argument works with f^{-1} in place of f. This finishes the proof.

I thank David Handelman for reminding me of the possible utility of the Kim-Roush example for the stochastic problem, and for helpful comments.

References

[BK] M. Boyle and W. Krieger, *Almost Markov and shift equivalent sofic systems*, Springer Lec. Notes 1342 (1988), 33-93.

[KR1] K.H. Kim and F.W.Roush, *Some results on decidability of shift equivalence*, J.Comb., Inf. and Sys. Sci **4** (1979), 123-146.

[KR2] _____, *An algorithm for sofic shift equivalence*, Erg.Th. and Dyn.Syst. **10** (1990), 381-393.

[KR3] _____, *Solution of two conjectures in symbolic dynamics*, Proc. AMS. **112** (1991), 1163-1168.

[KR4] _____, *William's conjecture is false for reducible subshifts*, Journal of the A.M.S. **5** (1992), 213-215.

[KRW] K.H.Kim, F.W. Roush and J.B. Wagoner, *Automorphisms of the dimension group and gyration numbers of automorphisms of the shift*, Journal of the A.M.S. **5** (1992), 191-212.

[MT] B. Marcus and S. Tuncel, *The weight-per-symbol polytope and scaffolds of invariants associated with Markov chains*, Erg.Th. and Dyn.Syst. **11** (1991), 129-180.

[PT] W. Parry and S. Tuncel, *On the stochastic and topological structure of Markov chains*, Bull. LMS **14** (1982), 16-27.

[PW] W. Parry and R.F. Williams, *Block coding and a zeta function for finite Markov chains*, Proc. London Math. Soc. **35** (1977), 483-495.

[W1] R.F. Williams, *Classification of one-dimensional attractors*, Proc. Sympos. Pure Math. **14** (1970), 341-361, Amer. Math. Soc., Providence, R.I..

[W2] _____, *Classification of subshifts of finite type*, Annals of Math. (1973), 120-153; Errata, Annals of Math, 99, (1974), 380-381.

MIKE BOYLE DEPARTMENT OF MATHEMATICS UNIVERSITY OF MARYLAND COLLEGE PARK, MD 20742 MMB@MATH.UMD.EDU

Contemporary Mathematics
Volume **135**, 1992

Predictions with Automata

ANNIE BROGLIO and PIERRE LIARDET

ABSTRACT. Reading sequencially the terms of an infinite string of digits, one can try to guess at each step the next digit that would be read. This is done by introducing the concept of predictions by maps, called predictors, defined on finite strings. We mainly consider the case of predictors depending on finite automata. Subsequences extracted by automata cannot have better predictions that the original sequence. Normal sequences are characterized as sequences having only one (and the lowest) prediction.

1. Introduction

1.1. The concept of random sequences has not yet find a single mathematical formulation and is strongly subordinate to the kind of problem we are working at. One of the main question is the existence of an effective process – in a suitable meaning – which recognizes randomness. For example, in the case of random sequences introduced by Martin-Löf [Mar], there exists a universal test of randomness. But the test is purely theoretical (based on the universal Turing machine) and we do not know how to built an efficient one. In this paper, random sequences are only considered from the point of view of A. Borel, which are also called normal sequences. We study the possibility to recognize or to generate such random sequences and we do this with the help of deterministic finite automata. Notice that such sequences exist which are not random according to Martin-Löf.

Let $\mathcal{A}$ be a finite set of q digits, $q \geq 2$, and let $u := (u_n)_{n \geq 0}$ be an infinite sequence of digits u_n. We also consider u as an infinite word. It is natural to think that if u is random, then no automaton is able, after reading $u_0, \ldots, u_{n-1}$ successively, to guess u_n with a probability of success exceeding $1/q$. We shall prove that effectively one can characterize normal sequences as the ones with the lowest automatic prediction. A previous work in the case of binary sequences has been done by M. O'Connor [OCn]. We generalize her work using a different method which replace a dichotomy argument by a probability one.

1991 Mathematics Subject Classification 11K16, 68Q68
Research partially supported under contract 901636/A000/DRET/DS/SR
This paper is in final form and no version of it will be submitted for publication elsewhere.

1.2. The paper is organized as the following. In section 2 we introduce basic definitions and notations. Section 3 and 4 give general properties of prediction ratios for an infinite word. We study the particular case of prediction for subwords extracted by automata. Roughly speaking, we show that a finite machine cannot create from any sequence a more complicated one. Section 5 is devoted to various characterizations of normal sequences. The first one that we can relate to automata was given for binary sequences by Ville [Vill], namely:

(**P**) *For any string $a_1 \ldots a_m$ of digits, and any digit $a = 0, 1$ one has*

$$\lim_{N \to \infty} \frac{\text{card}\{0 \le n < N; u_n \ldots u_{n+m} = a_1 \ldots a_m a\}}{\text{card}\{0 \le n < N; u_n \ldots u_{n+m-1} = a_1 \ldots a_m\}} = \frac{1}{2}.$$

This characterization can be extended to any sequence of q digits (see Proposition 3). We prove more (Theorem 3): Normal sequences are those whose prediction ratios (obtained by finite states automata) are all identical to $\frac{1}{q}$. The last section deals with subsequences of normal sequences extracted by automata. They are also normal. Our method gives a new proof of this result which was first anounced by Agafonov [Aga] in the case of binary sequences (see also T. Kamae and B Weiss [Ka-We]).

2. Notations and definitions

2.1. Let $\mathcal{A}$ be the set of q elements $0, 1, \ldots, q-1$ ($q \ge 2$). Throughout this paper, the set $\mathcal{A}$ is fixed and called *alphabet*. Its elements are called *letters* or *digits* and a word w of length $|w| := n$ is a finite string $w = w_1 \ldots w_n$ of letters w_j. The empty word (no letter) is denoted by $\wedge$. As usual, the set of words $\mathcal{A}^*$ is endowed with the concatenation law $(w, w') \mapsto ww'$ and $\mathcal{A}^n$ denotes the set of words of length n. The set of infinite strings (or infinite words, or sequences) $u := u_0 u_1 u_2 \ldots$ of letters is denoted by $\mathcal{A}^\infty$ and we put $u[n] := u_0 \ldots u_{n-1}$, $u[0] = \wedge$.

2.2. Definitions. (1) A *predictor* is a map $P : \mathcal{A}^* :\to \mathcal{A} \cup \{\#\}$, where $\#$ is a fixed symbol which does not belong to $\mathcal{A}$. If $P(w) = a$ ($\ne \#$) we say that P makes the prediction a on w. Otherwise $P(w) = \#$ and we say that P makes no prediction on w. If P never takes the value $\#$, then P is said to be a *complete* predictor.

Let $\mathcal{S} := (S, s_0, \Phi)$ be a semi-automaton. In the sequel we shall assume that $\mathcal{S}$ is finite and deterministic *i.e.*, S is a finite set (the set of states), s_0 is the initial state and Φ is the set of instructions $\phi_a : S \to S$ given for each letter a. For short, we denote by $\phi_{w_1 \ldots w_n}$ the composition $\phi_{w_n} \circ \phi_{w_{n-1}} \circ \ldots \circ \phi_{w_1}$. We also put $\phi_\wedge := id_{|_S}$ (the identity map on S). An automaton $\mathcal{S}^\tau := (S, s_0, \Phi, \tau)$ is defined by a semi-automaton $\mathcal{S} := (S, s_0, \Phi)$ and an output function $\tau : S \to \mathcal{B}$.

(2) A predictor P is called *automatic* if there exists an automaton $\mathcal{S}^\tau = (S, s_0, \Phi, \tau)$ with $\mathcal{B} := \mathcal{A} \cup \{\#\}$ such that

$$\forall \, w \in \mathcal{A}^*, \ P(w) = \tau \circ \phi_w(s_0).$$

Notice that P can be given by many different automata, but there exists a very natural one that we call *Nerode automaton*. In fact, any map $f : \mathcal{A}^* \to \mathcal{B}$ defines an equivalent relation $\sim_f$ (the well-known Nerode relation) given by

$$w \sim_f w' \iff \forall\, v \in \mathcal{A}^*,\ f(wv) = f(w'v).$$

Let $w \mapsto \overline{w}$ be the canonical map $\mathcal{A}^* \to \mathcal{A}^*/\sim_f$ and let $\overline{\phi}_a$ be the instruction given by $\overline{\phi}_a(\overline{w}) = \overline{wa}$. We consider the automaton $\Sigma_f := (\overline{\mathcal{A}^*}, \Lambda, \overline{\Phi}, \overline{f})$ with $\overline{\mathcal{A}^*} := \mathcal{A}^*/\sim_f$, $\overline{\Phi} := \{\overline{\phi}_a \,;\, a \in \mathcal{A}\}$ and $\overline{f} : \overline{\mathcal{A}^*} \to \mathcal{B}$ defined by $\overline{f}(w) := \overline{f(w)}$. The automaton Σ_f defines f and is minimal in the sense that if (S, s_0, Φ, τ) is any automaton which defines f, then there exists a map $\gamma : S \to \overline{\mathcal{A}^*}$ such that $\gamma \circ \phi_a = \overline{\phi}_a \circ \gamma$ for all $a \in \mathcal{A}$ and $\tau = \overline{f} \circ \gamma$.

(3) *Prediction ratio.* Let P be a predictor. For any word $w = w_0 \ldots w_{n-1}$ of length n the prediction ratio of w is by definition

$$\rho(P, w) := \frac{\sum_{j<n} |P(w[j]), w_j|}{n - \sum_{j<n} |P(w[j]), \#|}$$

where the sums run over the set of integers $j = 0, 1, \ldots, n-1$ and $|x, y|$ denotes the Kronecker symbol (equal to 1 if $x = y$ and 0 otherwise). If $P(w[j]) = \#$ for all possible j we put $\rho(P, w) := 0$. We extend the definition to an infinite word u by

$$\rho(P, u) := \limsup_{n \to \infty} \rho(P, u[n]).$$

and put

$$\mathcal{N}(P, n, u) := \mathrm{card}\{0 \leq j < n \,;\, P(u[j]) \in \mathcal{A}\}.$$

(4) We say that the predictor P is *regular for* u if $P(u[n]) \in \mathcal{A}$ infinitely often. According to the intuitive idea, a *good* predictor is one verifying $\rho(P, u) > \frac{1}{q}$. Regular but not complete predictors are usefull to study subsequences of u while complete predictors give a global prediction on u. For exemple, let $z = (z_n)_{n \geq 0}$ be the infinite word on $\mathcal{A} = \{0, 1\}$ defined by

$$z_{3n+j} := \begin{cases} 0 & \text{if } j = 0, \\ x_{2n} & \text{if } j = 1, \\ x_{2n+1} & \text{if } j = 2. \end{cases}$$

where $x = (x_n)_n$ is chosen such that

$$\lim_{M \to \infty} \frac{1}{N} \sum_{n < M} |x_{2n+i}, k| = 1/2$$

for all $i = 0, 1$, $k = 0, 1$. Then, for the semi-automaton $\mathcal{C} := \{\{0, 1, 2\}, 0, \Phi\}$ where $\phi_i(j) \equiv j + 1 \pmod 3$ we have $\rho(\mathcal{C}^\tau, z) := 1$ if $\tau : \{0, 1, 2\} \to \{0, \#\}$ with $\tau^{-1}(0) = \{2\}$. But for any $\sigma : \{0, 1, 2\} \to \{0, 1\}$ one has $\rho(\mathcal{C}^\sigma, z) = 1/3$ or $2/3$ according to $\sigma(2)$ is equal to 0 or 1.

3. Comparison of ratios

We first start with a simple but basic estimation:

Proposition 1. *For any* $u \in \mathcal{A}^{\infty}$, *define*

$$\beta(u) := \max_{a \in \mathcal{A}} \limsup_{N \to \infty} \frac{1}{N} \sum_{n < N} |u_n, a|.$$

and let $\mathcal{S} = (S, s_0, \Phi)$ *be any semi-automaton. There exists an output function* $\tau : S \to \mathcal{A}$ *such that* $\rho(P, u) \geq \beta (\geq \frac{1}{q},)$ *where* P *is the predictor given by* $\mathcal{S}^{\tau}$.

Proof. Let the maximum in the definition of β be obtained for $b \in \mathcal{A}$ and let $(N_k)_k$ be a non decreasing sequence of integers such that

$$\beta(u) = \lim_{k \to \infty} \frac{1}{N_k} \sum_{n < N_k} |u_n, b|.$$

Now for any semi-automaton $\mathcal{S} = (S, s_0, \Phi)$ choose the output function τ to be constant, equal to b. Then the predictor defined by $\mathcal{S}^{\tau}$ verifies $\rho(P, u) = \beta(u)$. ∎

Now we compare prediction ratios obtained from predictors and complete predictors.

Theorem 1. *Let* u *be an infinite word and let* P *be a regular predictor for* u *such that*

$$\liminf_{n \to \infty} \frac{\mathcal{N}(P, n, u)}{n} = h.$$

Then, there exists a complete predictor P' *and* $p \in [h, 1]$ *such that* $P(w) = P'(w)$ *if* $P(w) \neq \#$ *and*

$$\rho(P', u) \geq p\rho(P, u) + \frac{1 - p}{q}.$$

Moreover, if P *is automatic, we can choose* P' *automatic with the same underlying semi-automaton.*

Proof. For each letter a we define the complete predictor F_a on $\mathcal{A}^*$ by

$$F_a(w) := \begin{cases} P(w) & \text{if } P(w) \in \mathcal{A}, \\ a & \text{otherwise.} \end{cases}$$

If P is automatic, then F_a is also automatic with the same underlying semi-automaton (we just have to change the output function). We can choose an infinite subset J of $\mathbf{N}$ such that

$$\rho(P, u) = \lim_{\substack{j \to \infty, \\ j \in J}} \frac{1}{\mathcal{N}(P, j, u)} \sum_{n < j} |P(u[n]), u_n|$$

and all the following limits exist:

$$p := \lim_{\substack{j \to \infty \\ j \in J}} \frac{1}{j} \mathcal{N}(P, j, u), \quad \lambda_a := \lim_{\substack{j \to \infty \\ j \in J}} \frac{1}{j} \sum_{n < j} |F_a(u[n]), u_n|, \quad a \in \mathcal{A}.$$

One has $p \geq h$ and $\rho(F_a, u) \geq \lambda_a$. From the equality (considered for j large enough)

$$\sum_{a \in \mathcal{A}} \left(\frac{1}{j - \mathcal{N}(P, j, u)} \sum_{n < j} |P(u[n]), \#| \cdot |u_n, a| \right) = 1$$

we conclude that there exists a letter b and an infinite subset J_b of J such that the limit

$$\ell_b := \lim_{\substack{j \to \infty \\ j \in J_b}} \frac{1}{j - \mathcal{N}(P, j, u)} \sum_{n < j} |P(u[n]), \#| \cdot |u_n, b|$$

exists with $\ell_b \geq \frac{1}{q}$. But notice that

$$\sum_{n < j} |F_b(u[n]), u_n| = \sum_{\substack{n < j \\ P(u[n]) \neq \#}} |P(u[n]), u_n| +$$

$$\sum_{\substack{n < j \\ P(u[n]) = \#}} |P(u[n]), \#| \cdot |u_n, b|,$$

therefore

$$\lambda_b = p\rho(P, u) + (1 - p)\ell_b.$$

Choose $P' = F_b$ to end the proof. ∎

4. Subwords

4.1. Skew-product of automata. Let $\mathcal{K} := (K, k_0, \Psi, \kappa)$ be a given automaton with output function $\kappa : K \to \{1, \#\}$. We assume that $R := \kappa^{-1}(\{1\}) \neq \emptyset$. For any automaton $\mathcal{S} = (S, s_0, \Phi, \tau)$ we construct a new automaton $\mathcal{K} \times_R \mathcal{S} := (K \times S, (k_0, s_0), \Phi', \tau')$ called skew-product over R where the instructions are defined by

$$\phi'_a(k, s) := \begin{cases} (\psi_a(k), \phi_a(s)) & \text{if } k \in R \\ (\psi_a(k), s) & \text{otherwise.} \end{cases}$$

The output function τ' is given by

$$\tau'(k, s) := \begin{cases} \tau(s) & \text{if } k \in R \\ \# & \text{otherwise.} \end{cases}$$

From the definition we get for any word $\alpha = a_1 \ldots a_n$

$$\phi'_\alpha(k, s) = (\psi_\alpha(k), \phi_{a_{i_1} \ldots a_{i_m}}(s))$$

where $i_1 < \ldots < i_m$ are all ordered indices i such that $\psi_{a_1 \ldots a_{i-1}}(k) \in R$.

Let u be an infinite word and again consider the above automaton $\mathcal{K}$. Let J be the set of integers n such that $\kappa(\psi_{u[n]}(k_0)) = 1$. If J is infinite (in that case $\mathcal{K}$ is said to be regular for u) then we can define the following infinite word $E(\mathcal{K}, u) := (u_{j_m})_m$ where $(j_m)_{m \geq 0}$ is the increasing indexation of J. We denote by $E(\mathcal{K}, u)$ the infinite subword *extracted* by $\mathcal{K}$ from u.

Theorem 2. *Let u be an infinite word, let $\mathcal{K}$ be a regular automaton for u and let $v = E(\mathcal{K}, u)$. Then for any automatic predictor P regular for v, the prediction ratio $\rho(P, v)$ is also a prediction ratio $\rho(P', u)$ of u corresponding to some automatic predictor P' regular for u.*

Proof. With above notations, let the predictor P be defined by (S, s_0, Φ, τ). We claim that if P' is the predictor given by the skew-product $\mathcal{K} \times_R S :=$ $(K \times S, (k_0, s_0), \Phi', \tau')$ over $R = \kappa^{-1}(1)$ then $\rho(P', u) = \rho(P, v)$. Let v be the sequence $(v_{j_m})_m$ and for any natural number N let L be the unique integer such that $j_{L-1} < N \leq j_L$. Then

$$
\begin{aligned}
\rho(P', &u[N]) \\
&= \frac{1}{\text{card}\{n < N \,;\, P'(u[n]) \neq \#\}} \sum_{n < N} |\tau'(\phi'_{u[n]}(k_0, s_0)), u_n| \\
&= \frac{1}{\text{card}\{\ell < L \,;\, P(v[\ell]) \neq \#\}} \sum_{\ell < L} |\tau(\phi_{v[\ell]}(s_0)), v_\ell| \\
&= \rho(P, v[L]).
\end{aligned}
$$

Our claim follows. ∎

4.2. Remarks. (1) Theorem 1 says that we cannot increase automatic prediction by passing to any subsequence extracted by automaton.

(2) It is easy to see that the set of prediction ratios of $E(\mathcal{K}, u)$ can be strictly included in the corresponding set of prediction ratios of u (see example 2.2(4)).

(3) Theorem 2 furnishes another reason why we do not only consider regular predictors.

(4) Notice that in the above proof if P is a complete predictor then $P'(u[n]) = \#$ if and only if $\kappa(\psi_{u[n]}(k_0)) = \#$.

5. Normal words

5.1. Uniformly distributed infinite words.

We say that an infinite word u is uniformly $k-$distributed if

$$
\forall\, w \in \mathcal{A}^k, \quad \lim_{N \to \infty} \frac{1}{N} \sum_{n < N} |u_n \ldots u_{n+k-1}, w| = q^{-k}.
$$

Proposition 2. *Let u be an infinite word and assume that there exists an integer k, such that u is uniformly $k-$distributed but not $(k+1)-$distributed. Then there exists an automatic complete predictor P with less than $k+1$ states and $\rho(P, u) > \frac{1}{q}$.*

The proof is analogous to the one given by O'Connor [O'Co] in the case of binary infinite words.

5.2. We recall that an infinite sequence (or word) $u \in \mathcal{A}^\infty$ is said to be *normal* if it is uniformly $k-$distributed for all integers k. The result of J. Ville [Vill] given in the introduction has a straightforward generalization, namely:

Proposition 3. *An infinite word $u \in \mathcal{A}^{\infty}$ is normal if and only if for any word $w_1 \ldots w_m \in \mathcal{A}^m$, $m \geq 1$ and any letter $a \in \mathcal{A}$ one has*

(1) $$\lim_{N \to \infty} \frac{\operatorname{card}\{0 \leq n < N; \, u_n \ldots u_{n+m} = w_1 \ldots w_m a\}}{\operatorname{card}\{0 \leq n < N; \, u_n \ldots u_{n+m-1} = w_1 \ldots w_m\}} = \frac{1}{q}.$$

Proof. Property (1) follows directly from the assumption that u is normal. To see the converse, let J be any infinite subset of $\mathbf{N}$ such for that each letter a the limit $p_a := \lim_{N \to \infty} N^{-1} \operatorname{card}\{0 \leq n < N; \, u_n = a\}$ exists. Then $\sum_{a \in \mathcal{A}} p_a = 1$ and (1) implies that for all words $w \in \mathcal{A}^m$ and all letters a, the limit

$$p_{aw} := \lim_{N \in J, \, N \to \infty} N^{-1} \operatorname{card}\{0 \leq n < N; \, u_n \ldots u_{n+m} = aw\}$$

exists and is equal to $p_a q^{-m}$. Moreover for all letters b the equality $\sum_{a \in \mathcal{A}} p_{ab} = p_b$ holds, so that $p_b = \frac{1}{q}$. This implies that we can choose $J = \mathbf{N}$. Therefore the word u is normal. ∎

It is worth to notice that (1) is the prediction ratio of an automaton $\mathcal{L}_{wa} = (L, \ell_0, \Psi, \sigma)$ which recognizes the word $w_1 \ldots w_m$ and then predict a, otherwise $\mathcal{L}_{wa}$ makes no prediction. More precisely, take $L := \{\ell_0, \ldots \ell_m\}$ with $\ell_i = w[i]$ and for all $x \in \mathcal{A}$ take $\psi_x(\ell_i) = \ell_j$, where j is the greatest length of a suffix of $w[i]x$ which is a prefix of w. The output function σ is given by

$$\sigma(\ell_i) = \begin{cases} \# & \text{if } i \neq m \\ a & \text{otherwise.} \end{cases}$$

If f denotes the predictor defined by $\mathcal{L}_{wa}$ and if u is normal, then

$$\rho(f, u) = \frac{1}{q} = \limsup_{N \to \infty} \frac{1}{\mathcal{N}(f, N, u)} \sum_{n < N} |f(u[n]), u_n|$$

$$= \lim_{N \to \infty} \frac{\operatorname{card}\{0 \leq n < N; \, u_n \ldots u_{n+m} = wa\}}{\operatorname{card}\{0 \leq n < N; \, u_n \ldots u_{n+m-1} = w\}}.$$

Now we state our main result:

Theorem 3. *An infinite word u is normal if and only if for all automatic predictors P, regular for u, one has $\rho(P, u) = \frac{1}{q}$.*

The proof requires several steps.

STEP 1. Let P be a complete predictor. For any word w we define $[P, w]_n : \mathcal{A}^n \to \mathbf{N}$ by

$$[P, w]_n(v_1 \ldots v_n) := |P(w), v_1| + |P(wv_1), v_2| + \ldots$$
$$+ |P(wv_1 \ldots v_{n-1}), v_n|$$

and

$$[P]_n(v) = \max\{[P, w]_n(v) \, ; \, w \in \mathcal{A}^*\}, \quad (v \in \mathcal{A}^n).$$

It is worth to notice that $[P, w]_n(\cdot)$ only depends on the equivalence class $\overline{w}$ of w, modulo the Nerode equivalence $\sim_P$. For all $u \in \mathcal{A}^\infty$ and all $n \geq 1$, a straightforward computation gives

$$(2) \qquad \sum_{m<N} |P(u[m]), u_m| \leq n + \frac{1}{n} \sum_{k<N-n} [P]_n(u_k u_{k+1} \ldots u_{k+n-1}).$$

Proposition 4. *Assume that u is normal. Then for all complete predictors P and all integers $n \geq 1$, one has*

$$\rho(P, u) \leq \frac{1}{n} \int_{\mathcal{A}^n} [P]_n(v) d\lambda_n(v)$$

where λ_n denotes the uniform probability on $\mathcal{A}^n$.

Proof. From (2) we obtain

$$\rho(P, u) \leq \limsup_{N \to \infty} \frac{1}{Nn} \sum_{k<N} [P]_n(u_k u_{k-1} \ldots u_{k+n-1}).$$

Since u is normal, the upper limit is a limit and,

$$\lim_{N \to \infty} N^{-1} \sum_{k<N} [P]_n(u_k \ldots u_{k+n-1}) = \int_{\mathcal{A}^n} [P]_n(v) d\lambda_n(v).$$

This ends the proof. ∎

STEP 2. In order to estimate the bound in Proposition 4, we assume that P is given by a finite automaton. Let S be the set $\mathcal{A}^*/ \sim_P$. Then, by assumption, S is a finite set and for any $s \in S$ corresponding to the class of w modulo $\sim_P$, one can put $[P, s]_n := [P, w]$.

Lemma 1. *For $0 \leq k \leq n$, one has*
1) $\operatorname{card}\{v \in \mathcal{A}^n \,;\, [P, s]_n(v) = k\} = \binom{n}{k}(q-1)^{n-k}$,
2) $\operatorname{card}\{v \in \mathcal{A}^n \,;\, [P]_n(v) \geq c\} \leq \operatorname{card}(S) \sum_{k=c}^n \binom{n}{k}(q-1)^{n-k}$,
3) $\operatorname{card}\{v \in \mathcal{A}^n \,;\, [P]_n(v) \leq c\} \leq \sum_{k=0}^c \binom{n}{k}(q-1)^{n-k}$.

Proof. To construct $v = v_1 \ldots v_n$ such that $[P, s]_n(v) = k$ with $s = \overline{w}$, we notice whatever is the state $\overline{wv_1 \ldots v_{m-1}}$ the value $P(\overline{wv_1 \ldots v_{m-1}})$ gives only one choice for v_m in order to be a good prediction and $(q-1)$ different choices for a bad prediction. Hence there are exactly $\binom{n}{k}(q-1)^{n-k}$ words v of length n such that $[P, s]_n(v) = k$. This proves 1).

From the definition

$$\bigcup_{s \in S} \{v \in \mathcal{A}^n \,;\, [P, s]_n(v) \geq c\} = \{v \in \mathcal{A}^n \,;\, [P]_n(v) \geq c\}$$

and

$$\{v \in \mathcal{A}^n \,;\, [P]_n(v) \leq c\} = \bigcap_{s \in S} \{v \in \mathcal{A}^n \,;\, [P, s]_n(v) \leq c\}.$$

The inequalities in 2) and 3) follow easily from these equalities. ∎

Lemma 2. *Let P be a complete automatic predictor and let u be a normal word. Then*

$$\lim_{n \to \infty} \frac{1}{n} \int_{\mathcal{A}^n} [P]_n(v) d\lambda_n(v) = \frac{1}{q} = \rho(P, u).$$

Proof. Put $p_n := \int_{\mathcal{A}^n} [P]_n(v) d\lambda_n(v)$ for short. From the definition we obtain

$$[P]_{m+n}(v_1, \ldots, v_{m+n}) \leq [P]_m(v_1, \ldots, v_m) + [P]_n(v_{m+1}, \ldots, v_{m+n}),$$

so that the sequence $n \mapsto p_n$ is subadditive and consequently the sequence $n \mapsto p_n/n$ has a limit, namely $\lim_{n \to \infty} p_n/n = \inf_n p_n/n$. Now p_n can be written such as

$$\frac{p_n}{n} = \frac{1}{q^n} \sum_{k=0}^{n} \frac{k}{n} \operatorname{card}([P]_n^{-1}(k)).$$

Given any $\varepsilon > 0$, we cut the sum in three parts. The first is easy to estimate:

$$\sum_{|\frac{k}{n} - \frac{1}{q}| < \varepsilon} \frac{k}{n} \operatorname{card}([P]_n^{-1}(k)) \leq q^n (\varepsilon + \frac{1}{q}).$$

We use Lemma 2 to estimate the two remaining sums:

$$\sum_{\frac{k}{n} - \frac{1}{q} \leq -\varepsilon} \frac{k}{n} \operatorname{card}([P]_n^{-1}(k)) \leq \sum_{\frac{k}{n} - \frac{1}{q} \leq -\varepsilon} \binom{n}{k} (q-1)^{n-k}$$

and

$$\sum_{\frac{k}{n} - \frac{1}{q} \geq \varepsilon} \frac{k}{n} \operatorname{card}([P]_n^{-1}(k)) \leq \operatorname{card}(S) \sum_{\frac{k}{n} - \frac{1}{q} \geq \varepsilon} \binom{n}{k} (q-1)^{n-k}.$$

Hence

$$\frac{1}{n} \int_{\mathcal{A}^n} [P]_n(v) d\lambda_n(v) \leq \varepsilon + \frac{1}{q} + \operatorname{card}(S) D_n(\varepsilon),$$

where

$$D_n(\varepsilon) = \sum_{|\frac{k}{n} - \frac{1}{q}| \geq \varepsilon} \binom{n}{k} \left(\frac{1}{q}\right)^k \left(1 - \frac{1}{q}\right)^{n-k}.$$

The well-known Tchebichev inequality for the binomial law $(1/q, 1 - 1/q)$ gives

$$(3) \qquad\qquad D_n(\varepsilon) \leq \frac{q-1}{nq^2 \varepsilon^2}.$$

On the other hand we also get

$$\frac{1}{n} \int_{\mathcal{A}^n} [P]_n(v) d\lambda_n(v) \geq \sum_{|\frac{k}{n} - \frac{1}{q}| < \varepsilon} \frac{k}{n} \operatorname{card}([P]_n^{-1}(k))$$

$$\geq (\frac{1}{q} - \varepsilon)(1 - D_n(\varepsilon))$$

and consequently

$$(4) \qquad (\frac{1}{q} - \varepsilon)(1 - D_n(\varepsilon)) \leq \frac{1}{n} \int_{\mathcal{A}^n} [P]_n(v) d\lambda_n(v) \leq \frac{1}{q} + \varepsilon + D_n(\varepsilon).$$

Finally, (3) and (4) imply

$$\lim_{n \to \infty} \frac{1}{n} \int_{\mathcal{A}^n} [P]_n(v) d\lambda_n(v) = \frac{1}{q}.$$

In order to finish the proof we consider

$$[P]_n^-(v) = \min\{[P, w]_n(v) \; ; \; w \in \mathcal{A}^*\}, \quad (v \in \mathcal{A}^n),$$

and $r_n := \int_{\mathcal{A}^n} [P]_n^-(v) d\lambda_n(v)$. As above it is easy to see that $n \mapsto r_n$ is upperadditive and so $\lim_n \frac{r_n}{n} = \sup_n \frac{r_n}{n}$. Inequalities 2) and 3) in Lemma 1 have analogous counterparts, namely

$$\text{card}\{v \in \mathcal{A}^n \; ; \; [P]_n^-(v) \geq c\} \leq \sum_{k=c}^{n} \binom{n}{k} (q-1)^{n-k},$$

and

$$\text{card}\{v \in \mathcal{A}^n \; ; \; [P]_n(v) \leq c\} \leq \text{card}(S) \sum_{k=0}^{c} \binom{n}{k} (q-1)^{n-k}$$

which are consequence of

$$\{v \in \mathcal{A}^n \; ; \; [P]_n^-(v) \geq c\} = \bigcap_{s \in S} \{v \in \mathcal{A}^n \; ; \; [P, s]_n(v) \geq c\}.$$

and

$$\bigcup_{s \in S} \{v \in \mathcal{A}^n \; ; \; [P, s]_n(v) \leq c\} = \{v \in \mathcal{A}^n \; ; \; [P]_n^-(v) \leq c\}.$$

We proceed as above to obtain

$$\lim_{n \to \infty} \frac{1}{n} \int_{\mathcal{A}^n} [P]_n^-(v) d\mu_n(v) = \frac{1}{q}.$$

But $\rho(P, u) \leq \frac{p_n}{n}$ from Proposition 4 and in a quite similar way we have $\frac{r_n}{n} \leq \rho(P, u)$, therefore $\rho(P, u) = \frac{1}{q}$. ∎

STEP 3. This section is general; we do not assume that u is normal. Recall that an automaton (S, s_0, Φ, τ) is said to be connected if for all states s, s', there exists a word $w(\in \mathcal{A}^*)$ such that $\phi_w(s) = s'$.

Proposition 5. *Let P be an automatic predictor and let u be an infinite word. Assume that P is regular for u, then there exist a connected predictor P' and an integer t such that for the infinite word u' defined by $u'_n = u_{n+t}$ $(n \in \mathbf{N})$, P' is regular for u' and $\rho(P, u) = \rho(P', u')$.*

Proof. Let P be defined by the automaton (S, s_0, Φ, τ). If the sequence $s_n = \phi_{u[n]}(s_0)$ is ultimately constant then take for t any integer such that $s_{n+t} = s_{n+t+1}$ for all n. Then take the constant predictor P' given by $P'(v) = \tau(s_t)$. By the regularity of P for u we have $\tau(s_t) \neq \#$. Hence P' is regular and also $\rho(P, u) = \rho(P', u')$. If $(s_n)_n$ is not ultimately constant, we introduce the set $S_1 := \{s \in S \,;\, \exists\, a \in \mathcal{A}, \phi_a(s) \neq s\}$ and the equivalence relation $\sim$ on S_1 defined by

$$s \sim s' \iff \exists\, (w, w') \in (\mathcal{A}^*)^2, \phi_w(s) = s' \ \& \ \phi_{w'}(s') = s.$$

Let $c(s)$ denote the equivalence class of s. We avoid the case where $(s_n)_n$ is ultimately constant, hence s_n belongs to S_1 for any n. Now from the definition of $\sim$ if $c(s_m) \neq c(s_{m+1})$ for an m then $c(x_m) \neq c(x_{m+n})$ for all n. Therefore the sequence $(c(s_n))_n$ is ultimately constant. Let $S'(\subset S_1)$ be this constant class and let t be such that $c(s_{n+t}) = S'$ for all n. Then we define (S', s'_0, Φ', τ') by

$$s'_0 := s_t, \qquad \tau' := \eta_{|S'}, \qquad \phi_a(s) = \begin{cases} \phi_a(s) & \text{if } \phi_a(s) \in S' \\ s & \text{otherwise} \end{cases},$$

for all $a \in \mathcal{A}$. By construction this automaton is connected, the corresponding predictor P' is regular for $u' := (u_{n+t})_n$ and $\phi_{u[n+t]}(s_0) = \phi'_{u'[n]}(s_t)$. The equality $\rho(P, u) = \rho(P', u')$ follows at once. ∎

STEP 4. We go back to normal words:

Proposition 6. *Let u be a normal word and let $\mathcal{S} := (S, s_0, \Phi, \tau)$ be a connected automaton. Assume that the predictor P defined by $\mathcal{S}$ is regular for u, then for any state $s \in S$,*

$$\liminf_{m \to \infty} \frac{1}{m} \operatorname{card}\{n < m \,;\, \phi_{u[n]}(s_0) = s\} \geq \frac{1}{q^{d(d-1)+1}}$$

where $d := \operatorname{card}(S)$.

Proof. For any fixed state $s \in S$ we claim that there exists a word W which verifies the property:

(5) $\forall\, s' \in S, \exists\, p, 1 \leq p \leq |W| \text{ and } \phi_{W[p]}(s') = s.$

We construct W as a product $W_1 \ldots W_d$ of words W_i. There exists a letter a and a state s' such that $\phi_a(s') = s$. Now let S be indexed as $S = \{s_1, \ldots, s_d\}$ with $s_1 = s'$. Then we take $W_1 = a$. Assume that words $W_1, \ldots, W_i$ are already chosen for an $i < d$ such that $\phi_{W_1 \ldots W_j}(s_j) = s$ for $1 \leq j \leq i$. Since $\mathcal{S}$ is connected, for any states σ and σ' there exists a word w of length at most d such that $\phi_w(\sigma) = \sigma'$. In particular, there exists a word W_{i+1} of length at most

d such that $\phi_{W_{i+1}}(\phi_{W_1...W_i}(s_{i+1})) = s$. Hence, by finite induction, we construct W which verifies (5). By construction

$$\text{card}\{n < m \,;\, u_n \ldots u_{n+|W|-1} = W\}$$
$$\leq \text{card}\{n < m + |W| - 1 \,;\, \phi_{u[n]}(s_0) = s\}$$

and since u is a normal word, W occurs in u with the frequency $\geq 1/q^{d(d-1)+1}$.
∎

FINAL STEP. If $u \in \mathcal{A}^\infty$ is not normal then u is not uniformly k–distributed for some k and by Proposition 2 there exists an automatic complete predictor P with $\rho(P, u) > \frac{1}{q}$. Now assume that u is normal and let P be an automatic predictor regular for u. We have to prove that $\rho(P, u) = \frac{1}{q}$. From Proposition 4 we may assume that P is defined by a connected automaton (S, s_0, Φ, τ). Let s be a state such that $\tau(s) \neq \#$. Then by Proposition 6

$$h := \liminf_{n \to \infty} \frac{\mathcal{N}(P, m, u)}{m} \geq \liminf_{n \to \infty} \frac{\text{card}\{n < m \,;\, \phi_{u[n]}(s_0) = s\}}{m} > 0.$$

Theorem 1 says that there exist a complete predictor P' and $p \in [h, 1]$ such that $\rho(P', u) \geq p\rho(P, u) + (1 - p)/q$. But $\rho(P', u) = \frac{1}{q}$ by Lemma 2 and $p \neq 0$, therefore $\rho(p, u) \leq \frac{1}{q}$. At this state of the proof we have obtained

Lemma 3. *For all automatic predictors P, regular for a given normal word u, one has $\rho(P, u) \leq \frac{1}{q}$.*

We finish the proof of Theorem 3 using a combinatorial argument. Let P^τ be defined by the automaton (S, s_0, Φ, τ) and consider the partition of $\mathbf{N}$ into sets $N_a(\tau) = \{n \,;\, P^\tau(u[n]) = a\}$, $a = \#, 0, 1, \ldots, q$. The sum

$$E_m(\tau) := \sum_{n < m} |P^\tau(u[n]), u_n|$$

can be organized as the following:

$$E_m(\tau) = \sum_{a \in \mathcal{A}} \left(\sum_{\substack{n < m \\ n \in N_a(\tau)}} |a, u_n| \right).$$

Let T be the set of all maps $\sigma : S \to \mathcal{A} \cup \{\#\}$ such that $N_\#(\sigma) = N_\#(\tau)$ and

$$\forall a \in \mathcal{A}, \, \exists b \in \mathcal{A}, \, N_a(\sigma) = N_b(\tau).$$

Let ν be the number of non empty sets $N_a(\tau)$, $a \in \mathcal{A}$. Then $\text{card}(T) = q(q - 1)\ldots(q - \nu + 1)$ and

$$\sum_{\sigma \in T} E_m(\sigma) = \sum_{\sigma \in T} \sum_{a \in \mathcal{A}} \left(\sum_{\substack{n < m \\ n \in N_a(\sigma)}} |a, u_n| \right).$$

But, for every fixed n in $\mathbf{N}\backslash N_\#(\tau)$,

$$\sum_{\sigma\in T}\sum_{\substack{a\in\mathcal{A}\\n\in N_a(\sigma)}}|a,u_n|=\sum_{a\in\mathcal{A}}\sum_{\substack{\sigma\in T\\\sigma(s_n)=a}}[a,u_n]=(q-1)(q-2)\ldots(q-\nu+1),$$

where $s_n:=\phi_{u[n]}(s_0)$. Consequently

$$\sum_{\sigma\in T}E_m(\sigma)$$
$$=(m-\text{card}(N_\#(\tau)\cap\{0,1,\ldots,m-1\}))(q-1)\ldots(q-\nu+1).$$

Therefore

(6)
$$\limsup_{n\to\infty}\sum_{\sigma\in T}\frac{E_m(\sigma)}{(m-\text{card}(N_\#(\sigma)\cap\{0,1,\ldots,m-1\}))}$$
$$=(q-1)\ldots(q-\nu+1).$$

Let P^σ be the predictor defined by $(S,s_0\Phi,\sigma)$, then (6) gives

$$\sum_{\sigma\in T}\rho(P^\sigma,u)\ge(q-1)(q-2)\ldots(q-\nu+1).$$

Lemma 3 says that $\rho(P^\sigma,u)\le\frac{1}{q}$ for all σ so that $\rho(P^\sigma,u)=\frac{1}{q}$ and particularly the equality holds for P^τ. This ends the proof of Theorem 3. ■

Corollary 1. *For any infinite word u the following properties are equivalent:*
(i) u is normal,
(ii) For any automatic predictor P, regular for u, $\rho(P,u)=\frac{1}{q}$,
(iii) For any automatic complete predictor P, $\rho(P,u)=\frac{1}{q}$,
(iv) For any automatic predictor P, regular for u, $\rho(P,u)\le\frac{1}{q}$.

Notice that *(iv)* is a by-product of the combinatorial part in the above final step.

6. Words extracted from normal words

An easy application of Theorem 2 and Theorem 3 is the following result first obtained by V. N. Agafonov [Aga] in the case of binary sequences:

Theorem 4. *Let u be a normal sequence and let $\mathcal{K}$ be a regular automaton for u. Then the subsequence $E(\mathcal{K},u)=(u_{j_m})_m$ extracted by $\mathcal{K}$ from u is normal and moreover*

$$\limsup_{m\to\infty}\frac{j_m}{m}<+\infty.$$

A simple consequence of this theorem is:

Corollary 2. *Let $u\in\mathcal{A}^\infty$ be normal. Let $F\subset\mathcal{A}^\infty$ be a finite set and define*

$$J_F:=\{m\in\mathbf{N};\exists w\in F,u_{m-|w|}\ldots u_{m-1}=w\}.$$

Then the extracted sequence $(u_m)_{m\in J_F}$ is normal.

REFERENCES

[Aga] AGAFONOV V. N., Normal sequences and finite automata, *Dokl. Akad. Nauk SSSR*, **179**, No 2 (1968); *Soviet Math. Dokl.* **9**, No 2 (1968), 324–325.

[Ka-We] KAMAE T. and WEISS B., Normal numbers and selection rules, *Israel J. Math.*, **21** (1975), 101–110.

[Mar] MARTIN-LÖF P., The definition of random sequences, *Information and Control*, **9** (1966), 602–619.

[O'Co] O'CONNOR M., An unpredictability approach to finite state randomness, *J. Comp. System Science* **37** (1988), 324–336.

[Vill] VILLE J., *Etude critique de la notion de collectif*, Gauthier-Villars, Paris (1939).

Annie BROGLIO
Université de Grenoble 1
Institut Fourier, BP 74
F-38402 St-Martin-d'Heres

Pierre LIARDET
Université de Provence
UFR-MIM, case 96
F–13331 Marseille cedex 3
E-mail: liardet@ frccup51.bitnet

Contemporary Mathematics
Volume **135**, 1992

Common Closing Extensions and Finitary Regular Isomorphism for Synchronized Systems

DORIS FIEBIG

ABSTRACT. Two synchronized systems are shown to be finitary regular isomorphic iff they admit a common synchronized extension with factor maps which are right closing a.e. and 1-1 a.e. A complete invariant in terms of the Fischer cover is presented. Some computable invariants are discussed.

1. Introduction

The object of this paper is to show that two synchronized systems are finitary regular isomorphic (with respect to ergodic measures with full support which have a Markov block) iff they admit a common extension by a synchronized system with right closing almost everywhere, 1-1 a.e. factor maps. The proof of this theorem shows that the existence of such an extension defines an equivalence relation on the set of synchronized systems. We prove that this equivalence relation, when restricted to sofic systems, coincides with almost topological conjugacy. We present some complete and derive some non complete, but computable, invariants.

We fix some notation. Let N^Z be endowed with the product topology of the discrete topology on N. Let $\sigma : N^Z \to N^Z$ denote the shift map. A subshift S is a closed, shift invariant subset of N^Z, not necessarily compact. We deal only with transitive subshifts. For a point $x = (x_i)_{i \in Z}$ in S and $-\infty < m \le n < \infty$ we denote by $x[m, n]$ the subblock $x_m x_{m+1} \ldots x_n$ and by $x[n, \infty)$ and $x(-\infty, n]$ the right infinite ray $x_n x_{n+1} \ldots$, respectively left infinite ray $\ldots x_{n-1} x_n$. An S-*block* is a finite sequence of symbols which occurs as a subblock of some point in S. For an S-block m and $n \in Z$ denote by $[m]_n$ the set $\{x \in S \mid x[-|m|+1+n, n] = m\}$ and by $_n[m]$ the set $\{x \in S \mid x[n, n+|m|-1] = m\}$, where $|m|$ denotes the length of m. We denote the shift map restricted to S by σ, too. A point of S is right (left) transitive if its forward (backward) orbit is dense. A point of S is doubly transitive if it is right and left transitive.

1991 Mathematics Subject Classification. Primary 54H20, 28D20.
This paper is in final form and no version of it will be submitted for publication elsewhere.

Let S and T be subshifts. By a *factor map* $\varphi : S \to T$ we mean a shift commuting, uniformly continuous map such that $\varphi(S)$ is dense in T. By uniform continuity, there is $n \in \mathbf{Z}^+$ such that for all $x \in S$ $\varphi(x)_0$ is determined by $x[-n,n]$. We call $2n+1$ a coding length for φ. If S is compact, then $\varphi(S)$ is closed and thus φ is onto.

A factor map $\varphi : S \to T$ is

- *1-1 a.e.* if any doubly transitive point in T has exactly one preimage under φ.
- *right closing (r.c.)* if there is $n \in \mathbf{N}$ such that for any two points $x, y \in S$ with $x(-\infty, 0] = y(-\infty, 0]$ and $\varphi(x)(-\infty, n] = \varphi(y)(-\infty, n]$ it holds $x_1 = y_1$. In this case φ is called *n-step right closing*.
- *right closing almost everywhere (r.c.a.e.)* if there is $n \in \mathbf{N}$ such that for any two left transitive points $x, y \in S$ with $x(-\infty, 0] = y(-\infty, 0]$ and $\varphi(x)(-\infty, n] = \varphi(y)(-\infty, n]$ it holds $x_1 = y_1$. Here φ is called *n-step right closing a.e.*

We say two subshifts S and T *admit a common extension* if there is a subshift R and factor maps $\varphi : R \to S$, $\psi : R \to T$. S and T are *conjugate* if there are factor maps $\varphi : S \to T$, $\psi : T \to S$ with $\psi\varphi = $ id and $\varphi\psi = $ id.

Now we describe certain subclasses of subshifts. Let $S \subset \mathbf{N}^\mathbf{Z}$ be a subshift. An S-block m is a *synchronizing block* if for any two S-blocks am and mb it holds that amb is an S-block, too (see also [BH], [FF]). A subshift $S \subset \mathbf{N}^\mathbf{Z}$ is a *Markov shift* if there is $n \in \mathbf{N}$ such that any S-block of length n is synchronizing. The compact Markov shifts are exactly the shifts of finite type (SFT). A *sofic shift* is a factor of a compact Markov shift. A *coded system* is a compact factor of a Markov shift ([BH], [FF]). A *synchronized system* is a compact factor of a Markov shift which has a synchronizing block ([BH], [FF]).

Let S be a subshift. A measure on S is an invariant Borel probability measure. An S-block m is a *Markov block* for a measure μ on S if for all S-blocks a, b, c with $\mu([am]_0) > 0$ and $\mu([bm]_0) > 0$ it holds $\mu(_1[c] \mid [am]_0) = \mu(_1[c] \mid [bm]_0)$. We denote by $M(S)$ the set of ergodic measures on S which have full support and a Markov block. For a synchronized systems $M(S)$ is always non empty (see § 3).

Now let S and T be subshifts endowed with measures μ_S and μ_T. A measure theoretic homomorphism $\Phi : (S, \sigma, \mu_S) \to (T, \sigma, \mu_T)$ is *regular* if there is an $k \in \mathbf{Z}^+$ (*the anticipation of* Φ) such that for μ_S-almost all $x \in S$ $\Phi(x)_0$ is determined by $x(-\infty, k]$. Φ is *finitary* if for μ_S-almost all $x \in S$ there is an $n = n(x) \in \mathbf{Z}^+$ such that $\Phi(x)_0$ is determined by $x[-n, n]$. A *finitary regular isomorphism* $\Phi : (S, \sigma, \mu_S) \to (T, \sigma, \mu_T)$ is a measure theoretic isomorphism such that Φ and Φ^{-1} are both finitary and regular. Our main result is:

THEOREM 1.1. *Let S and T be synchronized systems. Then S and T have a common extension by a synchronized system, where the factor maps are r.c.a.e. and 1-1 a.e. iff there are measures $\mu_S \in M(S)$ and $\mu_T \in M(T)$ such that (S, σ, μ_S) and (T, σ, μ_T) are finitary regular isomorphic.*

The difficult part of this theorem is the construction of the desired common extension given a finitary regular isomorphism. For SFT´s, endowed with Markov measures, this was done by [BT]. To extend this result we have to give a different construction essentially for two reasons. In the synchronized case the Fischer cover is not compact and r.c.a.e. factor maps do not preserve topological entropy as right closing maps do.

I would like to thank Ulf Fiebig for his extremely helpful suggestions for the presentation of this paper.

2. Proof of the main theorem

We begin by showing that r.c.a.e., 1-1 a.e factor maps have a strong decoding property when the domain is a synchronized system.

DEFINITION 2.1. *Let* $\varphi : S \rightarrow T$ *be a factor map between arbitrary subshifts* S *and* T. *A* T-*block* w *is a decoder block for* φ *if there is* $k \in \mathbf{Z}^+$, *the anticipation of* w, *such that for all* $n \in \mathbf{N}$ *and all points* x, y $\in$ S *with* $\varphi(x)[-|w|+1, 0] = \varphi(y)[-|w|+1, 0] = w$ *and* $\varphi(x)[1, n+k] = \varphi(y)[1, n+k]$ *it holds that* x[1, n] = y[1, n].

REMARK 2.2. *Let* $\varphi : S \rightarrow T$ *be a factor map with a decoder block. Let* μ *be an ergodic measure on* S *with full support. Then* $\varphi : (S, \mu) \rightarrow (T, \varphi\mu)$ *is a finitary regular isomorphism, where* $\varphi\mu$ *is the measure* μ *transported by* φ.

LEMMA 2.3. *Let* S *be synchronized system and* $\varphi : S \rightarrow T$ *a factor map. Then* φ *is r.c.a.e., 1-1 a.e. iff* φ *has a decoder block.*

PROOF. " $\Rightarrow$ ": Fix $k \in \mathbf{Z}^+$ such that φ is k-step r.c.a.e. Fix a synchronizing S-block m with length greater than a coding length for φ. Let $z \in S$ be a doubly transitive point such that z(-∞, 0] ends with m. Because φ is 1-1 a.e. and S is compact, metric there is $p \in \mathbf{N}$ such that for $x \in S$ with $\varphi(x)[-p, p] = \varphi(z)[-p, p]$ it holds that x(-∞, 0] ends with m. We shall show that w:= $\varphi(z)[-p, p]$ is a decoder block with anticipation k. For that let x, y $\in$ S with $\varphi(x)[-|w|+1, 0] = \varphi(y)[-|w|+1, 0] = w$ and $\varphi(x)[1, n+k] = \varphi(y)[1, n+k]$. Then, by the choice of w, x(-∞, -p] and y(-∞, -p] both end with m. Thus there are points $\bar{x}, \bar{y} \in S$ with $\bar{x}$(-∞, -p] = z(-∞, 0], $\bar{y}$(-∞, -p] = z(-∞, 0], $\bar{x}$[-p+1, ∞) = x[-p+1, ∞) and $\bar{y}$[-p+1, ∞) = y[-p+1, ∞), because m is synchronizing. $\bar{x}, \bar{y}$ are left transitive points and $\varphi(\bar{x})$(-∞, n+k] = $\varphi(\bar{y})$(-∞, n+k]. This shows that $\bar{x}$[-p+1, n] = $\bar{y}$[-p+1, n], because φ is k-step r.c.a.e.. By definition of $\bar{x}, \bar{y}$ we get x[1, n] = y[1, n].

" $\Leftarrow$ ": Let w be a decoder block with anticipation k. Let $y \in T$ be such that for some $i \in \mathbf{Z}$ y[i-|w|+1, i] = w. By the decoding property of w all coordinates j > i of the preimages of y are uniquely determined. This shows that φ is 1-1 a.e. A coordinate j > i of the preimages of y is determined by y[i-|w|+1, j+k]. This shows that φ is k-step r.c.a.e.. $\square$

LEMMA 2.4. *Let* $\Phi : (S, \sigma, \mu_S) \rightarrow (T, \sigma, \mu_T)$ *be a finitary regular homomorphism, where* $\mu_S \in M(S)$. *Let* $k \in \mathbf{Z}^+$ *denote the anticipation of* Φ. *Then there is a measurable set* $M \subset S$ *with* $\mu_S(M) = 1$ *and the property that for all points* $x \in M$ *and all* $i \in \mathbf{Z}$ *there is* $n \in \mathbf{N}$ *such that* $\Phi(x)_i$ *is determined by* x[-n+i, i+k].

PROOF. It suffices to show that for μ_S-almost all $x \in S$ there is $n \in N$ such that $\Phi(x)_0$ is determined by $x[-n, k]$. Because μ_S is ergodic with full support and Φ is finitary there is a measurable set M, contained in the doubly transitive points of S, with $\mu_S(M) = 1$ such that Φ is defined and continuous on M. Fix $x \in M$. Then there is $n \in N$ such that $\Phi(x)_0$ is determined by $x[-n, n]$ and $w = x[-n, k]$ is a Markov block for μ_S. Let $W = [w]_k$, $A = {}_{k+1}[x[k+1,n]]$ and $B = \{y \in S \mid \Phi(y)_0 = \Phi(x)_0\}$. We may assume that B is in the σ-algebra generated by the coordinate projections with indices less or equal to k, because Φ is regular with anticipation k. Because μ_S has full support and $A \cap W \cap M \subset B$, we have $\mu_S(A \cap B \cap W) = \mu_S(A \cap W \cap M) > 0$. Thus $\mu_S(B \mid W) = \mu_S(B \mid W \cap A)$ $\mu_S(A \mid W)\mu_S(A \mid W \cap B)^{-1} = \mu_S(B \mid W \cap A) = 1$, because w is a Markov block and because $A \cap W \cap M \subset B$. Therefore $W \subset B$ almost surely. This proves the lemma. $\qquad\square$

LEMMA 2.5. *Let S be a compact subshift and $\varphi : S \to T$ be a 1-1 a.e. factor map. Let μ be a measure which has a Markov block. Then $\varphi\mu$, the measure transported by φ, has a Markov block, too.*

PROOF. Let m be a Markov block for μ with length greater than a coding length for φ. Choose a doubly transitive point $x \in S$ such that $x(-\infty, 0]$ ends with m. Because φ is 1-1 a.e and S is compact, metric there is $n \in N$ such that for all $y \in S$ with $\varphi(y)[-n, n] = \varphi(x)[-n, n]$ it holds that $y(-\infty, 0]$ ends with m. Then $\varphi(x)[-n, n]$ is a Markov block for $\varphi\mu$. $\qquad\square$

PROOF OF THEOREM 1.1. "$\Rightarrow$": Let R be synchronized and let $\varphi : R \to S$ and $\psi : R \to T$ be r.c.a.e., 1-1 a.e. factor maps. Let $\mu_R \in M(R)$. By Lemma 2.3. and Remark 2.2. $\psi\varphi^{-1}$ is a finitary regular isomorphism from $(S, \sigma, \varphi\mu_R)$ to $(T, \sigma, \psi\mu_R)$. $\varphi\mu_R \in M(S)$ and $\psi\mu_R \in M(T)$, by Lemma 2.5.

"$\Leftarrow$": Let S and T be synchronized systems. Let $\mu_S \in M(S)$, $\mu_T \in M(T)$ and let $\Phi : (S, \sigma, \mu_S) \to (T, \sigma, \mu_T)$ be a finitary regular isomorphism.

If S consists of a single orbit and then T, too, and S is conjugate to T.

Otherwise there is a conjugate system $\overline{S}$ such that in $\overline{S}$ there is a synchronizing symbol, say $*$, and for each n there is a block $*w*$ in $\overline{S}$ of length greater than n and such that $*$ is not a subblock of w. We call $\overline{S}$ again S. Composing Φ with a suitable power of the shift we may assume that Φ is finitary with anticipation 0 and Φ^{-1} is finitary with anticipation k, for some $k > 0$. By Lemma 2.4. there are shift invariant dense subsets M and $\overline{M}$ of the doubly transitive points of S and T such that $\Phi : M \to \overline{M}$ is shift commuting, for all $x \in M$ there is $n = n(x) \in Z^+$ such that $\Phi(x)_0$ is determined by $x[-n, k]$ and for all $y \in \overline{M}$ there is $n = n(y) \in Z^+$ such that $\Phi^{-1}(y)_0$ is determined by $y[-n, k]$. From now on Φ denotes always this homeomorpism from M to $\overline{M}$.

Because the construction of the desired extension is rather technical, we give a brief outline of the underlying idea.

Coded systems are exactly those transitive, compact subshifts which are generated by a (countable) set of blocks in the sense that the coded system is the closure of all biinfinite concatenations of those blocks, [BH]. This point of view shows how to find a coded system R which is a common extension of S and T. Namely, let $\{b_n, n \in N\}$ be a generating set of blocks for S and $\{\overline{b}_n, n \in N\}$ a generating set of blocks for T, with $|b_n| = |\overline{b}_n|$ for all n. Then the set of blocks $X :=$ $\{(b_n, \overline{b}_n), n \in N\}$ generates a coded system $R \subset S \times T$ which factors onto S and T by coordinatewise projections. To make R synchronizing we construct the set $\{b_n, n \in N\}$ such that b_1 cannot overlap itself, all blocks b_n have length at least as b_1 and b_1 occurs in a block b_n at most at the end of b_n. Then $(b_1, \overline{b}_1)$ is a synchronizing block for R (see part III). We ensure that b_1 is a decoder block for the projection map onto S by the following two properties. b_1 is a marker in the sense that if x is a point in S which sees the block b_1 twice, say $x(-\infty, 0]$ and $x(-\infty, n]$ end with b_1, then for any preimage $(x, y) \in R$ under the projection onto S it holds that $(x, y)[1, n]$ is a finite concatenation of blocks of X. Furthermore X has a certain kind of rigidity, namely, if for some b_n and b_m the beginning blocks of length, say p, coincide then the beginning blocks of length p of $\overline{b}_n$ and $\overline{b}_m$ coincide. The main problem in getting a decoder block for the projection onto T is to construct a marker for the map. $(b_1\overline{b}_1b_1\overline{b}_1$ will do this job as Claim 4 proves). This explains the tedious construction in part I.

Part I Construction of the blocks b_1 and $\overline{b}_1$, which yield marker blocks.

LEMMA. *There are S-blocks* u *and* w *and T-blocks* $\overline{u}$ *and* $\overline{w}$ *and a set* $E \subset \{\overline{b} \mid \overline{w}\overline{b}$ *is a T-block and* $|\overline{b}| = k\}$ *(k is the anticipation of* Φ^{-1}*) such that:*

(1) w *ends with a block* $*v_1...v_s*$
     where $s > 2k$ *and* $v_i \neq *$ *for all* $1 \leq i \leq s$.
(2) $\overline{w}$ *is synchronizing for* T.
(3) u *ends with* w *and* $\overline{u}$ *ends with* $\overline{w}$.
(4) *If* $x \in M$ *and* $x(-\infty, 0]$ *ends with* u *then* $\Phi(x)(-\infty, 0]$ *ends with* $\overline{u}$.
(5) *If* $y \in M$, $y(-\infty, 0]$ *ends with* $\overline{u}$ *and* $y[1, k] \in E$ *then* $\Phi^{-1}(y)(-\infty, 0]$ *ends with* w.
(6) *If* $x \in M$ *and* $x(-\infty, 0]$ *ends with* w *then* $\Phi(x)(-\infty, 0]$ *ends with* $\overline{w}$
     and $\Phi(x)[1, k] \in E$.
(7) u *begins with* $*$.

PROOF Let $\{e^1, ..., e^d\} = \{e \mid *e$ is an S-block and $|e| = k\}$. Fix a block $*v_1...v_s*$ as in (1). Fix $x^1 \in M$ such that $x^1[-s-1, 0] = *v_1...v_s*$ and $x^1[1, k]$ $= e^1$. Because $\Phi(x^1)$ is doubly transitive, there is $p > s+1$ such that $\Phi(x^1)[-p, 0]$ is synchronizing for T. Let $\overline{w} := \Phi(x^1)[-p, 0]$. Because Φ is continuous without anticipation, there is $n_1 > p$ such that $x^1[-n_1, k]$ determines $\Phi(x^1)[-p, k]$. Inductively, for $1 \leq i \leq d - 1$, fix $x^{i+1} \in M$ such that $x^{i+1}[-n_i, 0] = x^1[-n_i, 0]$ and $x^{i+1}[1, k] = e^{i+1}$. Because Φ is continuous without anticipation, there is $n_{i+1} \geq n_i$ such that $x^{i+1}[-n_{i+1}, k]$ determines $\Phi(x^{i+1})[-p, k]$. Let $\overline{E} :=$ $\{\Phi(x^i)[1,k] \mid 1 \leq i \leq d\}$. Then $w := x^d[-n_d, 0]$ satisfies (6). Because Φ^{-1} is continuous with anticipation k, there is $n \geq n_d$ such that $\Phi(x^d)[-n, k]$ determines $\Phi^{-1}(\Phi(x^d))[-n_d, 0]$. Let $\overline{u} := \Phi(x^d)[-n, 0]$. Use again the continuity of Φ and that x^d is doubly transitive to find $m \geq n$ such that $x^d[-m, 0]$ determines $\Phi(x^d)[-n, 0]$

and $x^d(-\infty, -m-1]$ ends with $*$. Then $u := x^d[-m-1, 0]$ satisfies (4) and (7). By construction (1) - (3) are satisfied. It remains to show that (5) is satisfied. Let $y \in \overline{M}$ such that $y(-\infty, 0]$ ends with $\overline{u}$ and $y[1, k] = \Phi(x^i)[1, k]$ for some $1 \le i \le d$. Let $r > n$ be such that for $x \in M$ with $\Phi(x)[-r, k] = y[-r, k]$ it holds $x[-n_d, 0] = \Phi^{-1}(y)[-n_d, 0]$. Let $t > r$ be such that for $x \in M$ with $x[-t, 0] = \Phi^{-1}(y)[-t, 0]$ it holds $\Phi(x)[-r, 0] = y[-r, 0]$. By definition of $\overline{u}$ we get $\Phi^{-1}(y)[-n_d, -s-1]$ ends with $*$. Thus, because $*$ is synchronizing for S, there is $x \in M$ with $x[-t, -s-1] = \Phi^{-1}(y)[-t, -s-1]$, $x[-n_d, 0] = w$ and $x[1, k] = e^i$. Then $\Phi(x)[-r, 0] = y[-r, 0]$ and $\Phi(x)[1, k] = y[1, k]$, by choice of t and by definition of w. Thus, by definition of r, $w = x[-n_d, 0] = \Phi^{-1}(y)[-n_d, 0]$, which shows (5). $\square$

Let a, c be S-blocks such that $*aucu$ is an S-block, where a and c do not see $*$, $|c| > 2|u|$ and $|a| > 2|ucu|$. Then let $b_1 := aucu$. Fix some point $x^{(0)} \in M$ such that $x^{(0)}(-\infty, 0]$ ends with u and $x^{(0)}[1, \infty)$ begins with b_1. Let $y^{(0)} := \Phi(x^{(0)})$ and let $\overline{b}_1 := y^{(0)}[1, |b_1|]$.

OBSERVATIONS.
a) Let $\overline{b}_1 = \overline{abcd}$ with $|\overline{a}| = |a|$, $|\overline{b}| = |\overline{d}| = |u|$.
    Then $\overline{a}[1, k], \overline{c}[1, k] \in \overline{E}$ and both $\overline{b}$ and $\overline{d}$ end with $\overline{u}$.
b) b_1 cannot overlap itself other than trivial (use that u end with $*$).
c) $(\overline{b}_1)^\infty \in \overline{T}$ and has least period $|\overline{b}_1|$. (because (5) implies that in $\overline{a}$ never occurs $\overline{b} \circ \overline{c}[1, k]$ and $|\overline{a}| > 2|\overline{bcd}|$)

Part II Inductive construction of the code X.
From now on $x^{(0)}$, $y^{(0)}$, b_1 and $\overline{b}_1$ are fixed as in the definition above.
The first codeword is $(b_1, \overline{b}_1)$. To $(b_1, \overline{b}_1)$ is assigned a memory block $(w_1, \overline{w}_1)$, where $(w_1, \overline{w}_1) := (x^{(0)}, y^{(0)})[-p, 0]$ with $p > |u|$ such that
a) $x \in \overline{M}$, $x[-p, |b_1|] = w_1 b_1 \Rightarrow \Phi(x)[1, |b_1|] = \overline{b}_1$.
b) $y \in \overline{M}$, $y[-p, |b_1|] = \overline{w}_1 \overline{b}_1$, $y[|b_1|+1, |b_1|+k] \in \overline{E} \Rightarrow \Phi^{-1}(y)[1, |b_1|] = b_1$.
For a finite set of blocks, say A, denote by A^* the set of blocks which are finite concatenations of blocks of A.
    Then define inductively, for $n > 1$, $(b_n, \overline{b}_n)$, $(w_n, \overline{w}_n)$ in the following way (the following remark shows that the required choices can be made in each step of the construction):
If n is even, then
    I(n):   Let $s = s(n) \in N$ be maximal such that $x^{(0)}[1, s] \in \{b_1, b_2,...,b_{n-1}\}^*$.
    II(n): Let $r = r(n) \in Z^+$ maximal such that there is $1 \le i \le n-1$ with
            $x^{(0)}[s+1, s+r] = b_i[1,r]$.
    III(n): Let $t = t(n) \ge 0$ such that $(x^{(0)}, y^{(0)})(-\infty, t]$ ends with $(w_{n-1}, \overline{w}_{n-1})$ and
            $x^{(0)}[t+1,t+r+1] = x^{(0)}[s+1,s+r+1]$.
If n is odd, then
    I(n):   Let $s = s(n) \in N$ be maximal such that $y^{(0)}[1, s] \in \{\overline{b}_1, \overline{b}_2,...,\overline{b}_{n-1}\}^*$
            and $y^{(0)}[s+1, s+k] \in \overline{E}$.
    II(n): Let $r = r(n) \in Z^+$ maximal such that there is $1 \le i \le n-1$ with
            $y^{(0)}[s+1, s+r] = \overline{b}_i[1,r]$.
    III(n): Let $t = t(n) \ge 0$ such that $(x^{(0)}, y^{(0)})(-\infty, t]$ ends with $(w_{n-1}, \overline{w}_{n-1})$ and

$$y^{(0)}[t+1,t+r+k] = y^{(0)}[s+1,s+r+k].$$

Then for any n

IV(n): Let $d = d(n) > 0$ be minimal such that $x^{(0)}[t+1, t+d]$ ends with b_1 or
$(x^{(0)}[t+1, t+d]$ ends with w and $y^{(0)}[t+1, t+d]$ ends with $\overline{b_1})$

V(n): $(b_n, \overline{b_n}) := (x^{(0)}, y^{(0)})[t+1, t+d]$.

VI(n): Let $p = p(n) > |w_{n-1}|$ such that for $(w_n, \overline{w_n}) := (x^{(0)}, y^{(0)})[t-p, t]$ holds

  a) $x \in \underline{M}$, $x[-p, |b_n|] = w_n b_n \Rightarrow \Phi(x)[1, |b_n|] = \overline{b_n}$.

  b) $y \in \underline{M}$, $y[-p, |b_n|] = \overline{w_n} \overline{b_n}$, $y[|b_n|+1, |b_n|+k] \in \overline{E} \Rightarrow$
    $\Phi^{-1}(y)[1, |b_n|] = b_n$.

Then let $X := \{(b_n, \overline{b_n}) \mid n \in N\}$. Let R be the closure of all biinfinite
concatenations of blocks of X.

REMARK There is always s satisfying I(n), because in S there are arbitrary
long blocks of the form $*m_1...m_t*$, where $m_i \neq *$ for all i and $x^{(0)}$ is doubly
transitive. $t \geq 0$ satisfying III(n) always exists, because $x^{(0)}$ is doubly transitive
and $y^{(0)}[s+1, s+k] \in \overline{E}$. Furthermore d satifying IV(n) always exists, because
$x^{(0)}[1, \infty)$ sees b_1 infinitely often.

I, II and IV will ensure the generating property of $\{b_m, m \in N\}$ in the even
case and that of $\{\overline{b_m}, m \in N\}$ in the odd case, thus the surjectivity of the
projection maps. III and VI yields the above described rigidity property. IV forces
R to be synchronized and gives the desired marker blocks.

We list a collection of properties of X, which we will use in part III.

CLAIM 0 (*How b_n and $\overline{b_n}$ look like.*)
For all $n \in N$ the following holds:

  a) b_n *and* $b_n[1, |b_n| - |cu|]$ *both end with* w.

  b) $\overline{b_n}$ *ends with* $\overline{u}$.

  c) $\overline{b_n}[1, |b_n| - |cu|]$ *ends with* $\overline{u}$ *and* $\overline{b_n}[|b_n| - |cu| + 1, |b_n| - |cu| + k] \in \overline{E}$.

  d) $\overline{u}$ *is not a subblock of* $\overline{b_n}[|b_n| - |b_1|+1, |b_n| - |b_1| + |a|]$.

  e) *If for some* $p \in N$, $|b_n| - |cu| \leq p < |b_n| - |c|$, $\overline{b_n}[1, p]$ *ends with* $\overline{u}$ *and*
    $\overline{b_n}[p + 1, p + k] \in \overline{E}$ *then* $p = |b_n| - |cu|$.

PROOF a) - c) follow from the definition of $(b_n, \overline{b_n})$ and (4) - (6). d) follows
from (5) and (1) and that a does not see a $*$. We prove e). The assumption
implies, by (5), that b_n sees a $*$ at position p. The block c does not see a $*$
and thus, if b_n ends with b_1 then $p = |b_n| - |cu|$. Otherwise $\overline{b_n}$ ends with $\overline{b_1}$,
which shows that $\overline{b_1}[1, |au| + q]$ ends with $\overline{u}$ and $\overline{b_1}[|au| + q + 1, |au| + q + k] \in$
$\overline{E}$, where $q = p - |b_n| - |cu|$. Therefore, by (5), b_1 sees a $*$ at position $|au| + q$,
which shows $q = 0$, because c does not see a $*$. $\qquad\square$

CLAIM 1 (*The marker property of b_1.*)
If $b_n[|b_n|-p+1, |b_n|] = b_1[1, p]$ *for some* $1 \leq p \leq |b_1|$, *then* $p = |b_1|$.

PROOF If b_n ends with b_1 we are done, because of observation b). In the other
case $\overline{b_n}$ ends with $\overline{b_1}$. By Claim 0 a) and because w ends with $*$, we get that

b_n and $b_n[1, |b_n| - |cu|]$ both end with $*$. Because b_1 does not see a $*$ in the first $|a|$ positions, this shows $p \geq |a| + |cu|$. Therefore, $|b_n| - p + |au| < |b_n|$ and $b_n[1, |b_n| - p + |au|]$ ends with u, thus by (4), $\overline{b}_n[1, |b_n| - p + |au|]$ ends with $\overline{u}$ and $\overline{b}_n[|b_n| - p + |au|+1, |b_n| - p + |au|+k] \in \overline{E}$. Claim 0 e) yields $p = |b_1|$. □

CLAIM 2 *If for some* $p < |\overline{b}_n|$ $\overline{b}_n[1, p - |\overline{cd}|]$ *ends with* $\overline{u}$ *and* $\overline{b}_n[p - |\overline{cd}| + 1, p - |\overline{cd}| + k] \in \overline{E}$ *then* $p < |\overline{b}_n| - k$. *In particular, if for some* $p < |\overline{b}_n|$ $\overline{b}_n[1, p]$ *ends with* $\overline{b}_m$ *then* $p < |\overline{b}_n| - k$.

PROOF By (5), $\overline{b}_n[1, p - |\overline{cd}|]$ ends with w. By Claim 0 a) $\overline{b}_n$ and $\overline{b}_n[1, |\overline{b}_n| - |\overline{cd}|]$ end with w. From (1) it follows that $|\overline{b}_n| - p > k$. □

CLAIM 3 *(The marker property of* $\overline{b}_1 \overline{b}_1 \overline{b}_1 \overline{b}_1$.)
If $\overline{b}_{i_1}[q, |b_{i_1}|] \overline{b}_{i_2} ... \overline{b}_{i_{m-1}} \overline{b}_{i_m}[1, p] = \overline{b}_1 \overline{b}_1 \overline{b}_1 \overline{b}_1$ *for some* $1 \leq q \leq |b_{i_1}|$, $1 \leq p \leq |b_{i_m}|$, *then* $p = |\overline{b}_1|$ *and* $(b_{i_{m-1}}, b_{i_{m-1}}) = (b_1, \overline{b}_1)$.

PROOF By Claim 2 we get $\overline{b}_{i_1}[q, |b_{i_1}|] = \overline{b}_1[1, s]$ for some $1 \leq s \leq |b_1|$, because $\overline{b}_1[1, k] \in \overline{E}$ and (5). If $s = |b_1|$, then we are done. For $s < |b_1|$, by Claim 2, $\overline{b}_{i_2} = \overline{b}_1[s+1, |b_1|] \overline{b}_1[1, t]$ for some $s \leq t \leq |b_1|$. Thus the assertion follows from:
$$\overline{b}_n = \overline{b}_1[s, |b_1|] \overline{b}_1[1, t] \text{ for some } s > 1, 1 \leq t \leq |b_1| \Rightarrow t = |b_1|.$$
To prove this: It is $\overline{b}_n[|b_n| - |b_1|+1, |b_n|] = \overline{b}_1[t+1, |b_1|] \overline{b}_1[1, t]$. Write $b_1 = aucu$ and $b_1 = \overline{abcd}$ as in observation a). We first show that $t \geq |u|$. For that assume that $t < |u|$. Then $t + |au| + k < |b_1|$ and $\overline{b}_n[|b_n| - |b_1|+1, |b_n| - |cu|+k] = \overline{b}_1[t+1, t+|au|+k]$. This block ends with $\overline{u} \overline{e}$ for some $\overline{e} \in \overline{E}$, by Claim 0 c). $t < |u|$ implies $t + |au| + k < |au| + |c|$, which contradicts Claim 0 e), because $t > 0$. Thus, we have shown $t \geq |u|$. It is $\overline{b}_n[|b_n| - t+1, |b_n|] = \overline{b}_1[1, t]$. Therefore, $t \geq |u|$ and Claim 0 b) and d) show $t > |a|$. Which now implies, by Claim 0 c) and d), that $t > |a| + |uc|$. By observation a), this shows that $\overline{b}_n[|b_n| - t+1, |b_n| - t+|au|+k]$ ends with $\overline{u} \overline{c}[1, k]$, which proves $t = |b_1|$, by Claim 0 e) and observation a). □

CLAIM 4 *(*$\{\overline{b}_n, n \in N\}$ *is decipherable.)*
Let $\overline{b}_{i_1} \overline{b}_{i_2} ... \overline{b}_{i_{m-1}} \overline{b}_{i_m}[1, p] = \overline{b}_{j_1} \overline{b}_{j_2} ... \overline{b}_{j_{d-1}} \overline{b}_{j_d}[1, q]$ *for some* $1 \leq p \leq |\overline{b}_{i_m}|$, $1 \leq q \leq |\overline{b}_{j_d}|$. *Then* $p \geq k + 1$ *implies* $m = d$, $i_s = j_s$, *for* $1 \leq s \leq m-1$ *and* $b_{i_m}[1, p-k] = b_{j_d}[1, p-k]$.

PROOF Induction on m.

m = 1: Assume d > m. Then $|\overline{b}_{j_1}| < p \leq |\overline{b}_{i_1}|$. By VI($i_1$) b) and VI($j_1$) b) and $\overline{b}_{i_1}[1, |\overline{b}_{j_1}|] = \overline{b}_{j_1}$ we get $b_{i_1}[1, |b_{j_1}|-k] = b_{j_1}[1, |b_{j_1}|-k]$. By $\overline{b}_{j_1} = \overline{b}_{i_1}[1, |b_{j_1}|]$, Claim 2 implies that $|b_{i_1}| > |b_{j_1}| + k$ and thus $\overline{b}_{i_1}[|b_{j_1}|+1, |b_{j_1}|+k] = b_{j_2}[1, k] \in \overline{E}$, which shows that $b_{i_1}[1, |b_{j_1}|]$ ends with w, by (5). Thus $(b_{i_1}, \overline{b}_{i_1})[1, |b_{j_1}|] = (b_{j_1}, \overline{b}_{j_1})$ a contradiction to the fact that X is a prefix code, see part III. Thus $d = m = 1$ and $q = p$. The assertion now follows from VI(i_1) b) and VI(j_1) b).

m-1 → m: Let $\overline{b}_{i_1} \overline{b}_{i_2} ... \overline{b}_{i_{m-1}} \overline{b}_{i_m}[1, p] = \overline{b}_{j_1} \overline{b}_{j_2} ... \overline{b}_{j_{d-1}} \overline{b}_{j_d}[1, q]$ for some $k + 1 \leq p \leq |b_{i_m}|$, $1 \leq q \leq |b_{j_d}|$. Then $\overline{b}_{i_1} \overline{b}_{i_2} ... \overline{b}_{i_{m-1}} = \overline{b}_{j_1} \overline{b}_{j_2} ... \overline{b}_{j_{s-1}} \overline{b}_{j_s}[1, t]$

for some $1 \le s \le q$, $1 \le t \le |\overline{b}_{j_s}|$. Because $|\overline{b}_{i_{m-1}}| > k + 1$, we get $s = m-1$ and $i_r = j_r$ for $1 \le r \le m - 2$ and $b_{i_{m-1}}[1, |\overline{b}_{i_{m-1}}|-k] = b_{j_{m-1}}[1, |\overline{b}_{i_{m-1}}|-k]$, by induction hypothesis. The argument from the case $m = 1$ shows that $t = |\overline{b}_{j_s}| = |\overline{b}_{i_{m-1}}|$ and $b_{i_{m-1}} = b_{j_{m-1}}$. Thus we have $\overline{b}_{i_m}[1, p] = \overline{b}_{j_m}...\overline{b}_{j_{d-1}}\overline{b}_{j_d}[1, q]$. Apply again induction hypothesis to see that $d = m$ and $q = p$ and $b_{i_m}[1, p-k] = b_{j_m}[1, p-k]$. $\qquad\square$

Part III Verification.

First we check that each step of construction yields a new word of X and that X is prefix, that is, no block is the beginning of another block.

The condition IV(n) in the definition of the code X forces that no $(b_n, \overline{b}_n)$ can be a proper prefix of a $(b_j, \overline{b}_j)$. It remains to show that $(b_j, \overline{b}_j) = (b_n, \overline{b}_n)$, $j \le n$ implies that $j = n$. If $|b_n| < r + 1$, then for n even we get $b_j = b_n = x^{(0)}[s+1, s+|b_n|]$ and thus the definition of $s = s(n)$ shows that $j = n$. For n odd, $|b_n| < r + 1$ implies $b_j = b_n = y^{(0)}[s+1, s+|b_n|]$. If $j < n$ then, by the definition of $s = s(n)$, it follows that $y^{(0)}[s+|b_n|+1, s+|b_n|+k] \notin \overline{E}$, therefore, by III(n), $y^{(0)}[t+|b_n|+1, t+|b_n|+k] \notin \overline{E}$, a contradiction to IV(n). Thus $j = n$. If $|b_n| \ge r + 1$, then for n even we have $b_j[1, r+1] = b_n[1, r+1] = x^{(0)}[s+1, s+r+1]$ and for n odd we have $\overline{b}_j[1, r+1] = \overline{b}_n[1, r+1] = y^{(0)}[s+1, s+r+1]$. Now the definition of $r = r(n)$ shows $j = n$.

Thus X is a prefix code and each step of the construction yields a new codeword.

By definition of $(b_n, \overline{b}_n)$, we have that $*b_n$ is an S-block, b_n ends with $*$ and $\overline{w}\overline{b}_n$ is a T-block, $\overline{b}_n$ ends with $\overline{w}$ (using IV(n) and Claim 0 b). Thus for $(x, y) \in R$ we have $x \in S$ and $y \in T$.

Now look at the sequence $s(2n)$, defined by I(2n), $n \in \mathbf{N}$. Clearly $s(2n) \le s(2n+2)$. If $s(2n) \to \infty$, then any S-block appears as a subblock of some $b_{i_1} b_{i_2}...b_{i_m}$, because $x^{(0)}$ is doubly transitive. Thus the map $pr_1 : R \to S$, $(x, y) \to x$ is onto. Otherwise, there is some s such that $s(2n) \le s$ for all n and that implies $s(2n) = s$ for all large enough n. This shows $|b_{2n}| \ge r(2n) + 1$ for all large enough n and thus $x^{(0)}[s+1, s+r(2n)+1] = b_{2n}[1, r(2n)+1]$. But $s(2n) = s$ for all large enough n, implies also $r(2n) \to \infty$, which shows that $pr_1 : R \to S$ is onto. Looking at the sequence $s(2n+1)$, $n \in \mathbf{N}$, shows $pr_2 : R \to T$ is onto, too.

Clearly R is coded, [BH]. Now we check that R is synchronized. For that we show that $(b_1, \overline{b}_1)$ is synchronizing for R: Let $(a, \overline{a})$ and $(b, \overline{b})$ be R-blocks such that $(ab_1, \overline{ab}_1)$ is an R-block and $(b_1b, \overline{b}_1\overline{b})$ is an R-block. That means $(ab_1, \overline{ab}_1)$ is a subblock of some $(b_{j_1}b_{j_2}...b_{j_m}, \overline{b}_{j_1}\overline{b}_{j_2}...\overline{b}_{j_m})$ and $(b_1b, \overline{b}_1\overline{b})$ is a subblock of some $(b_{j_1}b_{j_2}...b_{j_s}, \overline{b}_{j_1}\overline{b}_{j_2}...\overline{b}_{j_s})$. But Claim 1 shows that b_{i_m} ends with b_1 and thus also $\overline{b}_{i_m}$ ends with $\overline{b}_1$. Furthermore Claim 1 shows that b_{j_1} ends with b_1 and thus also $\overline{b}_{j_1}$ ends with $\overline{b}_1$, which implies that $(b, \overline{b})$ is the beginning of $(b_{j_2}...b_{j_s}, \overline{b}_{j_2}...\overline{b}_{j_s})$. Therefore $(b_{i_1}b_{i_2}...b_{i_m}b_{j_2}...b_{j_s}, \overline{b}_{i_1}\overline{b}_{i_2}...\overline{b}_{i_m}\overline{b}_{j_2}...\overline{b}_{j_s})$ contains $(ab_1b, \overline{ab}_1\overline{b})$ as a subblock. Thus $(b_1, \overline{b}_1)$ is a synchronizing block for R.

Next we will show that $pr_1 : R \to S$ is r.c.a.e., 1-1 a.e.. For that we prove that b_1 is a decoder block of pr_1 with anticipation 0.

Let $(x, y), (\overline{x}, \overline{y}) \in R$ and $n \in N$ with $x[-|b_1|+1, 0] = \overline{x}[-|b_1|+1, 0] = b_1$ and $x[1, n] = \overline{x}[1, n]$. Then there are $i_1, ..., i_m$ such that $(x, y)[-|b_1|+1, n] = (b_{i_1}, \overline{b}_{i_1})[q, |b_{i_1}|] (b_{i_2}, \overline{b}_{i_2}) ... (b_{i_{m-1}}, \overline{b}_{i_{m-1}})(b_{i_m}, \overline{b}_{i_m})[1, p]$ for some $1 \le q \le |b_{i_1}|$, $1 \le p \le |b_{i_m}|$. By Claim 1 $q = |b_{i_1}| - |b_1| + 1$ and thus $(x, y)[1, n] = (b_{i_2}, \overline{b}_{i_2}) ... (b_{i_{m-1}}, \overline{b}_{i_{m-1}})(b_{i_m}, \overline{b}_{i_m})[1, p]$. In an analogue way, $(\overline{x}, \overline{y})[1, n] = (b_{j_2}, \overline{b}_{j_2}) ... (b_{j_{d-1}}, \overline{b}_{j_{d-1}})(b_{j_d}, \overline{b}_{j_d})[1, q]$ for some $1 \le q \le |b_{j_d}|$.
Using VI(i_2) a), VI(j_2) a) and IV shows that $\overline{b}_{i_2} = \overline{b}_{j_2}$. Repeating this argument m times shows $y[1, n] = \overline{y}[1, n]$ and proves that b_1 is a decoder block for pr_1.

Now we will show that $pr_2 : R \to T$ is r.c.a.e., 1-1 a.e., by proving that $b_1 \overline{b}_1 \overline{b}_1 b_1$ is a decoder block for pr_2.

Let $(x, y), (\overline{x}, \overline{y}) \in R$, $n \in N$ with $y[-4|b_1|+1, 0] = \overline{y}[-4|b_1|+1, 0] = b_1 \overline{b}_1 \overline{b}_1 b_1$ and $y[1, n+2k] = \overline{y}[1, n+2k]$. Then $(x, y)[-4|b_1|+1, n+2k]$ is a subblock of some concatenation of blocks of X. By Claim 3 it follows that for some indices $i_1, ..., i_m$ $(x, y)[-|b_1|+1, n+2k] = (b_{i_1}, \overline{b}_{i_1})(b_{i_2}, \overline{b}_{i_2}) ... (b_{i_{m-1}}, \overline{b}_{i_{m-1}})$ $(b_{i_m}, \overline{b}_{i_m})[1, p]$, $1 \le p \le |b_{i_m}|$. For $p \le k$ we look only at $(x, y)[-|b_1|+1, n+2k-p] = (b_{i_1}, \overline{b}_{i_1})(b_{i_2}, \overline{b}_{i_2}) ... (b_{i_{m-1}}, \overline{b}_{i_{m-1}})$. As above we find $j_1, ..., j_d$ and some $1 \le q \le |b_{j_d}|$ such that $(\overline{x}, \overline{y})[-|b_1|+1, n+2k-p] = (b_{j_1}, \overline{b}_{j_1}) ... (b_{j_{d-1}}, \overline{b}_{j_{d-1}})(b_{j_d}, \overline{b}_{j_d})[1, q]$. Apply Claim 4 to see that $x[-|b_1|+1, n+k-p] = \overline{x}[-|b_1|+1, n+k-p]$, which proves $x[1, n] = \overline{x}[1, n]$, because $p \le k$. For $p > k$ we find $j_1, ..., j_d$ and some $1 \le q \le |b_{j_d}|$ such that $(\overline{x}, \overline{y})[-|b_1|+1, n+2k] = (b_{j_1}, \overline{b}_{j_1}) ... (b_{j_{d-1}}, \overline{b}_{j_{d-1}})(b_{j_d}, \overline{b}_{j_d})[1, q]$. By Claim 4 it follows $x[1, n] = \overline{x}[1, n]$. Thus we have shown that $b_1 \overline{b}_1 \overline{b}_1 b_1$ is a decoder block with anticipation 2k. $\square$

3. Complete invariants

To any synchronized system S is associated in a canonical way a Markov shift Σ_S and a factor map $\rho_S : \Sigma_S \to S$, the so-called Fischer cover of S. We recall its definition. The follower set of an S-block m is the set of blocks b such that mb is an S-block. The pair (Σ_S, ρ_S) is given by a directed, labeled graph, which we call, in accordance to [FF], (X_0^+, ρ_0^+). The vertices of the graph X_0^+ are the follower sets of synchronizing blocks. There is an edge from vertex i to vertex j with ρ_0^+-label a (a is an S-block of length 1) iff i is the follower set of the synchronizing block m and j is the follower set of the block ma. Let Σ_S be the space of biinfinite paths of edges in the graph X_0^+ and $\rho_S(z) := (\rho_0^+(z_i))_{i \in Z}$ for $z = (z_i)_{i \in Z} \in \Sigma_S$. From now on (Σ_S, ρ_S) denotes always the Fischer cover of S. It is a canonical, by [FF, Theorem 2.12.]. For further properties of the Fischer cover see [FF, Remark 2.13., Theorem 2.16.].

Without mentioning further we will make use of the fact that any factor map $\phi : \Gamma \to S$, where Γ is a Markov shift and S is compact, can be presented by a labeled graph, by passing to a higher block system of Γ.

By [FF, Theorem 2.16. iv)], $\rho_S : \Sigma_S \to S$ has a decoder block. Now we can use the argument of Lemma 2.5 to see that any Markov measure on Σ_S is mapped to a measure with a Markov block by ρ_S. Thus, because there are Markov measures with full supprt on Σ_S, [G], we have that $M(S)$ is non empty.

THEOREM 3.1. *Let* S *and* T *be synchronized. Equivalent are the following:*
(1) S *and* T *have a common extension by a synchronized system where the factor maps are r.c.a.e., 1-1 a.e..*
(2) *There is a homeomorphism with bounded anticipation between a subset of the doubly transitive points of* S *and a subset of the doubly transitive points of* T.
(3) Σ_S *and* Σ_T *have a common extension by a Markov shift where the factor maps have decoder blocks.*
(4) S *and* T *have a common extension by a Markov shift where the factor maps have decoder blocks.*

PROOF (1) $\Leftrightarrow$ (2): By the proof of Theorem 1.1.

(1) $\Rightarrow$ (3): Let R be synchronized and $\varphi : R \to S$, $\psi : R \to T$ r.c.a.e., 1-1 a.e factor maps. Then $\varphi \circ \rho_R : \Sigma_R \to S$ is a right closing factor map (use the same argument as in [BMT, Proposition 4.10]). Because $\rho_R(\Sigma_R)$ contains all doubly transitive points of R (see [FF]) and φ is 1-1 a.e., we get that $(\varphi \circ \rho_R)(\Sigma_R)$ is a residual subset of S. Applying [FF, Theorem 2.17.] yields a factor map $\hat{\varphi} : \Sigma_R \to \Sigma_S$ such that $\rho_S \circ \hat{\varphi} = \varphi \circ \rho_R$. By the proof of Lemma 2.3. there is a synchronizing R-block m and a synchronizing S-block w and a number k such that the following holds:

x, y $\in$ R, $\varphi(x)[-p, p] = \varphi(y)[-p, p] = w$, $\varphi(x)[p+1, p+n+k] = \varphi(y)[p+1, p+n+k]$ implies x(-∞, 0] and y(-∞, 0] end with m and x[1, n] = y[1, n].

Let $\overline{w}$ be an Σ_S-block such that x $\in \Sigma_S$, x[-p, p] = $\overline{w}$ implies $\rho_S(x)[-p, p] = w$. We show that $\overline{w}$ is a decoder block for $\hat{\varphi}$ with anticipation k. For that let x, y $\in \Sigma_R$, $\hat{\varphi}(x)[-p, p] = \hat{\varphi}(y)[-p, p] = \overline{w}$, $\hat{\varphi}(x)[p+1, p+n+k] = \hat{\varphi}(y)[p+1, p+n+k]$. Then $\varphi(\rho_R(x))[-p, p] = \varphi(\rho_R(y))[-p, p] = w$, $\varphi(\rho_R(x))[p+1, p+n+k] = \varphi(\rho_R(y))[p+1, p+n+k]$, because $\rho_S \circ \hat{\varphi} = \varphi \circ \rho_R$ and ρ_S has coding length 1. Thus, $\rho_R(x)(-∞, 0]$ and $\rho_R(y)(-∞, 0]$ end with m and $\rho_R(x)[1, n] = \rho_R(y)[1, n]$. Synchronizing blocks are decoder blocks for ρ_R with anticipation 0 and thus x[1, n] = y[1, n].

(3) $\Rightarrow$ (4): Let Γ be a Markov shift and $\varphi : \Gamma \to \Sigma_S$, $\psi : \Gamma \to \Sigma_T$ be factor maps with a decoder block. Then $\rho_S \circ \varphi : \Gamma \to S$, $\rho_T \circ \psi : \Gamma \to T$ are factor maps with decoder block, because ρ_S, ρ_T and φ, ψ have decoder blocks.

(4) $\Rightarrow$ (1): Let Γ be a Markov shift and $\varphi : \Gamma \to S$, $\psi : \Gamma \to T$ be factor maps with a decoder block. Let R be the closure of $(\varphi, \psi)(\Gamma)$. Then $\varphi : R \to S$, $(x, y) \to x$ and $\psi : R \to T$, $(x, y) \to y$ are factor maps, R is synchronizing, because $(\varphi, \psi) : \Gamma \to R$ has a decoder block and thus $(\varphi, \psi)(\Gamma)$ is a residual subset of R, apply [FF, Theorem 1.1.]. Let w be a decoder block with anticipation k for φ. Let (x, y), $(\overline{x}, \overline{y}) \in R$ with x[-|w|+1, 0] = $\overline{x}$[-|w|+1, 0] = w, x[1, n+k] = $\overline{x}$[1, n+k]. Let u, $\overline{u} \in \Gamma$ with $(\varphi, \psi)(u)[-|w|+1, n+k] = (x, y)[-|w|+1, n+k]$ and $(\varphi, \psi)(\overline{u})[-|w|+1, n+k] = (\overline{x}, \overline{y})[-|w|+1, n+k]$. Then u[1, n] = $\overline{u}$[1, n], because w is a decoder block for φ with anticipation k. Thus, $\psi(u)[i, n-i] = \psi(\overline{u})[i, n-i]$, where i is a coding length for ψ. Therefore, we get y[i, n-i] = $\overline{y}$[i, n-i]. This shows that any S-block wb with length of b greater than i is a decoder block for φ. Apply Lemma 2.3. to see that φ is r.c.a.e. and 1-1

a.e. Same argument for ψ.   ☐

For two synchronized systems S and T a r.c.a.e., 1-1 a.e. factor map induces a homeomorphism with bounded anticipation between the doubly transitive points of S and the doubly transitive points of T, by Lemma 2.3. Thus, by Theorem 3.1., having a common extension by a synchronized system with r.c.a.e., 1-1 a.e. factor maps defines an equivalence relation on the set of synchronized systems, which we denote by $\approx$.

COROLLARY 3.2. *Let* S *and* T *be synchronized systems where* Σ_S *and* Σ_T *are conjugate. Then* $S \approx T$.

4. Some non complete invariants

If two SFT´s admit a common r.c., 1-1 a.e extension by a SFT, then their topological entropies coincide. We show first that the topological entropy is not an invariant for the equivalence relation $\approx$ on the set of synchronized systems:

Consider the subshift $S = X$ defined in [P, Example 3.2.]. By [FF, Theorem 2.16.], the defining labeled graph is actually the Fischer cover. Replacing M by another minimal shift yields another synchronized system, say T, with $\Sigma_S = \Sigma_T$, thus by Corollary 3.2., $S \approx T$. By choosing the second minimal shift with a topological entropy that is different from that of M, we get that S and T have different topological entropies.

We are now going to define a new invariant which shall play the role of toplogical entropy for the equivalence relation $\approx$. Let $h(\mu)$ be the entropy of the measure μ.

DEFINITION 4.1. *Let* S *be a synchronized system. Let* $0 \le R_S \le 1$ *be defined by* $-\log R_S = \sup\{h(\mu) \mid \mu$ *is an ergodic measure on* Σ_S *with full support*}.

REMARK 4.2. a) By [G], $-\log R_S = \sup\{h(\mu) \mid \mu$ is a measure on $\Sigma_S\}$.
b) The graph X_0^+ has bounded out-degree, because ρ_0^+ is 1-step r.c.. Thus the graph X_0^+ has a positive radius of convergence which equals R_S, since X_0^+ is irreducible, [G]. Thus $0 < R_S < 1$, if S does not consist of a single orbit.
c) R_S coincides with the radius of synchronization ρ [BH].
d) R_S is a conjugacy invariant for S, by [FF, Theorem 2.12.].

There is a measure μ on Σ_S with $h(\mu) = -\log R_S$ iff the defining graph, here X_0^+, is positive recurrent. In this case such a measure is unique and it is a Markov measure with full support, which can be calculated in an analogue way as the Parry measure for SFT´s, [G].

DEFINITION 4.3. *A synchronized system* S *is positiv recurrent if there is a measure* μ *on* Σ_S *with* $h(\mu) = -\log R_S$.

THEOREM 4.4. *Let* S *and* T *be synchronized and* $\varphi : S \to T$ *be a r.c.a.e., 1-1 a.e. factor map. Then* $R_S = R_T$ *and* S *is positiv recurrent iff* T *is positiv recurrent.*

PROOF Any synchronizing S-block is a decoder block for ρ_S, [FF], thus ρ_S induces a homeomorphism between the doubly transitive points of Σ_S and of S. Thus, for any ergodic measure μ on Σ_S with full support (Σ_S, μ) is finitary isomorphic to $(\Sigma_T, (\rho_S \circ \varphi \circ (\rho_T)^{-1})\mu)$, by Remark 2.2. This proves the theorem. □

REMARK 4.5. Using decoder blocks it is not hard to see that $per(S) := gcd\{n \mid$ there is $z \in \Sigma_S$ with $\sigma^n z = z\}$ is invariant for $\approx$. $per(S)$ can be different from $gcd\{n \mid$ there is $x \in S$ with $\sigma^n x = x\}$.

The following is a generalization of [BMT, Proposition 8.2.] to synchronized systems.

COROLLARY 4.6. *Let* S *and* T *be positive recurrent, synchronizing. Let* $\mu_S = \rho_S(\mu)$ *and* $\mu_T = \rho_T(\nu)$, *where* μ *and* ν *are the measures of maximal entropy on* Σ_S *resp. on* Σ_T. *Then* S $\approx$ T *iff* (S, μ_S) *is finitary regular isomorphic to* (T, μ_T).

PROOF If S $\approx$ T, then there is a synchronizing system R and r.c.a.e, 1-1 a.e factor maps $\varphi : R \to S$ and $\psi : R \to T$. By Theorem 4.4. $-\log R_S = -\log R_R = -\log R_T$ and R is positive recurrent. Let $\mu_R = \rho_R(\lambda)$, where λ is the measure of maximal entropy on Σ_R. By Lemma 2.3. and Remark 2.2., it follows that (R,μ_R) is finitary regular isomorphic to (S, μ_S) and (R, μ_R) is finitary regular isomorphic to (T, μ_T), because μ_R, μ_S and μ_T are ergodic measures with full support. □

The next corollary shows that the equivalence relation $\approx$, when restricted to sofic systems, coincides with almost toplolgical conjugacy.

COROLLARY 4.7. *Let* S *and* T *be sofic systems with* S $\approx$ T. *Then there is a SFT* R *and factor maps* $\varphi : R \to S$, $\psi : R \to T$, *both r.c. and 1-1 a.e.*

PROOF Sofic systems are synchronized, [BH], and $-\log R_S$ equals the topological entropy. Thus S and T have the same topological entropy and μ_S and μ_T are the measures of maximal entropy on S and T. By Corollary 4.6. (S, μ_S) is finitary regular isomorphic to (T, μ_T). Because $\rho_S : \Sigma_S \to S$ is a regular isorphism, we get that the Fischer cover of S and of T are regular isomorphic with respect to their measures of maximal entropy. Apply [BT, Theorem 1] to see that the Fischer covers of S and T admit a common r.c., 1-1 a.e. extension by a SFT. This proves the corollary, because the maps ρ_S and ρ_T are r.c., 1-1 a.e. □

QUESTIONS.
1) For sofic shifts the period, the topological entropy and an ideal class form a complete invariant for $\approx$, by Corollary 4.7. and [BMT, Proposition 8.2.]. Is there an analogon to the ideal class for synchronized systems?

2) Is Theorem 1.1. true with "finitary regular" replaced by "regular"?
3) Can the construction in the proof of Theorem 1.1. be used to generalize [BT, Theorem 1] to synchronized systems, i.e., can one find additionally a measure $\mu_R \in M(R)$ such that $\varphi\mu_R = \mu_S$ and $\psi\mu_R = \mu_S$?

REFERENCES

[BH]. F. Blanchard & G. Hansel, Systèmes codés, Theor. Com. Sci. **44** (1986), 17-49.

[BMT]. M. Boyle, B. Marcus & P. Trow, Resolving maps and the dimension group for shifts of finite type, Mem. Amer. Math. Soc. **377** (1987).

[BT]. M. Boyle & S. Tuncel, Regular isomorphism of Markov chains is almost topological, Ergod. Th. & Dynam. Sys. **10** (1990), 89-100.

[FF]. D. Fiebig & U. Fiebig, Covers for coded systems, Preprint.

[G]. B. M. Gurevic, Shift entropy and Markov measures in the path space of denumerable graph, Dokl. Akad. Nauk SSSR **192** (1970). Soviet Math. Dokl. **11** (1970), 744-747.

[P]. K. Petersen, Chains, entropy, coding, Ergod. Th. & Dynam. Sys. **6** (1986), 415-448

Institut für Angewandte Mathematik, Universität Heidelberg, 6900 Heidelberg, Germany

Contemporary Mathematics
Volume **135**, 1992

Covers for Coded Systems

DORIS FIEBIG AND ULF-RAINER FIEBIG

ABSTRACT. A coded system is a subshift (with finite alphabet) which is the closure of an uniformly continuous image of a countable irreducible Markov chain. Such a Markov chain together with the uniformly continuous map is called a "cover". The coded systems are a natural generalization of transitive sofic systems. Our aim is to generalize parts of the sofic theory to certain subclasses of coded systems, the so-called synchronized and half-synchronized systems. We give alternate definitions for (half-)synchronized systems which show that any subshifts with at most countably many follower sets (of left infinite rays) is synchronized.

We define a canonical Fischer cover for (half-)synchronized systems and discuss its minimality properties in the space of all covers. We characterize the Fischer cover in various ways and develop an analogue of the sofic almost Markov theory. A synchronized system will be "almost Markov" if its Fischer cover is uniformly left closing. Then its Fischer cover intercepts any other cover by a uniformly continuous factor map.

We characterize those Fischer covers which continuously intercept any other cover.

We show that a Fischer cover has no proper uniformly continuous factors and characterize those Fischer covers which have no proper continuous factors.

0. Background and Definitions

We give an informal introduction followed by precise definitions.

Sofic systems are the continuous images of shifts of finite type ([W]). Some fundamental properties of a sofic system S are: (i) there is a finite directed graph with labeled edges such that reading off the labels along the bi-infinite paths in the graph gives the points of S (and the graph can be chosen to be irreducible if the sofic system is transitive) [F1, Theorem 1], (ii) each sofic system has a "magic" or "synchronizing" word m, i.e. whenever u,v are words such that um and mv are allowed in S, then also umv is allowed, (iii) sofic systems are intrinsically characterized as those subshifts which have a finite number of follower sets [W].

We shall consider only transitive sofic systems, resp. subshifts.

1991 Mathematics Subject Classification. Primary 54H20. Secondary 68Q45,68Q48.

This paper is in final form and no version of it will be submitted for publication elsewhere.

The above properties suggest the following three ways to generalize transitive sofic systems. One could consider transitive subshifts (with finite alphabet A) which
(1) are given by a labeled directed graph which is irreducible but may have countably many vertices ("*coded systems*", [BH], where "given" means that the subshift is the closure in A^Z of all the bi-infinite sequences of labels occurring along bi-infinite paths in the graph), or
(2) have a synchronizing word ("*synchronized systems*", [BH]), or
(3) have at most countably many follower sets (of left infinite rays) [P].

We shall see that (2) implies (1) ([BH]), and prove that (3) implies (2).

In the following any countable, irreducible, labeled graph that represents a coded system S in the sense of (1) is called a "*cover*" for S. A finite cover is a cover whose graph has a finite number of vertices.

Thus a transitive subshift is coded iff it has a cover, and it is sofic iff it has a finite cover.

As a first example consider the transitive subshift S which is the closure (in $\{0,1\}^Z$) of all bi-infinite concatenations of the words $0^n 1^n$, $n \in N$. This system is not sofic since it has infinitely many follower sets (a left infinite ray $1^{\infty} 0^n 1$, $n \in N$, can be followed by the word $1^k 0$ if and only if $k = n-1$). But S is coded, because attaching to one initial vertex, say α, for each $n \in N$ a loop path of length $2n$ whose edges are labeled $0^n 1^n$ (all paths disjoint except for α) defines a cover for S. Moreover, S has the special property that the number of follower sets of left infinite rays is countable (it could be uncountable). Thus, by our result that (3) implies (2), the system has to be synchronized. Which, of course, is straightforward in this example since 101 is a synchronizing word (there are many others). And it is the existence of a synchronizing word that will allow us to define a canonical cover for S, a "Fischer cover" (which is different from the one given above).

Thus our aim is to generalize parts of the theory for sofic systems to synchronized systems, and to discuss possible generalizations to the larger class of "*half-synchronized*" systems (introduced by W.Krieger, oral communication), which include the Dyke-system and the β-shifts. The half-synchronized systems will often serve as a landmark within the set of coded systems, showing the difficulties in extending the class of synchronized systems and keeping a satisfactory analogon of the sofic theory.

Our results deal with the existence and the canonicity of a "Fischer cover" for synchronized and half-synchronized systems (which we have to define yet), and its minimality properties in the set of all covers of the system.

We recall some results about (finite) covers for sofic systems.
Since a sofic system S has many different finite covers, one was looking for finite covers for S which are more closely related to S than others, or, which are minimal in some sense. The covers in [CK] had the property that the entropy of the

given sofic system was computable from the directed graph (discarding the labels). R.Fischer defined covers which had a minimal number of vertices among all covers [F1], [F2]. W.Krieger showed that these "Fischer covers" are in fact canonically associated to the sofic system [Kr1]. Following work dealt with the so-called almost finite type sofic systems, those systems which are factors of subshifts of finite type by a bi-closing factor map (for example [M],[N]). It has been shown that a system is almost finite type iff the right Fischer cover is left closing ([BKM],[Kr2],[N]). Then [BKM] investigated further the minimality properties of the Fischer cover. The (right) Fischer cover is shown to be minimal within the set of right closing covers (with respect to factor maps), which led to an alternate proof for its canonicity. They showed that a sofic system is almost finite type iff any other cover factors through the Fischer cover. Furthermore, a sofic system is AFT iff it has a cover that intercepts any other cover.

What is new when considering infinite labeled graphs instead of finite ones?

Closing properties play an important role in the study of sofic systems via labeled graphs ("almost finite type sofic systems", [M]). Now there are many different closing properties an infinite graph may have. For example, we have to distinguish between uniformly left closing, left closing, and "weakly left closing" (Definition 3.1). A labeling of a finite graph will always be weakly left closing, and left closing iff it is uniformly left closing.

Given a finite labeled graph, which defines a sofic system S, the map Φ which maps bi-infinite paths to bi-infinite sequences of labels, is onto. For infinite graphs the image of Φ is usually properly contained in its closure S (which is then a coded system). Still, Φ may have a topologically small or large image in S. The latter case will lead to an alternate definition of synchronized systems.

Continuous factor maps between labeled graphs (or their associated Markov chains), are no longer automatically uniformly continuous (as in the setting of finite graphs), since the associated Markov chains are no longer compact. Thus at least two types of factor maps will be studied: continuous and uniformly continuous factor maps.

The paper is organized as follows.

The remaining part of §0 provides the necessary fundamental definitions (see below), in particular, we define the Fischer cover for (half-)synchronized systems. In §1 alternate definitions for (half-)synchronized systems are given. They imply that (3) is contained in (2). We show that any coded system has a bi-resolving cover (which contrasts the fact that not any sofic system has a finite bi-closing cover). In §2 we give proofs for the canonicity of the Krieger graph and the Fischer cover, via "saturated" covers. Various characterizations of the Fischer cover yield an easy way to construct examples of Fischer covers (for synchronized systems), also with certain pre-assigned properties. A first minimality result is that the Fischer cover intercepts any other uniformly right closing cover with topologically large image, which leads to an alternate canonicity proof. In §3 we start the discussion of synchronized systems whose Fischer covers have additional closing properties. In particular, §4 extends the almost finite type (also called almost Markov) theory from sofic to synchronized systems. Almost all results about covers valid in the sofic setting have an appropriate generalization to the almost Markov synchronized case. In §5 we consider a continuous analogue of the almost Markov theory. We show that the Fischer cover intercepts any other cover by a continuous (not necessarily uniformly continuous) factor map iff it is locally finite and left closing. In §6 we first consider general coded systems. We show that any cover (of any coded system) has a locally finite factor. We investigate "rigid" covers, and then apply these general results to show that a Fischer cover of a (half-)synchronized system has no proper continuous factors if and only if it is locally finite and "weakly left closing".

As special cases of our work we obtain and slightly extend all the mentioned sofic results (for example the Fischer cover of an AFT sofic system does not only intercept all finite but even all countable covers), by arguments that (often necessarily) differ from those for the sofic case (for example we give a category argument for the existence of synchronizing words, and a canonicity proof for the Fischer cover via "saturated" covers). Thus we hope that even for the purely sofic case there are some interesting new aspects in this work.

We like to thank W.Krieger for introducing us to the subject and many helpful discussions.

Now we provide the relevant definitions. We start with labeled graphs and their associated topological Markov shifts. Then coded and (half-)synchronized systems, the Krieger graph of an arbitrary subshift, the (right) Fischer cover of a half-synchronized system, and factor maps between labeled graphs are defined.

DEFINITIONS 0.1. *A (directed) graph* Γ *is a pair* (V_Γ, E_Γ) *where* V_Γ *is an arbitrary set of vertices and* E_Γ *an arbitrary set of directed edges, such that each edge* $e \in E_\Gamma$ *has its initial vertex* $i(e)$ *and its terminal vertex* $t(e)$ *in* V_Γ. *We allow multiple edges between two vertices.We always assume* $V_\Gamma = \{i(e) \mid e \in E_\Gamma\}$ $= \{t(e) \mid e \in E_\Gamma\}$.
A sequence of edges $e_1...e_n$ *is called a path (of length* n) *in* Γ *if* $t(e_k) = i(e_{k+1})$ *for* $1 \le k \le n-1$. *A graph* Γ *is irreducible if for each pair of vertices* α_1, α_2 *there is a finite path leading from* α_1 *to* α_2.
A graph has finite out-degree (resp. finite in-degree) at a vertex α *if the set of edges* e *with* $\alpha = i(e)$ *(resp. with* $\alpha = t(e)$) *is finite. We call a graph locally finite if it has finite out-degree and finite in-degree at any vertex.* □

DEFINITION 0.2. *A labeled graph is a pair* (Γ, Φ), *where* Γ *is a graph together with a labeling function* $\Phi : E_\Gamma \to A$, *where* A *is a finite set of symbols (the "alphabet"). We shall assume that for each symbol* $s \in A$ *and any two vertices* α_1, α_2 *there is at most one edge from* α_1 *to* α_2 *labeled* s. *Thus* E_Γ *is countable iff* V_Γ *is countable. The* Φ*-label of a path* $e_1...e_n$ *i n* Γ *is the block* $\Phi(e_1...e_n) = \Phi(e_1)...\Phi(e_n) \in A^n$. *Analogously the label of a right* *(resp. left) infinite path is defined.*

The follower set of a vertex α *of* Γ *is* $\omega_+(\alpha) = \{ (\Phi(e_i))_{i \geq 1} \mid (e_i)_{i \geq 1}$ is a path in Γ with $i(e_1) = \alpha\}$. *Analogously let* $\omega_-(\alpha) = \{\Phi$-labels of all left-infinite paths terminating at $\alpha\}$ *and also* $F_+(\alpha) = \{\Phi$-labels of all finite paths starting at $\alpha\}$, $F_-(\alpha) = \{\Phi$-labels of all finite paths terminating at $\alpha\}$. *If* $\omega_+(\alpha) \neq \omega_+(\beta)$ *for any pair* $\alpha, \beta \in V_\Gamma$, $\alpha \neq \beta$, *then* (Γ, Φ) *is a labeled graph with distinct follower sets.*

The labeling Φ *is called right closing if there are no distinct right-infinite paths which are equally labeled and have the same initial vertex. We call* Φ *uniformly right closing if there is some* $n \in N$ *such that for any two paths* $x_1 \ldots x_n$ *and* $y_1 \ldots y_n$ *in* Γ *with the same initial vertex and the same* Φ-label *it holds* $x_1 = y_1$. *This will also be called* n-*step right closing. If the above holds for* $n = 1$ *then we call* Φ *right resolving.*

Analogously, (uniformly) left closing and left resolving are defined. $\square$

Note that for a labeled graph with finite out-degree $\omega_+(\alpha)$ is determined by $F_+(\alpha)$.

DEFINITIONS 0.3. *Given a directed graph* Γ, *let* Σ_Γ *be the set of bi-infinite paths in* Γ, *considered as a closed shift invariant subset of the space* $(E_\Gamma)^Z$ *endowed with the product topology (discrete topology on* E_Γ), *and let* σ_Γ *be the left shift. We call* $(\Sigma_\Gamma, \sigma_\Gamma)$ *the associated topological Markov chain. For a path* $e_1 \ldots e_n$ *in* Γ, $N \in Z$, *let* $_N[e_1 \ldots e_n]$ *denote the cylinder set* $\{(x_i)_{i \in Z} \in \Sigma_\Gamma \mid x_{N+k} = e_{1+k}, 0 \leq k \leq n - 1\}$. *We will consider* Σ_Γ *as being endowed with the complete metric given by* $d(x, y) = 0$ *if* $x = y$ *and for* $x \neq y$, $d(x, y) = 1/(k+1)$ *where* k *is the least non-negative integer such that* $x_k \neq y_k$ *or* $x_{-k} \neq y_{-k}$.

For a labeled graph (Γ, Φ) *the labeling* Φ *induces a uniformly continuous, shift invariant map* $(x_i)_{i \in Z} \rightarrow (\Phi(x_i))_{i \in Z}$ *from* Σ_Γ *to* A^Z, *which we denote also by* Φ. *Let* $\Phi(\Gamma) = \Phi(\Sigma_\Gamma) \subset A^Z$, *which we call the image of* (Γ, Φ). $\square$

Note that Σ_Γ is compact iff Γ is a finite graph, and that Σ_Γ is locally compact (i.e. any cylinder set is a compact) iff Γ is locally finite.

Any topological Markov chain with a uniformly continuous shift invariant map into A^Z is uniformly continuous conjugate to a pair (Σ_Γ, Φ) associated to a labeled graph.

DEFINITIONS 0.4. *Let* A *be a finite alphabet. A subshift is a closed, shift invariant subset of* A^Z. *An element* $m = m_1 \ldots m_n \in A^n$, $n \in N$, *is called a word of the subshift* $S \subset A^Z$ *(or an allowed word or an* S-word*) if there is an* $x \in S$ *with* $x_i = m_i$ $1 \leq i \leq n$. *A subshift* S *is transitive if for any pair of* S-words u, v *there is an* S-word w *such that the concatenation* uwv *is an* S-word.

Let $P_{[0,\infty)}$ *be the projection which maps* $(x_i)_{i \in Z} \in A^Z$ *onto its future* $(x_i)_{i \in Z+}$ $\in A^{Z+}$. *For a subshift* $S \subset A^Z$ *let* $S^+ = P_{[0,\infty)}(S)$. *Analogously,* $P_{(-\infty,0)}$ *projects* $(x_i)_{i \in Z}$ *onto* $(x_i)_{i < 0}$ *and* $S^- = P_{(-\infty,0)}(S)$. *Elements* $x_+ = (x_i)_{i \in Z+} \in S^+$ *(resp.* $x_- = (x_i)_{i<0} \in S^-$) *are called right (resp. left) infinite* S-rays.

Given $x_+ \in S^+$ *and an* S-word $u = s_0 \ldots s_n$, *then the concatenation* ux_+ *will be considered as an element of* A^{Z+}, *which may or may not be an* S-ray.

Analogously x_-u *is defined. For* $x_+ \in S^+$, $x_- \in S^-$, *let* x_-x_+ *denote the element* $(y_i)_{i \in \mathbb{Z}} \in A^{\mathbb{Z}}$ *with* $y_i = x_i$ *for all* i, *which may or may not be a point of* S.

For an left infinite S-ray x_- *its follower set is* $\omega_+(x_-) = \{x_+ \in S^+ \mid x_-x_+$ *is a* point in S } *and its block follower set is* $F_+(x_-) = \{S\text{-blocks } u \mid x_-u \in S^-\}$.

Analogously, we define predecessor sets $\omega_-(x_+) = \{x_- \in S^- \mid x_-x_+ \in S\}$, *and block predecessor sets* $F_-(x_+) = \{S\text{-blocks } u \mid ux_+ \in S^+\}$ *for* $x_+ \in S^+$.

We say that S *has finitely many (resp. countably many) follower sets if the map* $x_- \to \omega_+(x_-)$ *from* S^- *to the power set of* S^+ *has finite (resp. countable) range. By compactness this holds iff the map* $x_- \to F_+(x_-)$ *has finite (resp. countable) range.*

There also is a notion of follower sets for S-blocks. But if not specified otherwise, the term "follower set" will always mean the follower set of a left infinite S-ray.

For an S-block m *its follower set is* $\omega_+(m) = \{x_+ \in S^+ \mid mx_+ \in S^+\}$, *and its block follower set is* $F_+(m) = \{S\text{-blocks } u \mid mu \text{ is an S-block}\}$.

Analogously, predecessor sets $\omega_-(m) = \{x_- \in S^- \mid x_-m \in S^-\}$ *and block predecessor sets* $F_-(m) = \{S\text{-blocks } u \mid um \text{ is an S-block}\}$ *are defined.*   □

Note that the number of different follower sets (of rays) may be uncountable, whereas the number of different follower sets of S-blocks is necessarily finite or countable.

DEFINITION 0.5. *A generating (labeled) graph for a subshift* S *of* $A^{\mathbb{Z}}$ *is any labeled graph* (Γ, Φ) *with* $S = \overline{\Phi(\Gamma)}$, *where* $\overline{\Phi(\Gamma)}$ *denotes the closure of* $\Phi(\Gamma)$ *in* $A^{\mathbb{Z}}$. *If* (Γ, Φ) *is countable and irreducible, then we call* (Γ, Φ) *a cover for* S.

DEFINITION 0.6. *A subshift which has a cover (i.e. it admits some irreducible countable generating graph) is called a coded system* ([BH]).

A coded system is transitive and has a dense set of periodic points.

The irreducibility condition is the important feature in Definition 0.6 whereas the countability condition may be omitted. For that assume that S is a subshift which admits an irreducible, but not necessarily countable, generating graph (Γ, Φ). Then choose any vertex α of Γ and a countable collection of loops from α to α such that any finite S-block occurs as a label block along one of these labeled loops. This defines a cover for S, i.e. S is coded.

And the irreducibility assumption is essential, since any subshift has a countable generating graph, which is easily seen starting with a countable dense subset $D \subset S$.

REMARK 0.7. We want to point out that, given a cover (Γ, Φ) for a coded system $S = \overline{\Phi(\Gamma)}$, then $\Phi(\Gamma)$ is either a topologically small subset of S ("of first category") or topologically large ("residual" which means that $S - \Phi(\Gamma)$ is of first category in S).

If (Γ, Φ) is a finite cover and thus S is irreducible sofic, then $S = \Phi(\Gamma)$, i.e. the covering map is surjective. Now consider a general coded system S. Let (Γ, Φ) be a countable generating graph (Γ, Φ) with $S = \overline{\Phi(\Gamma)}$ (here we do not need irreducibility).

Then the associated Markov chain Σ_Γ is a separable (here we use countability) topologically complete space (see Definition 0.3). Thus $\Phi(\Gamma) = \Phi(\Sigma_\Gamma)$ is an analytic subset of S [Ku, §38]. It follows that $\Phi(\Gamma)$ has the Baire-property , i.e. $\Phi(\Gamma) = O \Delta P$ with O open and P of first category, and is contained in the completion of the Borel σ-algebra of S with respect to any probability measure. Thus, by shift invariance,

$$\Phi(\Gamma) = \bigcup_{n \in Z} S^n (O \Delta P) \supset \bigcup_{n \in Z} S^n (O) - \bigcup_{n \in Z} S^n (P).$$

If $\Phi(\Gamma)$ is not of first category, then O is non-empty, thus the first union is a non empty dense (by topological transitivity) open set and the second is of first category, thus $\Phi(\Gamma)$ is residual.

One can give examples of countable labeled graphs (even covers) where $\Phi(\Gamma)$ is not a Borel set. But often $\Phi(\Gamma) = \bigcup_{e \in E_\Gamma} \Phi([e]_0)$ is a Borel set. For example, if Φ is biclosing, then $\Phi([e]_0)$ is the injective continuous image of the Borel set $[e]_0$, thus again a Borel set [Ku, p.487]. Or, if (Γ, Φ) is locally finite, then each $[e]_0$ is a compact subset of Σ_Γ, thus the sets $\Phi([e]_0)$ are closed and $\Phi(\Gamma)$ is a F_σ-set.

Each sofic system has a "magic" or "synchronizing" word. This property is used to single out a special subclass of coded systems, the synchronized systems. The example given at the beginning is synchronized but not sofic. For more see Examples 5.8.

DEFINITION 0.8. *A transitive subshift* S *is synchronized if there is a block* m *of* S *such that for any pair of blocks* $a \in F_-(m)$ *and* $b \in F_+(m)$ *also* amb *is a block of* S. *Any such block* m *is called a synchronizing block for* S.

DEFINITION 0.9. *A transitive subshift* S *is half-synchronized iff there is a block* m *of* S *and a left transitive point* $x \in S$ *such that* $x[-|m|+1,0] = m$ *and* $\omega_+(x(-\infty,0]) = \omega_+(m)$. *Then* m *is called a half-synchronizing block of* S.

Any synchronized system is half-synchronized since any synchronizing block is a half-synchronizing block. If S is synchronized then S^{-1} is synchronized, but there are half-synchronized systems S such that S^{-1} is not half-synchronized.

That half-synchronized systems are coded will follow from Definition 0.14 which yields a special cover.

EXAMPLES 0.10. a) A well known example for a non-synchronized coded system is the *Dyke-system* S ([Kr3],[BH]). Its alphabet A consists of four brackets, say [,], (, and). A point $(x_i)_{i \in Z} \in A^Z$ is a point in S iff any finite subblock $x_i...x_k$, $j \le k$, obeys the standard bracket rules.

For any block b of S, there are blocks $a \in F_-(b)$, $c \in F_+(b)$ such that abc does not obey the standard bracket rules, thus S is not synchronized. But still the Dyke-system is half-synchronized, in fact any block happens to be half-synchronizing.

b) Any β-shift S is a coded system ([B1, p.136]) and even half-synchronized (it is not hard to see that any block is half-synchronizing). It is synchronized iff there is a block m of S which does not occur in the defining sequence ([B-M]).   ☐

As in the sofic case we shall define the "Fischer-cover" of a (half-)synchronized system as a special component of the Krieger graph.

For that we recall the construction of the Krieger graph, which can be carried out for any subshift, but usually leads to non-irreducible and often even uncountable graphs.

DEFINITION 0.11. (Essentially [Kr1]) *Let* S *be a subshift. Consider the collection of all follower sets* $\omega_+(x_-)$ *of left infinite S-rays* x_- *as the set of vertices of a graph* X^+. *There is an edge from* α_1 *to* α_2 *labeled* s *iff there is an S-ray* x_- *such that* x_-s *is an S-ray and* $\alpha_1 = \omega_+(x_-)$, $\alpha_2 = \omega_+(x_-s)$. *This labeled graph* (X^+, ρ^+) *is called the Krieger graph for* S.

It is easily seen that the Krieger graph is in fact a right resolving generating graph with $S = \rho^+(X^+)$. We will show in §2 that it is canonically associated to the given subshift (for the sofic case this is [Kr2,Theorem 2.2]).

Let $\alpha = \omega_+(x_-)$ be a vertex of X^+ and let $\omega_+(\alpha)$ be the follower set of the vertex α in (X^+,ρ^+). Then $\omega_+(x_-) = \omega_+(\alpha)$ as subsets of S^+. Thus (X^+,ρ^+) has distinct follower sets by construction. Moreover, for any $\alpha = \omega_+(x_-) \in V_{X^+}$ there is actually a left infinite path in X^+, terminating at α and labeled by x_-, i.e. $x_- \in \omega_-(\alpha)$. Thus, if α is connected to a vertex β by some finite path labeled u, then $\omega_+(\beta) = \omega_+(x_-u)$.

In general (X^+,ρ^+) is neither countable nor irreducible. An *irreducible component* (Γ,Φ) of (X^+,ρ^+) is a maximal sub-graph Γ with the property that for any two vertices $\alpha_1, \alpha_2 \in V_\Gamma$ there is a path from α_1 to α_2, and a path from α_2 to α_1. The labeling function Φ is then the restriction of ρ^+ to the edges of Γ. Note that irreducible components are countable labeled graphs. An irreducible (Γ,Φ) component is called *forwardly* (resp. *backwardly*) *closed* if $e \in E_{X^+}$, $i(e) \in V_\Gamma$ implies $t(e) \in V_\Gamma$ (resp. if $t(e) \in V_\Gamma$ implies $i(e) \in V_\Gamma$).

DEFINITION 0.12. *Let* S *be a synchronized (resp. half-synchronized) system and* m *a synchronizing (resp. half-synchronizing) block of* S. *Let* (X_0^+,ρ_0^+) *be the irreducible component of the Krieger graph* (X^+,ρ^+) *containing the vertex* $\omega_+(m)$. *Then the definition of* (X_0^+,ρ_0^+) *is independent of the particular choice of* m *(see below), it is forwardly closed and* $\rho_0^+(X_0^+) \subset S$ *is dense. Thus* (X_0^+,ρ_0^+) *is a cover for* S *, which is called the (right) Fischer cover of* S.   ☐

Since (X_0^+,ρ_0^+) is a cover, any half-synchronized system is coded.

We show that the definition of (X_0^+,ρ_0^+) is independent of the particular choice of m. Let m, m' be two half-synchronizing words of a half-synchronized system S. There are $x_-, x_-' \in S^-$ with terminal segment m and m', respectively, such that $\omega_+(x_-) = \omega_+(m)$ and $\omega_+(x_-') = \omega_+(m')$. So we may view $\omega_+(m)$, $\omega_+(m')$ as vertices

α, α' of the Krieger graph. We show that α and α' are in the same irreducible component. There is a word u such that mum' is the terminal segment of x'_{-}. Since $\omega_{+}(\alpha) = \omega_{+}(m)$, we have a finite path in X^{+} with initial vertex α and labeled um'. Let β denote its terminal vertex. We show that $\omega_{+}(\beta) = \omega_{+}(x'_{-})$ which implies $\beta = \alpha'$. One inclusion holds because $\omega_{+}(\beta) = \omega_{+}(x_{-}um') \subset \omega_{+}(m') = \omega_{+}(x'_{-})$, and the other because for any $x_{+} \in \omega_{+}(x'_{-})$ we have $x_{+} \in \omega_{+}(mum')$, thus $um'x_{+} \in \omega_{+}(m)$, thus $um'x_{+} \in \omega_{+}(x_{-})$, thus $x_{+} \in \omega_{+}(x_{-}um') = \omega_{+}(\beta)$.

We add some remarks on the structure of the right Fischer cover.

The right Fischer cover is right resolving and has distinct follower sets.

For any vertex α of the right Fischer cover of a synchronized (resp. half-synchronized) system there is some synchronizing (resp. half-synchronizing) word m $\in F_{-}(\alpha)$ such that $\omega_{+}(m) = \omega_{+}(\alpha)$. On the other hand, if m is a synchronizing (resp. half-synchronizing) word then there is a unique vertex α of X_0^+ with the properties $\omega_{+}(\alpha) = \omega_{+}(m)$ and $m \in F_{-}(\alpha)$.

If for some S-word m there is a unique vertex α of X_0^+ with the property $m \in F_{-}(\alpha)$, then m is called a *magic word for the right Fischer cover*. Any magic word is obviously synchronizing. Thus only Fischer covers for synchronized systems can have magic words. On the other hand, if S is synchronized, then any synchronizing S-word is a magic word for the right Fischer cover of S.

As in the sofic case, a synchronized system S has also a *left Fischer cover* $(\dot{X_0}, \dot{\rho_0})$. This uses predecessor sets instead of follower sets. The left Fischer cover for S can be obtained in constructing the right Fischer cover for S^{-1} and then reversing the directions of all the edges in $X_0^+(S^{-1})$. A half-synchronized system may not have a left Fischer cover since S^{-1} may not be half-synchronized.

Thus, whenever we use the term *Fischer cover* for a (half-)synchronized system we mean the *right Fischer cover*.

To investigate the special properties of the Fischer cover in the set of all covers, we shall consider different types of factor maps between covers or generating graphs.

DEFINITION 0.13. *Let (Γ, Φ) and (Λ, Ψ) be generating graphs for a subshift S. We say that (Λ, Ψ) is a continuous factor of (Γ, Φ) if there is a continuous, shift commuting map π from Σ_{Γ} to Σ_{Λ}, with dense image in Σ_{Λ}, such that $\Phi(x) = \Psi\pi(x)$ for all $x \in \Sigma_{\Gamma}$. Any such π will be called a continuous factor map.*

If π can be chosen to be uniformly continuous (with respect to the metric given in Definition 0.3), then we call (Λ, Ψ) a uniformly continuous factor, or a block factor of (Γ, Φ). Then any number $N \in \mathbb{N}$ such that $(\pi(x))_0$ is determined by $x_{-N}, \dots, x_N$ for all $x \in \Sigma_{\Gamma}$ is called a coding length for π. If $(\pi(x))_0$ is determined by x_0 for all $x \in \Sigma_{\Gamma}$ then π is a one-block factor map.

If π is continuous (resp. uniformly continuous) and continuously (resp. uniformly continuously) invertible, then we call (Γ, Φ) and (Λ, Ψ) continuously conjugate (resp. block conjugate). Instead of "conjugate" we also use the term *"isomorphic"*.   □

If (Γ,Φ) is a finite labeled graph then any continuous factor (Λ,Ψ) is necessarily a finite labeled graph and π is surjective and uniformly continuous.

In a few cases we shall consider factor maps between covers for different coded systems (Definition 0.14). But in general, whenever two labeled graphs are considered as generating graphs for one given subshift, then the term *(uniformly) continuous factor* is to be understood as in Definition 0.13.

DEFINITION 0.14. *Let* A *and* B *be two finite sets of symbols (alphabets). Let* (Γ,Φ) *and* (Λ,Ψ) *be graphs labeled by symbols from* A *and* B, *respectively.*

We say that (Λ,Ψ) *is a continuous factor (resp.uniformly continuous factor, or a block factor) of* (Γ,Φ) *if there is a continuous (resp. uniformly continuous) shift commuting map* π *from* Σ_Γ *to* Σ_Λ, *with dense image in* Σ_Λ, *and a continuous, shift commuting map* φ *from* $S = \overline{\Phi(\Gamma)} \subset A^Z$ *onto* $T = \overline{\Psi(\Lambda)} \subset B^Z$ *such that* $\varphi\Phi(x) = \Psi\pi(x)$ *for all* $x \in \Sigma_\Gamma$.

If π *and* φ *are invertible with inverses satisfying the same conditions, then* (Γ,Φ) *and* (Λ,Ψ) *are continuously conjugate (resp. block conjugate).*   □

1. Alternate definitions for (half-)synchronized systems and bi-resolving covers for coded systems

We characterize synchronized systems as those transitive subshifts S which admit a countable (not necessarily irreducible) generating graph with a "topologically large" image in S (Theorem 1.1, see also Remark 0.7). The proof gives a new argument for the existence of synchronizing words, even in the sofic case. Our characterization shows that transitive subshifts with at most countably many follower sets are synchronized (Corollary 1.3). For half-synchronized systems we have a one-sided version of this characterization (Theorem 1.4).

For sofic systems the existence of a finite bi-closing cover means something special, the system is then "almost finite type" [M, Definition 4],[N, §6].

This is contrasted by the fact that any coded system admits a uniformly bi-closing (even bi-resolving) cover (Theorem 1.7).

Nevertheless, Theorem 1.1 indicates a way to get an analogue of the almost finite type theory, at least for synchronized systems. The clue is to consider covers which are not only uniformly bi-closing, but also have "topologically large" image. This will lead to a proper subclass of synchronized systems, studied in §3 and §4.

THEOREM 1.1. *The synchronized systems are those transitive subshifts* S *that admit a countable (not necessarily irreducible) generating graph* (Γ,Φ) *such that* $\Phi(\Gamma)$ *is a residual subset of* S. *(See also Remark 0.7).*

Proof. " $\Rightarrow$ ": Let (X_0^+,ρ_0^+) be the Fischer cover of a synchronized system S. Let m be a synchronizing block of S. Then $\rho_0^+(X_0^+)$ contains all points of S which see m infinitely often in the past. This is a dense G_δ-set, so $\rho_0^+(X_0^+) \subset S$ is residual.

" $\Leftarrow$ ": Let $\alpha_1,\alpha_2,...$ be an enumeration of the vertices of Γ. Let $A_j = \{ x \in \Sigma_\Gamma$

with $i(x_n) = \alpha_j$}. Not all the sets $\Phi(A_j)$ can be nowhere dense in S. Thus there is an j' and a cylinder $M = _{-n}[m_{-n},...,m_0,...,m_n]$ of S such that $\Phi(A_{j'})$ is dense in M. Let $m = m_{-n},...,m_0,...,m_n$. Since $\Phi(A_{j'})$ is dense in M, we have that for any u $\in F_-(m)$ there is some finite path terminating at $\alpha_{j'}$ and labeled $um_{-n}...m_{-1}$, and for any $v \in F_+(m)$ there is some finite path starting at $\alpha_{j'}$ and labeled $m_0...m_n v$. Concatenating these two paths shows that umv is a word of S. Thus m is a synchronizing block. □

REMARK 1.2 For later use we observe that if Φ is right resolving, and α denotes the terminal vertex of the unique path starting at $\alpha_{j'}$ and labeled $m_0...m_n$, then $F_+(\alpha) = F_+(m)$ and $um \in F_-(\alpha)$ whenever um is a block in S.

A transitive subshift is sofic iff it has finitely many follower sets [W]. In [P] it has been suggested to consider transitive subshifts with at most countably many follower sets. We have the following

COROLLARY 1.3. *A transitive subshift with at most countably many follower sets is a synchronized system. The converse does not hold.*

Proof. If S has at most countably many follower sets, then the Krieger graph is countable. Since $S = \rho^+(X^+)$, the Krieger graph has residual image. Apply Theorem 1.1 (note that the Krieger graph is not necessarily irreducible).

We modify the Dyke system (see Example 0.10) such that it becomes synchronized, but still has uncountably many follower sets. Add a new symbol "∗" (which will be a synchronizing symbol) to the set of four brackets and let S be the subshift which consists of all bi-infinite sequences of these five symbols such that any finite subblock which does not contain a "∗" obeys the standard bracket rules.□

We recall that $P_{[0,\infty)}$ denotes the projection $(x_i)_{i \in Z} \to (x_i)_{i \in Z^+}$.

THEOREM 1.4. *A coded system S is half-synchronized iff it admits a cover (Γ,Φ) such that $P_{[0,\infty)}(\Phi(\Gamma))$ is residual in S^+.*

Proof. "⇒": We show that the image of the Fischer cover (X_0^+,ρ_0^+) is residual in the one-sided shift S^+. There is a word m and a vertex α of X_0^+ such that $m \in F_-(\alpha)$ and $\omega_+(\alpha) = \omega_+(m)$. Let $e_0...e_n$ be a finite path in X_0^+ with terminal vertex α such that $m = \rho_0^+(e_0)...\rho_0^+(e_n)$. Let $A = \{ x \in \Sigma_{X_0^+} \mid x[i,i+n] = e_0...e_n$ for some i $\geq 0\}$. Then $P_{[0,\infty)}(\Phi(A))$ is obviously dense in S^+. We are going to show that $P_{[0,\infty)}(\Phi(A))$ is open and thus residual. Consider a point $y_+ = P_{[0,\infty)}(\Phi(x))$, $x \in A$. Fix an $i \geq 0$ with $x[i,i+n+1] = e_0...e_n$. Let z_+ be any other ray in S^+ such that $z_+[0,i+n+1] = y_+[0,i+n+1]$. Then $z_+[i+n+2,\infty) \in \omega_+(m)$, thus there is a right infinite path w_+ in X_0^+, starting at α and labeled by $z_+[i+n+2,\infty)$. But then the concatenation of $x(-\infty,i+n+1]$ with w_+ gives a path x' in A with $z_+ = P_{[0,\infty)}(\Phi(x'))$. Thus $P_{[0,\infty)}(\Phi(x))$ contains an open neighborhood of y_+.

"⇐": Let $\alpha_1,\alpha_2,...$ be an enumeration of the vertices of Γ. Let $A_j = \{x \in \Sigma_\Gamma$ with $i(x_0) = \alpha_j\}$. Not all the sets $P_{[0,\infty)}(\Phi(A_j))$ can be nowhere dense in S^+. Thus there is a j' and a cylinder $M = {}_0[m_0,...,m_n]$ of S^+ such that $P_{[0,\infty)}(\Phi(A_{j_i}))$ is dense in M. Let m be the word $m_0,...,m_n$. Then $mu \in F_+(\alpha_{j_i})$ for all $u \in F_+(m)$. Choose any transitive, left infinite path in Γ which terminates at α_{j_i}, say labeled w_-. Then $u \in F_+(w_-m)$ whenever $u \in F_+(m)$, i.e. m is a half-synchronizing block. $\qquad\square$

REMARK. At the end of the above proof we had to use irreducibility and it can be shown that, in contrast to Theorem 1.1, this assumption cannot be omitted.

REMARK 1.5. The above proof shows that if (Γ,Φ) is a right resolving cover, then $P_{[0,\infty)}(\Phi(\Gamma))$ is residual in S^+ iff there is a vertex α and $m \in F_-(\alpha)$ such that $\omega_+(\alpha) = \omega_+(m)$ (which, for right resolving covers, is equivalent to $F_+(\alpha) = F_+(m)$). Note that any such m is a half-synchronizing word.

Requiring additional properties often forces a half-synchronized system to be already synchronized. We give a first example. Related results are Theorems 3.5, 4.6.

THEOREM 1.6. *A half-synchronized system* S, *which admits a cover* (Γ,Φ) *such that* $P_{[0,\infty)}(\Phi(\Gamma)) \subset S^+$ *and* $P_{(-\infty,0)}(\Phi(\Gamma)) \subset S^-$ *are residual, is synchronized.*

Proof. If $P_{[0,\infty)}(\Phi(\Gamma)) \subset S^+$ is residual, then, by the proof of Theorem 1.4, there is a vertex α and a block $m \in F_+(\alpha)$, such that $mu \in F_+(\alpha)$ whenever $u \in F_+(m)$. Analogously, if $P_{(-\infty,0)}(\Phi(\Gamma))$ is residual in S^-, there is a vertex β and a block $m' \in F_-(\beta)$ such that $vm' \in F_-(\beta)$ whenever $v \in F_-(m')$. Let m'' be the label of some finite path from β to α. Then $m'm''m$ is a synchronizing block for S. $\qquad\square$

In the sofic setting closing properties of finite covers play an important role. How about coded systems which admit a uniformly bi-closing (or even bi-resolving) cover?

THEOREM 1.7. *Any coded system* S *admits a bi-resolving cover.*

Proof. We shall construct a bi-resolving cover such that for any vertex α there are at most two edges terminating at α and at most two edges starting at α.

Let (Γ,Φ) be some cover for S. Let α be some vertex of Γ. Choose a shortest loop path from α to itself. Let c be its label. Then $c^\infty \in \omega_+(\alpha)$ (where c^∞ denotes infinite concatenation of c). Assume that $\omega_+(\alpha) = \{c^\infty\}$. Then, by irreducibility, any block of S would be a subblock of c^∞, and the above chosen loop path would be a bi-resolving cover for S. Thus we may assume that $\omega_+(\alpha)$ contains at least two elements. Analogously, we may assume that $|\omega_-(\alpha)| \geq 2$.

So there are $k \geq 1, n \geq 0$ and two paths $e_{-k} ... e_0 ... e_n$ and $f_{-k} ... f_0 ... f_n$ with $i(e_0) = i(f_0) = \alpha$, $\Phi(e_j) = \Phi(f_j)$ for $-k+1 \leq j \leq n-1$ and $\Phi(e_{-k}) \neq \Phi(f_{-k})$, $\Phi(e_n) \neq \Phi(f_n)$. Let $u = \Phi(e_n)u'\Phi(e_{-k})$ where u' is the label of some finite path from $t(e_n)$ to $i(e_{-k})$. Let $v = \Phi(f_n)v'\Phi(f_{-k})$ where v' is the label of some finite path from $t(f_n)$ to $i(f_{-k})$. Let w_i', $i \in \mathbb{Z}$ be an enumeration of the labels of all finite paths from α

to α. Let $w_i = \Phi(e_{-k+1})...\Phi(e_{-1})w_i'\Phi(e_0)...\Phi(e_{n-1})$, $i \in \mathbf{Z}$.

Denote the infinite concatenation $...uw_{-n}uw_{-n+1}...uw_{-1}uw_0uw_1...uw_n....$ by $(x_i)_{i \in \mathbf{Z}}$, say with x_0 equal to the first symbol of w_0. By construction the orbit of $(x_i)_{i \in \mathbf{Z}}$ in S is dense.

To define the bi-resolving cover (Λ, ψ) we start with a vertex set equal to $\mathbf{Z}$ and one edge from i to $i+1$, labeled x_i, $i \in \mathbf{Z}$. For each $n \in \mathbf{N}$ add a finite path labeled by v from the "terminal vertex of w_n" (which is length($w_0 uw_1...uw_n$) $\in$ $\mathbf{Z}$) to the "initial vertex of w_{-n}" (which is -length($w_{-n}...uw_{-1}u$)). Then (Λ, ψ) is irreducible and has dense image in S. The first symbols of u and v being different, as well as their last symbols, shows that ψ is bi-resolving. $\qquad\square$

REMARK 1.8. With a little more work the covering map can be made injective, then the first return codes are circular [B2]. In particular, this proves a result of W.Krieger (oral communication) that a coded system is the closure of an increasing sequence of irreducible shifts of finite type.

The cover constructed above is very particular and probably to far apart from the coded system it describes. For example, for synchronized S (provided they are not the orbit of one periodic point) the image of this cover is only of first category in S.

But if S is synchronized then it has covers with residual image (Theorem 1.1). And we shall see that in fact *not* any synchronized systems admits a uniformly bi-closing cover with residual image (see §3). Those who do will be called "almost Markov synchronized systems", and will be studied in §4.

We round off our first discussion of the impact of closing properties with two results.

If we weaken uniformly bi-closing to bi-closing, then the additional requirement of a residual image does not single out a proper subclass of synchronized systems.

THEOREM 1.9. *Each synchronized system* S *has a locally finite, bi-closing cover* (Γ, Φ) *with* $\Phi(\Gamma)$ *residual in* S *and* Φ *is injective and in particular 1-1 on an open non-empty set.*

Proof. Let $m = m_1...m_n$ be a synchronizing block for S. Assume first for simplicity that m has only trivial overlaps, i.e. $m_1...m_i \neq m_{n-i+1}...m_n$ for all $1 \leq i < n$. Fix two vertices α and β. Add a loopless path p_0 from α to β, labeled by m. Now, for each word u such that mum is a word of S, add a path from β to α with label u. Denote this path by p_u. So, in general, we added infinitely many disjoint (except for their initial and terminal vertices) paths from β to α. This cover already has all the properties claimed, except for not being locally finite at β (where it has infinite out-degree) and at α (where it has infinite in-degree). Now, split each p_u of length ≥ 3 into two paths p_u^1 and p_u^2 (i.e. p_u first traverses p_u^1 and then p_u^2) which have about the same length. Let γ_u denote the terminal vertex of p_u^1. Delete everything from (Γ, Φ) except for the path p_0 and the vertices γ_u. Add a right resolving labeled tree with root β such that for each path p_u^1 there is a path in the tree starting at β and equally labeled as p_u^1. For any vertex δ of this tree such that the path from β to δ is equally labeled as p_u^1 for some (unique) u, delete the edge e which terminates at δ and add an edge from i(e) to γ_u labeled by the previous label of e. Do a symmetric construction at α. Now it is not hard to see that the resulting irreducible component which contains α and β gives the desired cover. The essential clue is that for each length k there were only finitely many paths p_u of length $\leq k$.

Finally , if m has non-trivial overlaps just add some paths from β to suitable vertices of p_0. This does not change the claimed properties of the graph. $\qquad\square$

REMARK 1.10. Each half-synchronized system admits a uniformly bi-closing cover with residual image in the one-sided shift.

Since it is rather tedious we only indicate the construction. Given some right resolving cover (Γ, Φ) for some coded system S one constructs a bi-resolving cover (Λ, Ψ) such that there is a vertex α of Γ and a vertex β of Λ with $\omega_+(\alpha) = \omega_+(\beta)$. The existence of these vertices with the same follower set immediately implies $P_{[0,\infty)}(\Psi(\Lambda)) = P_{[0,\infty)}(\Phi(\Gamma))$ by irreducibility. Then the claim of this remark follows since the (right) Fischer cover of a half-synchronized system has residual image in the one-sided shift and is right resolving.

It can be shown that for strictly half-synchronized (i.e. not synchronized) systems any uniformly bi-closing cover with residual image in the one-sided shift has the property that for each vertex α there is a sequence of different vertices α_i, $i \in \mathbf{N}$, with $\omega_+(\alpha_i) = \omega_+(\alpha)$ for all i.

2. Canonical covers for synchronized and half-synchronized systems

A right resolving labeled graph is "saturated" if the follower set of each vertex is as large as it possibly can be with respect to the predecessor set (Definition 2.6). Saturatedness is related to the property that a cover has a topologically large image (Theorem 2.9). Saturated, right resolving labeled graphs with all follower sets distinct have good properties. For example, the identity map is their only continuous endomorphism (Lemma 2.10).

The Krieger graph (of any subshift) and the Fischer cover (of any half-synchronized systems) are saturated. This leads to a modified proof for the canonicity of the Krieger graph [Kr2, Theorem 2.2] and to a proof for the canonicity of the Fischer cover.

We give various characterizations of the Fischer cover of a (half-)synchronized system. Especially magic words (Theorem 2.16iv) give a handy tool to construct labeled graphs which are the Fischer cover for some synchronized system.

Finally, we note that the Fischer cover of a half-synchronized system is a block factor of any other uniformly right closing cover with topologically large image. This is well known in the sofic setting [BKM, Proposition 4], and leads to a second proof for the canonicity of the Fischer cover.

First we collect some basic facts about isomorphic covers and right resolving factors, which are well known for finite covers and carry directly over to the countable case.

DEFINITION 2.1. *Let* $C^+(S)$ *denote the set of all covers* (Γ,Φ) *for a coded system* S *which are right resolving and have distinct follower sets (see Definition 0.2).*

Two C^+-covers with a common follower set are shown to be essentially the same.

LEMMA 2.2. *Two covers* (Γ,Φ) *and* (Λ,Ψ) *in* $C^+(S)$ *are block isomorphic iff there are vertices* $\alpha \in V_\Gamma$ *and* $\beta \in V_\Lambda$ *such that* $\omega_+(\alpha) = \omega_+(\beta)$. *Furthermore, then there is a one-block isomorphism* $\pi: (\Gamma,\Phi) \to (\Lambda,\Psi)$ *with a one-block inverse* π^{-1}, *thus there is a graph isomorphism* $\Gamma \to \Lambda$ *that respects the edge-labeling.*

Proof. Let π be a block isomorphism $(\Gamma,\Phi) \to (\Lambda,\Psi)$. Let N be a coding length for π and π^{-1}. Let $x \in \Sigma_\Gamma$ and $\alpha = t(x_N)$, $\beta = t(\pi(x)_N)$. Consider $y \in \Sigma_\Gamma$ with $y(-\infty,N] = x(-\infty,N]$. Then $\pi(y)_0 = \pi(x)_0$, thus, since Ψ is right resolving, $\pi(y)_N = \pi(x)_N$. So $\Phi(y)[N+1,\infty) \in \omega_+(\beta)$. This shows $\omega_+(\alpha) \subset \omega_+(\beta)$. Using π^{-1} we obtain the reversed inclusion and thus equality.

Now assume that there are $\alpha \in V_\Gamma$ and $\beta \in V_\Lambda$ with $\omega_+(\alpha) = \omega_+(\beta)$. We define a one-block map π. Let $\alpha' \in V_\Gamma$, and let $e_1,...,e_n$ be a path from α to α'. By assumption, there is a path in Λ labeled $\Phi(e_1)...\Phi(e_n)$ with initial vertex β. It is unique since Ψ is right resolving. Let β' denote its terminal vertex. It is easy to see that $\omega_+(\alpha') = \omega_+(\beta')$. Thus, since (Γ,Φ) has distinct follower sets and (Λ,Ψ), too, this gives a well defined bijection $\varphi: V_\Gamma \to V_\Lambda$ such that $\omega_+(\alpha') = \omega_+(\varphi(\alpha'))$ for all $\alpha' \in V_\Gamma$. Let π map $e \in E_\Gamma$ to the unique $f \in E_\Lambda$ with $i(f)=\varphi(i(e))$ and $\Phi(f) = \Phi(e)$. Since Φ and Ψ are right resolving, π is an isomorphism of labeled graphs. $\square$

LEMMA 2.3. ([BKM, Proposition 1], [K]) *Any uniformly right closing cover is block isomorphic to a right resolving cover.*

LEMMA 2.4. ([F2, proof of Lemma 3]) *Any right resolving cover* (Γ,Φ) *has a one-block factor* (Λ,Ψ) *which is right resolving and has all follower sets distinct.*

Proof. For later use we recall the construction. The vertices of Λ are the different follower sets of (Γ,Φ), i.e. $V_\Lambda = \{\omega_+(\alpha) \mid \alpha \in V_\Gamma\}$. For $\alpha', \beta' \in V_\Lambda$ draw one edge from α' to β' labeled s if there are $\alpha, \beta \in V_\Gamma$ with $\alpha' = \omega_+(\alpha)$, $\beta' = \omega_+(\beta)$ and an edge from α to β labeled s. The one-block factor map $(\Gamma,\Phi) \to (\Lambda,\Psi)$ maps an edge e of Γ to the unique edge f in Λ with $i(f) = \omega_+(i(e))$ and $\Psi(f) = \Phi(e)$. $\square$

LEMMA 2.5. ([N, p.103]) *Any uniformly bi-closing cover* (Γ,Φ) *of a coded system* $\overline{S}$ *is block conjugate to a bi-resolving cover* (Λ,Ψ) *of a conjugate coded system* $\overline{S}$.

DEFINITION 2.6. *A right resolving labeled graph* (Γ,Φ) *is saturated at a vertex*

α *if for any* $u_+ \in S^+$ *with the property that* $x_- u_+ \in S$ *for every* $x_- \in \omega_-(\alpha)$, *it holds* $u_+ \in \omega_+(\alpha)$. *If* (Γ, Φ) *is saturated at each vertex, then* (Γ, Φ) *is called saturated*.

Thus a right resolving labeled graph is saturated if for each vertex the follower set is maximal with respect to the predecessor set.

If (Γ, Φ) is saturated and (Λ, Ψ) is another right resolving labeled graph which is block conjugate to (Γ, Φ), then it is not hard to see that (Λ, Ψ) is saturated, too. Thus a uniformly right closing labeled graph could be called saturated if it is block conjugate to a right resolving saturated graph, but for our purposes the given definition suffices.

EXAMPLES 2.7. The Fischer cover of a half-synchronized system is saturated. For each vertex α there is a (half-synchronizing) word m with $m \in F_-(\alpha)$ and $\omega_+(\alpha)$ $= \omega_+(m)$. Thus, if $u_+ \in S^+$ is compatible with all $x_- \in \omega_-(\alpha)$, then in particular with the block m, thus $u_+ \in \omega_+(\alpha)$.

The Krieger graph for any subshift S is saturated, since for any vertex α there is a left infinite path terminating at α and labeled say u_-, such that $\omega_+(\alpha) = \omega_+(u_-)$. Now use the above argument. The same holds for any irreducible component of the Krieger graph which is forwardly and backwardly closed.

If a right resolving labeled graph is a cover, saturatedness may be checked easier.

LEMMA 2.8. *A right resolving cover is saturated if it is saturated at one vertex* α.

Proof. Let β be another vertex and u_+ a right infinite ray with $x_- u_+ \in S$ for any $x_- \in \omega_-(\beta)$. By irreducibility there is a path from α to β, say labeled u. Then $y_- u \in \omega_-(\beta)$ for all $y_- \in \omega_-(\alpha)$. Thus $y_- u u_+ \in S$ for all $y_- \in \omega_-(\alpha)$. Saturatedness at α implies $u u_+ \in \omega_+(\alpha)$, thus $u_+ \in \omega_+(\beta)$ by right resolving. $\square$

THEOREM 2.9. *Let* (Γ, Φ) *be a right resolving cover for a coded system* S. *If* (Γ, Φ) *has residual image in the one-sided shift then it is saturated. The converse holds for synchronized systems* S.

REMARKS. a) Each of the uncountably many irreducible components of the Krieger graph for the Dyke-system is saturated (since they are all forwardly and backwardly closed). But only the Fischer cover has residual image in the one-sided shift.

b) A saturated C^+-cover of a synchronized system has residual image (Theorem 2.16).

Proof. "$\Rightarrow$": By Remark 1.5 there are $\alpha \in V_\Gamma$ and $m \in F_-(\alpha)$ such that $\omega_+(\alpha) = \omega_+(m)$. Thus, as in Example 2.7, (Γ, Φ) is saturated at α. Apply Lemma 2.8.

"$\Leftarrow$": Choose a finite path in Γ from some vertex α to some vertex β and labeled by some synchronizing block m. Let $u_+ \in \omega_+(m)$. Then $x_- m u_+ \in S$ for any $x_- \in \omega_-(\alpha)$, since m is synchronizing. Saturatedness at α implies $m u_+ \in$

$\omega_+(\alpha)$. So $u_+ \in \omega_+(\beta)$ by right resolving. Thus $\omega_+(m) = \omega_+(\beta)$. Apply Remark 1.5. □

The following property of saturated labeled graphs will be used to show the canonicity of the Krieger graph.

LEMMA 2.10. *Let* (Γ, Φ) *be a saturated, right resolving labeled graph with all follower sets distinct. Let* π *be a continuous, shift commuting map* $\Sigma_\Gamma \to \Sigma_\Gamma$ *with* $\Phi(x) = \Phi\pi(x)$ *for all* $x \in \Sigma_\Gamma$ *(we do not assume that* π *has dense image). Then* π *is the identity map.*

Proof. Let $x \in \Sigma_\Gamma$. Let $\alpha = t(x_0)$ and $\beta = t(\pi(x)_0)$. It suffices to show $\alpha = \beta$. All follower sets are distinct, so it suffices to show $\omega_+(\alpha) = \omega_+(\beta)$.

Choose $y \in \Sigma_\Gamma$ which is eventually periodic in the past and satisfies $y_0 = x_0$ and $\pi(y)_0 = \pi(x)_0$. Then, since Φ is right resolving,

(*) there is an $N \in \mathbb{N}$ such that for any $z \in \Sigma_\Gamma$ with $z[-N,0] = y[-N,0]$ it holds $\pi(z)_0 = \pi(y)_0$. This implies $\omega_+(\alpha) \subset \omega_+(\beta)$. Now let $u_+ \in \omega_+(\beta)$. Then, for any $v_- \in \omega_-(i(y_{-N}))$ it holds $v_-\Phi(y)[-N,0]u_+ \in S$ by (*). Saturatedness at $i(y_{-N})$ implies $\Phi(y)[-N,0]u_+ \in \omega_+(i(y_{-N}))$. Thus, by right resolving, $u_+ \in \omega_+(t(y_0)) = \omega_+(\alpha)$. □

THEOREM 2.11. (Essentially [Kr2, Theorem 2.2]) *Let* $\varphi : S \to \bar{S}$ *be an isomorphism of subshifts. Then there is a continuous factor map* $\psi : \Sigma_{X^+(S)} \to \Sigma_{X^+(\bar{S})}$ *of the Markov chains associated to their Krieger graphs, with the property* $\rho^+(\bar{S})\psi = \varphi\rho^+(S)$. *Furthermore, there is only one such factor map* ψ *and it is an uniformly continuous isomorphism, i.e. the labeled Krieger graphs are block conjugate.*

Proof. W.Krieger has shown that there is a block map $\psi : \Sigma_{X^+(S)} \to \Sigma_{X^+(\bar{S})}$ with $\rho^+(\bar{S})\psi = \varphi\rho^+(S)$ [Kr1]. For completeness, we recall the definition. Fix a common coding length N for φ and φ^{-1}. As a sliding block code let ψ map a path $e_{-3N},...,e_{3N}$ in $X^+(S)$ to an edge f in $X^+(\bar{S})$ in the following way. Choose a left infinite path x_- in $X^+(S)$ with terminal vertex $i(e_{-3N})$ and $\omega_+(\rho^+(x_-)) = \omega_+(i(e_{-3N}))$. Choose a point y in S with $y(-\infty, -3N-1] = \rho^+(x_-)$ and $y[-3N,3N] = \rho^+(e_{-3N}),...,\rho^+(e_{-3N})$. Now let f be the unique edge in $X^+(\bar{S})$ with initial vertex $\omega_+((\varphi(y)(-\infty,-1])$ which is labeled $\varphi(y)_0$.

We give an alternate argument that ψ is unique and a block isomorphism. Here we shall only use the fact that ψ is a well-defined block map [Kr1], but neither the explicit definition of ψ nor that it has dense image.

For that fix a block map $\bar{\psi} : \Sigma_{X^+(\bar{S})} \to \Sigma_{X^+(S)}$ with $\varphi^{-1}\rho^+(S) = \rho^+(S)\bar{\psi}$ (which exists by [Kr1]). Then $\bar{\psi}\psi$ and $\psi\bar{\psi}$ satisfy the hypothesis of Lemma 2.10. The Krieger graphs are saturated (see Example 2.7), thus $\bar{\psi}\psi = $ id and $\psi\bar{\psi} = $ id by Lemma 2.10. Thus $\bar{\psi} = \psi^{-1}$. This shows that ψ is a uniformly continuous isomorphism. But for any other continuous factor map $\psi': \Sigma_{X^+(\bar{S})} \to \Sigma_{X^+(S)}$ with

$\rho^+(S)\psi' = \phi\rho^+(S)$ we also have $\psi' = \psi^{-1}$, by the same reasoning. Thus ψ is uniquely determined, even in the set of continuous factor maps. $\square$

The canonicity of the Fischer cover of half-synchronized systems has also been shown by W.Krieger (oral communication).

THEOREM 2.12. *Let* S *be a synchronized or half-synchronized system. The Fischer cover* (X_0^+,ρ_0^+) *is the unique irreducible component of* $(X^+(S),\rho^+(S))$ *with residual image in the one-sided shift. Furthermore,* (X_0^+,ρ_0^+) *is canonical.*

Proof. For any irreducible component (Γ,Φ) of X^+ with residual image in the one-sided shift there are $\alpha \in V_\Gamma$ and a halfsynchronizing word $m \in F_-(\alpha)$ with $\omega_+(\alpha) = \omega_+(m)$, by Remark 1.5. But then α is a vertex of X_0^+ by definition, thus $\Gamma = X_0^+$.

Given an isomorphism ϕ between two half-synchronized systems S and $\bar{S}$, the image of $X_0^+(S)$ under the isomorphism ψ from Theorem 2.11. is some irreducible component Γ of $X^+(S)$. Let N be a coding length for ϕ and ϕ^{-1}. Choose a path $e_0,...,e_n$ as in the proof of Theorem 1.4, with $n \geq N$. Let $A = \{ x \in \Sigma_{X_0^+} \mid x[i,i+n] = e_0...e_n$ for some $i \geq N\}$. Now $\rho^+(A)$ is dense in S, thus $\phi\rho^+(A)$ is dense in S, thus $P_{[0,\infty)}(\phi\rho^+(A))$ is dense in S^+. As in the proof of Theorem 1.4 one sees that $P_{[0,\infty)}(\phi\rho^+(A))$ is open in S^+. Thus $P_{[0,\infty)}(\rho^+(\Gamma))$, which contains $P_{[0,\infty)}(\phi\rho^+(A))$, is a residual subset of S^+. Thus, by the preceding paragraph, $\Gamma = X_0^+(S)$. $\square$

REMARK 2.13. The Fischer cover of a synchronized system S is the unique irreducible component of the Krieger graph which has residual image, or, where a synchronizing word can be read off, or, which is a cover for S.

THEOREM 2.14. *Let* S *be half-synchronized and* $(\Gamma,\Phi) \in C^+(S)$. *Then* (Γ,Φ) *is block isomorphic to* (X_0^+,ρ_0^+) *iff* (Γ,Φ) *has a residual image in the one-sided shift.*

Proof. By Remark 1.5 there is a vertex $\alpha \in V_\Gamma$ such that $\omega_+(\alpha) = \omega_+(m)$ for some half-synchronizing word m. Apply the definition of (X_0^+,ρ_0^+) and Lemma 2.2. $\square$

DEFINITION 2.15. *A finite sequence* $m = m_1...m_n$ *of labels is called a magic word for a right resolving cover if there is one and only one vertex* α *such that* $m \in F_-(\alpha)$.

Thus there might be many realizations of m in the given cover, but all of them have to terminate at the same vertex. Note that a magic word is always synchronizing.

We characterize the Fischer cover in the set of C^+-covers for a synchronized system.

Note that in any of the following cases (Γ,Φ) actually equals (X_0^+,ρ_0^+) up to a graph isomorphism that respects the labels (see Lemma 2.2).

THEOREM 2.16. *Let* S *be synchronized. Let* (X_0^+,ρ_0^+) *denote its (right) Fischer cover. The following are equivalent for a cover* $(\Gamma,\Phi) \in C^+(S)$.
i) (Γ,Φ) *is block isomorphic to* (X_0^+,ρ_0^+)
ii) $\Phi(\Gamma)$ *is residual in* S
iii) (Γ,Φ) *is saturated*
iv) (Γ,Φ) *has a magic word*
v) (Γ,Φ) *has residual image in the one-sided shift.*

Proof. i) implies ii) - v) since by Lemma 2.2 (Γ,Φ) is actually equal to (X_0^+,ρ_0^+) up to a graph isomorphism that respects the labels, and (X_0^+,ρ_0^+) has these properties.
ii) $\Rightarrow$ i) By Remark 1.5, the definition of (X_0^+,ρ_0^+) and Lemma 2.2.
iv) $\Rightarrow$ i) Let m be a magic word and let α be the unique vertex with $m \in F_-(\alpha)$. Then $F_+(\alpha) = F_+(m)$. But for right resolving graphs this implies $\omega_+(\alpha) = \omega_+(m)$, again apply the definition of (X_0^+,ρ_0^+) and Lemma 2.2.
iii) $\Rightarrow$ v) by Theorem 2.9. v) $\Rightarrow$ i) by Theorem 2.14. $\square$

In the sofic case, with all covers finite, the following result is [BKM, Prop.4].

THEOREM 2.17. *Let* S *be synchronized or half-synchronized and* (Γ,Φ) *a uniformly right closing cover with residual image in the one-sided shift. Then* (X_0^+,ρ_0^+) *is a block factor of* (Γ,Φ).

Proof. By Lemma 2.3 and Lemma 2.4 (Γ,Φ) has a C^+-block factor which necessarily has still residual image in the one-sided shift, and is thus block isomorphic to (X_0^+,ρ_0^+) by Theorem 2.14. $\square$

Inspired by [BKM, Theorem 6], we give a second proof for the canonicity of the Fischer cover for half-synchronized systems. We cannot use the sofic arguments since ρ_0^+ is not 1-1 a.e. if the system is strictly half-synchronized.
Recode $(X_0^+(S),\varphi\rho_0^+(S))$ such that it becomes a cover of $\overline{S}$ with a one-block label function. Since φ is an isomorphism, it follows that the new cover is uniformly right closing. By Theorem 2.17 there is a block map $\psi\colon X_0^+(S) \to X_0^+(\overline{S})$ with $\varphi\rho_0^+(S) = \rho_0^+(\overline{S})\psi$. By the same arguments as used in the proof for the canonicity of the Krieger graph it follows that ψ is in fact a block isomorphism. $\square$

3. Subclasses of synchronized systems defined via closing properties of the Fischer cover

The Fischer cover of a synchronized system is right resolving by definition, but it may have additional closing properties as (increasingly ordered)
- locally finiteness,
- locally finiteness and weakly left closing (see Definition 3.1),
- locally finiteness and left closing,

- uniformly left closing.

By canonicity (Theorem 2.12), these properties are isomorphism invariants. Using Theorem 2.16iv it is easy to construct Fischer covers which have one of these closing properties but not the next stronger one. So these closing properties define different classes of synchronized systems, which will be discussed in more detail in the remaining sections.

Here we show that the existence of some "strong" cover with one of the above closing properties implies that the Fischer cover has this closing property (Theorem 3.2).

As the above list shows, we favor locally finiteness of the Fischer cover. Although left closing (or weakly left closing) by itself is an isomorphism invariant, we do not have good results if the Fischer cover lacks locally finiteness (Example 3.3).

Finally we discuss possible closing properties for Fischer covers of strictly half-synchronized systems.

The following "weakly left closing" property will become important in §6 (when characterizing Fischer covers without proper continuous factors). A labeled graph is left closing if any left infinite path is uniquely determined by its terminal vertex and its labels. "Weakly left closing" only requires this for left infinite paths which additionally diverge to infinity in the past, i.e. which eventually leave any finite set of vertices.

DEFINITION 3.1. *A labeled graph* (Γ, Φ) *is called weakly left closing if for any two paths* $x, y \in \Sigma_\Gamma$ *with* $t(x_0) = t(y_0)$, $x_0 \neq y_0$, *and* $\Phi(x_i) = \Phi(y_i)$ *for all* $i \leq 0$, *it holds that there is some vertex* α *such that* $t(x_i) = \alpha$ *or* $t(y_i) = \alpha$ *for infinitely many* $i \leq 0$.

Note that left closing implies weakly left closing but not vice versa.

THEOREM 3.2. *Let* S *be a synchronized system which admits a cover* (Γ, Φ) *that is uniformly right closing, has residual image in* S, *and has additionally one of the following closing properties (which are increasingly ordered):*
a) Γ *is locally finite,*
b) (Γ, Φ) *is locally finite and weakly left closing,*
c) (Γ, Φ) *is locally finite and left closing,*
d) Φ *is uniformly left closing.*
Then the (right) Fischer cover (X_0^+, ρ_0^+) of S has the same additional closing property.

If we weaken the assumption "uniformly right closing" to "locally finite and right closing", then the theorem fails, since, by Theorem 1.9, any synchronized system has such a cover which satisfies c), but in general the Fischer cover of a synchronized system is not even locally finite. Remark 1.10 shows that the theorem fails if we weaken "residual image" to "residual image in the one-sided shift".

Proof. By Lemma 2.3 we may assume that (Γ, Φ) is right resolving. The proof of Theorem 2.17 shows that there is a block factor map $\pi: (\Gamma, \Phi) \to (X_0^+, \rho_0^+)$ which is a graph homomorphism, i.e. it maps vertices to vertices and connecting edges between two vertices to connecting edges (with the same labels) of the image-vertices. Thus we can talk about preimages of vertices or finite paths. By Remark 1.2 there is a synchronizing block m and a vertex α' such that $F_+(\alpha') = F_+(m)$ and $um \in F_-(\alpha')$ whenever um is a block of S. Let α be the image of α'. Then $\omega_+(\alpha) = \omega_+(\alpha') = \omega_+(m)$, thus α is the unique vertex of X_0^+ with $m \in F_-(\alpha)$.

a) We shall show that if β is a vertex of X_0^+ and u a block with $um \in F_+(\beta)$, then β has some preimage vertex β' in Γ which leads to the special vertex α' by a path labeled um. If Γ is locally finite, then for fixed u there are only finitely many such vertices in Γ, thus there are only finitely many vertices β of X_0^+ with $um \in F_+(\beta)$, which will imply that X_0^+ is locally finite.

Thus assume $um \in F_+(\beta)$. Since any path labeled um and starting at β has to terminate at α, there is a loop path at α, labeled wum for some block w. Then $mwum$ is a block in S, thus there is a path in Γ terminating at α' and labeled $mwum$. Let β' be the terminal vertex of the initial segment of this path labeled mw. Since the π-image of this initial segment has to be in α after $|m|$ steps, by right resolving the π-image of β' has to be β.

The other parts of the theorem will follow from the following observation.

Let (Γ, Φ) be locally finite. Let x_- be some left infinite path in X_0^+ whose last $|m|$ edges are labeled by m. By the above considerations we know that any finite final segment of x_- has a preimage path that terminates at α'. Since Γ is locally finite a compactness argument gives a left infinite path Γ, terminating at α' and being mapped onto x_-. (This implies also that π is surjective, in particular $\Phi(\Gamma) = \rho_0^+(X_0^+)$).

b) Now assume that (Γ, Φ) is additionally weakly left closing. Let $x_- \neq y_-$ be two equally labeled left infinite paths in X_0^+, terminating at the same vertex. Choose some common finite extension to the right with the last $|m|$ edges labeled by m. These new left-infinite paths have preimages terminating at α'. (Γ, Φ) is weakly left closing, so one of these preimage paths hits some vertex β' infinitely often. Since (Γ, Φ) is locally finite this holds also for its image (the set of sequences hitting β' at time zero is compact, so its image, which can thus be in only finitely many vertices at time zero).

c) Proceed as in b), by left closing the preimages are equal, thus also $x_- = y_-$.

d) Assume that (Γ, Φ) is N-step left closing. Consider two equally labeled finite paths $x_1 \ldots x_{N+1}$ and $y_1 \ldots y_{N+1}$ terminating at the same vertex of X_0^+. Choose a common finite extension to the right, labeled by vm for some block v. These two extended paths have preimages terminating at α'. By N-step left closing these preimages coincide in the last $|vm|+1$ edges. Thus also their images, so $x_{n+1} = y_{n+1}$. $\qquad\square$

EXAMPLE 3.3. If we choose only "left-closing" (i.e. without locally finiteness) as one of the additional properties in Theorem 3.2, then ρ_0^+ has not to be left closing. For that consider the following (countable) cover. Its vertex set is $\mathbf{Z}_+$. Draw an edge labeled d from 0 to 1, an edge labeled c from 1 to 2, edges labeled a from i to i+1 for all $i \geq 2$, edges labeled b from each $i \geq 1$ to 0, and finally a loop at

1, labeled a. This cover (Γ,Φ) is right resolving and left closing by inspection.

The coded system it defines is a sofic system S whose right Fischer cover is given by the finite labeled graph with "transition matrix" $M = \begin{bmatrix} a & c & b \\ 0 & a & b \\ d & 0 & 0 \end{bmatrix}$, i.e.

1,2,3 is the set of vertices and there is an edge from i to j labeled s if $M_{i,j} = s$. Although the image of (Γ,Φ) is residual in S (in fact equal) , the right Fischer cover is not left closing. □

For half-synchronized systems all the above mentioned closing properties of the Fischer cover are invariants, too. But which of them can really occur for strictly half-synchronized (i.e. non-synchronized) systems ? We have a question and a theorem.

QUESTION 3.4. Are there strictly half-synchronized systems which have a locally finite and left-closing Fischer cover ? For each of these two properties separately we have examples (the Dyke system works for the first) but not for their combination.

THEOREM 3.5. *Any half-synchronized system* S *with a uniformly left closing Fischer cover* (X_0^+,ρ_0^+) *is synchronized (compare Theorem 1.6).*

Proof. Assume ρ_0^+ is N-step left closing. Choose $\alpha \in V_{X_0^+}$ and a block m of S such that $m \in F_-(\alpha)$ and $F_+(\alpha) = F_+(m)$. We may assume that m has length greater than N (if necessary extend m to the right). We will show that α is the unique vertex with $m \in F_-(\alpha)$. This obviously implies that m is a synchronizing block.

So assume $m \in F_-(\beta)$ for some vertex β. Choose a path from β to α labeled vm for some block v. Then $vm \in F_+(m)$, thus $vm \in F_+(\alpha)$. Consider the unique path starting at α and labeled vm. Let δ denote its terminal vertex. Obviously $F_+(\delta) \subset F_+(m) = F_+(\alpha)$. And for $w \in F_+(\alpha)$ we have $vmw \in F_+(m)$, thus $vmw \in F_+(\alpha)$, thus $w \in F_+(\delta)$. So $F_+(\delta) = F_+(\alpha)$ which shows that $\alpha = \delta$. Thus we have a loop at α, labeled vm ,and an equally labeled path from β to α. Since $m \in F_-(\beta)$ and $m \in F_-(\alpha)$, we obtain $\beta = \alpha$ by N-step left-closing. □

REMARK 3.6. We note one consequence of Theorem 3.5. It is not possible to extend Theorem 3.2 to the class of half-synchronized systems in replacing "residual" by "residual in the one-sided shift". By Remark 1.10 the strongest hypothesis d) of such a theorem would be satisfied for any half-synchronized system, but by Theorem 3.5 the conclusion fails if the system is not already synchronized.

4. Almost Markov synchronized systems

We shall generalize the notion of "almost Markov" (also called "almost finite type") from sofic to synchronized systems. Many properties of the Fischer cover of an almost Markov sofic system ([BKM, §3]) have their analogue in the setting of almost Markov synchronized systems (Theorems 4.2, 4.3). We also prove an analogue of [Wi, Theorem 4.1] for non-almost Markov synchronized systems

(Theorem 4.4).

Finally we indicate some obstructions when looking for an appropriate notion of "almost Markov" for half-synchronized systems.

We generalize the definition of almost Markov sofic systems as given in [N, §6].

DEFINITION 4.1. *A synchronized system* S *is called almost Markov iff it has a uniformly bi-closing cover* (Γ,Φ) *such that* $\Phi(\Gamma) \subset S$ *is residual.*

REMARK. Theorem 1.7 shows that it is crucial to require that $\Phi(\Gamma)$ is residual in S to select a proper subclass of synchronized systems by this definition.

Curious enough, extending the original definition of almost Markov sofic systems as given by B. Marcus [M, Definition 4] (S has a (finite) cover which is 1-1 on an open non-empty set) would not give a proper subclass of synchronized systems as Theorem 1.9 shows.

The following theorem extends some results [BKM, §3] from almost Markov sofic systems to almost Markov synchronized systems. Even in the almost Markov sofic setting it contains some additional information. For example, that the Fischer cover intercepts all covers, not only the finite ones. We shall give a direct construction of the factor map to show ii) ⇒ iii), since we cannot use fiber products and entropy arguments as in the sofic setting [BKM, Theorem 9].

THEOREM 4.2. *Let* S *be synchronized with Fischer cover* (X_0^+,ρ_0^+). *Equivalent are*

  i)   S *is almost Markov.*
  ii) (X_0^+,ρ_0^+) *is uniformly left-closing.*
  iii) (X_0^+,ρ_0^+) *is a block factor of any (!) other cover.*
  iv) S *has a cover which is a block factor of any other cover.*
  v) (X_0^+,ρ_0^+) *is block-conjugate to* (X_0,ρ_0)
  vi) S *admits a uniformly bi-closing cover which is* 1-1 *on an non-empty open set.*

Proof. i) ⇒ ii) by Theorem 3.2d.

ii) ⇒ iii) : By Lemma 2.5 we may assume that (X_0^+,ρ_0^+) is bi-resolving. Choose a synchronizing block m of S and let α be the unique vertex of X_0^+ with $m \in F_-(\alpha)$. Now let (Γ,Φ) be some cover for S. Since $\Phi(\Gamma)$ is dense in S, there is an $\alpha` \in V_\Gamma$ with $m \in F_-(\alpha`)$. Define a one-block map $\pi : (\Gamma,\Phi) \to (X_0^+,\rho_0^+)$ in the following way. Given an edge e in Γ choose a path from $\alpha`$ to i(e). Let v be the label of this path. Let $\gamma = \gamma(v)$ denote the unique terminal vertex in X_0^+ of the path starting at α and labeled v. Map e to the edge in X_0^+ which starts at γ and has label $\Phi(e)$. All we have to show is that γ is independent of the particular choice of the path from $\alpha`$ to i(e). So let w be the label of some other path from $\alpha`$ to i(e). Fix a path from i(e) to $\alpha`$ with label um for some u. Then there is also a path labeled um starting at $\gamma(v)$ and a path labeled um starting at $\gamma(w)$.

Since α is the unique vertex in X_0^+ with $m \in F_-(\alpha)$, both these two paths terminate at α, thus $\gamma(v) = \gamma(w)$ by left resolving.

Finally, $\pi(\Sigma_\Gamma)$ is dense in $\Sigma_{X_0^+}$, since (X_0^+, ρ_0^+) has a magic word.

iii) $\Rightarrow$ iv): trivial.

iv) $\Rightarrow$ v): If (Γ, Φ) is a cover which is a block factor of any other cover, then it is in particular a block factor of (X_0^+, ρ_0^+). We shall see in Theorem 6.2 that (X_0^+, ρ_0^+) does not have proper block factors, i.e. we have that (X_0^+, ρ_0^+) is block conjugate to (Γ, Φ). The same holds for (X_0, ρ_0), thus (X_0^+, ρ_0^+) and (X_0, ρ_0) are block conjugate.

v) $\Rightarrow$ vi): If (X_0^+, ρ_0^+) and (X_0, ρ_0) are block conjugate then (X_0^+, ρ_0^+) has to be uniformly left closing since (X_0, ρ_0) is left resolving. Then ρ_0^+ is 1-1 on the open set of sequences in S which contain some specified synchronizing word m.

vi) $\Rightarrow$ i): Since the image of a cover is shift invariant we have that it contains a dense open set once it contains a non empty open set. Thus there is a uniformly bi-closing cover with residual image and S is almost Markov by definition. $\square$

The characterization [BKM, Prop.7, Cor.10] of the Fischer cover of almost Markov systems carries over.

THEOREM 4.3. *Let S be almost Markov. A cover (Γ, Φ) is block-conjugate to (X_0^+, ρ_0^+) iff (Γ, Φ) is uniformly bi-closing and Φ is 1-1 a.e. (1-1 a.e. means that any doubly transitive point of S has precisely one preimage).*

Proof." $\Rightarrow$ ": follows from Theorem. 4.2.ii. and the fact that ρ_0^+ is 1-1 a.e.

" $\Leftarrow$ ": That Φ is 1-1a.e. implies that $\Phi(\Gamma) \subset S$ is residual. By Remark 1.2 there is a synchronizing block m and a vertex α such that $F_+(\alpha) = F_+(m)$ and $um \in F_-(\alpha)$ whenever um is a word of S. We show first that α is the only vertex with $m \in F_-(\alpha)$. Let $e_0, ..., e_r$ be a finite path in Γ labeled m. Extend this path to a doubly transitive labeled infinite path. Since (Γ, Φ) is bi-resolving, there is an infinite path with the same sequence of labels which is in α at time zero. By 1-1 a.e this infinite path is unique, thus $t(e_r) = \alpha$.

By Lemma 2.3 we may assume that (Γ, Φ) is right resolving and N-step left closing for some $N \in \mathbb{N}$. Let π be the one-block factor map as in the proof of Lemma 2.4. Then the image of (Γ, Φ) under π is block-conjugate to (X_0^+, ρ_0^+) by Theorem 2.16ii. Thus it suffices to show that π is block invertible. Let $x, y \in \Sigma_\Gamma$ with $\pi(x)[-N, N] = \pi(y)[-N, N]$. We show that $x_0 = y_0$, which implies that π has an inverse π^{-1} with coding length N. Since $\pi(x)_N = \pi(y)_N$ we have by the definition of π that $F_+(x_N) = F_+(y_N)$. For some block v we have $vm \in F_+(x_N)$. Thus also $vm \in F_+(y_N)$. By the preceding paragraph the two finite path labeled vm and starting at $t(x_N)$ and $t(y_N)$, respectively, both terminate at α. Since also $\Phi(x)[-N, N] = \Phi(y)[-N, N]$, N-step left closing now implies $x_0 = y_0$. $\square$

S.Williams constructed for any non-almost Markov sofic system a family of countably many finite covers such that no two of them have a common block factor [Wi, Theorem 4.1]. Thus there cannot be a finite family of finite covers such that any other finite cover would factor through at least one of them.

THEOREM 4.4. *Let* S *be synchronized but not almost Markov. Then there are uncountably many covers, all with residual image in* S, *and no two of them have a common block factor.*

Proof. Fix a vertex $\alpha \in V_{X_0^+}$ and a loop path at α, labeled by m such that α is the unique vertex with $m \in F_-(\alpha)$.

Because ρ_0^+ is not uniformly left closing, we can choose for each $n \in \mathbb{N}$ two paths in X_0^+ of length n and equally labeled, say by the block w_n, terminating at the same vertex, but with different last edges. For each of these paths we choose a shortest path from α to its initial vertex, one with label, say s_n, the other with label $\bar{s}_n$. Furthermore we choose a shortest path from the common terminal vertex to α, labelled, say t_n.

By passing to a subsequence we may assume that

(1) $|w_{n+1}| > 2 \cdot \max(|m^2 s_n w_n t_n m^2|, |m^2 \bar{s}_n w_n t_n m^2|)$ and $|w_1| > 2 \cdot |m|$. Here m^2 denotes the concatenation of m with itself.

Let $I \subset \mathbb{N}$, $|I| = \infty$.

We construct a cover $(\Gamma, \Phi) = (\Gamma_I, \Phi_I)$ which contains (X_0^+, ρ_0^+) as a subgraph.

Take (X_0^+, ρ_0^+) and add for each $n \in I$ first a new loop path at α labeled $m^2 s_n w_n t_n m^2$, and then a new path, labeled $m^2 \bar{s}_n$, from α to the terminal vertex of the initial segment labeled $m^2 s_n$ of the path just constructed. Thus for each $n \in I$ we created two new loops at α, labeled $m^2 s_n w_n t_n m^2$ and $m^2 \bar{s}_n w_n t_n m^2$, which have the last $|w_n t_n m^2|$ edges in common. Let (Γ, Φ) be this new labeled graph.

Obviously (Γ, Φ) is a cover for S, and since $\Phi(\Gamma) \supset \rho_0^+(X_0^+)$, $\Phi(\Gamma) \subset S$ is residual.

We leave the proofs of the following two claims to the reader. Hint: use (1).

Claim 1. m^2 is not a subblock of $s_n w_n t_n$ or $\bar{s}_n w_n t_n$.

Claim 2. The number of $x \in \Sigma_\Gamma$ with $\Phi(x)(-\infty, 0] = m^\infty$ and $\Phi(x)[1, \infty) = m^2 s_n w_n t_n m^\infty$ ($m^2 \bar{s}_n w_n t_n m^\infty$) is 1 if $n \notin I$ and 2 if $n \in I$.

Now let $\pi: (\Gamma, \Phi) \to (\Lambda, \Psi)$ be a block factor with coding length N.

Because m^4 is magic in (Γ, Φ) and π is a block map, we get

(2) Let $y \in \Sigma_\Lambda$, $\Psi(y)(-\infty, 0] = m^\infty$, $\Psi(y)[1, \infty) = m^2 s_n w_n t_n m^\infty$ ($m^2 \bar{s}_n w_n t_n m^\infty$) then there is an $x \in \Sigma_\Gamma$ such that $\pi(x) = y$.

For $n \in I$, there are four points $x, \bar{x}, y, \bar{y} \in \Sigma_\Gamma$ such that $\Phi(x) = \Phi(\bar{x})$, $\Phi(y) = \Phi(\bar{y})$ with the labels as in Claim 2. Using that π is a block map shows that for $n > 2N+1$ $\pi(x) \neq \pi(\bar{x})$ or $\pi(y) \neq \pi(\bar{y})$. Thus we get:

Claim 3. Let $n \in I$, $n > 2N + 1$. Then there are two points y in Σ_Λ with $\Psi(y)(-\infty, 0] = m^\infty$ and $\Psi(y)[1, \infty) = m^2 s_n w_n t_n m^\infty$ *or* there are two points y in Σ_Λ with $\Psi(y)(-\infty, 0] = m^\infty$ and $\Psi(y)[1, \infty) = m^2 \bar{s}_n w_n t_n m^\infty$.

Putting Claim 2, (2) and Claim 3 together shows that if $I, J \subset \mathbb{N}$, $|I \triangle J| = \infty$ then

(Γ_I, Φ_I) and (Γ_J, Φ_J) do not have a common block factor. $\square$

Remark. If S is not almost Markov, but sofic, then the idea of adding two loops to the Fischer cover easily shows that there are infinitely many finite covers for S such that no two of them have a common block factor. This yields a shorter and more direct proof of [Wi, Theorem 4.1].

REMARK 4.5. For strictly half-synchronized systems no analogon of the almost Markov theory for synchronized systems is known to us. Generalizing Definition 4.1. in replacing "residual" by "residual in the one-sided shift" does not work, since any half-synchronizing system admits such a cover, see Remark 1.10. Also there are no strictly half-synchronized systems with uniformly left-closing Fisher cover (Theorem 3.5). Thus these two natural attempts fail.

One more result along these lines is the following easy consequence of Theorem 1.6.

THEOREM 4.6. *Let* S *and its inverse* S^{-1} *be half-synchronized. In this case* S *has also a left Fischer cover* (X_0, ρ_0). *Assume* (X_0^+, ρ_0^+) *and* (X_0, ρ_0) *have a common continuous factor. Then* S *is synchronized.*

5. Fischer covers which intercept continuously any other cover

We have seen in §4 that for an almost Markov synchronized system its Fischer cover (X_0^+, ρ_0^+) intercepts any other cover by a block map.

Here we shall consider continuous factor maps, rather than uniformly continuous ones, and ask the following questions.

1) For which synchronized systems does (X_0^+, ρ_0^+) intercept continuously any other cover ? A natural guess (because of Theorem 4.2) is to consider those systems where (X_0^+, ρ_0^+) is locally finite and left closing, which is the right guess by Theorem 5.6.

2) Which synchronized systems have some cover that intercepts continuously any other cover ? If it exists, this cover is unique up to continuous isomorphism by Corollary 5.2. Theorem 5.5 gives some sufficient conditions and Theorem 5.7 a characterization under the additional hypothesis that X_0^+ is locally finite.

As a bi-product we obtain that whenever X_0^+ is locally finite, then its image under ρ_0^+ contains (as a set) the images of all other covers. (Lemma 5.3).

Finally, a couple of examples show that not all statements of Theorem 4.2 have their valid counterparts in the continuous setting (Examples 5.8).

A series of lemmata will lead us to our main results. First we show that some properties of a Fischer cover are preserved under continuous factor maps.

LEMMA 5.1. *Let* S *be a synchronized system. Let* (Γ, Φ) *be a cover for* S *which is a continuous factor of* (X_0^+, ρ_0^+). *Then* (Γ, Φ) *has finite out-degree and is right closing.*

Furthermore, (Γ, Φ) *is 1-1 a.e. (see Theorem 4.3 for the definition).*

REMARK. Continuous factors of the Fischer cover do not have to be uniformly right closing (in Example 5.8a (X_0^+, ρ_0^+) is a continuous factor of the left Fischer cover, but not uniformly left closing).

Proof. Assume that δ is a vertex of Γ where (Γ, Φ) has infinite out-degree or is not right-closing. Then for each $N \in \mathbb{N}$ there are two path $e_1^N, ..., e_N^N$ and

$f_1^N,...,f_N^N$ of length N starting at δ which have different initial edges $e_1^N \neq f_1^N$, but are equally labeled by some word $w(N)$ of length N. A compactness argument in S shows that we may assume $w(N) = w_+[1,N]$ for all $N \in \mathbb{N}$ for some left-infinite ray $w_+ = w_1, w_2,... \in S_+$ (which is trivial in case (Γ,Φ) is not right closing at δ).

Choose some $m \in F_-(\delta)$ that contains some magic word of (X_0^+, ρ_0^+) as a subword. Let $\alpha` \in V_{X_0^+}$ be the unique vertex with $m \in F_-(\alpha')$. Then $F_+(\alpha') = F_+(m)$.

Now we construct an $x \in \Sigma_{X_0^+}$ which is in $\alpha`$ at time zero. Since $w(N) \in F_+(\delta)$ implies $w(N) \in F_+(\alpha`)$ for each N, we have $w_+ \in \omega_+(\alpha`)$ by right resolving. Let the (unique) future of $x$ be labeled by $w_+$. Now fix a loop at $\delta$ with label $um$ for some block $u$. Then analogously $um \in F_+(\alpha')$, i.e. there is a loop at $\alpha`$ with label um. Let the (unique) past of x be labeled $(um)^\infty$.

Let π denote the given continuous factor map $(X_0^+, \rho_0^+) \to (\Gamma,\Phi)$.

Now fix some $N \in \mathbb{N}$. Consider two paths $y, z \in \Sigma_{X_0^+}$ whose π-images are both in δ at time zero, who both run in the past through the above chosen loop at δ for N-times, but in the future $\pi(y)$ runs through $e_1^N,...,e_N^N$ where $\pi(z)$ runs through $f_1^N,...,f_N^N$. Then necessarily at least $y[-N,N] = x[-N,N]$ and $z[-N,N] = x[-N,N]$. Since π is continuous at x, for large enough N we have $e_1^N = f_1^N$, a contradiction.

We show that Φ is 1-1 a.e. Any doubly transitive point in S has a preimage under Φ since it has a preimage under ρ_0^+. Furthermore, by the existence of magic words in (X_0^+, ρ_0^+) and since $\pi(\Sigma_{X_0^+})$ is dense in Σ_Γ, it follows that any $y \in \Sigma_\Gamma$ which is labeled doubly transitive has a preimage in $\Sigma_{X_0^+}$. Since in X_0^+ any doubly transitive sequence of labels occurs only once, this shows that the same holds for (Γ,Φ). $\qquad\square$

COROLLARY 5.2. *If a synchronized systems has a cover (Γ,Φ) that intercepts continuously any other cover, then (Γ,Φ) is unique up to continuous isomorphism.*

Proof. Let (Λ,Ψ) be any other cover with this property. Then there are continuous factor maps $\pi: (\Gamma,\Phi) \to (\Lambda,\Psi)$ and $\tau: (\Lambda,\Psi) \to (\Gamma,\Phi)$. Consider an $x \in \Sigma_\Gamma$ which is doubly transitive. Then the sequence of labels of x is doubly transitive. By Lemma 5.1. (Γ,Φ) is 1-1 a.e., thus x is uniquely determined by its sequence of labels, thus $\tau\pi(x) = x$. Since the doubly transitive points are dense in Σ_Γ, we get $\tau\pi = \mathrm{id}$. Analogously $\pi\tau = \mathrm{id}$, i.e. π is a continuous isomorphism of covers. $\qquad\square$

LEMMA 5.3. *Let S be synchronized and assume that (X_0^+, ρ_0^+) is locally finite. Let (Γ,Φ) be any cover for S, let $y \in \Sigma_\Gamma$. Then there is an $x \in \Sigma_{X_0^+}$ with $\rho_0^+(x) = \Phi(y)$ and $\omega_+(t(x_0)) \supset \omega_+(t(y_0))$. In particular, $\Phi(\Gamma) \subset \rho_0^+(X_0^+)$ for all covers (Γ,Φ) of S.*

REMARK 5.4. There are examples where $\rho_0^+(X_0^+) = S$, thus $\Phi(\Gamma) \subset \rho_0^+(X_0^+)$ for all covers (Γ,Φ) of S, but (X_0^+, ρ_0^+) is not locally finite.

Proof of the Lemma. Let m be a magic word for (X_0^+, ρ_0^+) and let $\alpha` \in V_{X_0^+}$ be the unique vertex with $m \in F_-(\alpha`)$. Let $\alpha, \beta \in V_\Gamma$ be the initial- and terminal vertex, respectively, of some fixed path in Γ which is labeled by m.

First fix one path from $t(y_0)$ to α with label, say u. Now for each $n \in \mathbf{N}$ choose a path from β to $i(y_{-n})$ with label, say v_n. Then $v_n \Phi(y)[-n,0]um \in F_+(m)$. Thus, for each $n \in \mathbf{N}$ there is a (unique) path in X_0^+, starting at $\alpha`$ and labeled $v_n\Phi(y)[-n,0]um$. Let $\delta`(n)$ denote the terminal vertex of the initial segment of this path labeled $v_n\Phi(y)[-n,0]$. Since X_0^+ is locally finite and each $\delta`(n)$ leads to $\alpha`$ by a path labelled um, there is a vertex $\delta`$ such that $\delta`(n) = \delta`$ for infinitely many $n$. Then $\Phi(y)(-\infty,0] \in \omega_-(\delta`)$, since X_0^+ is locally finite.

On the other hand for any $u_+ \in \omega_+(t(y_0))$ also $v_n\Phi(x)[-n,0]u_+ \in \omega_+(m)$ for each $n \in \mathbf{N}$, thus by right resolving $u_+ \in \omega_+(\delta`(n))$ for each $n \in \mathbf{N}$, thus $u_+ \in \omega_+(\delta`)$, thus $\omega_+(t(y_0)) \subset \omega_+(\delta`)$. This shows the existence of x. $\qquad \square$

THEOREM 5.5. *Let S be synchronized and (X_0^+, ρ_0^+) locally finite. Assume (X_0^+, ρ_0^+) has a left closing factor (Γ, Φ). Then any cover factors continuously through (Γ, Φ).*

Proof. Let $\pi : X_0^+ \to \Gamma$ be a continuous factor map from (X_0^+, ρ_0^+) to (Γ, Φ). Let (Λ, Ψ) be any other cover.

We want to construct a continuous factor map τ from (Λ, Ψ) to (Γ, Φ).

We first construct a word w of S such that there is a unique vertex α in Γ with the property that $w \in F_-(\alpha)$ and $w \in F_+(\alpha)$. Let m be a magic word for (X_0^+, ρ_0^+) and α' the unique vertex of X_0^+ with $m \in F_-(\alpha')$. Choose a loop path at α' labeled um for some block u. Let x be the periodic point in $\Sigma_{X_0^+}$ which is labeled $(um)^\infty$ and is in α' at time zero. Choose k such that $x[-k,k]$ determines $\pi(x)_0$. Let w be the block $(um)^{k+1}$. Now let $e_1...e_{|w|}...e_{2|w|}$ be any finite path in (Γ, Φ) labeled ww. There has to be an $y \in \Sigma_{X_0^+}$ whose π-image runs through the path $e_1...e_{|w|}...e_{2|w|}$ and is in $e_{|w|}$ at time zero. But then because of the labeling $y[-k,k] = x[-k,k]$, which forces $e_{|w|} = \pi(y)_0 = \pi(x)_0$. Thus the vertex $\alpha = t(e_{|w|})$ is uniquely determined.

Now let $z = (z_j)_{j \in \mathbf{Z}} \in \Sigma_\Lambda$. By Lemma 5.3 there is an $x \in \Sigma_{X_0^+}$ with $\rho_0^+(x) = \Psi(z)$ and $\omega_+(t(x_0)) \supset \omega_+(t(z_0))$.

Define a shift commuting map τ by specifying $\tau(z)_0 := \pi(x)_0$.

We first show that $\tau(z)_0$ does not depend on the particular choice of x, then that τ actually maps bi-infinite paths to bi-infinite paths.

So let $\bar{x} \in \Sigma_{X_0^+}$ be any other point with the above properties. Fix an $N \in \mathbf{N}$ such that $x[-\infty, N]$ and $\bar{x}[-\infty, N]$ determine $\pi(x)_0$ and $\pi(\bar{x})_0$, respectively.

Let v be the label of the path $z_1,...,z_N$.

Choose $u_+ \in \omega_+(t(z_0))$ with $u_+[1,|v|] = v$ and such that the above constructed ww occurs as a subblock of u_+. Then by hypothesis $u_+ \in \omega_+(t(x_0)) \cap \omega_+(t(\bar{x}_0))$, thus there are $y, \bar{y} \in \Sigma_{X_0^+}$ such that $y(-\infty, 0] = x(-\infty, 0]$, $\bar{y}(-\infty, 0] = \bar{x}(-\infty, 0]$, and $\rho_0^+(y)[1, \infty) = u_+ = \rho_0^+(\bar{y})[1, \infty)$. Note that $u_+[1,|v|] = v$ forces even $y(-\infty, N] = x(-\infty, N]$, $\bar{y}(-\infty, N] = \bar{x}(-\infty, N]$, thus $\pi(x)_0 = \pi(y)_0$ and $\pi(\bar{y})_0 = \pi(\bar{x})_0$.

On the other hand, the label ww occurs along the future of the equally labeled paths $\pi(y)$ and $\pi(\bar{y})$ in Γ, thus $i(\pi(y)_k) = i(\pi(\bar{y})_k)$ for some $0 \leq k$. Since Φ is left-closing we have also $\pi(y)_0 = \pi(\bar{y})_0$. Thus $\pi(x)_0 = \pi(\bar{x})_0$, so $\tau(z)_0$ is well-defined.

We show that $\tau(z) \in \Sigma_\Gamma$. For that we show $i(\tau(\sigma z)_0) = t(\tau(z)_0)$, where σ denotes the shift map in $\Sigma_{X_0^+}$. By definition we have $\tau(z)_0 = \pi(x)_0$ where $x \in \Sigma_{X_0^+}$ is any point with $\rho_0^+(x) = \Psi(z)$ and $\omega_+(t(x_0)) \supset \omega_+(t(z_0))$. Since ρ_0^+ is right-resolving, σx works for σz. Thus $i(\tau(\sigma z)_0) = i(\pi(\sigma x)_0) = t(\pi(x)_0) = t(\tau(z)_0)$.

We show that $\tau(\Sigma_\Lambda) \subset \Sigma_\Gamma$ is dense. Given a finite path $e_1 \ldots e_n$ in Γ choose a loop path at the above defined vertex α with contains $e_1 \ldots e_n$ as a subpath and is labeled wvw for some block v. Now for any $z \in \Sigma_\Lambda$ with $\Psi(z)[1, |wwvwwl|] = wwvww$ its τ-image has to be in α after $|w|$ and also after $|wwvwl|$ steps, thus it traverses a loop at α with label wvw. Since Φ is left closing there is only one loop path at α labeled wvw, so $\tau(z)$ has to run through the path $e_1 \ldots e_n$.

We show that τ is continuous. Assume not. Then there is $z \in \Sigma_\Lambda$ such that for all n there is $z^{(n)} \in \Sigma_\Lambda$ with $z^{(n)}[-n, n] = z[-n, n]$, but $\tau(z^{(n)})_0 \neq \tau(z)_0$.

Thus, by the definition of τ, there are points $x^{(n)}, x \in \Sigma_{X_0^+}$ such that $\rho_0^+(x^{(n)}) = \Psi(z^{(n)})$, $\rho_0^+(x) = \Psi(z)$, $\omega_+(t(x_0^{(n)})) \supset \omega_+(t(z_0^{(n)}))$, $\omega_+(t(x_0)) \supset \omega_+(t(z_0))$, and $\pi(x^{(n)})_0 \neq \pi(x)_0$ for all n.

Note that $t(z_0^{(n)}) = t(z_0)$. Fix a path starting in $t(z_0)$ with label um, for some block u.

Since m is magic in (X_0^+, ρ_0^+) and X_0^+ is locally finite, the number of vertices $\alpha \in V_{X_0^+}$ with $um \in F_+(\alpha)$, is finite. Thus there are only finitely many vertices $\alpha \in V_{X_0^+}$ such that $\omega_+(\alpha) \supset \omega_+(t(z_0))$. So there is an edge e and a subsequence $x^{(n_k)}$ of $x^{(n)}$ with $x_0^{(n_k)+} = e$ for all k. Since X_0^+ is locally finite, we may as well assume that $x^{(n_k)}$ converges to a point $\bar{x} \in \Sigma_{X_0^+}$. Then $\omega_+(t(\bar{x}_0)) \supset \omega_+(t(z_0))$, and since $\rho_0^+(x^{(n_k)})[-n_k, n_k] = \rho_0^+(x)[-n_k, n_k]$ for all k, we also have $\rho_0^+(\bar{x}) = \rho_0^+(x)$. Thus by the part of the proof that $\tau(z)_0$ is well defined, we obtain $\pi(\bar{x})_0 = \pi(x)_0$. On the other hand $\pi(x^{(n_k)})_0 \neq \pi(x)_0$ for all k implies $\pi(\bar{x})_0 \neq \pi(x)_0$, a contradiction. $\square$

THEOREM 5.6. *Let* S *be a synchronized system. Equivalent are*
i) *Any cover factors continuously through* (X_0^+, ρ_0^+)
ii) (X_0^+, ρ_0^+) *is locally finite and left closing.*

Proof. i) $\Rightarrow$ ii): In particular, (X_0^+, ρ_0^+) is a factor of (X_0, ρ_0). Apply Lemma 5.1.
ii) $\Rightarrow$ i): Apply Theorem 5.5 with $(\Gamma, \Phi) = (X_0^+, \rho_0^+)$. $\square$

REMARK. a) Any continuous factor map that maps some cover to (X_0^+, ρ_0^+) is regular (i.e. it has bounded anticipation).

b) By Theorem 3.2c the synchronized systems with the property that (X_0^+, ρ_0^+) intercepts continuously any other cover, are those which admit some uniformly right closing cover with residual image that is locally finite and left closing.

THEOREM 5.7. *Let* S *be synchronized and* (X_0^+, ρ_0^+) *locally finite. Equivalent are*
i) *There is a cover which is a continuous factor of any other cover.*
ii) (X_0^+, ρ_0^+) *has a left closing factor.*

iii) (X_0^+, ρ_0^+) *and* (X_0^-, ρ_0^-) *have a common continuous factor.*

Proof. i) $\Rightarrow$ iii): trivial. iii) $\Rightarrow$ ii): Lemma 5.1. ii) $\Rightarrow$ i) : Theorem 5.5. □

So far our results parallel those obtained in §4 for almost Markov systems.

EXAMPLES 5.8. But some statements of Theorem 4.2 are no longer equivalent when translated into the continuous setting (i.e. replacing "uniformly left closing" by "left closing" etc.). In the following examples the synchronized systems are the closures of all bi-infinite concatenations of the given block systems. These examples have been found in defining labeled graphs that yield Fischer covers by Theorem 2.16iv, but for expository reasons we often have to leave it to the reader to reconstruct these Fischer covers. The symbol "1" will always be magic in (X_0^+, ρ_0^+).

a) A synchronized system where (X_0^+, ρ_0^+) is locally finite and left-closing (thus intercepts continuously any other cover), but (X_0^-, ρ_0^-) is not even locally finite, thus these two covers cannot be (continuously) conjugate, is given by the set of blocks $1ab^n ab^{n-1} ab^{n-2}...ab^2 aba$, $n \geq 1$. Thus Theorem 4.2ii $\Leftrightarrow$ v has no equivalent here.

b) A synchronized system S which has a cover (Γ, Φ) that intercepts continuously any other cover, but (Γ, Φ) is not continuously isomorphic to (X_0^+, ρ_0^+) or (X_0^-, ρ_0^-), is given by the set of blocks $0^{2^n} 1$, $n \geq 1$ (this contrasts Theorem 4.2: iv $\Leftrightarrow$ iii).

Here (X_0^+, ρ_0^+) is a graph with vertex set, say $\mathbf{N}$, and one edge labeled 0 from i to i+1, $i \in \mathbf{N}$, and one edge labeled 1 from vertex 2^i+1 to vertex 1, $i \in \mathbf{N}$. Since $S = S^{-1}$, (X_0^-, ρ_0^-) is obtained in simply reversing the edges of (X_0^+, ρ_0^+), both Fischer covers are not locally finite. The minimal cover (Γ, Φ) is locally finite and constructed as in the proof of Theorem 1.9 with $m = 1$. (hint: No cover can have the fixed point 0^∞ in its image, thus any cover factors through (Γ, Φ) using the magic word 1).

c) A synchronized systems where (X_0^+, ρ_0^+) and (X_0^-, ρ_0^-) have a common continuous factor, but this factor does not intercept continuously any other cover, is given by the set of blocks $0^{2n} 1$, $n \geq 1$ and $0^{2^n+1} 1$, $n \geq 1$. A common factor of (X_0^+, ρ_0^+) and (X_0^-, ρ_0^-) is the cover (Γ, Φ) constructed in Theorem 1.9 with $m = 1$. This cover does not intercept continuously any other cover since $0^\infty \notin \Phi(\Gamma)$ but there are covers (Λ, Ψ) with $0^\infty \in \Psi(\Lambda)$. In fact this system has no cover which intercepts continuously any other cover (hint: by Remark 5.9 such a cover would have residual image and would be bi-closing. There is no such cover whose image contains 0^∞).

REMARK 5.9. Let S be synchronized and assume that there is a cover (Γ, Φ) which is a continuous factor of any other cover. Then it is in particular a continuous factor of the right and the left Fischer cover, thus by Lemma 5.1: $\Phi(\Gamma)$ is residual in S, Γ is locally finite, Φ is bi-closing and 1-1 a.e. But Example 5.8c shows that a cover with these properties does not have to be a continuous factor of any other cover.

6. Covers without proper factors

The Fischer cover of a sofic system is "minimal" in many different ways (see §0). For example, it has no proper factors ([BKM], Prop.5).

Here we consider arbitrary coded systems and investigate covers which have no proper block (resp. continuous) factors, or, even stronger, for which any block (resp. continuous) factor map is invertible.

First we generalize [BKM, Prop.5] and show that any block factor map defined on the Fischer cover of a half-synchronized system is block invertible (Theorem 6.2).

Then we consider continuous factor maps, rather than uniformly continuous ones.

We show that any cover of a coded system has a locally finite factor (via a continuous factor map) (Theorem 6.4). Thus covers without proper continuous factors have to be locally finite.

We introduce the intermediate concept of "rigid covers" (Definition 6.6). Rigid covers are locally finite, and any cover without proper continuous factor maps is rigid and weakly left closing (Theorem 6.7). We give some sufficient (and manageable) conditions for a locally finite cover to be rigid (Theorem 6.9). In particular any locally finite Fischer cover (of a half-synchronized system) is rigid. The class of rigid covers is shown to be closed under continuous factors (Lemma 6.10). A right resolving cover with all follower sets distinct has no proper continuous factor maps if and only if it is rigid and weakly left closing (Theorem 6.12).

This leads to our main result.

The Fischer cover of a half-synchronized system has no proper continuous factors iff it has no proper continuous factor maps iff it is locally finite and weakly left closing (Theorem 6.13). Finally we extend this result to factors of the Fischer cover.

If not stated otherwise all covers in this section are covers of some coded system S.

DEFINITION 6.1. *Let* (Γ, Φ) *be a cover of a coded system* S. *We say that* (Γ, Φ) *has no proper block factors (resp. no proper continuous factors) iff any other cover which is a block factor (resp. continuous factor) of* (Γ, Φ), *is necessarily block isomorphic (resp. continuously isomorphic) to* (Γ, Φ). *We say* (Γ, Φ) *has no proper block factor maps (resp. no proper continuous factor maps) iff any block factor map (resp. continuous factor map) to any other cover of* S *is necessarily a block (resp. continuous) isomorphism.* ☐

The next theorem is a generalization of [BKM, Prop.5], but we do not use minimality arguments (as Theorem 2.17) or that the Fischer cover is 1-1 a.e. (which is not even true for strictly half-synchronized systems). We prove a general result about saturated (see Definition 2.6) C^+-covers (see Definition 2.1) for coded systems.

THEOREM 6.2. *A saturated* C^+-*cover* (Γ, Φ) *does not have proper block factor maps. In particular, the Fischer cover of a half-synchronized system has no proper block factor maps.*

REMARK. Such covers also have no proper continuous endomorphisms (Lemma 2.10).

Proof. Let $\pi : (\Gamma,\Phi) \to (\Lambda,\Psi)$ be a block-factor map with coding length n. We will show that for any pair x, y $\in \Sigma_\Gamma$ with $\pi(x)[-3n,3n] = \pi(y)[-3n,3n]$ it holds $x_0 = y_0$. This easily implies that π has an inverse π^{-1} with coding length 3n.

For $x_0 = y_0$ it suffices to prove $\omega_+(i(x_0)) = \omega_+(i(y_0))$, since all follower sets in (Γ,Φ) are distinct. Fix $u_+ \in \omega_+(i(x_0))$. Choose any $v_- \in \omega_-(i(y_{-3n}))$.

Choose $\bar{x} \in \Sigma_\Gamma$ such that $\bar{x}(-\infty,-1] = x(-\infty,-1]$ and $\Phi(\bar{x})[0,\infty) = u_+$.

Choose $\bar{y} \in \Sigma_\Gamma$ such that $\bar{y}[-3n,\infty) = y[-3n,\infty)$ and $\Phi(\bar{y})(-\infty,-3n-1] = v_-$.

Let $w = \Phi(x)[-3n,-1] = \Phi(y)[-3n,-1]$.

Since $\pi(\bar{x})_{-n} = \pi(\bar{y})_{-n}$ there is a path labeled $v_- w u_+$ in Λ, in particular $v_- w u_+$ is a point in S. Thus $w u_+$ is compatible with any $v_- \in \omega_-(i(y_{-3n}))$. Because (Γ,Φ) is saturated this implies $w u_+ \in \omega_+(i(y_{-3n}))$. By right resolving we obtain $u_+ \in \omega_+(i(y_0))$. This shows $\omega_+(i(x_0)) \subset \omega_+(i(y_0))$ and equality follows by symmetry. $\square$

A Fischer cover may well have proper continuous factors (it certainly has if it is not locally finite, as we will see in Theorem 6.4.). We show that uniformly left closing Fischer covers do not even have proper continuous factor maps. We shall see (Theorem 6.13) that the hypothesis of uniformly left closing can be considerably weakened.

THEOREM 6.3 *A saturated uniformly left closing* C^+-*cover* (Γ,Φ) *does not have proper continuous factor maps.*

Proof. Let Φ be L-step left closing and $\pi: (\Gamma,\Phi) \to (\Lambda,\Psi)$ a continuous factor map.

Choose any $z \in \Sigma_\Gamma$ and fix an $n \geq L$ such that $z[-n,n]$ determines $\pi(z)_0$. We show that for any other $x \in \Sigma_\Gamma$ with $\pi(x)[-n,n] = \pi(z)[-n,n]$ we have $x_0 = z_0$. For that fix $k \geq n+1$ such that $x[-k,k]$ determines $\pi(x)[-n,n]$. Let $w = \Phi(x)[n+1,k]$. Let $u_+ \in \omega_+(t(x_k))$. Then, for any $v_- \in \omega_-(i(z_{-n}))$ we have that $v_-\Phi(z)[-n,n]w u_+$ is a point in S since it is the label of a path in Λ. Thus $\Phi(z)[-n,n]w u_+ \in \omega_+(i(z_{-n}))$ since (Γ,Φ) is saturated. Thus, by right resolving, $w u_+ \in \omega_+(t(z_n))$. Now let α denote the terminal vertex of the path labeled w and starting at $t(z_n)$. We have shown that $\omega_+(t(x_k)) \subset \omega_+(\alpha)$. But, again by saturatedness and right resolving, it is easily seen that also $\omega_+(\alpha) \subset \omega_+(t(x_k))$. Thus $\alpha = t(x_k)$ since all follower sets are distinct. Now, since $\pi(x)[-n,n] = \pi(z)[-n,n]$ implies $\Phi(x)[-n,n] = \Phi(z)[-n,n]$ and since there is a path from $t(x_n)$ to $t(x_k)$, labeled w, as well as a path from $t(z_n)$ to $t(x_k)$, labeled w, L-step left closing implies $x_0 = z_0$.

Now let $e_0,...,e_L$ be some path of length L+1 in Λ. Let $x \in \Sigma_\Gamma$ with $\pi(x)[0,L] = e_0,...,e_L$. We will show that x_L is uniquely determined by $e_0,...,e_L$. Since L is a constant this implies that π is even block invertible.

Choose a path from $t(e_L)$ to $i(\pi(z)_{-n})$, say $f_1,...,f_j$. By continuity there is some point y in Σ_Γ with $y[0,L] = x[0,L]$, $\pi(y_L) = \pi(x_L)$, and for some $s \leq -2n-1$ it holds $\pi(y)[s-n,s+n] = \pi(z)[-n,n]$. Now consider any point y' in Γ with $\pi(y')[s-n,L] = \pi(y)[s-n,L]$ and $\pi(y')[L+1,L+k+2n+1] = f_1,...,f_k,\pi(z)_{-n},..,\pi(z)_0,..,\pi(z)_n$.

Then $y'_s = z_0$ by step 1, thus by right resolving $y'[0,L] = y[0,L] = x[0,L]$. But also $y'_{L+k+n+1} = z_0$ by step 1. Thus $x_0,..,x_L$ is the initial segment of a path labeled $\Psi(e_1),...,\Psi(e_1),\Psi(f_1),...,\Psi(f_j),\Phi(z_{-n}),...,\Phi(z_{-1})$ leading from $i(x_0)$ to $i(z_0)$. Thus we have by L-step left closing that x_L is uniquely determined by $e_0,...,e_L$. □

Now we are aiming at a characterization of those Fischer covers which do not have proper continuous factors. First we show that locally finiteness is necessary. The proof of this theorem is given at the end of this section.

THEOREM 6.4. *Any cover has a locally finite factor, i.e. given any cover* (Γ,Φ) *of any coded system* S *, then there is a locally finite cover* (Λ,Ψ) *for* S *and a continuous factor map* $\pi\colon (\Gamma,\Phi) \to (\Lambda,\Psi)$.

Since the cover (Γ,Φ) is locally finite iff Σ_Γ is locally compact, which is invariant under continuous conjugacy, we obtain

COROLLARY 6.5. *A cover which has no proper continuous factor maps is locally finite.*

Now we introduce the intermediate concept of rigid covers. Later, Theorem 6.9 gives some manageable sufficient conditions for a cover to be rigid.

DEFINITION 6.6. *Let* $\pi\colon (\Gamma,\Phi) \to (\Lambda,\Psi)$ *be a continuous factor map between two covers of a coded system* S. *Then* π *is called rigid iff for every edge* e *in* Λ *the set of edges* f *in* Γ *with* $\pi([f]_0) \cap [e]_0 \neq \varnothing$ *is finite.*
A cover (Γ,Φ) *is called rigid if any continuous factor map* $\pi\colon (\Gamma,\Phi) \to (\Lambda,\Psi)$, *to any possible factor* (Λ,Ψ), *is rigid.*

Now rigidity is shown to be stronger than locally finiteness, but still necessary for a cover to have no proper continuous factor maps.

THEOREM 6.7. a) *A rigid cover is locally finite.*
b) *A cover* (Γ,Φ) *which has no proper continuous factor maps is rigid and weakly left closing (weakly left closing is defined in §3).*

Proof. a) Let (Γ,Φ) be rigid. By Theorem 6.4 there is a locally finite cover (Λ,Ψ) and a continuous factor map $\pi\colon (\Gamma,\Phi) \to (\Lambda,\Psi)$. Fix $x \in \Sigma_\Gamma$. Fix n such that $x[-n,n]$ determines $\pi(x)_0$. Let α be any vertex of Γ and F the set of edges f with $t(f) = \alpha$. Fix some path $e_1,...,e_k$ from α to $i(x_{-n})$. For each $f \in F$ choose a point $y_f \in \Sigma_\Gamma$ with $(y_f)_0 = f$ and $y_f[1,k+2n+1] = e_1,...,e_k,x_{-n},...,x_n$. The paths $\pi(y_f)[0,k+n]$ in Λ have the same terminal vertex $i(\pi(x)_0)$ for all f. Since Λ is locally finite, the set of edges $\pi(y_f)_0$, $f \in F$, is finite. By rigidity, also F is finite. Similarly the out-degree at α is shown to be finite. Thus (Γ,Φ) is locally finite.
b) By Corollary 6.5 (Γ,Φ) is locally finite. Let $\pi\colon (\Gamma,\Phi) \to (\Lambda,\Psi)$ be any continuous factor map. By hypothesis π is invertible, thus (Λ,Ψ) is locally finite.

Thus, for any edge e in Λ the set $[e]_0$ is compact, thus its image under π^{-1} is compact, thus consists of a finite union of cylinders, thus π is rigid.

Now assume that (Γ, Φ) is not weakly left closing. Then there are points $x, y \in \Sigma_\Gamma$ with $\Phi(x) = \Phi(y)$, $x[0, \infty) = y[0, \infty)$, $x_{-1} \neq y_{-1}$, and for each vertex β there are only finitely many $j \leq 0$ such that $i(x_j) = \beta$ or $i(y_j) = \beta$.

We define a sequence of finite subsets $V_i \subset V_\Gamma$, $i \in \mathbb{N}$. Let $V_1 = \{i(x_0)\}$. Fix $N_1 > 1$ such that $i(y_j), i(x_j) \neq i(x_0)$ for all $j \leq -N_1$. Now, assuming that $V_n \subset V_\Gamma$, $N_n \in \mathbb{N}$ have been chosen for $n \leq i$, then let $V_{i+1} = \{i(x_j), i(y_j) \mid N_i \leq j \leq 0\}$ and choose $N_{i+1} > N_i$ such that $i(y_j), i(x_j) \notin V_{i+1}$ for $j \leq -N_{i+1}$.

We define two functions $f, g : V_\Gamma \to \mathbb{N}$. Let $f(i(x_0)) = g(i(x_0)) = 1$ and $f(\beta) = g(\beta) = N_{i-1}$ for $\beta \in V_i - V_{i-1}$, $i \geq 2$. Extend f, g on the remaining vertices such that $\mathrm{card}\{\alpha \in V_\Gamma \mid f(\alpha) = n\} < \infty$ and $\mathrm{card}\{\alpha \in V_\Gamma \mid g(\alpha) = n\} < \infty$ for all $n \in \mathbb{N}$. Let π be the continuous factor map constructed in the proof of Theorem 6.4. Our choice of f and g implies $\pi(x) = \pi(y)$, so π is a proper continuous factor map. $\square$

We give two sufficient and manageable conditions for rigidity.

DEFINITION 6.8. *We say a cover* (Γ, Φ) *has a maximal follower set at a vertex* α *of* Γ *if* $\omega_+(\alpha) \subset \omega_+(\beta)$, β *any vertex of* Γ, *implies* $\beta = \alpha$.

THEOREM 6.9. a) *A locally finite* C^+-*cover* (Γ, Φ) *with residual image in the one-sided shift is rigid.* b) *A locally finite, saturated, right resolving cover* (Γ, Φ) *which has a maximal follower set is rigid.*

Proof. By Theorem 2.9 any cover in a) is saturated, and by Remark 1.5 there are $\alpha \in V_\Gamma$, $w \in F_-(\alpha)$ such that $F_+(\alpha) = F_+(w)$. We prove a) and b) at the same time.

Let $\pi : (\Gamma, \Phi) \to (\Lambda, \Psi)$ be a continuous factor map. Let in case a) α be the above vertex and let in case b) α be a vertex with a maximal follower set. In case a) α has a maximal follower set among all vertices β with $w \in F_-(\beta)$. Now let $e \in E_\Lambda$. Fix a doubly transitive $x \in \Sigma_\Gamma$ with $\pi(x)_0 = e$. Choose an $n \in \mathbb{N}$ with the properties $t(x_n) = \alpha$, $x[-n,n]$ determines $\pi(x)_0$, and in case a) $\Phi(x)[0,n]$ should end with w.

Now consider any $y \in \Sigma_\Gamma$ with $\pi(y)_0 = \pi(x)_0$. Let $u_+ \in \omega_+(\alpha) = \omega_+(t(x_n))$. We will show that $\Phi(x)[0,n]u_+ \in \omega_+(i(y_0))$. This implies that there is a path, labeled $\Phi(x)[0,n]$, which leads from $i(y_0)$ to some vertex β with the property $\omega_+(\alpha) \subset \omega_+(\beta)$. But then, by maximality, $\beta = \alpha$. Since (Γ, Φ) is locally finite this implies that y_0 has to belong to a finite set of edges, which shows that π is rigid.

For that choose m such that $y[-m,m]$ determines $\pi(y)_0$. Let $v_- \in \omega_-(i(y_{-m}))$. Let $z \in \Sigma_\Gamma$ with $z[-m,m] = y[-m,m]$ and $\Phi(z)(-\infty,-m-1] = v_-$. Considering $\pi(z)$ shows that $v_-\Phi(y)[-m,-1] \in \omega_-(i(e))$. Analogously $\Phi(x)[0,n]u_+ \in \omega_+(i(e))$. Thus $v_-\Phi(y)[-m,-1]\Phi(x)[0,n]u_+$ is a point of the system. By saturatedness $\Phi(y)[-m,-1]\Phi(x)[0,n]u_+ \in \omega_+(i(y_{-m}))$, thus $\Phi(x)[0,n]u_+ \in \omega_+(i(y_0))$ by right resolving. $\square$

The following two lemmata will lead to the converse of Theorem 6.7b for C^+-covers.

LEMMA 6.10. a) *Continuous factor maps defined on rigid covers are closed and thus surjective. So they are isomorphisms iff they are injective.*
   b) *Continuous factors of rigid covers are rigid.*

Proof. a) Let (Γ, Φ) be rigid and $\pi: (\Gamma, \Phi) \to (\Lambda, \Psi)$ a continuous factor map.

Let $A \subset \Sigma_\Gamma$ be closed. Let $z^{(n)} \in \pi(A)$, $n \in \mathbf{N}$, with $z^{(n)} \to z$ for some $z \in \Sigma_\Lambda$. Choose $x^{(n)} \in A$ such that $z^{(n)} = \pi(x^{(n)})$, $n \in \mathbf{N}$. For all large enough n it holds $(z^{(n)})_0 = z_0$. Thus, by rigidity, $\{(x^{(n)})_0 \mid n \in \mathbf{N}\}$ is a finite set. Since Γ is locally finite by Theorem 6.7, the $x^{(n)}$ have a subsequence converging to, say, $x \in A$, and $\pi(x) = z$. So $\pi(A)$ is closed. In particular, $\pi(\Sigma_\Gamma) \subset \Sigma_\Lambda$ is closed, thus $\pi(\Sigma_\Gamma) = \Sigma_\Lambda$.

b) Let $\pi: (\Gamma, \Phi) \to (\Lambda, \Psi)$ and $\pi': (\Lambda, \Psi) \to (\Lambda', \Psi')$ be continuous factor maps, π is onto by a). Let $e \in E_{\Lambda'}$. Since $\pi'\pi$ is rigid, $(\pi'\pi)^{-1}([e]_0)$ is contained in a finite union of zero cylinders $[f_1]_0 \cup ... \cup [f_n]_0$, $f_i \in E_\Gamma$, $1 \le i \le n$. By Theorem 6.7 Γ is locally finite, so $(\pi'\pi)^{-1}([e]_0)$ is compact. Thus $(\pi')^{-1}([e]_0) = \pi((\pi'\pi)^{-1}([e]_0))$ is compact, thus intersects only a finite number of zero cylinders in Λ. So π' is rigid. ◻

LEMMA 6.11. *Let (Γ, Φ) be a rigid C^+-cover and $\pi: (\Gamma, \Phi) \to (\Lambda, \Psi)$ any factor map. Let $x, y \in \Gamma$ with $\Phi(x) = \Phi(y)$ and $\pi(x)_0 = \pi(y)_0$. Then, for some $j \ge 0$, $x_j = y_j$. (And thus $x_i = y_i$ for all $i \ge j$).*

Proof. First we show that by rigidity and C^+ there is a finite path $e_{-N}, ... , e_N$ in Λ (a "magic path" for π) such that there is only one edge f in Γ with

(*) $\pi([f]_0) \cap {}_{-N}[e_{-N}, ... , e_N]_N \ne \varnothing$. For that choose an edge e_0 in Λ. By rigidity there are only finitely many edges $f_1, ..., f_k$ in Γ with $\pi([f_i]_0) \cap [e_0]_0 \ne \varnothing$. Since (Γ, Φ) is locally finite by Theorem 6.7, $\bigcup_{i=1}^{k} [f_i]_0 \subset \Sigma_\Gamma$ is compact. Thus $\pi^{-1}([e_0]_0) \subset \bigcup_{i=1}^{k} [f_i]_0$ is compact and open. So $\pi^{-1}([e_0]_0) = \bigcup_{i=1}^{m} {}_{-L}[g_L^{(i)}, ..., g_L^{(i)}]_L$ for some $L \in \mathbf{N}$, $m \in \mathbf{N}$ and certain paths $g_L^{(i)}, ..., g_L^{(i)}$ in Γ. Some of these finite paths may have the same terminal vertex. But since all follower sets in (Γ, Φ) are distinct and Φ is right resolving, there is a $1 \le j \le n$ and $u \in F_+(t(g_L^{(j)}))$ such that $u \notin F_+(t(g_L^{(i)}))$ for all i with $t(g_L^{(i)}) \ne t(g_L^{(j)})$. Now choose $\bar{z} \in \Sigma_\Gamma$ with $\bar{z}[-L, L] = g_L^{(j)}, ..., g_L^{(j)}$ and $\Phi(\bar{z})[L+1, L+|u|] = u$. Let $N = L + |u|$, then for $e_{-N}, ... , e_N := \pi(z)[-N,N]$ by right resolving there is only one edge f in Γ with $\pi([f]_{L+1}) \cap {}_{-N}[e_{-N}, ... , e_N]_N \ne \varnothing$. Extending and shifting to the left gives a cylinder with the property (*).

Now let $e_{-N}, ... , e_N$ be a path as in (*) and choose $n \in \mathbf{N}$ such that $x[-n,n]$ and $y[-n,n]$ determine $\pi(x)_0$ and $\pi(y)_0$, respectively. We will show that $F_+(t(x_n)) = F_+(t(y_n))$, which implies $x_{n+1} = y_{n+1}$.

Let $u \in F_+(t(x_n))$. Choose $\bar{x} \in \Sigma_\Gamma$ with $\bar{x}(-\infty, n] = x(-\infty, n]$ and $\Phi(\bar{x})[n+1, n+|u|] = u$. By continuity and dense image, there is some point z in Γ with $z_0 = y_0$, $\pi(z_0) = \pi(y_0)$, and for some $r \le -N$ it holds $\pi(z)[r-N, r+N] = e_{-N}, ... , e_N$. Now consider any point y' in Γ with $\pi(y')[r\text{-}N,0] = \pi(z)[r\text{-}N,0]$ and $\pi(y')[1,n+|u|] = \pi(\bar{x})[1,n+|u|]$. Then $(y')_r = f$ and thus by right resolving $(y')_n = y_n$. Thus $u \in F_+(t(x_n))$. This shows $F_+(t(x_n)) \subset F_+(t(y_n))$ and equality follows by symmetry. $\square$

THEOREM 6.12. *Let* (Γ,Φ) *be a* C^+*-cover. Equivalent are the following.*

a) (Γ,Φ) *has no proper continuous factor maps*

b) (Γ,Φ) *is rigid and weakly left closing.*

Proof. a)$\Rightarrow$b):Theorem 6.7b.

b)$\Rightarrow$a): Let $\pi: (\Gamma,\Phi) \to (\Lambda,\Psi)$ be a continuous factor map. By Lemma 6.10a it suffices to show that π is injective. Assume not.

Then there are $x,y \in \Sigma_\Gamma$, with $x_0 \neq y_0$, but $\pi(x) = \pi(y)$ and thus also $\Phi(x) = \Phi(y)$. Then, by Lemma 6.11, it follows that $x_n = y_n$ for some $n > 0$. Considering the rays $(x_j)_{j<n}$ and $(y_j)_{j<n}$,by weakly left closing one of the rays hits some vertex infinitely many times. Thus w.l.o.g. there is a vertex $\alpha \in V_\Gamma$ and a sequence $N_i \in \mathbf{N}$ such that $i(x_{-N_i}) = \alpha$ for all $i \in \mathbf{N}$. Since Γ is rigid, thus locally finite by Theorem 6.7a, we may assume there is an edge e in Γ with $x_{-N_i} = e$ for all $i \in \mathbf{N}$. Now the set $[e]_0$ is compact, thus $\pi([e]_0)$ is compact, thus there is a finite set of edges $E' \subset E_\Lambda$ such that $\pi([e]_0) \subset \cup_{e' \in E'}[e']_0$. Since π is rigid, there is a finite set $E \subset E_\Gamma$ such that $y_{-N_i} \in E$ for all $i \in \mathbf{N}$. Therefore, passing to a subsequence, we may assume that there is an edge $f \neq e$ with $y_{-N_i} = f$ for all $i \in \mathbf{N}$. Since $[e]_0 \cup [f]_0$ is compact, π restricted to $[e]_0 \cup [f]_0$ is uniformly continuous. So there is an $L \in \mathbf{N}$ such that $x[N_i\text{-}L,N_i+L]$ determines $\pi(x)_{-N_i}$ and $y[N_i\text{-}L,N_i+L]$ determines $\pi(y)_{-N_i}$ for all i. Choose three indices $i < j < k$ such that $x_i = x_j = x_k = e$, $y_i = y_j = y_k = f$, and $k - j > L$, $j - i > L$. Let p be the periodic point obtained from the infinite concatenation of the block $x[i,k\text{-}1]$, and let q be obtained from the infinite concatenation of $y[i,k\text{-}1]$. Then $\Phi(p) = \Phi(q)$ and $\pi(p)_j = \pi(q)_j$, but $p \neq q$ since $e \neq f$, this contradicts Lemma 6.11. $\square$

THEOREM 6.13. *Let* (X_0^+,ρ_0^+) *be the Fischer cover of a half-synchronized system. Equivalent are the following.*

a) (X_0^+,ρ_0^+) *has no proper continuous factors.*

b) (X_0^+,ρ_0^+) *has no proper continuous factor maps.*

c) (X_0^+,ρ_0^+) *is locally finite and weakly left closing.*

Proof. b) $\Leftrightarrow$ c):By Theorem 6.9a and Theorem 6.7a (X_0^+,ρ_0^+) is locally finite iff it is rigid. Apply Theorem 6.12.

b) $\Rightarrow$ a) is trivial.We show a) $\Rightarrow$ b). Let $\pi :(X_0^+,\rho_0^+) \to (\Gamma,\Phi)$ be some continuous factor map. By a) there is some isomorphism $\varphi : (X_0^+,\rho_0^+) \to (\Gamma,\Phi)$. Then $\varphi^{-1}\pi$ is a continuous endomorphism of (X_0^+,ρ_0^+). By Lemma 2.10, $\varphi^{-1}\pi$ has to be the identity. Thus π is injective. Since (X_0^+,ρ_0^+) has to be locally finite (otherwise it would have a proper continuous factor by Corollary 6.5), it is rigid by Theorem 6.9a. By Lemma 6.10a the injectivity of π implies that π is continuously invertible. $\square$

EXAMPLE 6.14. All the irreducible components of the Krieger graph for the Dyke system have no proper continuous factors, except for the Fischer cover.

Proof. The explicit construction of the Krieger graph for the Dyke system is left to the reader. The Fischer cover is not weakly left closing by inspection. All the other components are saturated (since they are forwardly and backwardly closed) C^+-covers and left resolving by inspection, thus Theorem 6.3 applies. □

We are going to extend Theorem 6.13 to factors of locally finite Fischer covers.

THEOREM 6.15. *Let* (Γ,Φ) *be a rigid* C^+-*cover and let* (Λ,Ψ) *be a continuous factor of* (Γ,Φ). *Equivalent are the following.*
  a) (Λ,Ψ) *has no proper continuous factor maps.*
  b) (Λ,Ψ) *is weakly left closing.*

Proof. a)⇒b): By Theorem 6.7b.
  b)⇒a): So let (Λ,Ψ) be weakly left closing and let $\pi\colon (\Gamma,\Phi) \to (\Lambda,\Psi)$ be a continuous factor map. Note that (Λ,Ψ) is rigid by Lemma 6.10b, and thus locally finite by Theorem 6.7.
  Assume there is a proper factor map ϑ from (Λ,Ψ) to (Λ',Ψ'). Since (Λ,Ψ) is rigid, the map ϑ is not injective by Lemma 6.10a. Thus there are points $x,y \in \Sigma_\Lambda$, $x \neq y$, with $\vartheta(x) = \vartheta(y)$, thus also $\Psi(x) = \Psi(y)$. By Lemma 6.10a π is surjective, so there are points $\bar{x}, \bar{y} \in \Sigma_\Gamma$ with $\pi(\bar{x}) = x$ and $\pi(\bar{y}) = y$. Then $\bar{x} \neq \bar{y}$ and so, since Φ is right resolving, there is $i_0 \in \mathbf{Z}$ such that $\bar{x}_i \neq \bar{y}_i$ for all $i \leq i_0$. On the other hand $(\pi\varphi)(\bar{x}) = (\pi\varphi)(\bar{y})$ implies by Lemma 6.11 that $\bar{x}_i = \bar{y}_i$ for all large enough i. By changing $\bar{x}, \bar{y}$ in the future to a periodic point, we may assume that $x_i = y_i$ for all large enough i. Thus, by weakly left closing, w.l.o.g., there is a vertex α and a subsequence $N_i \in \mathbf{N}$ such that $i(x_{-N_i}) = \alpha$ for all $i \in \mathbf{N}$. Since (Λ,Ψ) is locally finite we may assume there is an edge e in Λ with $x_{-N_i} = e$ for all $i \in \mathbf{N}$. Since (Γ,Φ) is rigid, the set of edges $\{(\bar{x})_{-N_i} \mid i \in \mathbf{N}\}$ is finite, thus by taking one more subsequence we may assume that for some edge $\bar{e}$ of Γ $(\bar{x})_{-N_i} = \bar{e}$ for all $i \in \mathbf{N}$. So $\bar{x}$ has this property and $\bar{y}$ is another point in Σ_Γ with $\bar{x}_j \neq \bar{y}_j$ for some j and $\vartheta\pi(\bar{x}) = \vartheta\pi(\bar{y})$. In the last paragraph of the proof for Theorem 6.12 it has been shown that this is cannot happen for a factor map $\vartheta\pi$ of a rigid C^+-cover (Γ,Φ). By this contradiction it follows that there is no proper factor map from (Λ,Ψ) to (Λ',Ψ'). □

Corollary 6.16. *Let* (X_0^+,ρ_0^+) *be the Fischer cover of a half-synchronized system and assume that* (X_0^+,ρ_0^+) *is locally finite. Let* (Λ,Ψ) *be a continuous factor of* (X_0^+,ρ_0^+). *Equivalent are the following.*
  a) (Λ,Ψ) *has no proper continuous factors.*
  b) (Λ,Ψ) *is weakly left closing.*

Proof. (X_0^+,ρ_0^+) is a C^+-cover with residual image in the one-sided shift (see §1), and locally finite by assumption, so it is rigid by Theorem 6.9a. Apply Theorem 6.15. □

EXAMPLE 6.17. A synchronized system, where (X_0^+, ρ_0^+) is locally finite and any continuous factor of (X_0^+, ρ_0^+) has a proper continuous factor.

In the Fischer cover (X_0^+, ρ_0^+) of the synchronized systems which is given as the closure of all infinite concatenations of the blocks $1a^n b^{n+k}$, $n \geq 1$, $k \geq 0$, there is a vertex α and an edge e from α to α, labeled b, and a point $x \in \Sigma_{X_0^+}$ with $t(x_0) = \alpha$, the vertices $t(x_j)$, $j \leq 0$, are pairwise distinct, and $\rho_0^+(x_j) = b$ for all $j \in$ **Z**. Considering the left infinite paths $(x_i)_{i \leq 0}$ and $(x_{i+1})_{i \leq 0}$ shows that (X_0^+, ρ_0^+) is not weakly left closing. Since (X_0^+, ρ_0^+) is locally finite and thus rigid, the image of x under any continuous factor map has the same behavior, i.e. in the past it leaves any finite set of vertices, and in the future it finally ends up in a loop edge. Thus no continuous factor is weakly left closing, apply Corollary 6.16 . $\square$

Finally, we give the proof of Theorem 6.4.

Proof of Theorem 6.4. Given any cover (Γ, Φ) of any coded system S we shall show that here is a continuous factor (Λ, Ψ) of (Γ, Φ) which is a cover for S and has finite in-degree, i.e. for each $\alpha \in V_\Lambda$ it holds card$\{e \in E_\Lambda \mid t(e) = \alpha\} < \infty$, and if (Γ, Φ) has finite out-degree then this cover (Λ, Ψ) has finite out-degree, too.

Then the theorem follows in first applying a time-reversal of the construction to obtain a factor of (Γ, Φ) with finite out-degree, and then applying the original construction to this factor to obtain the locally finite factor.

Choose two functions $f, g : V_\Gamma \to \mathbf{N}$ with
(1) card$\{\alpha \in V_\Gamma \mid f(\alpha) = n\} < \infty$ and card$\{\alpha \in V_\Gamma \mid g(\alpha) = n\} < \infty$ for all $n \in \mathbf{N}$.
Depending on (Γ, Φ), f, and g we shall construct a cover (Λ, Ψ) and a continuous factor map $\pi : (\Gamma, \Phi) \to (\Lambda, \Psi)$. In the proof of Theorem 6.7 we use the given freedom in the choice of f and g to define factor maps that identify pre-assigned points.
For a point $x \in \Sigma_\Gamma$ let $f(x) := f(i(x_0))$ and $g(x) := g(i(x_0))$.
Let $s(x) := \min \{f(\sigma^i x) + i \mid i \geq 0\}$, where σ denotes the shift-map σ_Γ on Σ_Γ.
Let $j(x) := \min \{0 \leq i \leq s(x) \mid g(\sigma^i x) \leq g(\sigma^j x)$ for all $0 \leq j \leq s(x)\}$.
Now we show that for each x there is an n with $j(\sigma^n x) = 0$. If $j(x) > 0$ then there is an $n_1 \leq f(x)$ such that $g(x) > g(\sigma^{n_1} x)$. If now $j(\sigma^{n_1} x) > 0$ then there is an $n_2 \leq f(\sigma^{n_1} x)$ such that $g(\sigma^{n_1} x) > g(\sigma^{n_1 + n_2} x)$. Since the function g is bounded by 1 from below, after at most $g(x) - 1$ iterations we have $j(\sigma^{n_1 + n_2 \cdots + n_k} x) = 0$.

Thus we can define a function $\vartheta : \Sigma_\Gamma \to \mathbf{Z}_+$ which assigns to a point $x \in \Sigma_\Gamma$ the number $\vartheta(x) := \min\{n \geq 0 \mid j(\sigma^n x) = 0\}$. It is not hard to see that
(2) $\vartheta(\sigma x) = \vartheta(x) - 1$ if $\vartheta(x) > 0$, since $\vartheta(x)$ depends only on $(x_i)_{i \geq 0}$.
(3) ϑ is continuous, since given $x \in \Sigma_\Gamma$ then $\vartheta(y) = \vartheta(x)$ for any $y \in \Sigma_\Gamma$ with $y[0, \vartheta(x) - 1 + s(\sigma^{\vartheta(x)} x)] = x[0, \vartheta(x) - 1 + s(\sigma^{\vartheta(x)} x)]$.

Although not needed it in the sequel, we found it interesting to remark that there is a function h: $V_\Gamma \to \mathbf{N}$ such that $\vartheta(x)$ is determined by $x[0, h(i(x_0))]$, i.e. the number of coordinates of x needed to determine $\vartheta(x)$ is bounded by a function of the initial vertex of x_0. In each step of the above iteration we had
$n_s \leq f(\sigma^{n_1 + n_2 \cdots + n_{s-1}} x)$ and $g(\sigma^{n_1 + n_2 \cdots + n_{s-1}} x) \leq g(x)$. Thus $\vartheta(x) \leq n_1 + n_2 \cdots + n_k \leq$

$(g(x)-1) \cdot \max\{f(\beta) \mid g(\beta) \leq g(x)\}$. As mentioned above, $\vartheta(x)$ is determined by $x[0, \vartheta(x) -1 + s(\sigma^{\vartheta(x)}x)]$. Inductively one shows $g(\sigma^{\vartheta(x)-m}x) \geq g(\sigma^{\vartheta(x)}x)$, $m = 0, \ldots, \vartheta(x)$. In particular, $g(x) \geq g(\sigma^{\vartheta(x)}x)$. Thus $s(\sigma^{\vartheta(x)}x) \leq f(\sigma^{\vartheta(x)}x) \leq \max\{f(\beta) \mid g(\beta) \leq g(x)\}$, and a choice for h is $h(\alpha) = g(\alpha) \cdot \max\{f(\beta) \mid g(\beta) \leq g(\alpha)\}$, $\alpha \in V_\Gamma$.

To each point x we assign a pair consisting of a vertex and a block of labels. Let $\Theta(x) := (i(x_{\vartheta(x)}), \Phi(x)_0 \ldots \Phi(x)_{\vartheta(x)-1})$, where in the case $\vartheta(x) = 0$ the block $\Phi(x)_0 \ldots \Phi(x)_{\vartheta(x)-1}$ is the empty word, which we also denote by "$*$".

We define the cover (Λ, Ψ). Let $V_\Lambda := \{\Theta(x) \mid x \in \Sigma_\Gamma\}$. We denote vertices in V_Λ also by (α, b). We define the set of labeled edges. Draw one edge e with label $\Psi(e) = c$ leading from (α_1, b_1) to (α_2, b_2) iff there is an $x \in \Sigma_\Gamma$ with $\Theta(x) = (\alpha_1, b_1)$, $\Theta(\sigma x) = (\alpha_2, b_2)$ and $\Phi(x_0) = c$. This defines a countable labeled graph (Λ, Ψ). It is irreducible since (Γ, Φ) was irreducible and because of (3). Thus (Λ, Ψ) is a cover. That (Λ, Ψ) is in fact a cover for the given coded system S, follows from the fact that there is a factor map from (Γ, Φ) to (Λ, Ψ) with dense image, see below.

(4) We show some of the structure of (Λ, Ψ). Let $\Theta(x) = (\alpha, b_0 \ldots b_{n-1})$. Then $\vartheta(x) = n \geq 1$, thus $\vartheta(\sigma x) = \vartheta(x) - 1$ and so $i(\sigma x_{\vartheta(\sigma x)}) = i(x_{\vartheta(x)}) = \alpha$. Thus $\Theta(\sigma x) = (\alpha, b_1 \ldots b_{n-1})$, with $b_1 \ldots b_{n-1} = *$ if $n = 1$. So any vertex $(\alpha, b_0 \ldots b_{n-1})$ with $n \geq 1$ leads deterministically via a path of length n, labeled $b_0 \ldots b_{n-1}$, to the vertex $(\alpha, *)$.

We define a continuous factor map $\pi : (\Gamma, \Phi) \to (\Lambda, \Psi)$. For $x \in \Sigma_\Gamma$ and $j \in \mathbb{Z}$ let $\pi(x)_j$ be the edge of Λ with initial vertex $\Theta(\sigma^j x)$, terminal vertex $\Theta(\sigma^{j+1}x)$, and label $\Phi(x_j) = \Phi((\sigma^j x)_0)$. This gives a well defined map, commuting with the shifts and labeling, which is continuous because of (3). (Note that $\pi(x)_0$ is determined by $x[0, h(i(x_0))]$, i.e. the number of coordinates needed to determine $\pi(x)_0$ can be bounded by a function that depends only on the initial vertex of x_0).

We show that $\pi(\Sigma_\Gamma) \subset \Sigma_\Lambda$ is dense.

Claim. Let $x, y \in \Sigma_\Gamma$ be given with $j(\sigma^{\vartheta(x)}y) = 0$ and $i(x_{\vartheta(x)}) = i(y_{\vartheta(x)})$. Let $z \in \Sigma_\Gamma$ be defined by $z(-\infty, \vartheta(x) - 1] = x(-\infty, \vartheta(x) - 1]$, $z[\vartheta(x), \infty) = y[\vartheta(x), \infty)$. Then
(5) $\pi(z)_i = \pi(x)_i$ for $i = 0, \ldots, \vartheta(x) - 1$ and $\pi(z)_i = \pi(y)_i$ for $i \geq \vartheta(x)$.
(The second equality is trivial since ϑ depends only on the presence and the future).

Proof. It suffices to show $\vartheta(z) = \vartheta(x)$, then (4) applies. Since $j(\sigma^{\vartheta(x)}y) = 0$, also $j(\sigma^{\vartheta(x)}z) = 0$, thus $\vartheta(z) \leq \vartheta(x)$. We use this to show $\vartheta(z) = \vartheta(x)$. If $s(\sigma^{\vartheta(z)}z) \leq \vartheta(x) - \vartheta(z)$ then $z[0, \vartheta(z) - 1 + s(\sigma^{\vartheta(z)}z)] = x[0, \vartheta(z) - 1 + s(\sigma^{\vartheta(z)}z)]$ by definition, thus $\vartheta(z) = \vartheta(x)$ by (3).

Now assume $s(\sigma^{\vartheta(z)}z) > \vartheta(x) - \vartheta(z)$. Then $\vartheta(z) \leq \vartheta(x)$ implies $g(i(z_{\vartheta(z)})) \leq g(i(z_k))$ for $\vartheta(z) \leq k \leq \vartheta(x)$. For $\vartheta(z) \leq k \leq \vartheta(x)$, this implies $g(i(x_{\vartheta(z)})) \leq g(i(x_k))$ since $i(x_k) = i(z_k)$. By the definition of $\vartheta(x)$ also $g(i(x_{\vartheta(x)})) \leq g(i(x_k))$ for $\vartheta(x) \leq k \leq \vartheta(x) + s(\sigma^{\vartheta(x)}x)$. Thus $g(i(x_{\vartheta(z)})) \leq g(i(x_k))$ for $\vartheta(z) \leq k \leq \vartheta(x) + s(\sigma^{\vartheta(x)}x)$, so $j(\sigma^{\vartheta(z)}x) = 0$, so $\vartheta(x) \leq \vartheta(z)$, so $\vartheta(x) = \vartheta(z)$. This proves the claim.

To show that $\pi(\Sigma_\Gamma) \subset \Sigma_\Lambda$ is dense, let $e_0, \ldots, e_n$ be a path in Λ, and assume that $0 = k_1 < \ldots < k_s$ are the indices where $i(e_{k_j}) = (\alpha, *)$ for some $\alpha \in V_\Gamma$. For each j choose a point $x^{(j)} \in \Sigma_\Gamma$ with $\pi(x^{(j)})_{k_j} = e_{k_j}$. (These points exist by definition of (Λ, Ψ). Now consider the concatenation j (or simply $x^{(1)}$ if $s = 1$)

$$x^{(1)}(-\infty, k_2 - 1] \, x^{(2)}[k_2, k_3 - 1] \ldots x^{(s-1)}[k_{s-1}, k_s - 1] \, x^{(s)}[k_s, \infty).$$

Now (4),(5) imply that the π-image of this point runs through the path $e_0, \ldots,$

e_n.

We show that (Λ, Ψ) has finite in-degree. Let $(\alpha, b) \in V_\Lambda$ and consider $e \in E_\Lambda$ with $t(e) = (\alpha, b)$. Then, by (4), either

a) $i(e) = (\alpha, cb)$ and c is a word of length 1, or b) $i(e) = (\beta, *)$ for some $\beta \in V_\Gamma$.

Only for finitely many different edges a) can hold. So assume that b) holds.

If b is the empty word $*$ then let $k=0$, otherwise let k denote the length of b and write $b = b_1 ... b_k$. Because of property (1) of f and g there is an $M \in \mathbb{N}$ such that

(6) $\gamma \in V_\Gamma$, $g(\gamma) \geq M$ implies $f(\gamma) \geq k+1$

The number of edges $e \in E_\Lambda$ such that b) holds and $g(b) < M$, is finite since the number of vertices $b \in V_\Gamma$ with $g(\beta) < M$ is finite by (1).

We show that the number of edges $e \in E_\Lambda$ such that b) holds and $g(\beta) \geq M$, is also finite. By the definition of (Λ, Ψ), for each such edge $e \in E_\Lambda$ there is $x \in \Sigma_\Gamma$ with $\Theta(x) = (\beta, *)$, $\Theta(\sigma x) = (\alpha, b)$ and $\Phi(x_0) = \Psi(e)$. By (4), $\Theta(\sigma^j x) = (\alpha, b_j ... b_k)$ for $1 \leq j \leq k$ and $\Theta(\sigma^{k+1} x) = (\alpha, *)$. Let $\alpha_j = i(x_j)$, $j \in \mathbb{Z}$. Then $\alpha_0 = \beta$, $\alpha_{k+1} = \alpha$. By assumption $g(\beta) \geq M$, thus $f(\beta) \geq k+1$ by (6) and also $g(\alpha_1) \geq M$ (otherwise $\Theta(x) \neq (\beta, *)$, since $s(x) \geq 1$). Assume we have shown $f(\alpha_j) \geq k+1$ and $g(\alpha_j) \geq M$ for all $0 \leq j \leq n$ for some $n < k$. Then $s(x) \geq n$, so $g(\alpha_n) \geq M$ (otherwise $\Theta(x) \neq (\beta, *)$), which implies $f(\alpha_n) \geq k+1$. Thus $f(\alpha_i) \geq k+1$ for $0 \leq i \leq k$. Thus $s(x) \geq k+1$ and $\Theta(x) = (\beta, *)$ implies $g(\beta) \leq g(\alpha_{k+1}) = g(\alpha)$. By (1), the set of β with $g(\beta) \leq g(\alpha)$ is finite for fixed α. So there are only finitely many edges $e \in E_\Lambda$ with $t(e) = (\alpha, b)$.

We show that (Λ, Ψ) has finite out-degree if (Γ, Φ) had finite out-degree. Let $(\alpha, b) \in V_\Lambda$. If b not the empty word, then card$\{e \in E_\Lambda \mid i(e) = (\alpha, b)\} = 1$ by (4).

Now consider $(\alpha, *) \in V_\Lambda$. Assume that card$\{e \in E_\Lambda \mid i(e) = (\alpha, b)\}$ is infinite. Let $e_1, e_2, ...$ be an enumeration of these edges. For each j there is an $x^{(j)} \in \Sigma_\Gamma$ with $\pi(x^{(j)})_0 = e_j$. By the definition of π we have $i(x^{(j)}_0) = \alpha$ for all j. Since $\pi(x)_0$ does not depend on the past of a point x, we may as well assume that all the points $x^{(j)}$ agree in the past. Then the $x^{(j)}$ have a convergent subsequence $x^{(j_k)}$, since (Γ, Φ) has finite out-degree. But by construction $\pi(x^{(j_k)})_0 \neq \pi(x^{(j_{k'})})_0$ for any pair $k \neq k'$, a contradiction. Thus card$\{e \in E_\Lambda \mid i(e) = (\alpha, *)\}$ is finite. $\square$

REFERENCES

[B1] Blanchard, F., B-expansions and symbolic dynamics, Theoret. Comput. Sci. **65** (1989), 131-141.

[B2] Blanchard, F., Codes engendrant certains systemes sofiques, Theoret. Comput. Sci. **68** (1989), 253-265.

[BH] Blanchard, F., Hansel, G., Systèmes codés, Theoret. Comput. Sci. 44 (1986),17-49.

[B-M] Bertrand-Mathis, A., Questions diverses relatives aux systèmes codés: applications au θ-shift, Preprint.

[BKM] Boyle, M., Kitchens, B., Marcus, B., A note on minimal covers for sofic systems, Proc.Amer.Math.Soc. **95** (1985), Nr. 3, 403-411.

[CK] Csiszár, I., Komlós, J., On the equivalence of two models of finite-state noiseless channels from the point of view of the output, Information Theory, Vol.1, Proc.Colloq.Math.Soc. Janos Bolyai 1968, 129-133.

[F1] Fischer, R., Sofic systems and graphs, Monatsh. Math. **80** (1975) ,179-186.

[F2] Fischer, R., Graphs and symbolic dynamics, Information Theory, Vol.16, Proc.Colloq.Math.Soc. Janos Bolyai 1975, 229-244.

[K] Kitchens, B., Ph. D. Thesis, University of North Carolina, Chapel Hill, N.C., 1981.

[Kr1] Krieger, W., On sofic systems I, Israel J. Math. **48** (1984), 305-330 (1984)

[Kr2] Krieger, W., On sofic systems II, Israel J. Math. **60** (1987), 167-176.

[Kr3] Krieger, W., On the uniqueness of the equilibrium state. Math. Systems Theory **8** (1974), 97-104.

[Ku] Kuratowski, K., *Topology*, Vol. 1.Academic Press New York and London, 1966.

[M] Marcus, B., Sofic systems and encoding data, IEEE - Information Theory **31** (1985), 366-377.

[N] Nasu, M., An invariant for bounded-to-one maps between sofic subshifts, Ergod.Th. & Dynam.Sys. **5** (1985), 89-107.

[P] Petersen, K., Chains, entropy, coding, Ergod.Th. & Dynam.Sys. **6** (1986), 415-448.

[W] Weiss, B., Subshifts of finite finite type and sofic systems, Monatsh. Math. **77** (1973), 462-474.

[Wi] Williams, S., Covers of non-almost-finite type sofic systems, Proc.Amer. Math.Soc. **104** (1988), Nr. 1, 245-252.

Institut für Angewandte Mathematik, Universität Heidelberg, 6900 Heidelberg, Germany

Contemporary Mathematics
Volume **135**, 1992

Z-NUMBERS AND β-TRANSFORMATIONS

LEOPOLD FLATTO

ABSTRACT. $\alpha > 0$ is a Z-number if $\left\{\alpha \left(\frac{3}{2}\right)^n\right\} < \frac{1}{2}$ for all integers $n \geq 0$, where $\{x\}$ denotes the fractional part of x. The definition is due to Mahler, who raised the problem of whether such numbers exists. Though the problem is still unsolved, Mahler proves that for every integer $g \geq 0$, $[g, g+1)$ contains at most one Z-number, and that the number of Z-numbers less than $x > 0$ is $O(x^{.7})$. We relate Z-numbers to β-transformations, these being the maps of the unit interval $f(x) = \{\beta x\}$, where $\beta > 1$. Exploiting the relation, we improve on Mahler's estimate. We also obtain analogous results for $\{\alpha \theta^n\}$, where θ is any rational number > 1. The methods used yield asymptotic formulas for the number of admissible sequences of length n in β-expansions and for the number of fixed points of f^n, the nth iterate of f.

1. INTRODUCTION

In his paper "An unsolved problem on the powers of 3/2" [6], Mahler poses the following question:

Do there exist real numbers $\alpha > 0$ such that $\left\{\alpha \left(\frac{3}{2}\right)^n\right\} < \frac{1}{2}$ for all integers $n \geq 0$, where $\{x\}$ denotes the fractional part of x?

Mahler calls such numbers *Z-numbers*. He was apparently motivated to study them by questions involving the value of $g(k)$ in Waring's problem. For any integer $k \geq 2$, $g(k)$ is the smallest number m such that every positive integer is the sum of at most m k^{th} powers of positive integers. It is known [2, P. 337] that $g(k) = 2^k - \left[\left(\frac{3}{2}\right)^k\right] - 2$ for $k \geq 6$ whenever the inequality

$$\left\{\left(\frac{3}{2}\right)^k\right\} \leq 1 - \frac{\left[\left(\frac{3}{2}\right)^k\right] + 1}{2^k}$$

holds, $[x]$ denoting the integral part of x. Mahler [5] proved that the inequality holds for all but a finite number of k, and it remains an unsolved problem to effectively determine the exceptional k, if any. Resolving this question motivates the study of the fractional parts $\left\{\left(\frac{3}{2}\right)^k\right\}$.

The distribution of the fractional parts $\left\{\left(\frac{3}{2}\right)^n\right\}$, $0 \leq n < \infty$, over the interval $[0, 1]$ is an intriguing and mysterious subject. Numerical work indicates that $\left\{\left(\frac{3}{2}\right)^n\right\}$ is uniformly distributed on $[0, 1]$, but the only definite result is that of Vijayaraghavan [17] stating that, for p and q relatively prime integers with $p > q \geq 2$,

1980 *Mathematics Subject Classification* (1985 *Revision*). Primary 26A18, 58F03, 58F11, 58F20.

This paper is in final form and no version of it will be submitted for publication elsewhere.

the sequence $\left\{ \left(\frac{p}{q} \right)^n \right\}$ has an infinite number of limit points. No limit points are known explicitly for any p and q. Pollington [10, p. 123] has proved that there are no $\alpha > 0$ for which $\left\{ \alpha \left(\frac{p}{q} \right)^n \right\} < \frac{1}{p}$ for all integers $n \geq 0$. We rederive this result in §7. Choosing $\alpha = \left(\frac{p}{q} \right)^k$, k any integer ≥ 0, we obtain $\left\{ \left(\frac{p}{q} \right)^{k+n} \right\} \geq \frac{1}{p}$ for some integer $n \geq 0$. Hence $\left\{ \left(\frac{p}{q} \right)^n \right\} \geq \frac{1}{p}$ infinitely often, in particular there is at least one limit point $\geq \frac{1}{p}$. Similarly, the non-existence of Z-numbers implies that $\left\{ \left(\frac{3}{2} \right)^n \right\} \geq \frac{1}{2}$ infinitely often, in particular there is at least one limit point $\geq \frac{1}{2}$.

Mahler does not prove that Z-numbers do not exist, which seems a very difficult problem. By a theorem of Weyl [18], the sequence $\left\{ \alpha \left(\frac{3}{2} \right)^n \right\}$ is uniformly distributed on $[0,1]$ for almost all $\alpha > 0$. Hence Z-numbers form at most a set of Lebesgue measure zero. Mahler improves on this fact and proves the following two results, which we call Mahler's first and second theorems.

Mahler's First Theorem.   *For any integer $g \geq 0$, there exists at most one Z-number in $[g, g+1)$. Since $\{\alpha\} < \frac{1}{2}$ for a Z-number, α lies in fact in $\left[g, g + \frac{1}{2} \right)$.*

Mahler's Second Theorem.   *For $x > 0$, let $Z(x)$ be the set of Z-numbers $\leq x$, and $|Z(x)|$ the cardinality of $Z(x)$. Then $|Z(x)| = O(x^\delta)$, where $\delta = \log_2 \left(\frac{1+\sqrt{5}}{2} \right) = .70 \ldots$.*

Let Z be the set of Z-numbers. The first theorem implies that Z is at most countable, and the second that it has density zero.

The proofs of the two theorems rely on notions from symbolic dynamics. Mahler's main idea is to show that α is a Z-number iff its integral and fractional parts have certain symbolic representations which are identical (Proposition P of §3). This criterion for Z-numbers is exploited to prove Mahler's two theorems. The symbolic representation for the fractional part is the so called β-expansion related to the β-transformation. We recall the definition of these concepts. The β-*transformation*, $\beta > 1$, is the map $f(x) = \{\beta x\}$, of $I = [0,1)$ to itself. The intervals $J_k = \{x : [\beta x] = k\}$, $0 \leq k \leq [\beta]$, form a partition of I. The β-*expansion* of x is the sequence $\{\epsilon_n(x)\}$, $n \geq 0$, defined by $f^n(x) \in J_{\epsilon_n}(x)$, where f^n is the nth iterate of f.

The graph of $f(x)$ is given by a sawtooth function consisting of straight line segments of slope β. The case $\beta = \frac{3}{2}$ is depicted in Figure 1i), where J_k is labeled by k. Over each J_k, $0 \leq k < [\beta]$, the y-coordinate of the line segment varies from 0 to 1. Over $J_{[\beta]}$, the y-coordinate varies from 0 to $\{\beta\}$. Similarly the graphs of the iterates f^n consist of line segments of slope β^n. For $\beta = \frac{3}{2}$, $f^2(x)$ is depicted in Figure 1ii).

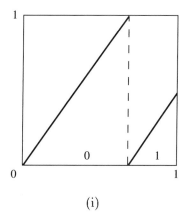

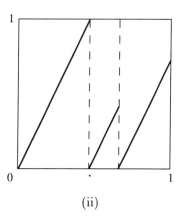

(i)   (ii)

Figure 1. Graphs of f and f^2.

The theory of β-transformations, in particular their ergodicity and invariant measure, is given in [9, 12].

In [6] the concepts of β-transformation and corresponding β-expansions, are mostly implicit. We elaborate these notions in the present paper. In doing so, we obtain new results on β-transformations (mainly Theorem 5.2 in §5), and the following improvement on Mahler's second theorem:

$$(1.1) \qquad |Z(x)| = O(x^\delta), \text{ where } \delta = \log_2\left(\frac{3}{2}\right) = .59\ldots$$

The plan of the paper is as follows. In §2, we review the proof of Mahler's first theorem. The proof depends on the easily proven fact that a number is determined by its β-expansion. The proof of the second theorem requires more knowledge. We need to estimate N_n, the number of distinct sequences $\{\epsilon_0, \ldots, \epsilon_{n-1}\}$, $n \geq 1$, arising in β-expansions. These sequences, called *admissible*, are studied in §3. In §4 we obtain asymptotic formulas for N_n and related quantities, as $n \to \infty$. The asymptotic formula for N_n has previously been obtained by Ito and Takahashi [4]. Aside from their application to Z-numbers, these formulas imply other interesting facts about β-transformations. We present these in §5, where we obtain the asymptotic behavior of $P_n(a, b)$, the number of fixed points of f^n in $[a, b)$, $0 \leq a < b \leq 1$. The asymptotics of $P_n(a, b)$ lead to another derivation of the invariant measure for the β-transformation, first obtained by Parry [9]. In §6, we return to Z-numbers and use the results of §4 to derive Mahler's second theorem and its improvement. Finally, in §7, we generalize the notion of Z-numbers by replacing 3/2 and 1/2 by $\theta > 1$ and $0 < t < 1$. For certain values of θ and t, our results on β-transformations give generalizations of Mahler's theorems. For these values of θ and t, the set

Z is at most countable. For certain other values of θ and t, we show that Z has cardinality C of the continuum. Pollington [10, 11] has obtained criteria of a similar kind guaranteeing that Z be at most countable and that Z have cardinality C. Tijdeman [15] has obtained criteria guaranteeing that Z be infinite. Combining his results with ours, we obtain examples for which Z is an infinite countable set.

The subject of this paper bears some relation to the papers of Milnor, Thurston [7] and Misiurewicz, Zlenk [8], where similar problems are considered for continuous maps of the unit interval, as opposed to the β-transformations which have discontinuities. Related work on β-expansions include [1, 3, 4, 13, 14].

2. Proof of Mahler's First Theorem

For every integer $n \geq 0$ let

$$(2.1) \qquad \alpha \left(\frac{3}{2} \right)^n = g_n + \frac{r_n}{2}$$

where g_n is an integer and $0 \leq r_n < 1$. g_n and $\frac{r_n}{2}$ are the integral and fractional parts of $\alpha \left(\frac{3}{2} \right)^n$. Suppose $0 \leq r_n < 1$. We investigate when $r_{n+1} < 1$. Multiplying both sides of 2.1) by $\frac{3}{2}$,

$$(2.2) \qquad g_{n+1} + \frac{r_{n+1}}{2} = \frac{3g_n}{2} + \frac{3r_n}{4}$$

We distinguish two cases.
A) g_n is even, hence $\frac{3g_n}{2}$ is an integer. Since $\frac{3r_n}{4} < \frac{3}{4}$ for $r_n < 1$, we equate integral and fractional parts in 2.2) to obtain

$$(2.3) \qquad g_{n+1} = \frac{3g_n}{2}, \; r_{n+1} = \frac{3r_n}{2}$$

The second equation of 2.3) shows that $r_{n+1} < 1$ iff $r_n < \frac{2}{3}$.
B) g_n is odd, hence both $\frac{3g_n \pm 1}{2}$ are integers. Rewrite 2.2) as

$$(2.4) \qquad g_{n+1} + \frac{r_{n+1}}{2} = \frac{3g_n - 1}{2} + \left(\frac{3r_n}{4} + \frac{1}{2} \right) = \frac{3g_n + 1}{2} + \left(\frac{3r_n}{4} - \frac{1}{2} \right)$$

Equating fractional parts, we get

$$(2.5) \qquad 1 \leq r_{n+1} = \frac{3r_n}{2} + 1 < 2, \text{ if } 0 \leq r_n < \frac{2}{3}$$

and

$$(2.6) \qquad r_{n+1} = \frac{3r_n}{2} - 1 < \frac{1}{2}, \text{ if } \frac{2}{3} \leq r_n < 1$$

2.5), 2.6) show that $r_{n+1} < 1$ iff $\frac{2}{3} \leq r_n < 1$, in which case

$$(2.7) \qquad g_{n+1} = \frac{3g_n + 1}{2}, \; r_{n+1} = \frac{3r_n}{2} - 1$$

We combine the two cases into the one statement 2.9). Let $\mathbb{N}$ be the set of non-negative integers and $\mathbb{N}_0, \mathbb{N}_1$ the sets of even and odd non-negative integers. Let $I = [0, 1)$, $I_0 = [0, \frac{2}{3})$, $I_1 = [\frac{2}{3}, 1)$. Define

$$(2.8) \quad T(g) = \begin{cases} \frac{3g}{2}, & g \in \mathbb{N}_0 \\ \\ \frac{3g+1}{2}, & g \in \mathbb{N}_1 \end{cases}, \quad f(r) = \left\{ \frac{3}{2}r \right\} = \begin{cases} \frac{3r}{2}, & r \in I_0 \\ \\ \frac{3r}{2} - 1, & r \in I_1 \end{cases}$$

T is a transformation on $\mathbb{N}$, and f is the β-transformation on I with $\beta = \frac{3}{2}$. From the above discussion,

(2.9) $\quad r_n$, $r_{n+1} < 1$ iff either $g_n \in \mathbb{N}_0$ and $r_n \in I_0$, or $g_n \in \mathbb{N}_1$ and $r_n \in I_1$.

Equations 2.3), 2.7) convert to

$$(2.10) \qquad\qquad g_{n+1} = T(g_n), \; r_{n+1} = f(r_n)$$

Repetition of the above argument for all $n \geq 0$ gives

Proposition P: *Let $\alpha = g_0 + \frac{r_0}{2}$, where g_0 is an integer ≥ 0 and $0 \leq r_0 < 1$. For $n \geq 0$, define $\epsilon_n = \epsilon_n(g_0)$ by $T^n(g_0) \in \mathbb{N}_{\epsilon_n}$, and $\delta_n = \delta_n(r_0)$ by $f^n(r_0) \in I_{\delta_n}$. α is a Z-number iff $\epsilon_n = \delta_n$ for $n \geq 0$. In this case the numbers g_n, r_n defined by 2.1) are given respectively by*

$$(2.11) \qquad\qquad g_n = T^n(g_0), \; r_n = f^n(r_0) .$$

The sequences $\{\epsilon_n(g_0)\}$ and $\{\delta_n(r_0)\}$ are called respectively the T- and β-*expansions* of g_0 and r_0. From $f^{k+1}(r_0) = \frac{3}{2}f^k(r_0) - \delta_k$, $0 \leq k \leq n$, we get

$$(2.12) \quad r_0 = \left(\frac{2}{3}\right)\delta_0 + \left(\frac{2}{3}\right)^2 \delta_1 + \ldots + \left(\frac{2}{3}\right)^{n+1} \delta_n + \left(\frac{2}{3}\right)^{n+1} f^{n+1}(r_0)$$

Letting $n \to \infty$,

$$(2.13) \qquad r_0 = \left(\frac{2}{3}\right)\delta_0 + \left(\frac{2}{3}\right)^2 \delta_1 + \ldots + \left(\frac{2}{3}\right)^{n+1} \delta_n + \ldots$$

Proposition P together with 2.13) imply the first theorem. For let $\{\epsilon_n(g_0)\}$ be the T-expansion of g_0. Then $g_0 + \frac{r_0}{2}$ is a Z-number iff $\{\epsilon_n(g_0)\}$ is also the β-expansion of r_0. But then r_0 is determined by 2.13), with δ_n replaced by $\epsilon_n(g_0)$. $\qquad\blacksquare$

3. Admissible Sequences

We introduce some concepts from symbolic dynamics and apply these to the study of the transformations T_r and f_β defined below. Just as T and f were used to obtain Mahler's first theorem, so T_r and f_β will be used in §7 to obtain a generalized version of this theorem.

Definition 3.1. i) Let U be a transformation on the space X and $P = \{X_0, \ldots, X_k\}$, $k \geq 1$, a finite partition of X. For $x \in X$ and $n \geq 0$, define $\epsilon_n = \epsilon_n(x)$ by $U^n(x) \in X_{\epsilon_n}$. $\{U^n(x)\}$ is called the U-*orbit* of x, and $\{\epsilon_n(x)\}$ the U-*expansion* of x with respect to the partition P.

ii) A finite sequence $\{\epsilon_i\}$, $0 \leq i \leq n$, where each $\epsilon_i \in \{0, 1, \ldots, k\}$, is called U-admissible iff $\epsilon_i = \epsilon_i(x)$, $0 \leq i \leq n$, for some $x \in X$. Infinite U-admissible sequences are defined in a similar manner.

We shall write admissible instead of U-admissible when this creates no ambiguity. For a finite sequence $\{\epsilon_0, \ldots, \epsilon_n\}$, each $\epsilon_i \in \{0, 1, \ldots, k\}$, let $X[\epsilon_0 \ldots \epsilon_n] = \{x \in X : U^k(x) \in X_{\epsilon_i}, 0 \leq i \leq n\}$. $X[\epsilon_0 \ldots \epsilon_n] \neq \emptyset$, the empty set, iff $\{\epsilon_0, \ldots, \epsilon_n\}$ is admissible. Similar notation is employed for infinite sequences.

Let $r = \frac{p}{q}$, p and q relatively prime positive integers with $p > q \geq 2$. Let $\mathbb{N}$ be the set of non-negative integers and $\mathbb{N}_i = \{g \in \mathbb{N} : pg + i \equiv 0 \pmod{q}, 0 \leq i \leq q - 1\}$. $\{\mathbb{N}_0, \mathbb{N}_1, \ldots, \mathbb{N}_{q-1}\}$ is a partition of $\mathbb{N}$. Define

$$(3.1) \qquad T_r(g) = \frac{pg + i}{q} \text{ for } g \in \mathbb{N}_i, \ 0 \leq i \leq q - 1$$

T_r is a transformation on $\mathbb{N}$. In particular $T_{3/2}$ is the transformation T of §2. From now on we write T for T_r, as this does not create ambiguity. The T-expansion of g will always be understood to be with respect to the partition $\{\mathbb{N}_0, \mathbb{N}_1, \ldots, \mathbb{N}_{q-1}\}$.

For given integers a and $d > 0$, the set of non-negative numbers $a + kd$, k an integer, is called a positive arithmetic sequence with difference d.

Theorem 3.1. For given integer $n \geq 0$, all q^{n+1} sequences $\{\epsilon_0, \ldots, \epsilon_n\}$, each $\epsilon_i \in \{0, 1, \ldots, q - 1\}$, are T-admissible. For each of these, $\mathbb{N}[\epsilon_0 \ldots \epsilon_n]$ is a positive arithmetic sequence with difference q^{n+1}. The converse also holds.

Proof: Let A be a positive arithmetic sequence with difference q^{n+1}. Choose $g_0 \in A$. A is the set of non-negative integer solutions g_0' to the congruence $g_0' \equiv g_0 \pmod{q^{n+1}}$. Let $g_k = T^k(g_0)$, $g_k' = T^k(g_0')$. From the definition of T, $g_{k+1}' - g_k = \frac{p}{q}(g_k' - g_k)$ whenever $g_k' \equiv g_k \pmod{q}$. Hence

$$g_1' \equiv g_1 \pmod{q^n}, \ g_2' \equiv g_2 \pmod{q^{n-1}}, \ldots, g_n' \equiv g_n \pmod{q} .$$

The congruences show that, for $0 \leq k \leq n$, g_k' and g_k belong to the same $\mathbb{N}_i$. Hence $A \subset \mathbb{N}[\epsilon_0 \ldots \epsilon_n]$, where $\epsilon_i = \epsilon_i(g_0)$, $0 \leq i \leq n$.

Conversely, let $g_0' \in \mathbb{N}[\epsilon_0 \ldots \epsilon_n]$ where $\epsilon_i = \epsilon_i(g_0)$, $0 \leq i \leq n$. We have $g_{k+1} = \frac{pg_k + \epsilon_k}{q}$, $0 \leq k \leq n$, with similar equations for g_k'. Solving for g_0,

$$(3.2) \qquad g_0 = -\frac{1}{p}\left\{\epsilon_0 + \frac{q}{p}\epsilon_1 + \ldots + \left(\frac{q}{p}\right)^n \epsilon_n\right\} + \left(\frac{q}{p}\right)^{n+1} g_{n+1}$$

with a similar equation for g_0'. Subtracting the two equations, we obtain $g_0' - g_0 = \left(\frac{q}{p}\right)^{n+1}(g_{n+1}' - g_{n+1})$. Hence $p^{n+1}(g_0' - g_0) = q^{n+1}(g_{n+1}' - g_{n+1})$ is divisible by q^{n+1}. As $(p, q) = 1$, we obtain $g_0' - g_0 \equiv 0 \pmod{q^{n+1}}$. Hence $\mathbb{N}[\epsilon_0 \ldots \epsilon_n] \subset A$, and so $A = \mathbb{N}[\epsilon_0 \ldots \epsilon_n]$. Since there are q^{n+1} distinct positive arithmetic sequences with difference q^{n+1}, we conclude that all q^{n+1} sequences $\{\epsilon_0, \ldots, \epsilon_n\}$, each $\epsilon_i \in \{0, 1, \ldots, q - 1\}$, are admissible. ∎

Theorem 3.1 shows that for an infinite admissible sequence $\{\epsilon_n\}$, $\mathbb{N}[\epsilon_0 \ldots \epsilon_n \ldots]$ consists of only one integer. For if $g_0, g_0' \in \mathbb{N}[\epsilon_0 \ldots \epsilon_n \ldots]$, then $g_0, g_0' \in \mathbb{N}[\epsilon_0 \ldots \epsilon_n]$ for $n \geq 0$. Hence $g_0' - g_0$ is divisible by q^{n+1} for all $n \geq 0$, implying that $g_0' = g_0$.

In distinction to the finite sequences, the description of the infinite T-admissible sequences is a mystery.

For $\beta > 1$, let $f_\beta(x) = \{\beta x\}$, $0 \le x < 1$. $f_\beta(x)$ is called the β-*transformation*. $f_\beta(x)$ maps $[0,1)$ onto $I = [0,1)$. In particular $f_{3/2}$ is f of §2. Let $J_k = \{x : [\beta x] = k\}$, $0 \le k \le [\beta]$. Then $J_k = \left\{x : \frac{k}{\beta} \le x < \frac{k+1}{\beta}\right\}$ for $k < [\beta]$, and $J_{[\beta]} = \left\{x; \frac{[\beta]}{\beta} \le x < 1\right\}$ ($J_{[\beta]} = \emptyset$ iff β is an integer). $\{J_0, J_1, \ldots, J_{[\beta]}\}$ is a partition of I.

From now on, we write f for f_β. The f-*expansion* $\{\epsilon_n(x)\}$ will always be understood to be with respect to the partition $\{J_0, \ldots, J_{[\beta]}\}$. Following the terminology used in the literature, we rename it the β-*expansion*. By definition, $\epsilon_n(x) = [\beta f^n x]$, so $f^{n+1}x = \{\beta f^n x\} = \beta f^n x - \epsilon_n$, $n \ge 0$. Solving these equations for $x = f^o(x)$, we get

$$(3.3) \qquad x = \sum_{n=0}^\infty \frac{\epsilon_n(x)}{\beta^{n+1}}$$

Thus x is determined by its β-expansion.

It seems natural and proves convenient to define $f(1)$ to be $\{\beta\}$, and the β-*expansion* of 1 to be $\{a_n\}$, $n \ge 0$, where $a_0 = [\beta]$ and $\{a_{n+1}\}$ is the β-expansion of $\{\beta\}$. Then $a_n = [\beta f^n 1]$, $n \ge 0$, and

$$(3.4) \qquad 1 = \sum_{n=0}^\infty \frac{a_n}{\beta^{n+1}}$$

β is called a β-*number* if the orbit $\{f^n 1\}$, $n \ge 0$, assumes only a finite number of distinct values. It is called a *simple* β-*number* if $f^{n_0} 1 = 0$ for some n_0, in which case $f^n 1 = 0$ for $n \ge n_0$. Equivalently, from $f^k 1 = \sum_{n=0}^\infty \frac{a_{n+k}}{\beta^{n+1}}$, we obtain that β is a β-number if $\{a_n\}$ is periodic from some n on, and a simple β-number if $a_n = 0$ from some n on (see [9]). We order finite or infinite sequences of integers of the same length lexicographically. I.e. $(b_0 \ldots b_n \ldots) < (c_0 \ldots c_n \ldots)$ when $b_n < c_n$ for the first n for which $b_n \ne c_n$.

Theorem 3.2 below describes the f-admissible sequences. The theorem follows essentially from results of Parry [9]. We do not prove it as it is not used in the sequel, but state it to contrast results for T and f.

Theorem 3.2. *i) If β is not simple, then $\{b_n\}$, $0 \le n < \infty$, is admissible iff $\{b_{n+k}\} < \{a_n\}$ for all integer $k \ge 0$.*

If β is simple with a_p the last non-zero term in $\{a_n\}$, then let $\{c_n\}$, $0 \le n < \infty$ be the periodic sequence of period $(p+1)$ defined by the initial block $a_0, a_1, \ldots, a_{p-1}, a_p - 1$. $\{b_n\}$, $0 \le n < \infty$, is admissible iff $\{b_{n+k}\} < \{c_n\}$ for all integer $k \ge 0$.

ii) If β is not simple, then the finite sequence $b_0 \ldots b_n$ is admissible iff:

$$b_0 b_1 \ldots b_n \leq a_0 a_1 \ldots a_n$$

$$b_1 \ldots b_n \leq a_0 a_1 \ldots a_{n-1}$$

$$\ldots \ldots$$

$$b_n \leq a_0$$

If β is simple, then the same result holds with the a_i's replaced by c_i's.

Comparing T and f, we observe that the finite T-admissible sequences have a simpler description. However, for infinite admissible sequences, the reverse is true.

We illustrate Theorem 3.2 ii) with the following example. For $\beta = \frac{3}{2}$, the f-orbit of 1 is: $1, \frac{1}{2}, \frac{3}{4}, \frac{1}{8}, \frac{3}{16}, \ldots$, so that $\{a_n\} = 1, 0, 1, 0, 0, \ldots$. From Theorem 3.2 we obtain that the admissible sequences of length n, $1 \leq n \leq 4$, are the 0, 1-sequences not containing the block 11. The admissible sequences of length 5 are the 0, 1-sequences distinct from 10101 and not containing the blocks 11. These facts can also be checked directly, without use of Theorem 3.2. Since $\epsilon_n(f^k x) = \epsilon_{n+k}(x)$ for any integer $k \geq 0$, we also conclude that the blocks $11, 10101$ do not appear anywhere in any β-expansion.

For the sequence $\{b_0 \ldots b_{n-1}\}$, each $b_i \in \{0, 1, \ldots, [\beta]\}$, let $I[b_0 \ldots b_{n-1}] = \{x : f^i(x) \in J_{b_i}, 0 \leq i \leq n-1\}$. For $n = 0$, $I[b_0 \ldots b_{n-1}]$ is interpreted to be I. Observe that $I[b_0] = J_{b_0}$. We collect the following facts about the sets $I[b_0 \ldots b_{n-1}]$, $n \geq 0$.

Theorem 3.3. *i) For any admissible $\{b_0, \ldots, b_{n-1}\}$, $I[b_0 \ldots b_{n-1}]$ is a half-open interval containing its left end point. These intervals are called n^{th} stage intervals and denoted generically by I_n.*
ii) Any I_n is a finite union of I_{n+1}'s.
iii) On each I_{n+1}, f is linear with slope β and maps I_{n+1} into an I_n. Except for the right most I_{n+1}, f maps I_{n+1} onto $I_n = f(I_{n+1})$.

iv) Let $\lambda(I_n)$ denote the length of I_n. Then $\lambda(I_n) \leq \frac{1}{\beta^n}$, so that $\lim_{n \to \infty} \left(\sup_{I_n} \lambda(I_n) \right) = 0$.

Proof: i) For $n = 0$, $I_0 = I$ and i) is obvious. Suppose i) holds for n. If $\{b_0, \ldots, b_n\}$ is admissible, then so is $\{b_1, \ldots, b_n\}$. From the definition,

$$(3.5) \qquad\qquad I[b_0 \ldots b_n] = J_{b_0} \cap f^{-1} I[b_1 \ldots b_n] \ .$$

f is linear with slope β on J_{b_0}, mapping it onto $[0, 1)$ for $b_0 < [\beta]$ and onto $[0, \{\beta\})$ for $b_0 = [\beta]$. By assumption, $I[b_1 \ldots b_n]$ is half-open containing its left end point. 3.5) then shows that the same is true for $I[b_0 \ldots b_n]$.
ii) By definition, $I_n = I[b_0 \ldots b_{n-1}]$ is the union of those $I[b_0 \ldots b_n]$ for which $\{b_0, \ldots, b_{n-1}, b_n\}$ is admissible.
iii) For $b_0 < [\beta]$, f maps J_{b_0} onto $[0, 1)$. 3.5) then shows that f maps $I[b_0 \ldots b_n]$ onto $I[b_1 \ldots b_n]$. For $b_0 = [\beta]$, f maps J_{b_0} onto $[0, \{\beta\})$. 3.5) then shows that f maps $I[b_0 \ldots b_n]$ onto $I[b_1 \ldots b_n]$, unless $\{\beta\}$ is in the interior of $I[b_1 \ldots b_n]$. In the latter case $I[b_0 \ldots b_n]$ is the right most I_{n+1}, and f maps it onto $I[b_1 \ldots b_n] \cap [0, \{\beta\})$.
iv) By iii), $\lambda(I_{n+1}) \leq \frac{\lambda(I_n)}{\beta}$. iv) follows by induction. ∎

4. Asymptotics of N_n

For any integer $n \geq 0$, let N_n be the number of f-admissible sequences of length n. Equivalently, N_n is the number of nth stage intervals. We obtain the asymptotic behavior as $n \to \infty$ of N_n. We show in Lemma 4.2 that over each nth stage interval I_n, the graph of f^n consists of a line segment of slope β^n whose y coordinate varies from 0 to $f^k 1$ for some integer $k \geq 0$. We first obtain an asymptotic formula for the number of I_n's corresponding to a given k. These formulas in turn lead to the asymptotics of N_n.

For simple β, we denote by $N = N(\beta)$ the smallest positive integer n for which $f^n 1 = 0$. For non-simple β, $N = N(\beta)$ is defined to be ∞. We recall that $a_k = [\beta f^k 1]$, $0 \leq k < \infty$.

Lemma 4.1. *Let f^n be linear with slope β^n on I_n, and mapping I_n onto $[0, f^k 1)$ for some $0 \leq k \leq N - 1$. Then:*

for $k \leq N - 2$, I_n is the union of $(a_k + 1) I_{n+1}$'s. On each of these f^{n+1} is linear with slope β^{n+1}. f^{n+1} maps the right most of these onto $[0, f^{k+1} 1)$ and all others onto $[0, 1)$.

for $k = N - 1$, I_n is the union of $a_k I_{n+1}$'s. On each of these, f^{n+1} is linear with slope β^{n+1} and maps it on $[0, 1)$.

Proof: If $k \leq N - 2$, then $f(f^k 1) = f^{k+1} 1 \neq 0$. As $f(x) = 0$ only for βx an integer, we conclude that $a_k < \beta f^k 1 < a_k + 1$. Thus $[0, f^k 1) = J_0 \cup \ldots \cup J_{a_k - 1} \cup K_k$ where $K_k = \left[\frac{a_k}{\beta}, f^k 1 \right) \neq \emptyset$. By assumption f^n is linear with slope β^n on I_n, mapping it onto $[0, f^k 1)$. Hence I_n is the union of $(n+1)^{th}$ stage intervals denoted by $I_{n\ell}$, $0 \leq \ell \leq a_k$, f^n mapping $I_{n\ell}$ onto J_ℓ for $0 \leq \ell < a_k$ and I_{na_k} onto K_k. Furthermore I_{na_k} is the right most of the $I_{n\ell}$. f is linear with slope β on each J_ℓ mapping it onto $[0, 1)$, and linear with slope β on K_k mapping it onto $[0, f^{k+1} 1)$. Hence $f^{n+1} = f \circ f^n$ is linear with slope β^{n+1} on each $I_{n\ell}$. f^{n+1} maps $I_{n\ell}$ onto $[0, 1)$ for $\ell < a_k$, and I_{na_k} onto $[0, f^{k+1} 1)$. ∎

Lemma 4.1 and induction on n give

Lemma 4.2. *For $n \geq 0$, f^n is linear with slope β^n on I_n and maps it onto $[0, f^k 1)$ for some $0 \leq k \leq \min[n, N - 1]$.*

If β differs from a non-simple β-number, then the numbers $f^k 1$, $0 \leq k \leq N - 1$, are distinct. In this case the integer k in Lemma 4.2 is uniquely determined by I_n. This suggests the following.

Definition 4.1. Let β differ from a non-simple β number. I_n is of *type k*, $0 \leq k < \infty$, iff f^n maps I_n onto $[0, f^k 1)$.

By Lemma 4.1, the type k of any I_n satisfies $0 \leq k \leq \min[n, N - 1]$. If β is a non-simple β-number, then the numbers $f^k 1$, $0 \leq k < \infty$, are not all distinct and Definition 4.1 does not apply. In this case we give the following

Definition 4.2. I_0 is of *type 0*. If I_n is of *type k*, where $0 \leq k \leq N - 2$, then the right most I_{n+1} contained in I_n is of type $(k + 1)$ and all other I_{n+1} contained in I_n are of type 0. If I_n is of type $(N - 1)$, then all I_{n+1} contained in I_n are of type 0.

Again by Lemma 4.1, we have $0 \leq k \leq \min[n, N-1]$. We remark that Definition 4.2 also applies when β differs from a non-simple β-number, but Definition 4.1 is simpler.

For $n, k \geq 0$, let F_{nk} be the number of nth stage intervals of type k. Then $F_{nk} = 0$ for $k > \min[n, N-1]$. N_n is related to F_{nk} by $N_n = \sum_{k=0}^{\min[n,N-1]} F_{nk}$.

Theorem 4.1. F_{nk} *satisfies:*

$$
\begin{aligned}
&i) && F_{00} = 1 \\
&ii) && F_{nk} = F_{n-k,0} \; for \; 0 \leq k \leq \min[n, N-1], \; n \geq 0 \\
&iii) && F_{n+1,0} = \sum_{k=0}^{n} a_k F_{n-k,0}, \; n \geq 0 \\
&iv) && F_{n0} \leq \beta^n, \; n \geq 0
\end{aligned}
$$

(4.1)

Proof: i) Follows from I_0 being of type 0.

ii) We may assume $1 \leq k \leq \min[n, N-1]$, $n \geq 1$. By Lemma 4.1, each I_{n-1} of type $(k-1)$ contains precisely one I_n of type k, and no other I_{n-1} contains an I_n of type k. Repetition of this argument gives $F_{nk} = F_{n-1,k-1} = \ldots = F_{n-k,0}$.

iii) Let $0 \leq k \leq \min[n, N-1]$, $n \geq 0$. By Lemma 4.1, each I_n of type k contains a_k I_{n+1}'s of type 0. Hence

$$
(4.2) \qquad F_{n+1,0} = \sum_{k=0}^{\min[n,N-1]} a_k F_{nk} = \sum_{k=0}^{\min[n,N-1]} a_k F_{n-k,0}
$$

For $0 \leq n \leq N-1$, 4.2) is 4.1). For $n > N-1$, 4.2) is $F_{n+1,0} = \sum_{k=0}^{N-1} a_k F_{n-k,0}$. As $a_k = 0$ for $k \geq N$, 4.2) is again 4.1).

iv) On I_n of type 0, f^n is linear with slope β^n and maps I_n onto $[0,1)$. Hence $\lambda(I_n) = \frac{1}{\beta^n}$. Summing over I_n of type 0, we get $\frac{F_{n0}}{\beta^n} \leq 1$. $\blacksquare$

We obtain the asymptotic behavior of F_{n0}, and hence of N_n, by studying the generating functions

$$
(4.3) \qquad F(z) = \sum_{n=0}^{\infty} F_{n0} z^n, \; \phi(z) = \sum_{n=0}^{\infty} a_n z^{n+1}
$$

From Theorem 4.1 iv), we conclude that $F(z)$ converges for $|z| < \frac{1}{\beta}$. Since $a_n \leq [\beta]$, $0 \leq n < \infty$, $\phi(z)$ converges for $|z| < 1$.

Theorem 4.2. $F(z)$ *can be continued analytically to* $|z| \leq r$ *for some* $\frac{1}{\beta} < r < 1$, *except for a simple pole at* $z = \frac{1}{\beta}$. *The residue at* $1/\beta$ *equals* $-1 / \sum_{n=0}^{\infty} \frac{f^n 1}{\beta^{n-1}}$.

Proof: Multiply both sides of 4.1) by z^n and sum over $0 \leq n < \infty$. Elementary manipulations yield

$$
(4.4) \qquad F(z) \cdot [1 - \phi(z)] = 1, \; |z| < \frac{1}{\beta}
$$

4.4) shows that $\phi(z) \neq 1$ for $|z| < \frac{1}{\beta}$. We show that also $\phi(z) \neq 1$ for $|z| = \frac{1}{\beta}$, except at $z = \frac{1}{\beta}$. We have $a_n = [\beta f^n 1] = \beta f^n 1 - \{\beta f^n 1\} = \beta f^n 1 - f^{n+1} 1$. Hence

(4.5)

$$\phi(z) = \sum_{n=0}^{\infty} \beta f^n 1 \cdot z^{n+1} - \sum_{n=0}^{\infty} f^{n+1} 1 \cdot z^{n+1} = 1 - (1 - \beta z) \sum_{n=0}^{\infty} f^n 1 \cdot z^n, \ |z| < 1$$

or

(4.6)

$$1 - \phi(z) = (1 - \beta z) \sum_{n=0}^{\infty} f^n 1 \cdot z^n, \ |z| < 1$$

Letting $z = \frac{1}{\beta}$ in 4.6) gives $\phi\left(\frac{1}{\beta}\right) = 1$, a fact already established in 3.4). From 4.3) we get

(4.7)

$$|\phi(z)| \leq \sum_{n=0}^{\infty} \frac{a_n}{\beta^{n+1}} = 1, \ \text{for } |z| = \frac{1}{\beta}.$$

Equality holds in 4.7) iff

(4.8)

$$z^{n+1} = \frac{e^{i\theta}}{\beta^{n+1}} \ \text{for some } 0 \leq \theta < 2\pi, \ \text{whenever } a_n \neq 0$$

Suppose $\phi(z) = 1$, for some $|z| = \frac{1}{\beta}$. As $a_0 \neq 0$, we conclude from 4.8) that $z = \frac{e^{i\theta}}{\beta}$. Also from 4.8), $\phi(z) = e^{i\theta} \phi\left(\frac{1}{\beta}\right) = 1$. As $\phi\left(\frac{1}{\beta}\right) > 0$, we conclude $\theta = 0$ so that $z = \frac{1}{\beta}$.

From 4.4), 4.6) we obtain that $F(z)$ has an analytic continuation to $|z| \leq r$ for some $\frac{1}{\beta} < r < 1$, except for a simple pole at $z = \frac{1}{\beta}$ where the residue equals $-1/\sum_{n=0}^{\infty} \frac{f^n 1}{\beta^{n-1}}$. ∎

Theorem 4.3. *Let $\sigma = \sum_{n=0}^{\infty} \frac{f^n 1}{\beta^n}$. There exists a $0 < \delta < \beta$ such that for $n \to \infty$*

(4.9)

$$F_{n0} = \frac{\beta^n}{\sigma} + O(\delta^n).$$

(4.10)

$$N_n = \frac{\beta(1 - \beta^{-N})}{\sigma(\beta - 1)} \beta^n + O(\delta^n)$$

where β^{-N} is defined to be 0 for $N = \infty$.

We remark that Theorem 4.1 i) and 4.9) imply for $0 \leq k \leq N - 1$

(4.11)

$$F_{nk} = \frac{\beta^{-k}}{\sigma} \beta^n + O(\delta^{n-k}), \ \text{for } n \geq k$$

the multiplicative constant hidden in the O-notation being independent of k.

Proof: From Theorem 4.2, we obtain for some $1 < \delta < \beta$

$$(4.12) \qquad\qquad F(z) = \frac{1}{\sigma} \frac{1}{1 - \beta z} + g(z) \,, \ |z| \le \frac{1}{\delta}$$

where $g(z) = \sum_{n=0}^{\infty} g_n z^n$ is analytic for $|z| \le \frac{1}{\delta}$, hence $g_n = O(\delta^n)$.

Expanding both sides of 4.12) into power series about 0 and equating coefficients, we get 4.9). 4.10) follows from 4.9) and $N_n = \sum_{k=0}^{\min[n,N-1]} F_{n-k,0}$. ■

We conclude the section with the following

Remarks. 1) From the definition of N_n, it follows readily that $N_{n+m} \le N_n \cdot N_m$ for n, m integers ≥ 0. Hence $\log N_n$ is a subadditive function, implying that $\lim_{n \to \infty} \dfrac{\log N_n}{n} = c$ for some c, or $N_n = e^{(c + o(n))}$. c is the topological entropy of the β-shift associated with the β-transformation [1, 4]. 4.10) improves on this and identifies the entropy as $\log \beta$.

2) The constant σ appearing in 4.9) can be explained as follows. Let I_n be of type k. f^n is linear with slope β^n on I_n, and maps it onto $[0, f^k 1)$. Hence $\lambda(I_n) = \frac{f^k 1}{\beta^n}$. Summing over all I_n,

$$\sum_{k=0}^{n} F_{nk} \frac{f^k 1}{\beta^n} = \sum_{k=0}^{n} \frac{F_{n-k,0}}{\beta^{n-k}} \frac{f^k 1}{\beta^k} = \sum_{I_n} \lambda(I_n) = 1 \,.$$

Suppose $F_{n0} \sim c\beta^n$ for some constant c. Insert this into the above sum and let $n \to \infty$. Interchanging limit and summation, justifiable by the dominated convergence theorem, we obtain $c = 1/\sigma$.

5. ASYMPTOTICS OF $P_n(a, b)$

For $n \ge 1$ and $0 \le a < b \le 1$, let $P_n(a, b)$ the *number of periodic points in* $[a, b)$ *of period n of f*, i.e. the points $0 \le x < 1$ satisfying $f^n(x) = x$. We obtain the asymptotic behavior of $P_n(a, b)$ from that of F_{n0}. We first prove a preliminary result that for fixed k, $0 \le k \le N - 1$, the nth stage intervals of type k become uniformly distributed over $[0, 1]$ as $n \to \infty$.

For $0 \le a < b \le 1$ and $0 \le k, n < \infty$, let $F_{nk}(a, b)$ be the number of nth stage intervals of type k contained in $[a, b)$, and $N_n(a, b)$ the number of nth stage intervals contained in $[a, b)$.

Theorem 5.1. i) For $0 \le k \le N - 1$, $\lim_{n \to \infty} \dfrac{F_{nk}(a, b)}{F_{nk}} = b - a$.

ii) $\lim_{n \to \infty} \dfrac{N_n(a, b)}{N_n} = b - a$.

Theorem 5.1 also appears in [4], where it is stated somewhat differently.

Proof: We prove Theorem 5.1 when $[a, b)$ is an mth stage interval I_m for some $m \ge 0$. The general result then follows by a standard approximation argument. We write $N_n(I_m)$, $F_{nk}(I_m)$ instead of $N_n(a, b)$, $F_{nk}(a, b)$. Consider first $k = 0$. The proof is by induction on m. For $m = 0$, $I_0 = [0, 1)$. Hence $F_{n0}(I_0) = F_{n0}, N_n(I_0) = N_n$, and Theorem 5.1 is obvious. Let $m > 0$ and suppose Theorem 5.1 holds for $m - 1$. Let K_m be the right most of all m^{th} stage intervals, and denote the other m^{th} stage intervals by L_m. By Theorem 3.3 iii), f is linear with slope β on L_m, mapping it into an $(m-1)^{th}$ stage interval $I_{m-1} = f(L_m)$. Hence $\lambda(L_m) = \frac{\lambda(I_{m-1})}{\beta}$. Similarly, for $n \geq m$ and $I_n \subset L_m$, f maps I_n onto $I_{n-1} = f(I_n)$. Since $f^n = f^{n-1} \circ f$ and f maps I_n onto I_{n-1}, I_n will be of type 0 iff I_{n-1} is of type 0. Hence for $n \geq m$,

$$(5.1) \qquad N_n(L_m) = N_{n-1}(I_{m-1}), \ F_{n0}(L_m) = F_{n-1,0}(I_{m-1})$$

From 4.10) and 5.1)

$$(5.2) \quad \lim_{n \to \infty} \frac{N_n(L_m)}{N_n} = \lim_{n \to \infty} \frac{N_{n-1}}{N_n} \cdot \lim_{n \to \infty} \frac{N_{n-1}(I_{m-1})}{N_{n-1}} = \frac{\lambda(I_{m-1})}{\beta} = \lambda(I_m)$$

As I is the union of K_m and the L_m's,

$$(5.3) \qquad N_n = N_n(K_m) + \sum_{L_m} N_n(L_m), \ 1 = \lambda(K_m) + \sum_{L_m} \lambda(L_m)$$

Hence

$$(5.4) \quad \lim_{n \to \infty} \frac{N_n(K_m)}{N_n} = 1 - \sum_{L_m} \lim_{n \to \infty} \frac{N_n(L_m)}{N_n} = 1 - \sum_{L_m} \lambda(L_m) = \lambda(K_m)$$

Thus for all m^{th} stage intervals I_m, $\lim_{n \to \infty} \frac{N_n(I_m)}{N_n} = \lambda(I_m)$. Similarly, we conclude from 4.9) and 5.1) that $\lim_{n \to \infty} \frac{F_{n0}(I_m)}{F_n} = \lambda(I_m)$.

For $k > 0$, Theorem 5.1 i) is proved by induction on k. Suppose it holds for k and I_m, where $0 \leq k \leq N-2$ and $m \geq 0$. By Lemma 4.1, we have for $n \geq m$,

$$(5.5) \qquad F_{n+1,k+1}(I_m) = F_{nk}(I_m), \ F_{n+1,k+1} = F_{nk}$$

Hence,

$$(5.6) \qquad \lim_{n \to \infty} \frac{F_{n,k+1}(I_m)}{F_{n,k+1}} = \lim_{n \to \infty} \frac{F_{nk}(I_m)}{F_n} = \lambda(I_m)$$

■

Theorem 5.2.

$$(5.7) \qquad P_n(a,b) \sim \frac{\int_a^b h(x)dx}{\sigma} \cdot \beta^n \ as \ n \to \infty,$$

where $h(x) = \sum_{f^k 1 > x} \beta^{-k}$

Proof: For each $n \geq 1$, $[a,b)$ is the union of the I_n's contained in it and at most two other intervals. Let $I_n = [a_n, b_n)$ be of type k and contained in $[a, b)$. If $f^n(x) = x$ for some $x \in I_n$, then $f^k 1 > f^n(x) = x \geq a$. If $f^k 1 > b$, then $f^n(a_n) - a_n = -a_n \leq 0$ and $f^n(b_n) - b_n = f^k 1 - b_n \geq f^k 1 - b > 0$, so that $f^n(x) = x$ for some $x \in I_n$. Furthermore, as f^n is linear with slope $\beta^n \neq 1$, $f^n(x) = x$ has a unique solution in I_n. Hence for $f^k 1 > b$, I_n contains precisely one fixed point of $f^n(x)$, and for $f^k 1 \leq a$, I_n does not contain a fixed point of $f^n(x)$. We conclude

$$(5.8) \quad \sum_{k=0}^{\infty} F_{nk}(a,b)\chi(f^k 1 > b) \leq P_n(a,b) \leq \sum_{k=0}^{\infty} F_{nk}(a,b)\chi(f^k 1 > a) + 2$$

where $\chi(y > x) = 1$ if $y > x$, and $\chi(y > x) = 0$ if $y \leq x$.

For $0 \leq k \leq N - 1$, we get from 4.11) and Theorem 5.1 i)

$$(5.9) \qquad \lim_{n \to \infty} \frac{F_{nk}(a,b)}{\beta^n} = \frac{\beta^{-k}}{\sigma}(b - a)$$

From Theorem 4.1,

$$(5.10) \qquad \frac{F_{nk}(a,b)}{\beta^n} \leq \frac{1}{\beta^k} \text{ for } n, k \geq 0$$

From 5.9), 5.10), and the dominated convergence theorem, we obtain for $0 \leq x \leq 1$

$$(5.11) \qquad \lim_{n \to \infty} \sum_{k=0}^{\infty} \frac{F_{nk}(a,b)}{\beta^n} \cdot \chi(f^k 1 > x) = \frac{h(x)}{\sigma}(b - a) \ .$$

Dividing 5.8) by β^n and using 5.11),

$$(5.12) \qquad \frac{h(b)}{\sigma}(b - a) \leq \varliminf_{n \to \infty} \frac{P_n(a,b)}{\beta^n} \leq \varlimsup_{n \to \infty} \frac{P_n(a,b)}{\beta^n} \leq \frac{h(a)}{\sigma}(b - a)$$

Let $x_i = a + \frac{(b-a)}{m}i$, $0 \leq i \leq m$. Then $P_n(a,b) = \sum_{i=0}^{m-1} P_n(x_i, x_{i+1})$, so that

$$(5.13) \qquad \varlimsup_{n \to \infty} \frac{P_n(a,b)}{\beta^n} \leq \sum_{i=0}^{m-1} \varlimsup_{n \to \infty} \frac{P_n(x_i, x_{i+1})}{\beta^n}$$

with the reversed inequality for $\varliminf$.

From 5.12), 5.13)

$$(5.14) \qquad \frac{1}{\sigma} \sum_{i=0}^{m-1} h(x_{i+1})(x_{i+1} - x_i) \ \leq \ \varliminf_{n \to \infty} \frac{P_n(a,b)}{\beta^n} \leq \varlimsup_{n \to \infty} \frac{P_n(a,b)}{\beta^n}$$

$$\leq \ \frac{1}{\sigma} \sum_{i=0}^{m-1} h(x_i)(x_{i+1} - x_i)$$

The sums in 5.14) are Riemann sums for $\int_a^b h(x)dx$. Hence, letting $m \to \infty$, we obtain $\lim_{n \to \infty} \frac{P_n(a,b)}{\beta^n} = \frac{1}{\sigma} \int_a^b h(x)dx$. ∎

Corollary. Let $P_n = P_n(0,1)$ be the number of periodic points of period n of the transformation f. Then

$$(5.15) \qquad P_n \sim \beta^n \text{ as } n \to \infty \ .$$

Proof: We have

$$(5.16) \qquad \int_0^1 h(x)dx \ = \ \int_0^1 \sum_{k=0}^{\infty} \beta^{-k} \chi(f^k 1 > x)dx$$

$$= \ \sum_{k=0}^{\infty} \beta^{-k} \int_0^1 \chi(f^k 1 > x)dx = \sum_{k=0}^{\infty} \beta^{-k} f^k 1 = \sigma \ .$$

The corollary then follows from 5.7) with $a = 0$, $b = 1$. ∎

We remark that 5.15) can also be derived from a formula for the zeta function of the transformation f. The *zeta function* of a transformation is defined by

$$\zeta(z) = \exp\left(\sum_{n=1}^{\infty} \frac{P_n}{n} z^n\right) \ , \ |z| \text{ small} .$$

Takahashi [14] has shown that for the transformation f

(5.17) $$\zeta(z) = \frac{1}{1 - \phi(z)}, \ |z| \text{ small}$$

where $\phi(z)$ is defined by 4.3). Taking logarithms

(5.18) $$\sum_{n=1}^{\infty} \frac{P_n}{n} z^n = \log\left(\frac{1}{1 - \phi(z)}\right) \ , \ |z| \text{ small}$$

where $\log\left(\frac{1}{1-\phi(z)}\right)$ is that analytic branch which equals 0 at $z = 0$. Use formula 4.6) for $(1 - \phi(z))$ and the series expansion

$$\log\left(\frac{1}{1 - \beta z}\right) = \sum_{n=1}^{\infty} \frac{\beta^n}{n} z^n \ , \ |z| < \frac{1}{\beta} .$$

Comparing coefficients in 5.18), we readily obtain 5.15).

Theorem 5.1 can be used to get another derivation of the formula for the invariant measure of f obtained by Parry.

Theorem 5.3. *Let $\mu(E) = \int_E h(x) \, dx$, E any measurable subset of I. $\mu(E)$ is an invariant measure for f, that is $\mu(f^{-1}E) = \mu(E)$.*

Proof: Let $\mathcal{P}_n(E)$ be the set of fixed points of f^n in E. It is a simple exercise to check that f^{n-1} is a one-to-one map from $\mathcal{P}_n(E)$ onto $\mathcal{P}_n(f^{-1}E)$. Hence $P_n(E) = P_n(f^{-1}E)$, where $P_n(E) = |\mathcal{P}_n(E)|$. If E is an interval, then $f^{-1}E$ is a finite union of disjoint intervals. 5.7) extends to finite unions of disjoint intervals. Hence

$$\mu(E) = \lim_{n \to \infty} \frac{\sigma P_n(E)}{\beta^n} = \lim_{n \to \infty} \frac{\sigma P_n(f^{-1}E)}{\beta^n} = \mu(f^{-1}E) .$$

We conclude that $\mu(E) = \mu(f^{-1}E)$ for E an interval. A standard extension argument shows that the formula persists for E measurable.   ∎

6. MAHLER'S SECOND THEOREM. AN IMPROVEMENT

Lemma 6.1. *Let $f_n, n \geq 0$, be the number of $0, 1$-sequences of length n not containing the block 11. Then $\{f_n\}$ is the Fibonacci sequence $1, 2, 3, 5, \ldots$*

Proof: Let $\mathcal{F}_n$ be the set of $0, 1$ sequences of length n not containing $1,1$; thus $|\mathcal{F}_n| = f_n$. Let $\{\epsilon_1, \ldots, \epsilon_n\}$ be an arbitrary $0, 1$-sequence of length $n \geq 2$. If $\epsilon_1 = 0$, then $\{\epsilon_1, \ldots, \epsilon_n\} \in \mathcal{F}_n$ iff $\{\epsilon_2, \ldots, \epsilon_n\} \in \mathcal{F}_{n-1}$. If $\epsilon_1 = 1$, then $\{\epsilon_1, \ldots, \epsilon_n\} \in \mathcal{F}_n$ iff $\epsilon_2 = 0$ and $\{\epsilon_3, \ldots, \epsilon_n\} \in \mathcal{F}_{n-2}$. Hence $f_n = f_{n-1} + f_{n-2}$, $n \geq 2$. By inspection, $f_0 = 1$ and $f_1 = 2$. Hence $\{f_n\}$ is the Fibonacci sequence.   ∎

From the well known formula $f_n = \frac{1}{\sqrt{5}} \left\{ \left(\frac{\sqrt{5}+1}{2} \right)^{n+2} + (-1)^{n+1} \left(\frac{\sqrt{5}-1}{2} \right)^n \right\}$, $n \geq 0$, we obtain

$$(6.1) \qquad f_n = O\left(\left(\frac{1+\sqrt{5}}{2} \right)^n \right) .$$

In the proof below, T and f are the transformations discussed in §2.

Proof of Mahler's Second Theorem. Let $x \geq 1$ and choose the integer $n \geq 1$ so that $2^{n-1} \leq x < 2^n$. By Proposition P of §2, for any integer $g_0 \geq 0$, the interval $[g_0, g_0+1)$ contains a Z-number iff the T-expansion $\{\epsilon_n(g_0)\}$ is identical with the f-expansion of a number $0 \leq r_0 < 1$, in which case $r_0 = \sum_{n=0}^{\infty} \epsilon_n(g_0) \left(\frac{2}{3} \right)^{n+1}$ and $g_0 + \frac{r_0}{2}$ is the desired Z-number. As remarked in §3, the f-expansion of any number $0 \leq r_0 < 1$ does not contain $1,1$. Thus if $[g_0, g_0+1)$ contains a Z-number, then the sequence $\{\epsilon_k(g_0)\}$, $0 \leq k < \infty$, does not contain $1,1$. By Theorem 3.1, the map $g_0 \rightarrow \{\epsilon_0(g_0), \ldots, \epsilon_{n-1}(g_0)\}$ is one-to-one from the integers $0, 1, \ldots, 2^n - 1$ onto all $0,1$-sequences of length n. We conclude, by use of Lemma 6.1, that

(6.2)
$$|Z(x)| \leq |\{0 \leq g_0 < 2^n : \{\epsilon_0(g_0), \ldots, \epsilon_{n-1}(g_0)\} \text{ does not contain } 1,1\}| = f_n =$$

$$\left(\left(\frac{1+\sqrt{5}}{2} \right)^x \right) = O(x^\delta)$$

where $\delta = \log_2 \left(\frac{1+\sqrt{5}}{2} \right)$.   ∎

As shown in [6], the above proof gives the following algorithm for deciding that $[g_0, g_0+1)$ does not contain a Z-number. We compute successively $g_0, g_1, \ldots$. If eventually for some n, g_n and g_{n+1} are both odd, then $[g_0, g_0+1)$ does not contain a Z-number. Using this algorithm, we have checked that there are no Z-numbers below 2000.

The results of §4 yield the following improvement on Mahler's second theorem.

Theorem 6.1. $|Z(x)| = O(x^\delta)$, where $\delta = \log_2(\frac{3}{2})$.

Proof: The proof is similar to that of Mahler's second theorem. Let $x \geq 1$ and choose $n \geq 1$ so that $2^{n-1} \leq x < 2^n$. By Proposition P of §2, if $[g_0, g_0+1)$ contains a Z-number then $\{\epsilon_0(g_0), \ldots, \epsilon_{n-1}(g_0)\}$ is an f-admissible sequence of length n. By 4.10), $N_n = O\left(\left(\frac{3}{2} \right)^n \right)$, where N_n is the number of f-admissible sequences of length n. Hence

(6.3)
$$|Z(x)| \leq |\{0 \leq g_0 < 2^n : \{\epsilon_0(g_0) \ldots \epsilon_{n-1}(g_0)\} \text{ is an } f\text{-admissible sequence}\}| = N_n$$

$$= O\left(\left(\frac{3}{2} \right)^x \right) = O(x^\delta) .$$

                                                                                    ∎

7. Generalizations

The notion of Z-numbers can be generalized to any $0 < t < 1$ and $\theta > 1$ by defining

$$Z_{t,\theta} = \{\alpha > 0 : \ 0 \leq \{\alpha\theta^n\} < t, \ 0 \leq n < \infty\}$$

$Z_{t,\theta}$ is called the set of Z-numbers for t, θ. In particular $Z_{\frac{1}{2},\frac{3}{2}}$ is the set Z of §1.

The following theorem can be proved without recourse to the continuum hypothesis.

Theorem 7.1. $Z = Z_{t,\theta}$ is either at most countable or has cardinality C of the continuum.

Proof: From the definition, $Z = \bigcap_{n=0}^{\infty} V_n$ where

$$V_n = \bigcup_{k=0}^{\infty} \left[\frac{k}{\theta^n}, \frac{k+t}{\theta^n} \right)$$

Let $\overline{V}_n = \bigcup_{k=0}^{\infty} \left[\frac{k}{\theta^n}, \frac{k+t}{\theta^n} \right]$. $\overline{V}_n$ is closed for all integer $n \geq 0$, and so is $Y = \bigcap_{n=0}^{\infty} \overline{V}_n$. It follows that Y is either at most countable or has cardinality C [15, P. 48-50]. $V_n \subset \overline{V}_n$ so that $Z \subset Y$. Furthermore,

$$(7.1) \quad Y - Z = \bigcap_{m=0}^{\infty} \overline{V}_m - \bigcap_{n=0}^{\infty} V_n = \bigcup_{n=0}^{\infty} \left(\bigcap_{m=0}^{\infty} \overline{V}_m - V_n \right)$$

$$\subset \bigcup_{n=0}^{\infty} (\overline{V}_n - V_n) = \bigcup_{n=0}^{\infty} \bigcup_{k=0}^{\infty} \left\{ \frac{k+t}{\theta^n} \right\}$$

7.1) shows that $Y - Z$ is denumerable. Hence Z has the cardinality of Y. ∎

The problem of deciding which of the alternatives to Theorem 7.1 holds for given t, θ seems difficult. We give sufficient conditions for each alternative. In Theorems 7.2)-7.4) p and q denote relatively prime integers with $p > q \geq 2$.

Theorem 7.2. i) Let $0 < t \leq \min\left(\frac{1}{q}, \frac{1}{\theta}\right)$, that is $t \leq \frac{1}{q}$ for $p < q^2$ and $t \leq \frac{q}{p}$ for $p > q^2$. Then for any integer $g \geq 0$, $[g, g+1)$ contains at most one Z-number.
ii) Let $p < q^2$, and assume the above condition $t \leq \frac{1}{q}$. Then $|Z(x)| = O(x^\delta)$, where $0 < \delta = \log_q \left(\frac{p}{q}\right) < 1$.

We illustrate Theorem 7.2 for $p = 3$, $q = 2$, $t = \frac{1}{2}$. In this case the hypotheses of both i) and ii) are satisfied. i) and ii) become respectively Mahler's first theorem and Theorem 6.1, the improved version of the second theorem.

Proof: i) For every integer $n \geq 0$, let

$$(7.2) \qquad\qquad \alpha \left(\frac{p}{q} \right)^n = g_n + \frac{r_n}{q}$$

where g_n is an integer and $0 \leq \frac{r_n}{q} < 1$. Suppose $0 \leq \frac{r_n}{q} < t$. We investigate when $\frac{r_{n+1}}{q} < t$. Multiplying both sides of 7.2) by $\frac{p}{q}$,

$$(7.3) \qquad g_{n+1} + \frac{r_{n+1}}{q} = \frac{p}{q}\, g_n + \frac{pr_n}{q^2}\ .$$

Let $T = T_{\frac{p}{q}}$, $f = f_{\frac{p}{q}}$, $\{\mathbb{N}_0, \ldots, \mathbb{N}_{q-1}\}$, $\{J_0, \ldots, J_{[\frac{p}{q}]}\}$, be defined as in §3. Let $g_n \in \mathbb{N}_i$, i.e. $p\,g_n + i \equiv 0 \pmod{q}$ and $0 \leq i \leq q - 1$. Multiply both sides of 7.3) by q and rearrange to get

$$(7.4) \qquad q(g_{n+1} - T(g_n)) + i + r_{n+1} = \frac{pr_n}{q}\ .$$

As $t \leq \frac{1}{q}$, 7.4) shows that $\frac{r_{n+1}}{q} < t$ iff

$$(7.5) \qquad \left[\frac{pr_n}{q}\right] = q(g_{n+1} - T(g_n)) + i$$

and

$$(7.6) \qquad \left(\frac{pr_n}{q}\right) < tq\ .$$

As $0 \leq i < q$ and $\frac{pr_n}{q} < pt \leq q$, 7.5) is equivalent to $\left[\frac{pr_n}{q}\right] = i$, i.e. $r_n \in J_i$. Hence $0 \leq \frac{r_{n+1}}{q} < t$ is equivalent to $r_n \in J_i$ and $f(r_n) < tq$. Furthermore, when the latter conditions hold, we conclude from 7.4), 7.5) that $g_{n+1} = T(g_n)$, $r_{n+1} = f(r_n)$. Repetition of the above argument gives the following generalization of Proposition P of §3.

Proposition P$'$: *Let $\alpha = g_0 + \frac{r_0}{q}$, where g_0 is a non-negative integer and $0 \leq r_0 < tq$. α is a Z-number iff: i) the T-expansion of g_0 is the β-expansion of r_0, where $\beta = \frac{p}{q}$, and*
ii) $f^n(r_0) < tq$ for all $n \geq 0$.

Since r_0 is determined by its β-expansion, Theorem 7.2 i) follows from Proposition P$'$.

ii) Since $Z_{t_1,\theta} \subset Z_{t_2,\theta}$ for $t_1 < t_2$, it suffices to consider $t = \frac{1}{q}$. In this case, condition ii) of Proposition P' is superfluous as $f(r_0) < 1$ for $0 \leq r_0 < 1$. Theorem 7.2 ii) then follows from Proposition P$'$ in the same way that Theorem 6.1) follows from Proposition P of §3. ∎

Proposition P$'$ implies Pollington's result quoted in the introduction, i.e. there are no Z-numbers for $t = \frac{1}{p}$, $\theta = \frac{p}{q}$. For let α be a Z-number. For our choice of t and θ, the condition $t < \min\left(\frac{1}{q}, \frac{1}{\theta}\right)$ of Theorem 7.2 is fulfilled. Since $f(r_0) = \frac{p}{q} r_0$ for $0 \leq r_0 < \frac{q}{p}$, condition ii) of Proposition P$'$ becomes $0 \leq r_0 < \left(\frac{q}{p}\right)^{n+1}$ for all $n \geq 0$. Hence $r_0 = 0$ and $g_0 = \alpha > 0$. All terms in the β-expansion of 0 equal 0. The T-expansion of g_0 contains non-zero terms, indeed the $(k + 1)^{th}$ term is not zero when q divides g_0 precisely k times. Thus condition i) of Proposition P$'$ fails, a contradiction.

Theorem 7.3 guarantees that Z is at most countable. The following result of Tijdeman [15] guarantees that Z is infinite.

Theorem 7.3. *Let $\theta = \frac{p}{q} > 2$ and $t > \frac{q-1}{p-q}$. Then for g an integer ≥ 0, $[g, g+1)$ contains at least one Z-number.*

Proof: We reproduce the proof in [14], where the discussion is restricted to $q = 2$. However, the same proof works for all q. Let $g_n = T^n_{p/q}(g)$, $n \geq 0$. Then $\theta g_n \leq g_{n+1} \leq \theta g_n + \frac{q-1}{q}$, from which

$$(7.7) \qquad 0 \leq \frac{g_{n+1}}{\theta^{n+1}} - \frac{g_n}{\theta^n} \leq \frac{q-1}{q\theta^{n+1}} \ .$$

Hence

$$(7.8) \qquad \frac{g_n}{\theta^n} \leq g + \left(\frac{q-1}{q}\right)\left(\frac{1}{\theta} + \frac{1}{\theta^2} + \dots\right) = g + \frac{q-1}{q(\theta-1)} \ .$$

The sequence $\left\{\frac{g_n}{\theta^n}\right\}$ is monotonically increasing and contained in $\left[g, g+\frac{1}{\theta-1}\right]$. Hence it has a limit $\alpha \in \left[g, g+\frac{1}{\theta-1}\right]$. From 7.7),

$$(7.9) \qquad 0 \leq \theta^n \alpha - g_n \ = \ \theta^n\left(\alpha - \frac{g_n}{\theta^n}\right) \leq \left(\frac{q-1}{q}\right)$$

$$\theta^n\left(\frac{1}{\theta^{n+1}} + \frac{1}{\theta^{n+2}} + \dots\right) = \frac{q-1}{q(\theta-1)} = \frac{q-1}{p-q}$$

■

Combining Theorems 7.2 and 7.3, we get Theorem 7.4, which gives examples where Z is an infinite countable set.

Theorem 7.4. *Let $p > q^2$, which is the same as $\frac{q-1}{p-q} < \frac{q}{p}$, and $\frac{q-1}{p-q} < t \leq \frac{q}{p}$. Then for any integer $g \geq 0$, $[g, g+1)$ contains precisely one Z-number.*

Finally, we give examples where Z has cardinality C.

Theorem 7.5. *Let $\theta > 3$, $\frac{2}{\theta-1} < t < 1$. For any integer $g \geq 0$, $[g, g+1) \cap Z$ has cardinality C.*

Proof: Let $\tau = \frac{2}{\theta-1}$, thus $0 < \tau < 1$, and

$$\overline{V}_n(\theta) = \bigcup_{k=0}^{\infty}\left[\frac{k}{\theta^n}, \frac{k+\tau}{\theta^n}\right], \ Y(\theta) = \bigcap_{n=0}^{\infty}\overline{V}_n(\theta) \ .$$

$Y(\theta) \subset Z$ for $\theta < t$, so that it suffices to show that $Y(\theta)$ has cardinality C. For an integer $k \geq 0$, let $j = 1 + [k\theta]$. Then $k\theta < j$, and

$$j + 1 + \tau \leq (1 + k\theta) + (1 + \tau) = (k + \tau)\theta \ .$$

We conclude that for all integers $n \geq 0$

$$(7.10) \qquad \frac{k}{\theta^n} < \frac{j}{\theta^{n+1}} < \frac{j+1+\tau}{\theta^{n+1}} \leq \frac{k+\tau}{\theta^n} \ .$$

From 7.10), we obtain the interval inclusions

$$(7.11) \qquad \left[\frac{r}{\theta^{n+1}}, \frac{r+\tau}{\theta^{n+1}}\right] \subset \left[\frac{k}{\theta^n}, \frac{k+\tau}{\theta^{n+1}}\right], \ r = j, j+1 \ .$$

Repeated use of 7.11) produces C sequences of nested intervals, all contained in $[k, k + \tau] \subset [k, k + 1)$. The intersection of the nested intervals in each sequence produces a distinct point of Z.   ■

Acknowledgments. I am grateful to Jeff Lagarias for helpful discussions and for bringing to my attention some of the literature related to the topic of this paper.

References

1. F. Blanchard, "β-Expansions and symbolic dynamics," *Theor. Comp. Sci.* **65** (1989), pp. 131–141.

2. G. H. Hardy and E. M. Wright, "An Introduction to the Theory of Numbers," Fourth Edition, Oxford Clarendon Press, 1960.

3. T. Hofbauer, "β-Shifts have unique maximal measures," *Monatsh. Math.* **85** (1978), pp. 189–198.

4. Sh. Ito and Y.Takahaski, "Markov subshifts and realizations of β-expansions," *J. Math. Soc. of Japan*, Vol. 26 (1974), pp. 33–56.

5. K. Mahler, "On the fractional parts of the powers of a rational number II," *Mathematika* **4** (1957), pp. 122–124.

6. K. Mahler, "An unsolved problem on powers of 3/2," *J. Australian Math. Soc.* **8** (1969), pp. 313–321.

7. J. Milnor and W. Thurston, "On iterated maps of the interval," *Lecture notes in Math.*, Springer Verlag, Vol. 1342 (1988), pp. 465–563.

8. M. Misiurewicz and W. Zlenk, "Entropy of piecewise monotone mappings," *Studia Math.* **67** (1980), pp. 45–63.

9. W. Parry, "On the β-expansion of real numbers," *Acta. Math. Acad. Sci. Hung.* **11** (1960), pp. 401–416.

10. A. D. Pollington, "Interval Constructions in the Theory of Numbers," Thesis, University of London, 1976.

11. A. D. Pollington, "On the density of sequences $\{n_k\xi\}$," *Illinois J. Math.* **23** (1979), pp. 511–515.

12. A. Renyi, "Representations for real numbers and their ergodic properties," *Acta. Math. Acad. Sci. Hung.* **8** (1957), pp. 472–493.

13. Y. Takahashi, "β-Transformations and symbolic dynamics," *Lecture Notes in Math*, Springer Verlag, Vol. 330 (1973), pp. 455–464.

14. Y. Takahashi, "Shift with free orbit basis and realization of one-dimensional maps," *Osaka J. Math.* **20** (1983), pp. 269–278.

15. R. Tijdeman, "Note on Mahler's 3/2 problem," *K. norske Vidensk. Selsk. Skr.* **16** (1972), pp. 1–4.

16. E. J. Townsend, "Functions of Real Variables," New York, Henry Holt and Co., 1928.

17. T. Vijayaraghavan, "On the fractional parts of the powers of a number I," *J. London Math. Soc.* **15** (1940), pp. 159–160.

18. H. Weyl, "Über die gleichverteilung von zahlen mod eins," *Math. Ann.* **77** (1916), pp. 313–352.

AT&T Bell Laboratories, 600 Mountain Avenue, Murray Hill, New Jersey 07974

Contemporary Mathematics
Volume **135**, 1992

QUASISYMMETRIC CONJUGACIES
FOR SOME ONE-DIMENSIONAL MAPS
INDUCING EXPANSION

M. JAKOBSON

ABSTRACT. We prove that topological conjugacies between certain one-dimensional maps are quasisymmetric by constructing induced maps which satisfy conditions of Adler's (Folklore) theorem.

1. INTRODUCTION

1.1. The idea of studying quasisymmetric classes of conjugacies between unimodal maps of different smoothness was introduced by Sullivan who in particular used it to study infinitely renormalizable dynamics (see [S]). A recent result of Yoccoz [Y] implies that topological and quasiconformal classes coincide for non-renormalizable quadratic polynomials.

In this paper we prove qs property for conjugacies between certain C^2-maps inducing expansion, in particular for S-unimodal maps satisfying Misiurewicz condition. Our methods are purely real, we do not assume that the maps under consideration are of the form $h \circ Q$ where h is a diffeomorphism and Q a quadratic polynomial. Our approach is based on the " Renyi property " (see below) which is also crucial for the existence of absolutely continuous invariant measures (see [A], [W], [J1]).

Since the first version of this work appeared in the preprint [J2] the author and Swiatek proved qs property of conjugacies for many topological classes of S-unimodal maps with non-renormalizable as well as with renormalizable dynamics(see [JS]). However these last results strongly use complexification arguments and can be only applied to the maps of the form $h \circ Q$. It would be interesting to generalize purely real arguments of this work beyond the class of Misiurewicz maps.

1.2. A homeomorphism of an interval is called *M-quasisymmetric* if for any x, y belonging to I and $z = (x + y)/2$ the inequality

$$\frac{|\varphi(x) - \varphi(z)|}{|\varphi(y) - \varphi(z)|} \leq M$$

1991 *Mathematics Subject Classification*. Primary 58F15, 30C60.
Key words and phrases. Unimodal maps, quasisymmetric conjugacy.
This work was supported by NSF grant DMS-901631
This paper is in final form and no version of it will be submitted for publication elsewhere.

holds.

In order to study quasisymmetric properties of a conjugacy between two uni-modal maps f and g we construct special induced maps F and G which have infinite number of expanding branches and then use that expansion and bounded distortion of F (G) implies that all iterates of F (G) have uniformly bounded distortions. We call it the *Renyi property* .

The proof of the following well known statement illustrates the method and contains somefacts which will be used later.

Lemma 1. *If f_1 , f_2 are expanding C^2 endomorphisms of the circle of the same degree then they are qs conjugate.*

Proof.

1. There exists a unique orientation preserving homeomorphism φ conjugating f_1 to f_2.
2. *Expansion property.* For any two points x, y there exists a nonnegative $n = n(x, y)$ such that $f^n | [x, y]$ is one-to-one and the distance between $f^n x$ and $f^n y$ is larger than some fixed constant d_0.
3. The uniform continuity of φ^{-1} implies the following

    Large scale property. If x, y and $z \in [x, y]$ are any three points such that $|x - y| > d_0$ and
    $$\theta_0^{-1} \; > \; \frac{|x - z|}{|y - z|} \; > \; \theta_0$$
    then
    $$M_0^{-1} \; < \; \frac{|\varphi(x) - \varphi(z)|}{|\varphi(y) - \varphi(z)|} \; < \; M_0$$
    where $M_0 = M_0(d_0, \theta_0)$.

4. C^2-smoothness and expansion imply

    Renyi property. There exist $M_{1i} > 0$ such that
    $$M_{1i}^{-1} \; < \; \frac{Df_i^n(u)}{Df_i^n(v)} \; < \; M_{1i}, \qquad i = 1, 2$$
    for any pair of points u, v such that $f_i^n | [u, v]$ is one-to-one.

5. As φ is equivariant 2, 3 and 4 imply qs property for any pair of points x, y with $M = M_0(d_0, M_{11}^{-1}) \cdot M_{12}$.

2. FULLY CHAOTIC MAPS

2.1. Let f be a C^2-smooth unimodal map of the unit interval, satisfying the following conditions

$f(0) = f(1) = 0$, the critical point c of f is nondegenerate, f has one more fixed point $q \in (c, 1)$, and both fixed points q and 0 are repelling.

In this section we shall study a particularly simple class of the so-called "fully-chaotic" maps which satisfy
$$f(c) \; = \; 1.$$

Let us consider the first return map F induced by f on the interval $[q_{-1}, q] \; = \; I$ bounded by the fixed point q and its preimage $q_{-1} \in (0, c)$. The map F is piecewise

C^2 with an infinite number of branches $F_i = f^i$, $i = 2, 3, \ldots$, which map their domains Δ_i onto I. The domains Δ_i accumulate toward c. It is straightforward that all but a finite number of F_i are expanding and all of them have uniformly bounded distortions. Namely $A_0 = df(0) > 1$ implies that there exist constants c_1, c_2, c_3 such that F_i satisfy

(1) $$C_2 A_0^{i/2} > |DF_i| > C_1 A_0^{i/2},$$

(2) $$\frac{|D^2 F_i|}{(DF_i)^2} < C_3.$$

Notice that (1) does not imply that F_i with small i are expanding, for example F can have an attracting periodic orbit of a small period. We introduce expanding property as an additional condition:

There exists a natural m_0 such that

(3) $$|DF^{m_0}(x)| > C_0' > 1.$$

where we use F^{m_0} to denote any composition of m_0 branches F_i.

Conditions (2) and (3) imply uniform bounded distortion property and consequently the Renyi property for all compositions of F_i (see [A]).

Remark

The condition (3) is satisfied for maps with negative Schwarzian derivative. For general C^2 - maps , because of (1), we need to check (3) only for a finite number of branches F_i. If (3) holds for some f, then it holds for any fully chaotic map f' which is sufficiently close to f in C^2-topology.

2.2. Now suppose that f' is another fully chaotic map satisfying (3). Then there exists a unique orientation-preserving homeomorphism φ conjugating f to f'.

Proposition 1. φ is quasisymmetric.

Proof.

Step 1. Reducing to $x, y \in I$.

Let us consider a sequence I_0^{-n} of the adjacent preimages of I converging to zero. As zero is a repelling fixed point, the maps

$$f^n : I^{-n} \to I$$

have distortions uniformly bounded by M_1. The same is true for the sequence of preimages I_1^{-n} converging to one.

We fix some small $d_0 \ll |I|$ and find $M_0(d_0, M_1^{-1})$ from the Large Scale property. Then our reasoning depends on the location of x and y .

a) If x and y are separated by I then qs property holds with $M = M_0$.

b) If x, y are outside of I and both belong to $(0, c)$ or to $(c, 1)$ then $[x, y]$ is mapped by an appropriate diffeomorphism f^n with uniformly bounded distortion onto $[x_1, y_1]$ such that either

b_1) $[x_1, y_1]$ is contained in I and we are done, or

b_2) $[x_1, y_1]$ contains a boundary point of I.

In the last case we use that q is a repelling fixed point. Namely if the distance between x_1 and y_1 is larger than d_0 then qs property for x, y holds with M_0 . And if the distance between x_1 and y_1 is small then under further iterates of f they will spiral around q until they become d_0 - separated. As these last iterates also have distortions uniformly bounded by some M_2 , qs property holds for x, y with $M_0(d_0, (M_1 \cdot M_2)^{-1})$.

2.3. Step 2. Now we assume $x, y \in I$ and respectively $\varphi x, \varphi y \in I'$ and prove φ is quasisymmetric on I. If x, y belong to the same domain Δ_i then we use uniformly bounded distortion of all compositions $F_{i_1} \circ F_{i_2}...$ and iterate until the images of x and y fall into different domains Δ_i , Δ_j. Let again denote these last iterates by x and y and the corresponding iterate of the middle point by z . Then we distinguish two cases.

a) The interval $[x, y]$ does not contain the critical point c_1 of f_1. In that case the points $X = fx, Y = fy$ belong to different preimages I^{-n} of I , and similarly to the Step 1 we obtain that the corresponding ratios on $[X, Y]$ and on $[\varphi X, \varphi Y]$ are uniformly comparable. The nondegeneracy of the critical point c implies the following

Bounded Critical Distortion property. For any $a > 0$ there is a $b > 0$ such that if u, s and $t \in [u, s]$ are any three points belonging to Imf and u_1, s_1, t_1 are their preimages with respect to the same branch of f^{-1} and if one of the ratios $\frac{\|[u,t]\|}{\|[u,s]\|}$ or $\frac{\|[u_1,t_1]\|}{\|[u_1,s_1]\|}$ is bounded away from zero by a then another one is bounded away from zero by b.

BCD property bounds ratios on $[\varphi x, \varphi y]$ in terms of ratios on $[\varphi X, \varphi Y]$ and we get a uniform estimate.

b) The interval $[x, y]$ contains c. We can assume that both x and y are in a small neighborhood of c, otherwise we were done by the Large Scale property with d_0 smaller than the size of that neighborhood.

Let assume for definiteness that z belongs to $[c, y]$. Quasisymmetric property is equivalent to the fact that both ratios

$$\frac{|\varphi(x) - \varphi(z)|}{|\varphi(y) - \varphi(x)|} \ and \ \frac{|\varphi(y) - \varphi(z)|}{|\varphi(y) - \varphi(x)|}$$

cannot be too small.

As the ratio $\frac{|x-z|}{|x-y|}$ is bounded away from zero, we obtain that either $|x - c|$ or $|c - z|$ is comparable to $|x - y|$. Let assume first that $|c - z|$ is comparable to $|x - y|$. The points z and y belong to the domain of monotonicity of f and we have

$$\frac{|z - c|}{|y - c|} > \frac{|z - c|}{|x - y|}$$

As BCD property holds for f and for f' it is enough to compare ratios on $(1, f(y))$ to ratios on $(1, f'(\varphi(y))$. But these intervals are mapped with uniformly bounded distortions onto the large intervals containing respectively $(0, q_{-1})$ and $(0, q'_{-1})$ .Using the Large Scale property we obtain that $\frac{|\varphi(z)-c'|}{|\varphi(y)-c'|}$ is bounded away from zero. But then the same is true for

$$\frac{|\varphi(x) - \varphi(z)|}{|\varphi(x) - \varphi(y)|} = \frac{|\varphi(x) - c'| + |\varphi(z) - c'|}{|\varphi(x) - c'| + |\varphi(y) - c'|}$$

Let assume now that $|x - c|$ is comparable to $|x - y|$. It is enough to prove that $\frac{|\varphi(x)-c'|}{|\varphi(x)-\varphi(y)|}$ is bounded away from zero. Let define $\bar{x}$ by $f\bar{x} = fx$. As c is a nondegenerate critical point, $|\bar{x} - c|$ is comparable to $|x - c|$. As $\bar{x}$ and y lie on the same side of c, the previous reasoning applies and we can estimate from below the ratio $\frac{|\varphi(\bar{x})-c'|}{|\varphi(y)-c'|}$ .Using that c' is a nondegenerate critical point we get for some $d_0 > 0$ $\frac{|\varphi(\bar{x})-c'|}{|\varphi(y)-c'|} > d_0$ and consequently

$$\frac{|\varphi(x) - c'|}{|\varphi(x) - \varphi(y)|} = \frac{|\varphi(x) - c'|}{|\varphi(x) - c'| + |\varphi(y) - c'|} > \frac{d_0}{1 + d_0}$$

Notice that the points z and y in the first estimate and the points $\bar{x}$ and y in the second estimate can be arbitrary close. It only means that the corresponding ratios are close to one.

In order to estimate the ratio $\frac{|\varphi(y)-\varphi(z)|}{|\varphi(y)-\varphi(x)|}$ we use that $\frac{|z-y|}{|x-y|}$ is uniformly bounded away from zero and so are

(i)
$$\frac{|y - c|}{|y - x|} > \frac{|z - y|}{|x - y|}$$

(ii)
$$\frac{|z - y|}{|y - c|} > \frac{|z - y|}{|x - y|}$$

Now, as y, z and c are in the domain of monotonicity of f, we can apply the previous arguments and obtain uniform estimates for $\frac{|\varphi(y)-c'|}{|\varphi(y)-\varphi(x)|}$ and for $\frac{|\varphi(y)-\varphi(z)|}{|\varphi(y)-c'|}$.

As a consequence the ratio

$$\frac{|\varphi(y) - \varphi(z)|}{|\varphi(y) - \varphi(x)|}$$

is uniformly bounded away from zero.

Combining steps 1 and 2 we get that qs property holds for any pair of points x, y in $[0, 1]$ with some uniform constant M depending on the constants in the above construction, which proves Proposition 1.

3. Misiurewicz maps

3.1. In this section we consider unimodal maps with aperiodic kneading invariants which satisfy Misiurewicz condition : critical point is non-recurrent . We concentrate on the classical case of maps with negative Schwarzian derivative (for definitions see for example [M]).The result can be generalized to C^2- maps which satisfy some additional expanding properties, similar to (3) from Section 2 , but in contrast to the case of fully chaotic maps there is no effective way to check these conditions.

Generally f can be a finitely renormalizable map. However the same arguments as above show that any two points can be iterated by a diffeomorphism with uniformly bounded distortion until they either will be d_0 - separated with some fixed d_0 , or fall into a periodic interval I such that $f^m|I$ is nonrenormalizable.

So we assume that f and g are non-renormalizable S-unimodal maps satisfying Misiurewicz condition which are topologically conjugate by a homeomorpism φ and prove

Theorem 1. φ *is quasisymmetric.*

3.2. The proof of Theorem 1 is based on the construction of an induced map F satisfying the Renyi property. For S-unimodal maps we use Koebe distortion property , which implies uniformly bounded distortions for the induced branches f^{n_i} whenever holds the following

Uniform Extendability property. Every branch $F_i = f^{n_i} : \Delta_i \to I$ can be extended as a diffeomorphism onto a larger domain Γ_i so that f^{n_i} maps Γ_i onto some fixed interval T which is a union of I with two adjacent intervals R and L. Then the distortions of F_i are bounded by a uniform constant. When all F_i are uniformly extendable so are all their iterates $F_{i_1} \circ F_{i_2} \circ \ldots \circ F_{i_s}$ and Renyi property follows.

3.3. We shall construct F by induction. We start by constructing the first return map F^1 on $I = [q_{-1}, q]$ as in Section 2. We assume that f is not a fully chaotic map and so the critical value of f is smaller than one. As a result F^1 has a finite number of monotone branches F_i^1, and a parabolic branch h^1 defined on the central domain δ^1 containing the critical point c. The branches F_i^1 as well as all their compositions $F_{i_1}^1 \circ F_{i_2}^1 \circ \ldots \circ F_{i_s}^1$ are uniformly extendable. The union of all preimages $\delta_{i_1 \ldots i_s}^1 = (F_{i_1}^1 \circ F_{i_2}^1 \circ \ldots \circ F_{i_s}^1)^{-1} \delta^1$ and the remaining Cantor set C^1 form a partition of I :

$$(\xi^1) \qquad\qquad I = \delta^1 \cup C^1 \cup \bigcup_{s=1}^{\infty} \bigcup_{i_1 \ldots i_s} \delta_{i_1 \ldots i_s}^1$$

Then we pull back the partition ξ^1 by $(h^1)^{-1}$ onto δ^1. The elements of that new partition are the domains of the first return map F^2 on δ^1. That map has an infinite number of monotone branches defined on preimages $(h^1)^{-1} \delta_{i_1 \ldots i_s}^1$. We denote these preimages by Δ_j where j are new composite indices and we denote the corresponding branches of F^2 by F_j^2.

The following argument depends on the position of the critical value $w = h^1(c)$ with respect to the partition ξ^1.

Let assume first that $w \in C^1$. Then we get a partition

$$(\eta^1) \qquad\qquad \delta^1 = \bigcup \Delta_j \cup C_0^1$$

As $\delta_{i_1 \ldots i_s}^1$ are preimages of δ^1 with respect to branches with uniformly bounded distortions, we obtain that all branches F_j^2 with the domains outside of a given neighborhood of c are uniformly extendable. In particular the domains of uniformly extendable branches fill in two intervals adjacent to the boundary points of δ^1. These domains satisfy the following

Boundary Scale property. There exists $c_1 > 0$ such that every domain Δ_j is separated from the boundary point of δ_1 by a distance larger than $c_1 |\Delta_j|$.

The last step of the construction is called

Boundary Refinement. We substitute non-extendable branches F_j^2 by compositions $F_{j_1}^2 \circ F_{j_2}^2$. BCD property for h^1 and Boundary Scale property imply that these compositions are extendable if and only if F_j^2 are. Then we substitute non-extendable branches $F_{j_1}^2 \circ F_{j_2}^2$ by triple compositions and so on. As a result we get a partition

$$(\zeta^1) \qquad\qquad \delta^1 = \bigcup \Delta_{j_1 \ldots j_l} \cup C_1^1$$

where C_1^1 is a Cantor set, and $\Delta_{j_1 \ldots j_l}$ are domains of uniformly extendable branches $F_{j_1}^2 \circ F_{j_2}^2 \circ \ldots \circ F_{j_l}^2$.

ζ^1 is our final partition in the case $w \in C^1$.

3.4. If w belongs to one of the preimages $\delta_{i_1 \ldots i_s}^1$ of δ^1, then we do more than one inducing before the boundary refinement.

In that case the induced map F^2 on δ_1 has a countable number of monotone branches and a central parabolic branch h^2 with the domain δ^2 containing c. The monotone branches of F^2 are uniformly extendable, although their extendability depends on the ratio $\frac{|\delta^2|}{|\delta^1|}$ and on the other parameters of f and can be much smaller than the extendability of the branches of F^1. As a consequence all compositions of monotone branches $F_{i_1}^2 \circ F_{i_2}^2 \circ \ldots \circ F_{i_k}^2$ are uniformly extendable.

Then we proceed as above: we delete preimages of δ^2 from δ^1 and obtain a partition

$$(\xi^2) \qquad\qquad \delta^1 = \bigcup (\delta^2)^{-k} \cup C^2$$

Then we check the position of $h^2(c) = w_2$ with respect to ξ^2. If $w_2 \in C^2$, then we have a case of depth 2. If $w_2 \in (\delta^2)^{-k}$ then the depth of induction is at least 3 and so on.

Misiurewicz case is characterized by a finite depth of induction. So eventually we obtain

$$(\xi^n) \qquad\qquad \delta^{n-1} = \bigcup (\delta^n)^{-k} \cup C^n$$

and $w_n \in C^n$. Then we finish our construction as in 3.3. We pullback ξ^n by the central parabolic branch h^n and obtain a partition η^n of δ^n into the domains of monotone branches F_α^n and a complementary Cantor set C_0^n. All branches F_α^n with the domains outside of a small neighborhood of the critical point are uniformly extendable. In particular the domains of extendable branches fill in two intervals adjacent to the boundary of δ^n and satisfy the Boundary Scale property.

For the remaining non-extendable branches we do boundary refinement , i.e. construct consecutive compositions $F_{\alpha_1}^n \circ F_{\alpha_2}^n \circ \ldots \circ F_{\alpha_s}^n$ and stop if and only if the last map in the composition $F_{\alpha_1}^n$ is extendable. BCD property for h^n and the Boundary Scale property imply that the whole composition is extendable. As a result we get a final partition ζ^n of δ^n into domains of uniformly extendable monotone branches F_β^n and the remaining Cantor set C_1^n.

3.5. Step 1. Reducing to $x, y \in \delta^n$.

Let consider the location of two arbitrary points $x, y \in I$ with respect to the partition ξ^1. If x, y belong to the same preimage $(\delta^1)^{-k}$ then the interval $[x, y]$ will be eventually mapped with uniformly bounded distortion onto δ^1. If this is not the case then either x, y belong to the different preimages of δ^1, or one of these points (or both) belongs to the Cantor set C^1. In all these cases there exists a branch $F_{i_1}^1 \circ F_{i_2}^1 \circ \ldots \circ F_{i_s}^1$ which maps $[x, y]$ onto $[x_1, y_1]$, such that the points x_1, y_1 belong to the domains of two different branches of the induced map F^1. As the boundary points of these domains are preimages of q we get as above that either x_1 and y_1 are at a distance larger than some fixed d_0, or they will be d_0 - separated after spiraling around q. So the problem reduces to $x, y \in \delta^1$.

In that case we can iterate x and y until either they both fall into δ^2 or they fall into different elements of ξ^2. The elements of ξ^2 are preimages of δ^2 with respect to compositions of the branches F_i^2. So we can iterate further until either the corresponding points, which we denote by x_1, y_1, fall into the domains of different monotone branches of F^2, or one of them falls into δ^2 and another one remains outside.

The first possibility means that $h^1(x_1), h^1(y_1)$ belong to the different elements of the partition ξ^1.

Now by the inductive assumption the critical values $h^i(c)$ for $i < n$ belong to some intervals δ_i^{-k}. That implies that $h^i|[x, y]$ are uniformly extendable for $[x, y] \subset \delta^{i-1} \setminus \delta^i$, $i \leq n$. So by applying a map with uniformly bounded distortion we come to the case which has been already considered.

It remains to consider the case when one of the points is inside δ^2 and another one is outside. As the boundary points q_i of δ^i, $i = 0, 1, \ldots, n$ are preimages of q, we can choose ϵ_0- neighborhoods U_i of q_i such that any interval $l \in U_i$ which contains q_i will be mapped by a diffeomorphism with uniformly bounded distortion onto an interval which contains q and has the size larger than some fixed ϵ_1. Using the Large Scale property with d_0 smaller than $min(\epsilon_0, \epsilon_1)$ we obtain a uniform estimate in that last case as well. So the problem reduces to $x, y \in \delta^2$.

Proceeding in this way we eventually either get for x, y a qs correspondence with some uniform M_0 or come to $x, y \in \delta^n$.

The previous argument proves the following fact that will be used later.

Lemma 2. *Let x, y be any two points belonging to the different elements of some partition ξ^i, $i \leq n$. Then for $z \in [x, y]$ uniformly bounded ratios $\frac{|x-z|}{|x-y|} > a > 0$ are transformed by φ into uniformly bounded ratios $\frac{|\varphi(x)-\varphi(z)|}{|\varphi(x)-\varphi(y)|} > b(a) > 0$.*

3.6. Step 2. Now the problem is reduced to $x, y \in \delta^n$. Here $h^n(c)$ belongs to C^n and some of the branches induced on δ^n can be non-extendable . Then we do boundary refinement as described in 3.4.

Now let take any two points $x, y \in \delta^n$. As above we can iterate them until they fall into the different elements of the partition ζ^n. By construction the elements of ζ^n either coincide to some extendable elements of $\eta^n = (h^n)^{-1}\xi^n$ or were obtained in the process of boundary refinement of non-extendable elements η^n. As extendable elements fill in two intervals adjacent to the boundary of δ^n we obtain that for any interval l contained in a domain of a composition of two branches $F_{j_1}^n \circ F_{j_2}^n$, which are both non-extendable, the image $F_{j_2}^n(l)$ is uniformly bounded away from the end points of δ^n and so $F_{j_2}^n|l$ is uniformly extendable.

This implies the following property of the iterates of two points x, y which belong to the different elements of ζ^n but to the same element of η^n. They are iterated with uniformly bounded distortion until they fall for the last time into a domain of some non-extendable branch (close to its boundary) and after that they fall into different extendable elements of η^n.

Because of BCD property of h^n the last iterate can distort ratios only by a finite amount. So the problem reduces to x, y which initially belong to the different elements of η^n. If $[x, y]$ does not contain c, then $h^n(x) = x_1$ and $h^n(y) = y_1$ belong to the different elements of ξ^n and Lemma 2 applies.

It remains to consider $x, y \in \delta^n$ which lie on different sides of the critical point. As c_1, c_2 are nondegenerate critical points of h_1^n (respectively of h_2^n) the arguments of Section 2.3 (b) reduce the problem to the triples of points lying on the same side of the critical point. For these triples we use Lemma 2 and get the result.

REFERENCES

[A]   R.L. Adler, *F-expansions revisited*, Lecture Notes in Math. **318** (1973), 1–5.

[J1] M. Jakobson, *Absolutely continuous invariant measures for one-parameter families of one-dimensional maps*, Commun. Math. Phys. **81** (1981), 39–81.

[J2] M. Jakobson, *Quasisymmetric conjugacy for some one-dimensional maps inducing expansion* (1989).

[JS] M.Jakobson and G. Swiatek, *Quasisymmetric conjugacies between unimodal maps* (1991).

[M]   M. Misiurewicz, *Absolutely continuous invariant measures for certain maps on an interval*, Publ. IHES **53** (1981), 17-51.

[S]   D. Sullivan, *Bounds, quadratic differentials and renormalization conjectures* (to appear).

[W]   P. Walters, *Invariant measures and equilibrium states for some mappings which expand distances*, Trans. Am. Math.Soc. **236** (1978), 127–153.

[Y]   J.C. Yoccoz, *On local connectivity of the Mandelbrot set* (to appear).

DEPARTMENT OF MATHEMATICS, UNIVERSITY OF MARYLAND AT COLLEGE PARK
E-mail address: mvy@lakisis.umd.edu

Contemporary Mathematics
Volume **135**, 1992

A MONOTONICITY PROPERTY IN
ONE DIMENSIONAL DYNAMICS (*)

JEAN MARC GAMBAUDO AND CHARLES TRESSER

(*) This paper is dedicated to Roy Adler, on the occasion of his 60th birthday.

ABSTRACT. For unimodal maps on the interval, we define a numerical topological invariant adapted from the classical theory of circle maps. For a class of continuous one-parameter families of interval maps containing, for any fixed real $\lambda \geq 1$, the quadratic-like families $q_a(\lambda, x) = 1 - a|x|^\lambda$ for $a \in [\frac{2\lambda+1}{\lambda+1}, 2]$, we give an elementary proof of the monotonicity of this number when varying the parameter. We also show that uncountably many distinct topological behavior cannot be C^k-structurally stable in the set of even C^k unimodal maps of the interval $[-1, 1]$, where $k \geq 1$.

1. INTRODUCTION AND PREPARATORY RESULTS

The present paper is motivated by (and brings a small contribution to) the following:

Conjecture 1. *For $k \geq 1$, in an open dense subset $S^k(I)$ of the space $C^k(I)$ of C^k endomorphisms of the interval I, all maps have a hyperbolic stable periodic orbit.*

Recall that for a given map f, the *orbit* of x is the sequence

$$(x, f(x), f^{\circ 2}(x), \dots),$$

where $f^{\circ n} = f \circ f^{\circ(n-1)}$. A periodic orbit with period p is *hyperbolic* if at any (hence at all) of its points q, $|(f^{\circ p})'(q)| \neq 1$. A hyperbolic orbit is *stable* if $|(f^{\circ p})'(q)| < 1$, and *unstable* otherwise.

Remark 1. Conjecture 1 has been proved by M. Jakobson when $k = 1$ [11], and we notice that it is precisely above $k = 1$ that one does not have a proof of the Closing Lemma. In fact the Closing Lemma is an alternate and more usual context to discuss our results; one would like to know if any non-wandering critical point can be C^k-perturbed to a periodic critical point.

The conjectured picture is that any piece of $U^k(I) = C^k(I) \setminus S^k(I)$ is of codimension at least one, the difficult case being when all periodic orbits are hyperbolic and unstable. Let then $\{f_a\}$ be a continuous one parameter family in $C^k(I)$, ample enough to cross both $S^k(I)$ and $U^k(I)$. We say that:
- $\{f_a\}$ is *monotonic* if any topological dynamics realized in $\{f_a\}$ corresponds to a single interval in the parameter space. It is said to be *monotonic at f_{a_0}* if the same holds true in a neighborhood of a_0.

1991 Mathematics Subject Classification. Primary 58F10, 54H20.

This paper is in final form and no version will be submitted for publication elsewhere.

The map f may belong to $U^k(I)$ only if $K(f)$ is not periodic, but stable periodic orbits can exist even when $K(f)$ is not periodic. Furthermore, $K(f)$ can be periodic while the critical orbit converges to a non-hyperbolic periodic point. Conjectures 1 and 2 for Unimodal maps have nevertheless nonequivalent counterparts in terms of Kneading theory, which are interesting conjectures by themselves. These read:

UK1. *For $k \geq 1$, $K(f)$ periodic holds in an open dense subset of the unimodal maps in $C^k(I)$.*

UK2. *For $k \geq 0$, any f in $C^k(I)$ belongs to a family $\{f_a\}$ such that $K(f_a)$ is increasing with a and strictly increasing when not periodic.*

The special versions of Conjectures 1 and 2 obtained by restricting the problem to the quadratic family $q_a(x) = 1 - ax^2$, and forgetting about the non-hyperbolic periodic points, are known to be equivalent to the conjectures **UK1** and **UK2** restricted to the quadratic family. These Conjectures are models for our main result in the next section (see Theorems 3 and 4).

Remark 4. Recognizing that a given family is fast or even monotonic is usually quite tricky even in the unimodal case, even if $k = 0$. An easy example is the tent family $t_a(x) = 1 - a|x|$. Another easy class of examples, which contains trapezoidal families (see [4,5] for other fast trapezoidal families), is provided by "cutting out families". More specifically, let $f_1 : [0,1] \rightarrow [0,1]$ with $f(0) = f(1) = 0$ be a unimodal map with critical set $J \subset]0,1[$ and $f(J) = 1$.

Define the *cutting out family* $\{f_R\}_{0 \leq R \leq 1}$ of f_1 by:

$$f_R(x) = \min(f_1(x), R).$$

Then $\{f_R\}_{0 \leq R \leq 1}$ is trivially a monotonic family. Furthermore, if f_1 has no *homterval* (i.e. no interval not converging to a stable or semi-stable periodic orbit, where all iterates are homeomorphisms), the family is easily seen to be fast (with some supplementary assumptions like f_1 trapezoidal, or f_1 with negative Schwarzian derivative, one even gets that the Lebesgue measure of the set $\{R|f_R \in U^0\}$ is zero). The basic fact in cutting out families is that any kneading sequence for a f_R can be read as an itinerary of some point under f_1, these points being precisely those which are the greatest in their orbit.

Remark 5. Conjectures similar to 1 and 2 (say 1' and 2') can be formulated for circle endomorphisms isotopic to identity or for circle endomorphisms with at least one critical point. Here again 1' is known to be true for $k = 1$ [11], and in contrast with the difficulty of these general problems for $k \geq 2$, we have:

Theorem 1. *Conjecture 2' holds true for any $k \geq 0$ when one restricts oneself to the space $\mathcal{E}^k$ of non-decreasing circle endomorphisms of degree one.*

This is in fact an easy extension of the same, but better known, result for circle homeomorphisms. The key points are that if f and g are lifts to $\mathbb{R}$ of non-decreasing circle endomorphisms of degree one, and if $\rho(.)$ stands for the rotation number, then:

C1: $f \geq g \Rightarrow \rho(f) \geq \rho(g)$,

C2: $\rho(g) \in \mathbb{R} \backslash \mathbb{Q}$ and $f > g$ everywhere $\Rightarrow \rho(f) > \rho(g)$,

C3: if $\rho(g) = p/q$, with p and q coprime, and $g \in S^k$, for all $\epsilon > 0$ there exists f ϵ-close to g with $\rho(f) = \rho(g)$ and $f \in S^k$. For ϵ small enough, when g^{oq} is not the identity, it is sufficient that, depending on g, either $f > g$ or $f < g$ everywhere.

For later use, we also record here the further well known property:

C4: $\rho : \mathcal{E}^k \to \mathbb{R}$ is continuous for $k \geq 0$.

The present status of Conjectures 1 and 2 in $C^k(I)$ is as follows for $k > 1$:

M: -The quadratic family $q_a(x) = 1 - ax^2$ is monotonic [6],[16], and similar results hold true for polynomial like families [6]. As a consequence, the topological entropy $h(q_a)$ is a monotonous function of a.

F: -No fast C^1 family is known, even in the case of a single critical point.

However, the quadratic family has been shown to be fast at the infinitely renormalizable maps of bounded type [20], and at the non infinitely renormalizable maps, the later result being in fact a real variable consequence of a more general result about the complex endomorphism $z \to c + z^2$ [23].

We recall (with just the amount of generality needed to make sense of the above report) that a unimodal map $f : I \to I$ is *n-renormalizable* if for some $n > 1$, f exchanges n intervals J_i, where $f^{on}|_{J_i}$ is a unimodal map, said to be *renormalized* from f. If the renormalized map is m-renormalizable, f is (n,m)-*twice renormalizable*, and so on. An infinitely renormalizable map is of *bounded type* if the sequence $n, m, \ldots$ is bounded.

It should also be noticed that under mild conditions, one can prove that aperiodic behavior with some expanding properties can be perturbed away in reasonable one-parameter families: see e.g., [2,12].

All known proofs of **M** and **F** rely on some aspect of the theory of quasi-conformal mappings. In fact, the main efforts to prove that the quadratic family is fast belong to a more general, but similar program for rational maps on the Riemann sphere: a main open problem in this context is to prove the *Fatou Conjecture* [7] that densely, all critical points of a rational map are in the basins of attraction of periodic sinks, i.e. that the Axiom A property is dense in the space of rational maps with degree d for any $d \geq 2$. Another way to (im)prove **M** and **F** would be to prove the well known:

Convex-Schwarzian Conjecture. *Let* $g : [-1, 1] \to [-1, 1]$ *be a* C^3 *convex function with negative Schwarzian derivative, with 0 as single critical point and such that* $g(0) = -1, g(-1) = g(1) = 1$. *Then the family* $f_R(x) = R - g(x)$ *(or* $h_a(x) = 1 - ag(x)$) *is fast.*

Remark 6. The negative Schwarzian condition cannot be simply dropped in this Conjecture; in the case g is only convex, Anna Zdunik constructed a counter example in [24]. Some results toward this Conjecture can be found in [3,9,13,15,16,24].

In §II of this paper, we will import in the theory of unimodal maps, a topological invariant (weaker than the topological entropy) coming from the theory of degree one circle endomorphisms. The corresponding monotonicity and fastness will lead us to weak versions of Conjectures 1 and 2 as expressed by Theorems 3 and 4. Using previously known results assembled in Theorem 2, the proof will essentially consist in a reduction to the key points **C1**, **C2** and **C3** mentioned above for circle maps. Some more material depending on the same constructions is assembled in §III.

2. ROTATION INTERVALS AND UNIMODAL MAPS ON THE INTERVAL

Let $f : [-1, 1] \to [-1, 1]$ be a continuous even unimodal map such that $f(0) = 1$, and let $\mathcal{C}^k, k \geq 0$, stand for the space of all such maps, equipped with the C^k topology. For an economy of notations, the circle will be defined as $\mathbb{T} = \mathbb{R}/2\mathbb{Z}$ with a realization as $[-1, 1]/(-1 \sim 1)$, universal cover $\mathbb{R}$, and canonical projection $\pi : \mathbb{R} \to \mathbb{T}$. Notice that this non conventional representation of the circle causes the rotation numbers to be the double of what they would usually be for the same topological dynamics.

To any map $f \in \mathcal{C}^0$, we associate a map $\overline{f} : \mathbb{R} \to \mathbb{R}$ as follows. We set:

$$f_0 = f \text{ on } [-1, 0] \text{ and } f_0 = 2 - f \text{ on } [0, 1[,$$

and we define:

$$f_n : [2n - 1, 2n + 1[\to [2n - 1, 2n + 3[,$$
$$x \mapsto f_0(x - 2n) + 2n ,$$

so that $\overline{f}$ is the real function satisfying (see Figure 1):

$$\overline{f}|_{[2n-1,2n+1[} = f_n .$$

$\overline{f}$ has monotonic upper and lower bounds $\overline{f}^+$ and $\overline{f}^-$ defined respectively by:

$$\overline{f}^+(x) = \sup_{y \leq x} \overline{f}(y) ,$$
$$\overline{f}^-(x) = \inf_{y \geq x} \overline{f}(y) ,$$

and we define its rotation interval as:

$$I(\overline{f}) = [\rho(\overline{f}^-), \rho(\overline{f}^+)] ,$$

where $\rho(g)$ is defined for any monotonic lift of a circle map as its (unique) rotation number. One can easily check that

$$\rho(\overline{f}^-) = 2 - \rho(\overline{f}^+) .$$

Hence we define the *rotation interval bound* $\omega(f) = \rho(\overline{f}^+)$. We denote by $h(f)$ the *topological entropy* of f. The extension of topological entropy to the special type of discontinuous maps we consider is discussed e.g. in [1]: we notice that $\overline{f}$ is in fact a special type of "old" map (old for Degree One Lifts [17]), a class already studied for its own sake for various reasons (see e.g., [1,8,10,17,21] for recent works and more references). We next list some simple and well known properties of $\omega(f)$:

Theorem 2.

- 1). $\omega(f)$ is a topological invariant (in fact determined by $h(f)$, which allows to generalize ω to any unimodal map).

- 2). $\omega(f) \in [1, 2]$, with:

$$\omega(f) = 1 \Leftrightarrow h(f) \leq \frac{\log 2}{2} ,$$
$$\omega(f) = 2 \Leftrightarrow h(f) \leq \log 2 ,$$

- 3). $\omega : \mathcal{C}^k \to [1, 2]$ is continuous for $k \geq 0$,

- 4). for any $\frac{p}{q}$ in $[1, \omega(f)]$, with p and q coprime, f has a periodic orbit with period q.

Proof of Theorem 2.

-3) follows directly from C4.

- The rest are particular cases (using the fact that $\rho(\overline{f}^{\,-}) = -\rho(\overline{f}^{\,+})$) of results which can be found e.g., in [1]. $\square$ (Theorem 2).

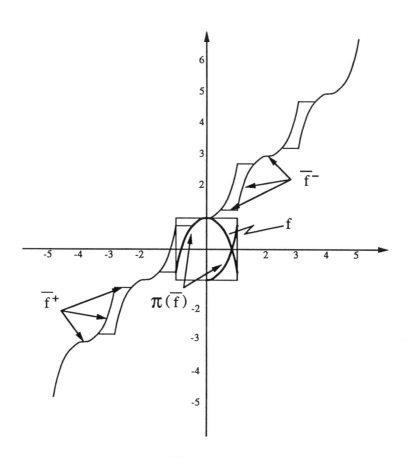

Figure 1.

- $\{f_a\}$ is *fast* if furthermore these intervals reduce to points for the dynamics in $U_k(I)$. It is said to be *fast at f_{a_0}* if no f_a is topologically conjugate to f_{a_0} for a in a neighborhood of a_0.

Remark 2. If the family is C^1-continuous, $\{f_a\}$ fast at f_{a_0} implies that there are a's arbitrarily close to a_0 such that f_a has a stable periodic point.

This motivates the:

Conjecture 2. *For $k \geq 0$, any $f \in C^k(I)$ belongs to a family which is fast at f, which would imply Conjecture 1.*

In the present paper, we shall deal only with *unimodal* endomorphisms of the interval $I = [a, b]$, i.e. maps which are both strictly increasing and strictly decreasing on a single interval that contains an end point of I, the interval (or point) in between (where the map is constant) being called in general the *critical set* to keep a terminology which is proper only in the smooth case, since the smooth case is our main concern. For definiteness, we shall assume that our maps are increasing on the left side of the critical set. For simplicity, we shall assume that:

E: zero is in the interior of I, and that all maps are *even*,

B: the right bound b of I is the critical value (i.e. the image of the critical set).

Remark 3. Since we will be concerned with topological properties, evenness seems an unreasonable restriction: any unimodal map $f : [a, b] \to [a, b]$ satisfying $f(a) = f(b)$ is transformed to an even one by a continuous change of coordinates. However, the topological invariant we want to introduce is most easily motivated for even maps. Furthermore, we will later be dealing with classes of deformations which are quite simply defined in the even case but out of our control otherwise.

We now recall some definitions and results from the kneading theory [16] for unimodal maps. Hence, let $f : I \to I$ be a unimodal map.

- The *address* $a(x)$ of $x \in I$ is L, R, or C according to wether x belongs to the left, right or critical interval.

- The *itinerary* of x is the sequence $I(x) = (a(x), a(f(x)), a(f^{o2}(x)), \dots)$.

- The *kneading sequence* $K(f)$ of f is then the itinerary of the image of the critical value.

It is easy to put an order on words written with the alphabet $\{R, C, L\}$, such that if $K(f) > K(g)$, the set of addresses for g is strictly contained in the set addresses of f, this statement being simplified by properties **E** and **B**. This order can be described as follows for words W_1 and W_2 which are either finite or not:

1- The order is almost lexicographical in that sense that if the first n letters of W_1 form a word bigger than the first n letters of W_2, then $W_1 > W_2$,

2- Set $R > C > L$,

3- If $W_1 = AX$ and $W_2 = AY$, where $X \neq Y$ both belong to $\{L, C, R\}$, then:

$$W_1 > W_2 \quad \Leftrightarrow \quad \begin{cases} \textit{either } A \textit{ has an even number of } R's \textit{ and } X > Y, \\ \textit{or } A \textit{ has an odd number of } R's \textit{ and } X < Y. \end{cases}$$

We can now state our main result:

Theorem 3. *A maps f with $\omega(f)$ irrational cannot be C^k- structurally stable in the set of even C^k unimodal maps of the interval $[-1,1]$, where $k \geq 1$.*

This gives the first uncountable set of topological behavior which cannot be C^k-structurally stable for $k \geq 1$ (but notice the mild evenness condition that we do not know how to drop for $k \geq 1$) and follows directly from the following more precise:

Theorem 4. *Any map f in C^k can be embedded in a C^k- continuous one parameter family $\{f_R\}_{R\in[A,B]}$ in C^k such that:*
- $\omega(f_R)$ is a monotonic function of R with $\omega(f_A) = 1$ and $\omega(f_B) = 2$,
- the family is fast at those f_{R_0} such that either $\omega(f_{R_0})$ is irrational, or f_{R_0} stands at an end point of a R-interval where $\omega(f_R)$ is a rational number.

Let us say that a C^k-continuous one parameter family $\{f_R\}_{R\in[A,B]}$ is n-ω-monotonic if $(\overline{f_S^+})^{on} \geq (\overline{f_T^+})^{on}$ for $S > T$, and *strictly n-ω-monotonic* if $(\overline{f_S^+})^{on} > (\overline{f_T^+})^{on}$ for $S > T$.

Proof of Theorem 4. For each f in C^k, we shall construct a family $\{f_R\}_{R\in[A,B]}$ in C^k such that $f = f_S$ for some $S \in [A, B]$, which will satisfy $\omega(f_A) = 1$ and $\omega(f_B) = 2$, and which will be 1-ω-monotonic and strictly 2-ω-monotonic. Using the properties **C1, C2** and **C3** of §II, this will be enough to prove the Theorem. The few inequalities below are obvious with a drawing pad at hand. We now proceed to the construction.
- We define the point x^* by $x^* > 0$ and $f(x^*) = x^*$.
- We next impose that:
-α) $f_{R_{|[-x^*,x^*]}} = f_{|[-x^*,x^*]}$ for $R \in [A, B]$.
- At last, keeping the required smoothness and the evenness, we deform f_R out of $[-x^*, x^*]$ by imposing that:
-β) $f_A(1) = -x^*, f_B(1) = -1$,
and:
-γ) $f_Q < f_P$ out of $[-x^*, x^*]$ for $Q > P$.
There is some freedom in this construction, left to the taste of the reader. Independently of these choices:
- From Theorem 2, it is plain that β) implies that $\omega(f_A) = 1$ and $\omega(f_B) = 2$.
- The 1-ω-monotonicity of the family is a direct consequence of α) and γ), and we notice that for $Q > P$, $\overline{f_Q^+} > \overline{f_P^+}$ everywhere in $[-1, 1[$, except on $[(\overline{f_Q^+})^{-1}(f_Q(1)), 0]$ where the maps are equal.
- Since $\pi(\overline{f}(-x^*)) = x^*$, and $(\overline{f_Q^+})^{-1}(f_Q(1)) > -x^*$, $[(\overline{f_Q^+})^{-1}$ $(f_Q(1)), 0]$ is mapped in a region where $\overline{f_Q^+} > \overline{f_P^+}$ by both $\overline{f_Q^+}$ and $\overline{f_P^+}$, and the strict 2-ω-monotonicity follows. $\square$ (Theorem 4)

3. MISCELLANEOUS COMMENTS

-1) Some deformations of even unimodal maps.
We have previously mentioned the fact that some precise one-parameter families have attracted special interest. The quite particular surgeries involved in the construction for the proof of Theorem 4 cannot be used to study families of

the form $h_a(x) = 1 - ag(x)$ as described in the Convex-Schwarzian Conjecture. Nevertheless, by introducing a general type of deformations of even unimodal endomorphisms, we shall see here that the rotation interval bound can be simply analyzed in parts of the quadratic-like families $q_a(\lambda, x) = 1 - a|x|^\lambda$. In particular:

Theorem 5. *For* $\lambda \geq 1$ *real and* $a \in \left[\frac{2\lambda+1}{\lambda+1}, 2\right]$, *the quadratic-like families* $q_a(\lambda, x) = 1 - a|x|^\lambda$ *is strictly 2-ω-monotonic.*

We now describe a class of deformation: $f \in C$ is a *(strict) regular deformation* of $g \in C$ if both the vertical and horizontal distances between the graphs of f and g increase (strictly) with $|x|$. $\{f_a\}$ is a *(strict) regular deformation family* if for $a \neq b$, f_a is a (strict) regular deformation of f_b. The surgeries in the proof of Theorem 4 give necessarily non-strict regular deformations families, while the families described in Theorem 5 are strict regular deformations families.

Let then $\{f_a\}$ be a continuous strict regular deformation family in C^k, with $k \geq 1$. With $\Delta_v(a, x)$ and $\Delta_h(a, x)$ standing for the length of the first order displacements of the graph of f_a, respectively in the vertical and horizontal directions, the following Lemma is a direct consequence of the definitions:

Lemma 1. *The family* $\{f_a\}$ *is strictly 2-ω-monotonic at* $a = a_0$ *if and only if*

$$\Delta_h(a, f_a(1)) > \Delta_v(a, x_0),$$

and

$$\Delta_h(a, x_0) < \Delta_v(a, x_1),$$

where x_0 *and* x_1 *are respectively defined by* $f_a(x_0) = -f_a(1)$ *with* $x_0 < 0$, *and* $f_a(x_1) = -x_0$ *with* $x_1 > 0$.

Proof of Theorem 5. For any map h_a of the form $h_a(x) = 1 - ag(x)$, we get

$$\Delta_v(a, x)dx = g(x)dx \text{ and } \Delta_h(a, x)dx = \frac{g(x)dx}{ag'(x)}.$$

Then apply Lemma 1. $\square$ (Theorem 5).

Remark 7. For the quadratic family $q_a(x) = 1 - ax^2$, the rotation interval bound when $a = \frac{5}{3}$ is $\frac{6}{5}$. It is $\frac{8}{7}$ at $a = \frac{2}{3}$ for the tent family $t_a(x) = 1 - a|x|$. From numerical tests, this lower rotation interval bound covered by Theorem 5 seems to increase monotonically with λ, until it saturates at $\frac{4}{3}$. Other tests suggest that for any λ, there is a n such that the a-family is stricly n-ω-monotonic for a rotation interval bound ranging all the way from 1 to 2. It seems that $n = 20$ is enough when $\lambda = 2$.

-2) A particular quadratic map.

From Theorem 5, it follows that there is a single quadratic map q_{a_G} whose rotation interval bound is equal to the Golden mean $\omega_G = \frac{\sqrt{5}-1}{2}$. We just report here the remark that the associated map $\overline{q_{a_G}^+}$ belongs to the class studied in [22], hence results originally obtained for circle maps find applications in the quadratic family. Combining this remark with the main result in [18], one gets that q_{a_G} has an invariant probability measure absolutely continuous with respect to Lebesgue. More general families and other rotation interval bounds can be approached similarly: we refer the interested reader to [22] and [18]. The map q_{a_σ} restricted to the closure of the orbit of its critical point is minimal. A different quadratic map with this property and not renormalizable has been studied in [14].

-3) A quick look at Sarkovskii's order.

A Theorem by Sarkovskii [19] asserts that:
- For the order $\ll$ on the natural numbers defined by:

$$1 \ll 2 \ll 4 \ll 8| \ldots |2^m.(2n+1) \ll \ldots \ll 2^m.7 \ll 2^m.5 \ll 2^m.3.$$
$$|\ldots 4n+2 \ll \ldots \ll 14 \ll 10 \ll 6| \ldots 2n+1 \ll \ldots \ll 7 \ll 5 \ll 3,$$

any continuous function from the interval to the real line having a periodic orbit with period p also has a periodic orbit with period q for each $q \ll p$.

The restriction of this order to the odd numbers has a simple meaning for functions in $\mathcal{C}$, since that part of the Theorem of Sarkovskii directly follows from statement 4 of Theorem 2 for f in $\mathcal{C}$.

REFERENCES

[1] L. Alseda, J. Llibre, M. Misiurewicz and C. Tresser, *Periods and entropy for Lorenz-like maps*, Ann. Inst. Fourier **39** (1989), 929–952.

[2] M. Benedicks and L. Carleson, *On iterations of $1 - ax^2$ on $(-1,1)$*, Ann. Math. **122** (1985), 1–25.

[3] W.A. Beyer, R.D. Mauldin, and P.R. Stein, *Shift-maximal sequences in function iteration: existence, uniqueness, and multiplicity*, J. Math. Anal. Appl. **115** (1986), 305–362.

[4] K.M. Brucks, *Uniqueness of aperiodic kneading sequences*, Proc. Amer. Math. Soc. **107** (1989), 223–229.

[5] K.M. Brucks, M. Misiurewicz and C. Tresser, *Monotonicity properties of the family of trapezoidal maps*, Commun. Math. Phys. **137** (1991), 1–12.

[6] A. Douady and J.H. Hubbard, *On the dynamics of polynomial like mappings*, Ann. Sci. Ecole Norm. Sup. (4) **18** (1985), 287–343 See also: Etude dynamique des polynômes complexes. Preprint Orsay (1984), and to appear.

[7] P. Fatou, *Sur les frontiéres de certains domaines*, Bull. Soc. Math. France **51** (1923), 16–22.

[8] J.M. Gambaudo and C. Tresser, *Dynamique régulière ou chaotique: applications à "une discontinuité" du cercle ou de l'intervalle*, C. R. Acad. Sc. Paris t.300 Série I (1985), 311–313.

[9] F. Hofbauer, *The topological entropy of the transformation $x \to ax(1-x)$*, Mh. Math. **90** (1980), 117–141.

[10] F. Hofbauer, *Periodic points for piecewise monotonic transformations*, Erg. Th. and Dynam. Sys. **5** (1985), 237–256.

[11] M.V. Jakobson, *On smooth mappings of the circle into itself*, Math. USSR Sbornik **14** (1971), 161–185.

[12] M. V. Jakobson, *Quasisymmetric conjugacy for some one-dimensional maps inducing expansion*, Preprint U. of Maryland (1989).

[13] L. Jonker, *A monotonicity theorem for the family $f_a(x) = a - x^2$*, Proc. A.M.S. **85** (1982), 434–436.

[14] M. Yu. Lyubich and J. Milnor, *The Fibonacci unimodal map*, Preprint Stony Brook No. 1991/15.

[15] S. Matsumoto, *On the bifurcation of periodic points of one dimensional dynamical systems of a certain kind*, Bull. Sci. Math. **107** (1983), 49–75.

[16] J. Milnor and W. Thurston, *On iterated maps of the interval*, Springer Lecture Notes No. 1342 (1988), 465–563.

[17] M. Misiurewicz, *Rotation intervals for a class of maps of the real line into itself*, Erg.Th. and Dynam. Sys. **6** (1986), 117–132.

[18] T. Nowicki and S. Van Strien, *Invariant measures exist under a summability condition for unimodal maps*, Preprint TU Delft (1990).

[19] A. Sarkovskii, *Coexistence of cycles of a continuous map of the line into itself*, Ukr. Mat. Z. **16** (1964), 61–71.

[20] D. Sullivan, *Bounds, quadratic differentials, and renormalization conjectures*, Preprint I.H.E.S. (1991). To appear in A.M.S. Centennial Publications, Volume 2: Mathematics into the Twenty-first Century.

[21] C. Tresser, *Nouveaux types de transitions vers une entropie topologique positive*, C. R. Acad. Sc. Paris t.296, Série I (1983), 729–732.

[22] J.J.P. Veerman and F. Tangerman, *Scallings in circle maps I and II*, Preprints Stony Brook, No. 1990/1 and No. 1990/8.

[23] J.C. Yoccoz, To appear.

[24] A. Zdunik, *Entropy of transformations of the unit interval*, Fundamenta Math. **124** (1984), 235–241.

I.N.L.N., B.P. no 71, 06018 Nice Cedex 2, FRANCE

E-mail: JMGA @ ECU.UNICE.FR

IBM, P.O. Box 218, Yorktown Heights, NY 10598, USA

E-mail: TRESSER @ IBM.COM

Contemporary Mathematics
Volume **135**, 1992

Finiteness of conjugacy classes of restricted block upper triangular matrices

In honour of Roy Adler

*David Handelman**

ABSTRACT. We prove finiteness of conjugacy or algebraic shift equivalence classes for some types of block upper triangular matrices, in response to a question of Paul Trow.

In answer to a question of Paul Trow (at the conference), a somewhat surprising finiteness result is presented, concerning the numbers of algebraic shift equivalence classes and conjugacy classes of certain block upper triangular matrices.

Let S be a commutative domain in which all ideals are principal and all proper factor rings are finite; for dynamical applications, the reader may as well take S to be **Z**. If A and B are square matrices with entries from S, we call them *algebraically shift equivalent* (the notion of shift equivalence translated to the algebraic setting, dropping the ordering) if there exist rectangular matrices X and Y of the appropriate dimensions such that the following equations hold:

$$(1) \qquad XA = BX \qquad AY = YB \qquad YX = A^l \qquad XY = B^l,$$

the latter for some integer l (the latter is known as the *lag*).

Two square matrices A and B of size n are *conjugate* if there exists Y in $\mathrm{GL}(n, S)$ such that $B = YAY^{-1}$.

Suppose B and C are square matrices over S of sizes m and n respectively, with characteristic polynomials f and g; suppose that f and g have no common divisors in $S[x]$ (equivalently, they have no common roots in the algebraic closure of the field of fractions of S). Then the set of matrices

$$(2) \qquad \left\{ \begin{bmatrix} B & X \\ \mathbf{0} & C \end{bmatrix} \mid X \in S^{m \times n} \right\}$$

(where the bold zero is used to denote a rectangular matrix of the appropriate size all of whose entries are zero, as opposed to the number 0) decomposes into only finitely

---

1991 Mathematics Subject classification. Primary 15A21, Secondary 13C05, 54H20.

*Supported in part by an operating grant from NSERC (Canada).

This paper is in final form and no version of it will be submitted elsewhere.

many algebraic shift equivalence classes and finitely many conjugacy classes (with respect to S, which we tacitly assume from now on). In fact for X and X' in $S^{m \times n}$, the matrices $\begin{bmatrix} B & X \\ 0 & C \end{bmatrix}$ and $\begin{bmatrix} B & X' \\ 0 & C \end{bmatrix}$ are conjugate if $X - X' = BZ - ZC$ for some matrix Z in $S^{m \times n}$—the conjugacy is implemented by $Y = \begin{bmatrix} I & Z \\ 0 & I \end{bmatrix}$.

Let $L_B, R_C : S^{m \times n} \to S^{m \times n}$ denote left multiplication by B and right multiplication by C, respectively, and set $T = L_B - R_C$. Then the matrices are conjugate if $X - X'$ belongs to the range of T. In particular, since B and C have disjoint spectra, $\operatorname{cok} T = S^{m \times n}/TS^{m \times n}$ is a finitely generated torsion S-module, and is thus finite (by the assumption that proper factor rings of S are finite). Hence there are only finitely many conjugacy classes of the matrices in the set in (2). More precise versions of this result are given in [ByH; Propositions 1.1 and 1.2].

If we now fix f (the characteristic polynomial of B) and let B vary over all matrices with characteristic polynomial f, then the same argument yields just finitely many conjugacy classes, provided the roots of f are distinct. However, if multiple roots occur, infinitely many conjugacy classes will arise when S is not a field. In the case that $S = \mathbf{Z}$ and the matrices are assumed primitive, see [By], where the corresponding discussion takes place with respect to shift equivalence. If we now let B and C vary over the matrices with characteristic polynomials f and g respectively, we can obtain infinitely many algebraic shift equivalence classes in (2), but only if either f or g has repeated roots.

Consider a matrix of the form $A = \begin{bmatrix} B & X \\ 0 & C \end{bmatrix}$. Recalling that S is a principal ideal domain, we can find a change of basis (a conjugacy) so that with respect to the new basis, A will have the form $\begin{bmatrix} C' & X' \\ 0 & B' \end{bmatrix}$ where the characteristic polynomial of the square matrices labelled with a B is always f and that of any matrix labelled with a C is g. This is well known (e.g., [N; Theorem III.12, p. 50]), but it is worthwhile to give the argument at least to establish notation. Let W' be the kernel of $g(A)$; this is an A-invariant subgroup of $S^n = H$ such that the quotient $H/W' = Y'$ is torsion free. As S is a principal ideal domain, W' is not only free, but is a direct summand (with complement necessarily free) of H. This means we can take a basis for W' and enlarge it to a basis for H. As the characteristic polynomial of A restricted to W' (subsequently written $A|W'$) is g, we have obtained our C'. Of course it does *not* follow that W' admits an invariant complement (this would force A to be conjugate to a block diagonal matrix, which it need not be—e.g., $S = \mathbf{Z}$ and $A = \begin{bmatrix} 1 & 1 \\ 0 & -1 \end{bmatrix}$). However, A acts as an endomorphism of the quotient module Y' via $\overline{A}(h + W') = Ah + W'$; we denote this $A\backslash Y'$. It is clear that the characteristic polynomial of the latter is f, so that $B' = A\backslash Y'$.

The dynamical question which led to the algebraic problems we will be considering here was kindly explained to me by Paul Trow. Let B be an integer matrix all sufficiently large powers of which are primitive ("IEP") whose characteristic polynomial is f. Let g be a monic integer polynomial. How many shift equivalence classes are there of (primitive integral) matrices A that eventually factor onto B in a right closing manner? (For the definition of eventual right closing maps, see [KMT].) If A eventually factors onto B in right closing fashion, then $A|\ker f(A)$ is algebraically shift equivalent to B. If g has no repeated roots, there are only finitely many choices for $C = A|\ker g(A)$ up to algebraic shift equivalence. A consequence of Theorem 1 is that there are only finitely many algebraic shift equivalence classes of matrices of the form, $\begin{bmatrix} B & X_\alpha \\ 0 & C_\alpha \end{bmatrix}$ where we let X_α and C_α vary subject to the restraints

$$\begin{bmatrix} B & X_\alpha \\ 0 & C_\alpha \end{bmatrix} \sim \begin{bmatrix} C & X'_\alpha \\ 0 & B_\alpha \end{bmatrix},$$

for some X'_α and B_α and all the B-matrices have f as their characteristic polynomial (away from zero), and all the C matrices have g as theirs. We prove more than this, by allowing C (the upper left corner of the right matrix) to vary as well, and concluding that only finitely many equivalence classes of B_α occur.

The referee also suggested the following dynamical interpretation of the result in Theorem 1. Let $S = \mathbf{Z}$ and let f g be monic polynomials in $\mathbf{Z}[x]$ having no (complex) roots in common. Let $\mathcal{M}$ denote the set of mixing subshifts, S, of finite type arising from (primitive) matrices A such that

- the characteristic polynomial of A is $x^k fg$ for some k;
- there is a left resolving quotient map from S onto a subshift of finite type given by a matrix which has characteristic polynomial f (modulo powers of x) and lies in a specified shift equivalence class.

Then there are only finitely many shift equivalence classes which contain matrices having characteristic polynomial f (modulo powers of x) and define shifts of finite type that are quotients of elements of $\mathcal{M}$ by right resolving maps.

The actual statement of the result corresponds to the transpose of Paul's question— thus, right resolving has been switched to left resolving. Implicit in the dynamical interpretation are the now classic results:

⋄ if A and A' are primitive integral matrices that are algebraically shift equivalent, then they are shift equivalent [**PW**];

⋄ any (square) matrix over a principal ideal domain is algebraically shift equivalent to a matrix with nonzero determinant [**Ef**] (see [**ByH**; 1.6]).

Theorem 1. Let S be a principal ideal domain all of whose proper factor rings are finite. Let α run over some index set. Let B and B_α be square matrices with characteristic polynomial f; let C_α and C'_α be square matrices with the same characteristic polynomial g, where f and g have no common divisors; let X_α and X'_α be rectangular matrices, all of these subject to the constraint

$$(3) \qquad \begin{bmatrix} C_\alpha & X_\alpha \\ 0 & B \end{bmatrix} \sim \begin{bmatrix} B_\alpha & X'_\alpha \\ 0 & C'_\alpha \end{bmatrix},$$

where $\sim$ denotes either algebraic shift equivalence or conjugacy with respect to S. Then only finitely many equivalence classes appear among the B_α satisfying these constraints, and an explicit bound (depending only on f and g) exists.

Proof: We prove the theorem first in the case of conjugacy, and then indicate what changes must be made for algebraic shift equivalence. Let A_α denote the matrix on the left of (3). Write $H = S^{n+m}$, the module consisting of columns; all the A_α act on the left. Let W be the submodule consisting of all elements for which the bottom n entries are zero. Note that W is A_α-invariant for all α, and $A_\alpha|W = C_\alpha$. Set $Y = H/W$, so that $B = A_\alpha\backslash Y$, and let $\pi : H \to Y$ denote the quotient map.

Now the element of S given by $k = \det f(C_\alpha)$ is nonzero (by the lack of common divisors of f and g) and clearly only depends on f and g, not on the choice of C_α. Let

K be the (finite) cardinality of S/kS (so if $S = \mathbf{Z}$, $K = |k|$). Let Y_α be the submodule of H, $\ker f(A_\alpha)$. Then B_α is, up to conjugacy, just $A_\alpha | Y_\alpha$. Obviously $Y_\alpha \cap W = (0)$ (again from f and g having no common divisors). We claim that $kH \subseteq W \oplus Y_\alpha \subseteq H$ (in fact, it is easy to see that for some nonzero element s_α of S, $s_\alpha H \subseteq W \oplus Y_\alpha$, by calculating their ranks; but we want something independent of α).

If U is any square matrix over S, then all the entries of the adjoint, $\mathrm{Adj}\, U$, belong to S. If $\det U$ is not zero, U^{-1} exists over the field of fractions of S and $\mathrm{Adj}\, U = U^{-1} \det U$. Hence $k f(C_\alpha)^{-1}$ is an endomorphism of W.

Choose h in H. We observe that $\pi f(A_\alpha) = f(B)\pi = 0$, so that $h_1 = f(A_\alpha)h$ belongs to W. Since $k f(C_\alpha)^{-1}$ is an endomorphism of W, $h_2 = k f(C_\alpha)^{-1} h_1$ belongs to W. We observe that

$$f(A_\alpha)(kh - h_2) = k f(A_\alpha)h - (f(A_\alpha|W)(k f(C_\alpha)^{-1}h_1)$$
$$= kh_1 - kh_1 = 0.$$

Hence $kh - h_2$ belongs to $\ker f(A_\alpha) = Y_\alpha$; as h_2 belongs to W, $kh \in Y_\alpha \oplus W$.

Let π_α be the restriction of π to Y_α; obviously π_α is an embedding, and by the preceding its image contains kY. Let $\pi^\alpha : kY \to Y_\alpha$ be defined by $\pi^\alpha(ky) = \pi_\alpha^{-1}(ky)$; this is well defined and one to one as π_α is one to one and its image contains kY. We have the commuting diagram

$$
\begin{array}{ccccc}
kY & \xrightarrow{\ \pi^\alpha\ } & Y_\alpha & \xrightarrow{\ \pi_\alpha\ } & Y \\
\downarrow{\scriptstyle B|kY} & & \downarrow{\scriptstyle A_\alpha|Y_\alpha} & & \downarrow{\scriptstyle B} \\
kY & \xrightarrow{\ \pi^\alpha\ } & Y_\alpha & \xrightarrow{\ \pi_\alpha\ } & Y
\end{array}
$$

In view of the earlier comment that B_α is conjugate to $A_\alpha | Y_\alpha$, the conjugacy class of B_α is thus determined by the invariant submodule of Y given by $\pi(Y_\alpha)$. Since there are only finitely many between kY and Y, only finitely many conjugacy classes appear. Explicitly, the number of such submodules (which is generally bigger than the number of conjugacy classes, as some pairs of submodules might be equivalent) is bounded by 2^{K^m} (the number of subsets of $(S/kS)^m$—recall that m is the size of B). Of course, much better bounds can be obtained for the number of orbits of subgroups of m copies of a finite abelian group of order K under the action of $\mathrm{GL}(m, S/kS)$.

The preceding argument works for conjugacy. To modify it so that it works for algebraic shift equivalence (and so as to obtain the same upper bound on the number of equivalence classes), we need only replace "conjugate" by "algebraically shift equivalent" in the remark that B_α is conjugate to $A_\alpha | Y_\alpha$. ✱

Paul's original question actually concerned matrices in the upper left rather than the lower right. We can obtain the finiteness result in this context (notice that interchanging the positions changes the status—restriction to an invariant submodule changes to an action on the quotient module, and *vice versa*) via the transpose:

$$
\begin{bmatrix} C & X \\ 0 & B \end{bmatrix} \sim \begin{bmatrix} B & 0 \\ X & C \end{bmatrix} = \begin{bmatrix} B^T & X^T \\ 0 & C^T \end{bmatrix}^T .
$$

Corollary 2. Let S be a principal ideal domain with all proper factor rings finite. Suppose the matrices B and C over S have characteristic polynomials

f and g respectively, and these have no common divisors. Let B_α, C_α, X_α and X_α' be matrices such that all the B_α have characteristic polynomial f, all the C_α have characteristic polynomial g, and

$$A_\alpha = \begin{bmatrix} B & X_\alpha \\ 0 & C_\alpha \end{bmatrix} \sim \begin{bmatrix} C & X_\alpha' \\ 0 & B_\alpha \end{bmatrix}.$$

Then only finitely many algebraic shift equivalence classes of B_α, C_α, and A_α appear.

Proof: Finiteness of the algebraic shift equivalence classes of the B_α is a direct consequence of Theorem 1 (applied to the transposes), as is finiteness of the set of equivalence classes of C_α. Finiteness of the equivalence classes of the A_α is now a consequence of [**ByH**; Proposition 1.2], as discussed earlier in connection with the difference of right and left multiplication operators, T.✸

The hypotheses on S, f, and g can be somewhat weakened in Theorem 1, if we are willing to accept a result with a more algebraic flavour. Let S be a commutative noetherian domain (so the property of being a principal ideal domain has been weakened to one of its consequences) all of whose proper factor rings are finite (so S could be any order in a number field). Let $\mathcal{P}$ be a class of finitely generated S-modules, and as in [**ByH**], define the equivalence relation, algebraic shift equivalence, on the collection of pairs of modules with an endomorphism

$$\{\, (\alpha, P) \mid P \in \mathcal{P}, \ \alpha : P \to P \,\},$$

as follows. The pairs (α, P) and (β, Q) are algebraically shift equivalent, if there exist module homomorphisms $X : Q \to P$ and $Y : P \to Q$ such that $XY = \alpha^l$, $YX = \beta^l$, $X\alpha = \beta X$, and $\alpha Y = Y\beta$. If $\mathcal{P}$ consists of finitely generated free modules, this reduces to the usual notion of algebraic shift equivalence. Conjugacy may similarly be defined. It is reasonable to insist that all finitely generated free modules are in $\mathcal{P}$.

Let f and g be monic polynomials in $S[x]$ with no common roots in the algebraic closure of the field of fractions of S (equivalently, no common divisors in $S[x]$—we do not have to go up to the field of fractions of S in this formulation, since the polynomials are monic). Consider all matrices A_α of size $m+n$ whose characteristic polynomial is fg, and such that the endomorphism of $W_\alpha = S^{m+n}/\ker g(A_\alpha)$ induced by A_α lies in a specific algebraic shift equivalence class (with respect to finitely generated S-torsion free modules). The argument in the proof of Theorem 1 yields that the number of algebraic shift equivalence classes of $(A_\alpha|\ker f(A_\alpha), \ker f(A_\alpha))$ is finite.

All that is really needed to make this work is that $S[x]/(f, g)$ be finite (this is weaker than finiteness of all factor rings of S; for example, (f, g) could be the improper ideal).✸

Roughly speaking, conjugacy corresponds to isomorphism as $S[A]$-modules, while algebraic shift equivalence corrresponds to $S[A, A^{-1}]$-module isomorphisms. To make this more precise, let f be a monic polynomial in $S[x]$, and let $\alpha : P \to P$ be an endomorphism of the finitely generated module P such that $f(\alpha) = 0$. Then, as is a standard construction, P admits the structure of an $S[x]/(f)$-module, via $(x + (f))(p) = \alpha(p)$. It is a tautology that (α, P) is conjugate to (β, Q) if and only if $P \simeq Q$ as $S[x]/(f)$-modules.

To view algebraic shift equivalence in a similar fashion, define $\overline{S} = S[x]/(f)$, and let $\overline{x}$ denote the image of x therein. We may formally invert $\overline{x}$, via the construction

$$\text{``}\overline{S}[\overline{x}^{-1}]\text{''} = \lim \overline{S} \xrightarrow{\ \times\,\overline{x}\ } \overline{S} \xrightarrow{\ \times\,\overline{x}\ } \ldots$$

This amounts to first factoring out the ideal of $\overline{S}$, $I = \{\,\sigma \in \overline{S} \mid \overline{x}^k \sigma = 0 \text{ for some } k\,\}$, and inverting the image of $\overline{x}$ in $\overline{S}/I$. In case $\overline{x}$ is not a zero divisor in $\overline{S}$ (as occurs if the constant term of f is not zero), this construction simplifies to $S[x, x^{-1}]/fS[x, x^{-1}]$— i.e., the ideal generated by f of the Laurent ring $S[x, x^{-1}]$ is factored out. Again, it is tautological that (α, P) is algebraically shift equivalent to (β, Q) if and only if $P \otimes_{\overline{S}} \overline{S}[\overline{x}^{-1}] \simeq Q \otimes_{\overline{S}} \overline{S}[\overline{x}^{-1}]$ as $\overline{S}[\overline{x}^{-1}]$-modules. (This is just another way of saying that the corresponding modules $\lim \alpha : P \to P$ and $\lim \beta : Q \to Q$ are so isomorphic.)

REFERENCES

[By] M. Boyle, Shift equivalence and the Jordan form away from zero, Ergodic Theory and Dynamical Systems **4** (1984) 367–369.

[ByH] M. Boyle and D. Handelman, Algebraic shift equivalence and primitive matrices, Trans. Amer. Math. Soc., (to appear).

[Ef] E. G. Effros, Williams' problem for positive matrices, unpublished manuscript (1981).

[KMT] B. Kitchens, B. Marcus, and P. Trow, Eventual factor maps and compositions of closing maps, Ergodic Theory and Dynamical Systems **11** (1991) 85–113.

[N] M. Newman, *Integral Matrices*, Academic Press, New York, 1978.

[PW] W. Parry and R. F. Williams, Block coding and a zeta function for Markov chains, Proc. L. M. S. **35** (1977) 483–495.

Mathematics Department, University of Ottawa, Ottawa, Ontario K1N 6N5 CANADA

E-mail address: dehsg%uottawa@ACADVM1.ca

Contemporary Mathematics
Volume **135**, 1992

Polynomials with a positive power

In honour of Roy Adler

David Handelman[*]

ABSTRACT. If some power of polynomial in several real variables has no negative coefficients, then all sufficiently high powers of either it or its negative have no negative coefficients.

Let p be a polynomial in d variables. I asked by Valerio deAngelis whether p^n having no negative coefficients for some n implies all sufficiently large powers of p have no negative coefficients. We show that this is indeed the case. Of course, we insist that $p(1, 1, \ldots, 1) > 0$ to remove the nuisance of multiplication by -1.

We use the notation of [H1], [H2], and [H3]. If $f = \sum \lambda_w x^w$ where λ_w are real and x^w is the usual monomial notation, define $\text{Log } f = \left\{ w \in \mathbf{Z}^d \mid \lambda_w \neq 0 \right\}$. The convex hull of a set S in $\mathbf{R}^d$ will be denoted cvx S. If S is a subset of $\mathbf{R}^d$ and n is a positive integer, nS will denote the set of sums of n elements of S.

Suppose that $p^n = P$ has no negative coefficients. Let v be a vertex of cvx $\text{Log } p$. Form *the cone at v for p*, defined as

$$\bigcup_{k \in \mathbf{N}} \left(\text{Log } p^k - kv \right).$$

This defines a cone in $\mathbf{Z}^d$. An *atom* is simply a nonzero element of the cone that is not a sum of nonzero elements in the cone. We can of course form the cone for P at nv. Since p^n has no negative coefficients, it follows immediately that the coefficient of x^{kv} and of x^a in p^k are positive if $a - kv$ is an atom in the cone of p. Hence all the atoms in the cone at v for p are also in the cone (and are atoms therein) at nv for P. In particular, the two cones are equal. Note that this does not imply that for example, $n\text{Log } p = \text{Log } P$, because cancellation may occur.

As a particular consequence, the groups generated by $\text{Log } p - \text{Log } p$ and by $\text{Log } P - \text{Log } P$ are equal; so we may assume they are both the standard copy of $\mathbf{Z}^d$ in $\mathbf{R}^d$.

Claim: If $v = 0$, then given M for all sufficiently large (depending on v) N, for all vertices w of cvx $\text{Log } p$,

$$(Mn - 1)w \in N\text{Log } P.$$

1991 Mathematics Subject classification. Primary 26C99, Secondary 54H20.

[*]Supported in part by an operating grant from NSERC (Canada).

This paper is in final form and no version of it will be submitted elsewhere.

Proof: We may write each w as a sum of atoms, say $k(w)$ of them (counting multiplicities) since 0 is a vertex. Hence we may take any N exceeding $\max_w(Mn - 1)k(w)/n$.✸

This result may be applied to an arbitrary vertex by the obvious translation.
Claim: For v a vertex of cvx Log p, there exists $l \equiv l(v)$ such that

$$\text{Log } p + (nl - 1)v \subset \text{Log } P^l,$$

and this holds for all larger values of l.
Proof: Via the translation argument, we may assume v is zero. Each nonzero u in Log p may be expressed as a sum of atoms from the cone of p at v, and by our assumption 0 is a vertex of Log p, so appears in Log p^t for all t. Let l be the maximum number of atoms required, going over Log p. Again since 0 is a vertex, any larger value of l will also do. ✸

Pick $M \geq \max \{ l(v) \mid v \in d_e\text{cvx Log } p \}$, and set $b_v = px^{(Mn-1)v}/P^M$. By the second claim, b_v is an element of R_P. Set $Q_M = \sum x^{(Mn-1)v}$. Thus $\sum b_v = pQ_M/P^M$. We note that Mnv belongs to Log pQ_M for all vertices v, and we already have Log $pQ_M \subseteq$ Log P^M. Since $p^{n-1}(pQ_M)$ has no negative coefficients, by either [H1; V.7] or the combination [H1; V.3A] and [H3; II.1], there exists K such that $P^K pQ_M$ has no negative coefficients.

Pick a vertex v_0 and translate so that this is zero. By the claim above, there exists k such that p/P^k belongs to R_P. Let N_v be an integer that will do in the first claim for an arbitrary vertex v. Set $a_v = x^{((N_v-M)n+1)v}Q_M/P^{N_v}$. Then a_v belongs to R_P (by the claim), it is positive (obviously), and by the preceding $a_v \cdot p/P^k$ belongs to R_P^+. The coefficient of $N_v nv$ in the numerator of a_v is not zero. Hence $h = \sum a_v$ is an order unit of R_P and $h \cdot p/P^k$ is positive in R_P. By [H3; Theorem II.5], p/P^k is positive in R_P, which is precisely what we want.✸

We have proved:

Theorem. Let p be a polynomial in several variables such that p^n has no negative coefficients for some n and $p(1, 1, \ldots, 1) > 0$. Then for all sufficiently large N, p^N has no negative coefficients.

By [H3; Corollary C2, p. 107], there exists k such that Log $P^k p^j =$ Log $P^k +$ jLog p for all positive integers j.

REFERENCES

[H1] D. E. Handelman, *Positive Polynomials and Product Type Actions of Compact Groups*, Memoirs of the A.M.S. **320** Amer. Math. Soc., 1985.

[H2] D. E. Handelman, Deciding eventual positivity of polynomials, Ergodic Theory and Dynamical Systems **6** (1985) 57–79.

[H3] D. E. Handelman, *Positive Polynomials, Convex Integral Polytopes, and a Random Walk Problem*, Springer Lecture Notes in Mathematics **1282** Springer-Verlag, New York, 1988.

Mathematics Department, University of Ottawa, Ottawa, Ontario K1N 6N5 CANADA

E-mail address: dehsg%uottawa@ACADVM1.ca

Contemporary Mathematics
Volume **135**, 1992

Spectral radii of primitive integral companion matrices and log concave polynomials

In honour of Roy Adler

*David Handelman**

ABSTRACT. A question arising from work of Goldberger, Lind, and Smorodinsky on renewal systems is answered; we show that the conjectured characterization of positive real numbers that are spectral radii of primitive integral companion matrices is correct. We also present an eventual strong unimodality result for repeated multiplication by a polynomial.

We answer a question raised by Doug Lind during his talk at the conference (both Mike Boyle and Jonathan Ashley had asked me the same question earlier). Which algebraic numbers appear as the large (Perron) eigenvalue of a primitive integral companion matrix? This question is natural in the context of renewal systems; in particular, in the determination of the spectral radii of uniquely decipherable ones—see Goldberger, Lind, and Smorodinsky [GLS]. We also obtain a result (necessary for the conclusion) concerning eventual strong unimodality of polynomials.

I would like to thank Valerio deAngelis and Alan Kelm for their careful reading of the manuscript; both pointed out numerous errors and made valuable suggestions.

Recall that an *algebraic integer* is a complex number λ satisfying a monic polynomial with integer coefficients. It is a *Perron number* if it is real, positive, and exceeds the absolute value of all of its algebraic conjugates. A real matrix is *primitive* if all of its entries are nonnegative and some power of it has only strictly positive entries. A consequence of the Perron Frobenius theorem is that the spectral radius of a primitive integral matrix is a Perron number and is an eigenvalue (and the only eigenvalue of its absolute value). The converse was established by Doug Lind [L] (see also [BMT]).

A *companion matrix* of a monic polynomial p of degree n is an $n \times n$ matrix obtained from the regular representation on the ring $\mathbf{Z}[x]/(p)$ (I have used the integers as the coefficient ring—this could be replaced by any commutative ring); alternately, it is the matrix whose action on the standard basis $\{e_i\}$ is given by

$$e_i \mapsto \begin{cases} e_{i+1} & \text{if } i < n \\ \sum t_j e_j & \text{if } i = n, \end{cases}$$

1991 Mathematics Subject classification. Primary 11R06, Secondary 15A36, 15A48, 54H20.

*Supported in part by an operating grant from NSERC (Canada).

This paper is in final form and no version of it will be submitted elsewhere.

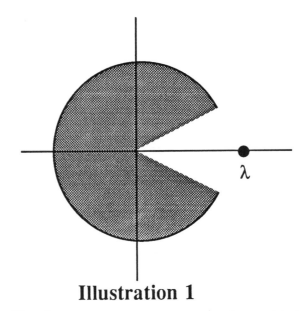

λ

Illustration 1

where $p = x^n - \sum t_j x^j$. The problem is not to be confused with that of determining the spectral radii of primitive integral matrices algebraically shift equivalent to a companion matrix—it is practically certain that *every* Perron number can be so realized.

The question we consider here is equivalent to finding the Perron numbers λ for which there is an equation of the form $\lambda^N = \sum_{0 \le i \le N-1} T_i \lambda^i$ where T_i are nonnegative integers, for some N. As Doug pointed out in his talk, we see that a necessary condition on λ is that none of its other algebraic conjugates can be real and positive (the polynomial $x^N - \sum_{0 \le j \le N-1} T_i x^i$ has at most one positive real root). We show in particular that this condition (together with being a Perron number) is sufficient Theorem 5. We actually work in a more general framework; we begin with a polynomial p whose roots have the necessary properties—there is exactly one positive real root and it exceeds the absolute values of all the other roots (see Illustration 1; the roots are either λ or appear in the pie-shaped figure whose radius is less than λ)—and we work over arbitrary subrings S of the reals rather than just the integers (the reader may simply set $S = \mathbf{Z}$ throughout).

A polynomial $p = \sum a_i x^i$ in $\mathbf{R}[x]$ is said to have *one sign change* if after deleting the zero coefficients, the resulting sequence of coefficients $(a_0, a_1, a_2, \ldots)$ has exactly one sign change. The polynomial p is *strongly unimodal* (also known as "log concave") if all the coefficients are nonnegative, there are no gaps in the coefficients (that is, if $i < j < k$ and $a_i a_k \ne 0$, then $a_j \ne 0$) and for all i, $a_i^2 \ge a_{i+1} a_{i-1}$. For a survey of results and examples relating to strong unimodality, see [S] (where the term *log concave* is used). Products of strongly unimodal polynomials are strongly unimodal.

In 1883, Poincaré [P] showed that if p is a real polynomial with exactly one positive real root, there exists a polynomial P such that the product Pp has one sign change. What we want (*inter alia*) is a *monic integral* polynomial P with the same property, and additionally P has no roots of modulus at least as large as that of p (Poincaré's construction for P would not generally be integral, nor would its roots be small). We show that in fact $P = (1 + x)^N$ will do for all sufficiently large N.

Meissner [M] in 1911 showed (in particular) that for all sufficiently large N, the polynomial $(1 + x)^N p/(x - \lambda)$ has no negative coefficients, but what we need (and show) is that for still larger N, the polynomial is strongly unimodal (it follows immediately that $(1 + x)^N p$ has one sign change, Lemma 2). I suspect this has been proved previously, but I could not find a reference for it. Of interest (but unfortunately not applicable here) is the beautiful result of Odlyzko and Richmond [OR], which asserts that for a real polynomial with no negative coefficients such that the two smallest and the two largest exponents have positive coefficients, all sufficiently large powers are strongly unimodal.

The following argument was suggested by a technique in [GLS].

Lemma 1.   Let S be a unital subring of $\mathbf{R}$, and suppose the positive real number λ satisfies a monic polynomial $p(\lambda) = 0$ where p belongs to $S[x]$ and $p(0) \neq 0$. Suppose that p has exactly one sign change, the degree of p is n, the multiplicity of λ is one, and all other roots of p have absolute value less than λ. Then there exists $N \geq n$ and positive T_i in S such that

$$\lambda^N = \sum_{i=0}^{n-1} T_i \lambda^i.$$

and p divides the polynomial $x^N - \sum_{i=0}^{n-1} T_i x^i$ (within $S[x]$).

Proof: Write $p(x) = x^n - \sum_{i=0}^{n-1} t_i x^i$. Let $\{e_i\}_{i=0}^{n-1}$ be the standard basis (of columns) for S^n, and let C be the companion matrix of p with respect to this basis. Thus $Ce_i = e_{i+1}$ for $0 \leq i \leq n - 2$, and $Ce_{n-1} = \sum t_i e_i$. Set $v = Ce_{n-1}$. We observe that the left eigenvector of C for the eigenvalue λ is $W = (1, \lambda, \lambda^2, \dots, \lambda^{n-1})$ (this is true for any companion matrix and eigenvalue λ).

The right eigenvector (unique up to scalar multiple) for λ is the column $V = (V_i)$, where $\sum_{i=0}^{n-1} V_{i+1} x^i = p(x)/(x - \lambda) := q(x)$. It is an easy exercise to verify that because p has exactly one sign change, the polynomial q has only positive coefficients—in fact the sequence of coefficients of $q(x\lambda)$ is unimodal.

Hence both W and V are strictly positive. By (for example) [H1; Lemma 1], all sufficiently large powers of C are strictly positive. Moreover, $Wv = WCe_{n-1} = \lambda We_{n-1} = \lambda^n > 0$. By the usual convergence arguments, there exists m so that $M \geq m$ entails $C^M v$ is a strictly positive column. This means $C^{m+1} e_{n-1} = \sum_{i=0}^{n-1} T_i e_i$ where T_i are positive (and of course belong to S). Thus $C^{m+n} e_{n-1} = \sum_{i=0}^{n-1} T_i C^i e_{n-1}$. Applying the left eigenvector W, we obtain $(\lambda^{m+n} - \sum T_i \lambda^i) W e_{n-1} = 0$. Since We_{n-1} is not zero, we deduce $\lambda^{m+n} = \sum_{i=0}^{n-1} T_i \lambda^i$. For $0 \leq j \leq n - 1$,

$$C^{m+n} e_j = C^{m+j+1} e_{n-1}$$
$$= C^j \sum T_i e_i$$
$$= \sum T_i C^i e_j.$$

Thus C satisfies the polynomial $x^{m+n} - \sum T_i x^i$.

To see that p divides the polynomial, just observe that p is also the minimal polynomial of C (true for any companion matrix) and C satisfies the polynomial, so p divides it. ✳

The entries of the right eigenvector, when read from the bottom up, can also be regarded the initial "segments" of p, divided by a suitable power of λ. So we require only that the initial segments of p evaluated at λ (λ^n, $\lambda^n - t_{n-1}\lambda^{n-1}$, $\lambda^n - t_{n-1}\lambda^{n-1} - t_{n-2}\lambda^{n-2}$, etc.) are all strictly positive. This need not occur even when p has only one positive real root exceeding the absolute value of all other roots. The technique in the lemma can be used to obtain sign patterns determined by the signs of the initial segment polynomials.

At this point, it is possible to give a short proof of the most important parts of Theorem 5, that p divides a polynomial whose companion matrix is primitive. This proof is due to Valerio de Angelis (who also pointed out a simplification of the original proof of Lemma 1, identifying V with the coefficients of a polynomial). As in the proof of Lemma 1, define q via $(x - \lambda)q(x) = p(x)$. Then q has no positive real roots and is monic; so $q(t) > 0$ for all positive real numbers t. By [M] or [H2; Theorem V.1], there exists M such that $(1 + x)^M q$ has no negative coefficients. By increasing M if necessary, we may assume that all the coefficients between the constant term and that of x^{M+n-1} in $Q = (1 + x)^M q$ are strictly positive. Set $P = (x - \lambda)Q$, and apply the argument of Lemma 1 with q replaced by Q and p by P. Since the coefficients of Q are all strictly positive, the right eigenvector, V, of the companion matrix of P, is strictly positive, and the argument yields the conclusion that P divides a polynomial of the desired form. Obviously, this forces the same about p, as well.

The next few results are devoted to proving that $(1 + x)^M q$ is actually strongly unimodal for all sufficiently large M (Proposition 4), from which Theorem 5 follows. Although it is computational, it is not difficult; and it appears to be new and interesting. This (using Lemma 1) was the original proof of the main result.

Lemma 2.   If the real polynomial Q is strongly unimodal, and λ is a positive real number, then the polynomial $(x - \lambda)Q$ has exactly one sign change.

Proof: We may write $Q = \sum a_i x^i$; set $r_i = a_{i+1}/a_i$ (with the usual conventions on zero in the denominator). Strong unimodality is equivalent (in the presence of no gaps and nonnegative coefficients) to $r_i \geq r_{i+1}$ for all i. If $r_i > 1/\lambda \geq r_{i+1}$, the coefficient of x^j in $(x - \lambda)Q$ is negative if $j \leq i$ and is nonnegative if $j > i$.✱

The corresponding result for two or more positive roots fails; as a simple example, the product $(x - 1)^2(x^2 + 1.9x + 3)$ has four sign changes, not two. On the other hand, it is easy to check that multiplication by terms of the form $x + a$ (where $a > 0$) does not increase the number of sign changes.

Let p be a monic polynomial satisfied by λ as in the statement of Lemma 1. Set $q = p/(x - \lambda)$, as in Lemma 1—this is a real polynomial (of course, it need not have coefficients from S). Our next result (Proposition 4) is that $(1 + x)^M q$ is strongly unimodal for all sufficiently large M. In some cases, q is already strongly unimodal (so M can be chosen to be zero). This occurs, for example, if the arguments θ of all the other roots of p lie in the interval $|\pi - \theta| \leq \pi/3$ [S].

Lemma 3.   Let d and e be real numbers with e positive satisfying $d^2 < 4e$, and let $f = x - d + ex^{-1}$. There exists N such that $(1 + x)^N f$ is strongly unimodal.

Proof: The condition on the coefficients ensures that $f|(0, \infty) > 0$. As we noted previously, for all sufficiently large n, all the coefficients of $(1 + x)^n f$ are nonnegative [M], [H2; Theorem V.1]. For unspecified (but large) n, let P_n denote $(1 + x)^n$, which we expand as $\sum a_i x^i$ (of course, $a_i = \binom{n}{i}$). Let (Q, x^k) denote the coefficient of x^k

in the polynomial Q. Then

$$(P_n f, x^k) = a_k \left(\frac{a_{k-1}}{a_k} - d + e \frac{a_{k+1}}{a_k} \right)$$

$$= a_k \left(\frac{k}{n-k+1} - d + e \frac{n-k}{k+1} \right).$$

Now $a_k^2 / a_{k+1} a_{k-1} = (1 + \frac{1}{k})(1 + \frac{1}{n-k})$. Set $t = k/n$ and define a function of t in $[0, 1]$, $G_n(t) = \frac{tn}{n(1-t)+1} - d + e \frac{n(1-t)}{nt+1}$. For sufficiently large n, $G_n(t) > 0$ for all t in the domain of G_n, from the positivity of the coefficients. We show to begin with that for any small $\delta > 0$, for all sufficiently large n for all t in the interval $[\delta, 1 - \delta]$ that

(1)
$$G_n^2(t)(1 + \frac{1}{k})(1 + \frac{1}{n-k}) > G_n(t + \frac{1}{n}) G_n(t - \frac{1}{n}).$$

This entails that for all sufficiently large n, for all integers k with $\delta n \leq k \leq (1 - \delta)n$, $(P_n f, x^k)^2 > (P_n f, x^{k+1})(P_n f, x^{k-1})$. This is log concavity of the coefficients over a wide range. Then we show the existence of small positive δ such that log concavity holds for $k < \delta n$ and $k > (1 - \delta)n$.

To start, assume δ is given and $2/n < \delta < t < 1 - \delta$ (we can increase n as much as we like). We calculate

$$G_n(t - \frac{1}{n}) - G_n(t) = \frac{tn - 1}{n(1-t)+2} - \frac{tn}{n(1-t)+1} + e \left(\frac{n(1-t)+1}{nt} - \frac{n(1-t)}{nt+1} \right)$$

$$= e \frac{n+1}{nt(nt+1)} - \frac{n+1}{(n(1-t)+2)(n(1-t)+1)}$$

$$= -\frac{1}{n} \left(\frac{1}{(1-t)^2} - \frac{e}{t^2} \right) + O \left(\frac{1}{n^2} \right).$$

(The big oh term is a consequence of t being bounded away from 0 and 1.) Thus $G_n(t - \frac{1}{n}) = G_n(t) - \frac{1}{n} \left(\frac{1}{(1-t)^2} - \frac{e}{t^2} \right) + O \left(\frac{1}{n^2} \right)$. Replacing t by $t + \frac{1}{n}$, we deduce $G_n(t + \frac{1}{n}) = G_n(t) + \frac{1}{n} \left(\frac{1}{(1-t)^2} - \frac{e}{t^2} \right) + O \left(\frac{1}{n^2} \right)$. As $G_n(t) > 0$, we deduce

$$G_n(t + \frac{1}{n}) G_n(t - \frac{1}{n}) = G_n^2(t) + O \left(\frac{1}{n^2} \right)$$

(conveniently, the $G_n(t)/n$ terms cancel and the terms involving $G_n(t) O \left(\frac{1}{n^2} \right)$ are themselves $O \left(\frac{1}{n^2} \right)$ since G_n are uniformly bounded on the interval $[\delta, 1 - \delta]$). Now G_n converges uniformly in n on the interval $[\delta, 1 - \delta]$ to $t/(1 - t) - d + e(1 - t)/t$, so that for sufficiently large n, G_n is bounded below away from zero on the interval. Thus

$$G_n^2(t)(1 + \frac{1}{tn})(1 + \frac{1}{n(1-t)}) > G_n^2(t)(1 + \frac{1}{tn}) > G_n^2(t) + O \left(\frac{1}{n^2} \right)$$

for all sufficiently large n (again this uses that the values of t are bounded below, and $\frac{1}{n}$ is infinitely bigger than $\frac{1}{n^2}$). This yields (1). With more care, this argument can be made to work given ε for all sufficiently large n, for all k with $n^\varepsilon \leq k \leq n - n^{1/2+\varepsilon}$.

Now we let t run over the interval $[0, \delta_0]$ where δ_0 is to be determined. For δ sufficiently small and n sufficiently large, $G_n(t)$ is monotone nonincreasing on $(0, \delta]$. Then

$$G_n(t - \frac{1}{n}) = G_n(t) + e\frac{n+1}{nt(nt+1)} - \frac{n+1}{(n(1-t)+2)(n(1-t)+1)}$$
$$\leq G_n(t) + e\frac{n+1}{nt(nt+1)} - \frac{1}{n(1-t)^2} + O\left(\frac{1}{n^2}\right).$$

(The second line follows from $1 - \delta$ being bounded below.) Similarly,

$$G_n(t + \frac{1}{n}) \leq G_n(t) - e\frac{n+1}{nt(nt+1)} + \frac{1}{n(1-t)^2} + O\left(\frac{1}{n^2}\right)$$

Therefore

$$G_n(t + \frac{1}{n})G_n(t - \frac{1}{n}) \leq G_n^2(t) - \left(e\frac{n+1}{nt(nt+1)} - \frac{1}{n(1-t)^2}\right)^2$$
$$+ O\left(\frac{1}{n^2}\right)\left(G_n(t) + \left|\frac{e(n+1)}{nt} - \frac{1}{n(1-t)^2})\right|\right)$$
$$\leq G_n^2(t) + O\left(\frac{1}{n^2}\right)G_n(t).$$

It is sufficient that $G_n^2(t)(1 + 1/nt) > G_n^2(t) + O\left(\frac{1}{n^2}\right)G_n(t)$, i.e., that $G_n(t)/nt > O\left(\frac{1}{n^2}\right)$. If we choose $t < \delta_0$ sufficiently small, then of course $\min_{t \leq \delta_0} G_n(t)$ will be arbitrarily large, so this last inequality is easy to arrange for sufficiently large n.

This yields that for all sufficiently large n, $(P_n f, x^k)^2 > (P_n f, x^{k+1})(P_n f, x^{k-1})$ for all $0 < k < \delta_0 n$. If we replace x by x^{-1} throughout (and multiply by a sufficiently large power of x), and deal with $x - d/e + x^{-1}/e$, we obtain δ_1 such that the same inequality holds (for sufficiently large n) for $k > (1 - \delta_1)n$. Combining the two results, we obtain the inequality for all k (for sufficiently large n); this is just strong unimodality of $P_n f$. ✳

Proposition 4. If f is a real polynomial with no positive real roots and $f(1) > 0$, there exists an integer N so that $(1 + x)^N f$ is strongly unimodal. *Proof:* Factor f as a product of quadratic polynomials and linear polynomials. The linear polynomials can only come from negative real roots, and we just apply Lemma 3 and the fact that a product of strongly unimodal polynomials is strongly unimodal. ✳

Let $\{P_i\}$ be a sequence of strongly unimodal polynomials. Define the *ratio* of a strongly unimodal polynomial $r(P)$ to be the ratio of the second largest coefficient to the largest. Suppose $\sum r(P_i) = \infty$. I conjecture that if f is a real polynomial with $f|(0, \infty) > 0$, there exists n such that $P_1 P_2 \ldots P_n f$ is strongly unimodal. (Divergence is necessary, as is easy to see.) In [BH; Theorems 1.9 and 2.3], we showed that the product eventually has no negative coefficients. In that paper, we did not use factorization into real quadratic and linear polynomials, because we felt that some of the results might generalize to several variables. In this paper, I had no qualms about using the factorization because there is no generalization of strong unimodality to two variables.

Theorem 5.   Let S be a unital subring of $\mathbf{R}$ and let p be a monic polynomial of degree n in $S[x]$ having a positive real root λ of multiplicity one that exceeds the absolute values of all other roots. Suppose additionally that $p(0) \neq 0$ and p has no other positive real roots.

(i) For all β in S with $0 < \beta < \lambda$, for all sufficiently large integers N, $(x+\beta)^N p$ has one sign change.

(ii) Using N, β from (i), for all sufficiently large M, there exist $T_1, T_2, \ldots, T_n$, positive elements in S, such that $\lambda^{M+N+n} = \sum_{i=0}^{N+n-1} T_i \lambda^i$.

(iii) The companion matrix of the polynomial $c = x^{M+N+n} - \sum_{i=0}^{N+n-1} T_i x^i$ (using the notation of (ii)) is primitive, has λ as its spectral radius, and p divides c (in $S[x]$).

Proof: Let $q = p/(x - \lambda)$ and set $q_\beta = q(x\beta)$. For all large N, $(1 + x)^N q_\beta$ is strongly unimodal. Replacing x by x/β does not affect strong unimodality, so that $(1 + x/\beta)^N q$ is strongly unimodal, and therefore so is $(\beta + x)^N q$. By Lemma 2, $(x - \lambda)(x + \beta)^N q = (x + \beta)^N p$ has one sign change.

Property (ii) follows from Lemma 1. Each of the statements in (iii) is now either obvious or follows from the previous results. ✸

The reason for considering all $\beta < \lambda$ (and not just $\beta = 1$, which will work in the case that $S = \mathbf{Z}$), is that λ need not exceed 1 when S is not $\mathbf{Z}$.

REFERENCES

[BH] B. M. Baker and D. E. Handelman, Positive polynomials and time dependent integer-valued random variables, Canad. J. Math. (to appear, 1992).

[BMT] M. Boyle, B. Marcus, and P. Trow, *Resolving maps and the dimension group for shifts of finite type*, Mem. Amer. Math. Soc. 377 Amer. Math. Soc., 1987.

[GLS] J. Goldberger, D. Lind, and M. Smorodinsky, The entropies of renewal systems, Israel J. of Math. (to appear).

[H1] D. E. Handelman, Positive matrices and dimension groups affiliated to C*-algebras and topological Markov chains, J. of Operator Theory 6 (1981) 55–74.

[H2] D. E. Handelman, *Positive Polynomials and Product Type Actions of Compact Groups*, Memoirs of the A.M.S. 320 Amer. Math. Soc., 1985.

[L] D. Lind, The entropies of topological Markov shifts and a related class of algebraic integers, Ergodic Theory and Dyn. Systems 4 (1984) 283–300.

[Me] E. Meissner, Über positive Darstellung von Polynomen, Math. Ann. 70 (1911) 223–235.

[OR] A. M. Odlyzko and L. B. Richmond, On the unimodality of high convolutions of discrete distributions, Annals of Prob. 13 (1985) 299–306.

[P] H. Poincaré, Sur les équations algébriques, C. R. Acad. Sci. Paris 97 (1883) 1418.

[S] R. P. Stanley, Log concave and unimodal sequences in algebra and combinatorics, in "Graph Theory and its Applications: East and West" Annals of the New York Academy of Sciences 576 (1989) 500–535.

Mathematics Department, University of Ottawa, Ottawa, Ontario K1N 6N5 CANADA

E-mail address: dehsg%uottawa@ACADVM1.ca

Contemporary Mathematics
Volume **135**, 1992

Self-Replicating Tilings

RICHARD KENYON

ABSTRACT. A tiling of $\mathbf{R}^n$ with one tile up to translation is called a **self-replicating tiling** (SRT) if there is a linear expanding map φ of $\mathbf{R}^n$ such that the image of any tile fits exactly over a union of tiles in the unexpanded tiling.

We show that a linear map φ can be the expansion of a quasiperiodic SRT if and only if φ is similar to an expanding integer-linear map. In addition, the translations between tiles in such an SRT generate a discrete rank-n subgroup of $\mathbf{R}^n$, invariant under φ. In $\mathbf{R}$ and $\mathbf{R}^2$, the expansion of *any* SRT is integral.

An SRT gives a radix representation of elements of $\mathbf{R}^n$, a "decimal system" in base φ with a finite set of digits in $\mathbf{R}^n$ depending on the tiling. We give necessary and sufficient conditions on the digits such that an SRT exists.

1. Introduction

A **tiling** $T = \{T_i\}_{i \in I}$ of $\mathbf{R}^n$ is a locally finite covering of $\mathbf{R}^n$ by compact sets T_i, which are closures of their interiors, such that the interiors of any two tiles are disjoint.

We say that a tiling T is **self-replicating** if it has the following two properties:

1. there is one tile type up to translation, i.e.

$$\forall i, j \in I, \ \exists v_{ij} \in \mathbf{R}^n \text{ such that } T_i = T_j + v_{ij}.$$

2. There is an expanding linear map φ of $\mathbf{R}^n$ (here expanding means the eigenvalues $\{\lambda_j\}$ of φ satisfy $|\lambda_j| > 1$) and for each $i \in I$, there is a finite subset $I_i \subset I$ such that

$$\varphi(T_i) = \bigcup_{j \in I_i} T_j.$$

1991 Mathematics Subject Classification. Primary 05B45.

Partially supported by an IBM graduate fellowship and a Chateaubriand fellowship.

This paper is in final form and no version of it will be submitted for publication elsewhere.

By property (1) we can and will re-index the tiles in an SRT so that the subscript of a tile represents the translation of it relative to a given tile T_0, that is, $T_v = T_0 + v$. Then the tiling is determined by a fixed tile T_0 and a set V of translations.

If we include the hypothesis of quasiperiodicity (see section 2), then SRTs are special cases of SSTs (self-similar tilings), which were introduced by Thurston [17] (an SST is a quasiperiodic SRT in which several tile shapes are allowed, not just one.) Their subdivision property is the same as the property of **inflation** of Robinson and Penrose [14, 13], who use this property to produce sets of aperiodic planar tiles.

The construction of Thurston of SSTs is related to the construction of Markov partitions of Sinai [15]; in particular, the construction although very general is not very constructive in practice, that is, usually requires a large number of tiles defined in a rather abstract way. This paper gives a more concrete construction for the (simpler) class of SRTs, and a theory of all possible tilings with given expanding map.

Several authors have studied SRTs in some way or another. In particular, Dekking [3, 4] has constructed some SRTs as 'recurrent sets', which are by definition sets with some sort of self-similar structure. Also see [7] for some background into the questions of more general "inflationary" tilings.

We will see that the tiles of an SRT are exactly the shapes which can be cut up into n similarly-oriented congruent shapes, each similar to the initial shape. This point of view is taken by Bandt [1], who constructs self-similar shapes for expanding $n \times n$ integer matrices. This is part of Theorem 12, below; the theorem also deals with the slightly non-trivial step from self-replicating shapes to self-replicating tilings.

Gilbert [6] discusses representing complex numbers in complex bases (where the base φ is a quadratic algebraic integer), shows that there are related SRTs of the plane, and obtains some geometric results about the tiles.

More recently, self-replicating tilings have found connections with the theory of wavelets, as described for example in [16].

In this paper we show that an SRT has the following properties.

Proposition 1 *All tiles subdivide in the same way, i.e. there is a finite set* $D \subset \mathbf{R}^n$ *such that for all* $v \in V$,

$$\varphi(T_v) = \bigcup_{d \in D} T_{\varphi(v)+d}.$$

Using this proposition and a rigidity result for planar tilings [9] allows us to prove:

Theorem 3 *An SRT of the line is periodic; an SRT of* $\mathbb{R}^2$ *is either quasiperiodic or half-periodic.*

A tiling is said to be **quasiperiodic** if there are a finite number of lo-cal "pictures" (arrangements of tiles), and each local picture occurs uniformly throughout the tiling, i.e. within a bounded distance of every point. (See section 2.) A tiling of $\mathbf{R}^2$ is **half-periodic** if it is invariant under some translation.

From Proposition 1 we see that the only information needed to determine an SRT is the expansion φ, and a finite set of translations D, called the set of **digits**, with $|D| = |\det \varphi|$.

Using a generalization of a result of Thurston [17], we prove in section 3:

Theorems 7, 9, 12 $\varphi \in M_n(\mathbf{R})$ *is the expansion of a quasiperiodic SRT (or any SRT if $n = 1$ or 2) if and only if φ is similar to an expanding element of $M_n(\mathbf{Z})$, that is, a $n \times n$ integer matrix with eigenvalues outside the unit circle. The set of translations V of a quasiperiodic SRT are a subset of a discrete rank-n φ-invariant lattice in $\mathbf{R}^n$.*

For D a finite subset of $\mathbf{R}^n$ let D^* be the set of finite sequences of elements of D. For $\varphi \in M_n(\mathbf{R})$ let R_φ be the map $D^* \to \mathbf{R}^n$ which sends $x = d_{i_0} d_{i_1} \ldots d_{i_k}$ to the point

$$R_\varphi(x) = d_{i_0} + \varphi(d_{i_1}) + \ldots + \varphi^k(d_{i_k}).$$

R_φ is the **representation map** in base φ. For example, in the decimal system, with $D = \{0, 1, \ldots, 9\}$, $R_{10}(D^*)$ is the non-negative integers.

Theorem 10 *Given an expanding $\varphi \in M_n(\mathbf{Z})$ and a set $D \subset \mathbf{R}^n$ lying on a discrete φ-invariant lattice, with $|D| = |\det \varphi|$ and $0 \in D$, there is a corre-sponding SRT of $\mathbf{R}^n$ iff R_φ is injective on D^*.*

For a survey of work related to the representation of numbers in unusual bases, see [10]. In particular, for real numbers with integral bases, Matula [11] and Odlyzko [12] discuss the case when D is a set of integers, with the condition that all integers should have a *finite* expansion. This condition is more restrictive than that which we need; we refer the reader to the examples after Theorem 15. Odlyzko [12] also finds a necessary condition for general digit sets, which turns out to be equivalent to having D lying on a discrete lattice; thus the second statement in Theorem 7 was shown, in the one-dimensional case and in an equivalent form, in [12]. W.J Gilbert [6] studied base representations for complex bases, using a particular set of digits $D = \{0, 1, \ldots, m\}$. He determined those quadratic algebraic integer bases for which the resulting representation is **simple**, in the same sense as Matula (all integral points have finite expansions).

For integer bases in the line, we prove:

Theorem 15 *If D is a set of b non-negative integers, then R_b is injective on D^* iff for all divisors $k > 1$ of b, either $\phi_k(x)|p_D(x)$ or there exists an $l = l(k) > 0$ such that*

$$\phi_b(x^{b^l/k})|p_D(x),$$

where

$$p_D(x) = \sum_D x^d,$$

and $\phi_k(x)$ is the minimal polynomial for the primitive kth roots of unity.

This paper is an extension of some results obtained in [8]. I would like to thank Bill Thurston and Yuval Peres for helpful discussions and David Epstein and Robert Strichartz for pointing out problems in earlier versions.

2. Quasiperiodicity for planar SRTs

Proposition 1 *Two tiles in an SRT $\cup_V T_v$ subdivide in the same way, that is, there is a finite set of vectors $D \subset \mathbb{R}^n$, such that for all $v \in V$,*

$$\varphi(T_v) = \bigcup_{d \in D} T_{\varphi(v)+d}.$$

Proof: Let $S = \varphi(T_\alpha)$ for some tile T_α. Choose a direction for which both S and T_α have a "farthest point" (x and y respectively), that is, there is a closed half-space C which contains only the point x of S, and a translate of C contains only the point y of T_α. Such a direction always exists for any finite collection of compact sets since the set of directions for a fixed compact set which have no farthest point is of measure zero (consider the convex hull of a compact set; it has a boundary which is differentiable a.e., and at every point of differentiability there is a unique support plane.)

In a covering of S by translates of T_α, one translate of T_α must move y exactly onto x. This determines the position of one subtile of S exactly. Remove this subtile from S, and repeat with the closure of the remaining region $S' = \text{closure}(S - (T_\alpha + x - y))$. □

A **local arrangement** of radius r in a tiling is that part of a tiling contained in a ball of radius r, up to translations of the whole figure. That is, two arrangements are considered equivalent if one is obtained from the other by a single translation. For our purposes here in which there is only one tile type, a local arrangement can be considered as simply a finite collection W of translation vectors containing 0, such that

$$B_r(0) \subset \bigcup_{w \in W} T_0 + w$$

and each $T_0 + w$ intersects $B_r(0)$.

A locally finite tiling of $\mathbb{R}^n$ is said to be **homogeneous** if for every local arrangement A there exists an $R > 0$ such that for all y the arrangement A occurs somewhere in $B_R(y)$. A tiling is **quasiperiodic** if it is homogeneous and if in addition for each $r > 0$ there are only a finite number of different local arrangements of radius r. Alternatively, for every $r > 0$ there is a $R > 0$ such that an arrangement of radius r occurs in every ball of radius R.

Both these definitions can be extended to the case of *coverings* of $\mathbf{R}^n$ with a finite number of compact sets up to translation.

We will call an SRT **pure** if any local arrangement has an ancestor which is a single tile, i.e. there is a tile T (possibly not contained in the tiling) such by successive expansion and subdivision of T we can find every possible local arrangement occurring in some subdivision. This property is automatic for an SRT if the origin is contained in the interior of a tile; however there are "impure" examples (see figure 1) which we wish to avoid discussing (in any case impure SRTs are always finite unions of 'purely' tiled regions, and all our results, with the appropriate generalizations, go through in each of these regions separately).

Lemma 2 *A pure SRT is homogeneous.*

Proof: For any arrangement, if it first occurs in the nth subdivision of a tile, then it occurs within $\|\varphi^n\|\mathrm{diam}(T)$ of any point x, using Proposition 1, since every point $\varphi^{-n}x$ is contained in a copy of T. $\square$

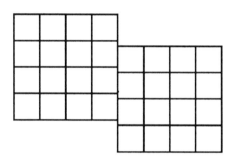

Figure 1: An impure SRT; the expansion is multiplication by 4.

From now on we will assume all SRTs are pure.

Theorem 3 *An SRT of the line is periodic. An SRT of $\mathbf{R}^2$ is half-periodic or quasiperiodic.*

The proof of Theorem 3 relies on a rigidity property of planar tilings whose tiles are closures of their interior; we state the result below (Theorem A). A *translation perturbation* of a tiling of a bounded region X is a discontinuous map from the tiling to another tiling covering a slightly smaller region X', the map being a (continuous) translation when restricted to a tile (and hence possibly multivalued on the tile boundaries).

Theorem A ([9]) *Any sufficiently small translation-perturbation (of a tiling covering a region X) in $\mathbf{R}^2$ is an earthquake, that is, the discontinuity set is* *a disjoint set of line segments ('fault lines'). If X is bounded, the fault lines extend to the boundary of X.*

This theorem basically says that you can't 'jiggle' the tiles in a tiling and still have a tiling, unless the tile boundaries contain a line, in which case you can shift the tiles relative to each other along the line, like an earthquake.

For an example of a non-quasiperiodic SRT, see figure 2, which displays part of a tiling in which $\varphi(z) = 3z$, with digits $\{-i, 0, i, -1-i, -1, -1+i, 1-i+ai, 1+ai, 1+i+ai\}$, where a is irrational. The relative y-coordinate of two adjacent tiles takes an infinite number of different values throughout the tiling. The fault lines in this case are the vertical lines at integer x-coordinates.

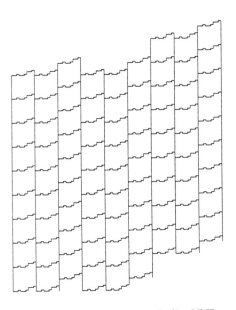

Figure 2: A non-quasiperiodic SRT.

Proof of Theorem 3:   The statement for a tiling of the line follows from Lemma 4 below.

Suppose we have an SRT in the plane. If there are only a finite number of different local arrangements, then the tiling is quasiperiodic by Lemma 2.

In the case where there are an infinite number of different local arrangements, there are convergent sequences of local arrangements, convergent in the sense that the relative translations between tiles converge. Any two arrangements S_1, S_2 in which the tiles occurring are sufficiently close to each other (so that one arrangement is a small perturbation of the other) must by Theorem A both contain locally a fault line or set of fault lines, such that S_2 is obtained from S_1 by sliding tiles along these lines (as in figure 2).

We will need:

Lemma 4 *Let $T \subset \mathbf{R}^2$ be a tile, $v \in \mathbf{R}^2, t_i \in \mathbf{R}$, so that $T, T + t_1 v, T + t_2 v, \ldots, T + t_n v \ldots$ is a set of translates of T; suppose the union of these translates tiles a region in the plane which contains an infinite strip (of positive thickness). Then the tiling of the strip is periodic.*

Proof. After tiling a sufficiently long portion, the placement of tiles at either end of the strip is predetermined (that is, the t_i are eventually determined). There are a finite number of local arrangements of any radius, by Theorem A. Hence there are two identical (long) local arrangements. Since each arrangement determines everything to its left and right, the two arrangements determine each other and so the tiling is periodic along the strip.   □

Lemma 5 *In a given SRT, a sufficiently small perturbation has whole fault lines (each fault line extends to infinity in both directions). That is, if two local arrangements are sufficiently close in the above sense, the union of the tile boundaries contains whole lines going through each of the arrangements.*

The difficulty in the proof of this lemma is that a fault line may not be isolated in the set of all fault lines. Since the proof is a bit long we will only sketch it.

Proof Sketch: Let ϵ be a small perturbation, so that all of its fault lines in some large neighborhood of the origin are longer than $10\mathrm{diam}(T_0)$. Let l be a long fault line of the perturbation adjacent to the region containing a component A of the interior of a tile, so that l is not a limit of other fault lines on at least one side. Assume l is vertical with the copy of A to its right. Since fault lines do not cross, if l has length R and A contains a ball of radius r, then there is a region N adjacent to l, on its right, of width r and height $R/2$ containing no point of any other fault line.

Take tiles intersecting N along its left edge, which are translated nontrivially relative to N by the perturbation. Subdivide these tiles if necessary to find tiles $\varphi^{-k} T_i$ which are adjacent to N, are translated non-trivially relative to N and touch N on their right-most point(s).

When we subdivide these small tiles at least two subtiles of each tile still touch N on their rightmost boundaries. Thus there are two distinct digits d_1, d_2 such that $\varphi^{-k}(d_1 - d_2)$ is parallel to l. Note that the smallest such $k \geq 0$ depends only on r, the thickness of N, which is fixed. Also, for $k' \geq k$, the same property holds: two subtiles of a tile of size $\varphi^{-k'}$ which touches N also touch N. Since $\{d_i - d_j | d_i, d_j \in D\}$ is a finite set, there is an $n \geq 0$ such that φ^n preserves the direction l (and is hence real).

Since there is a bound on the smallest k as above, there are only a finite number of possible directions for fault lines. By Lemma 4, tiles lined up along each of these directions line up with the same period depending only on the direction and not on the local arrangement.

Now for any sufficiently small perturbation, look at one of its fault lines; after a few subdivisions, the tiles line up periodically in a neighborhood of that fault line, so in the subdivided tiling the fault line must extend to infinity in both directions.

But now let l' be any complete fault line with a perturbation which is everywhere non-zero along it. If part of $\varphi^{-1}(l')$ is again a fault line, then the entire preimage must be fault line, since a fault line can only end at a point where the perturbation becomes continuous. $\square$

Continuation of proof of Theorem 3. Let l be a fault line of the small perturbation $S_1 \to S_2$, the perturbation being chosen sufficiently small so that l is a whole line. By homogeneity of the tiling, we can find a translated copy of either of the two arrangements S_1, S_2 at a bounded distance from any point. But each such arrangement gives rise to a new (whole) fault line parallel to l.

Thus by homogeneity there are lines parallel to l within a bounded distance from any point in the plane.

Choose $\epsilon > 0$ sufficiently small so that a tile is not contained entirely in disjoint parallel strips of width less than ϵ.

Consider the strips bounded by adjacent fault lines for which the width is at least ϵ. There is an upper bound to the width of such a strip by homogeneity, so there are either a finite number of widths of strips, or a convergent sequence of widths.

There are two possibilities: we have either a finite number of different widths of strips and the set of local arrangements (restricted to the strip) for all strips of a given width is finite, or else we have a convergent sequence of such strip-restricted arrangements.

In the latter case we must have from Theorem A again the existence of a fault line l' which cannot be parallel to l by definition of the strip, and yet cannot cross the boundary fault lines, since the perturbation must leave these lines invariant. This gives us a contradiction.

Thus there are a finite number of widths and the total number of different arrangements occurring between *any* two adjacent fault lines (of width at least ϵ) is finite.

By a similar argument as Lemma 4, in any strip, because there are a finite number of different arrangements, there is a *period* for the tiling of that strip. By multiplying all the possible periods for strips of all possible widths we find a period for the whole tiling. (We use here that each tile is at least partially contained in some strip of width larger than ϵ). Thus the tiling is at least half-periodic. $\square$

For SRTs of $\mathbf{R}^n$, $n \geq 3$ we have at present no rigidity result like Theorem A (in fact the theorem for general tilings is not true in dimension $n \geq 3$, see [9]). In what follows for SRTs of $\mathbf{R}^n$, $n > 2$ we will have to add in the hypothesis of quasiperiodicity. Thus our results are slightly more restricted in $\mathbf{R}^n$, $n \geq 3$.

3. Expansions

For what linear maps φ can we find a self-replicating tiling of $\mathbf{R}^n$, $n > 1$? We have the necessary conditions: φ is expanding (i.e. the eigenvalues have modulus greater than 1), since φ engulfs larger and larger regions of $\mathbb{R}^n$. We also have:

Lemma 6 $|\det \varphi|$ *is an integer, equal to the number of tiles that a tile divides into.*

Proof: Let S be a tile, and μ be Lebesgue measure. Then

$$\mu(\text{int}(\varphi(S))) \geq b \cdot \mu(\text{int}(S)),$$

where b is the number of tiles in the subdivision of S. Similarly

$$\mu(\text{int}(\varphi^k(S))) \geq b^k \mu(\text{int}(S)).$$

Let B_1 be a ball contained in S, and B_2 a ball containing S. Since

$$\mu(\varphi^k(B_i)) = |\det(\varphi)|^k \mu(B_i),$$

we have

$$b^k \mu(S) \geq \mu(\varphi^k S) \geq |\det \varphi|^k \mu(B_1)$$

and

$$|\det \varphi|^k \mu(B_2) \geq \mu(\varphi^k S) \geq \mu(\text{int}(\varphi^k S)) \geq b^k \mu(\text{int} S).$$

and so taking limits as k gets large we find $|\det \varphi| = b$. $\qquad\square$

A **self-similar tiling** (SST) is a generalization of a quasiperiodic SRT allowing a finite number of different tile shapes. Much of Theorem 7 below is based on the classification of expansions of general SSTs which was first proved for conformal SSTs in the plane by Thurston [17] and extended to $\mathbf{R}^n$, for SSTs with diagonalizable expansion in [8]. The proof below may also be regarded as an extension of the original proof of Thurston; the geometric ideas are all the same.

Fix a digit $d \in D$, and let T_0 be a tile containing the fixed point c_0 of the map $x \mapsto \varphi^{-1}(x - d)$. We say that c_0 is the **control point** of the tile T_0. The control point for a tile T_v is then defined to be $c_v = c_0 + v$. Note that the set C of control points maps into itself under φ. There is a similar definition of control points for SSTs, see [17].

We denote by $M_n(\mathbb{R}), M_n(\mathbb{Z})$ the set of $n \times n$ matrices with coefficients in $\mathbf{R}, \mathbb{Z}$.

Theorem 7 *If a linear map $\varphi \in M_n(\mathbf{R})$ is the expansion for a quasiperiodic SRT of $\mathbf{R}^n$ then φ is similar to an expanding element of $M_n(\mathbb{Z})$. The translations between the tiles lie on a φ-invariant discrete rank-n lattice.*

The construction of an SRT for any integer-linear expanding map is given in Theorem 12 in the next section.

Proof: Let $\cup_V T_v$ be an SRT of $\mathbf{R}^n$ with expansion $\varphi \in M_n(\mathbf{R})$.

By quasiperiodicity, a translation between control points of adjacent tiles in an SRT is one of a finite number of translations $\{v_1, v_2, \ldots, v_N\}$.

Since one can reach any control point from any other by hopping on a path between control points of adjacent tiles, we see that $V \subset \Gamma$, where Γ is the integer lattice generated by the set $\{v_1, v_2, \ldots, v_N\}$.

Since $\varphi V \subset V$, we see that for each v_i there are integers α_{ij} such that for $1 \le i \le N$,

$$\varphi v_i = \sum_{j=1}^N \alpha_{ij} v_j.$$

We define $\tilde{\varphi} \in M_N(\mathbb{Z})$ to be the transformation given by the matrix (α_{ij}).

Let π be the projection $\mathbf{R}^N \to \mathbf{R}^n$, which maps the ith basis vector, e_i, to v_i. Then $\pi\tilde{\varphi} = \varphi\pi$, and so $\mathbf{R}^N$ splits into two invariant spaces $\mathbf{R}^N = W \oplus \ker\pi$, where W is isomorphic to $\mathbf{R}^n$ and $\tilde{\varphi}|_W = \varphi$.

Let us first sort out the linear algebra. Denote by $C^{(l)}(\gamma) \in M_l(\mathbf{R})$ the matrix with λs on the diagonal and 1s on the upper diagonal (zeros elsewhere). We can decompose $\tilde{\varphi}$ over the complex numbers into Jordan form:

$$\tilde{\varphi} \sim \begin{pmatrix} C^{(l_1)}(\gamma_1) & 0 & \ldots & 0 \\ 0 & C^{(l_2)}(\gamma_2) & & \\ \vdots & & \ddots & \\ 0 & & & C^{(l_k)}(\gamma_k) \end{pmatrix},$$

where the γ_i are not necessarily distinct.

There is another decomposition which we shall be concerned with, which is the decomposition of $\tilde{\varphi}$ over the rationals. After conjugating by an element of $GL_N(\mathbf{Q})$, we can write $\tilde{\varphi}$ in the form

$$\tilde{\varphi} \sim \begin{pmatrix} D_1^{(r_1)} & 0 & \ldots & 0 \\ 0 & D_2^{(r_2)} & & \\ \vdots & & \ddots & \\ 0 & & & D_m^{(r_m)} \end{pmatrix},$$

where the $D_i^{(r)}$ are $r \times r$ matrices of the form

$$(1) \qquad\qquad D_i = \begin{pmatrix} A_i & I & & 0 \\ 0 & A_i & I & \\ \vdots & & \ddots & I \\ 0 & \ldots & 0 & A_i \end{pmatrix},$$

here A_i is some *irreducible* square integer matrix and I is the identity. Irreducible refers to the fact that the characteristic polynomial for A_i is irreducible over $\mathbf{Q}[x]$. We define W_{D_i} to be the subspace (of dimension r_i) on which $\tilde{\varphi}$ acts as the block matrix $D_i^{(r_i)}$.

Since W is a priori not a rational subspace, it may intersect the W_{D_i} non-trivially. We split W into invariant subspaces $W = W_1 \oplus W_2 \oplus \ldots \oplus W_m$, with $W_i = W \cap W_{D_i}$. Define $\varphi_s = \varphi|_{W_s}$, which is the restriction of $\tilde{\varphi}_s = \tilde{\varphi}|_{W_{D_s}}$. We claim that $W_s = \{0\}$ or $W_s = W_{D_s}$, which will complete the proof of the first part of the theorem.

Let λ_j, $1 \leq j \leq l$ be the eigenvalues of $\tilde{\varphi}_s$; by (1) they each have the same multiplicity k_s and are a complete set of Galois conjugates. By the invariance of W_s we can write

$$
(2) \qquad \varphi_s = \begin{pmatrix} C^{(k_1)}(\lambda_1) & 0 & \cdots & 0 \\ 0 & C^{(k_2)}(\lambda_2) & & \\ \vdots & & \ddots & \\ 0 & & & C^{(k_l)}(\lambda_l) \end{pmatrix},
$$

where all the λ_i are distinct and each k_i is either 0 or k_s (recall that $\ker \pi$ is invariant, so that $W_{D_s} \cap \ker \pi$ is also invariant).

Since W_{D_s} is a real subspace and π is a real projection, φ_s is a real linear map on W_s and so for each complex eigenvalue λ_j of φ_s, the eigenvalue $\overline{\lambda}_j$ occurs with the same multiplicity k_j. If all the k_i are equal, we are done.

Lemma 8 *If $k_i = k_s$ and $k_j = 0$, then $|\lambda_i| > |\lambda_j|$.*

Proof: Suppose not. Assume without loss of generality $i = 1$ and $j = 2$. Then $k_1 = k_s > 0$, $k_2 = 0$, and $|\lambda_1| \leq |\lambda_2|$. Let V_1 be the λ_1-eigenspace of W_{D_s}, (so that $\tilde{\varphi} - \lambda_1 I$ is nilpotent on V_1), and similarly define V_2. We construct a function $f : W \to V_2$ as follows.

Consider the control points $C \subset W$. The control points are projections of a set of lattice points $\tilde{C}$ sitting in $\mathbf{R}^N$. Because A_i is irreducible over $\mathbf{Q}$, we can lift C into the space V_2, by Galois conjugation of the coordinates of the points in V_1. Alternatively, let π_2 denote the projection of $\mathbf{R}^N$ (along complementary invariant subspaces) onto the space V_2. Then $f(c) = \pi_2(\tilde{c})$.

We then define f for each $n > 0$ on the set $\varphi^{-n}(C)$ by invariance:

$$
(3) \qquad f(\varphi^{-n} c) = \pi_2 \tilde{\varphi}^{-n}(\tilde{c}).
$$

The values of f on adjacent control points in C differ by only a finite number of possible values since this difference only depends on the vector joining the two points. Since we can get between any two control points by jumping along a path of adjacent control points, with at most a constant loss of efficiency, there is a *global* lipschitz constant for the function f on the original control points C.

Because φ^{-1} is contracting and λ_2^{-1} is contracting (a simple calculation shows that the absolute values of the matrix entries of $C^{(k)}(\lambda_2)^{-n}$ are all less than $Kn|\lambda_2^{-n}|$ for some constant K), the function f thus defined on the dense set $\cup_n \varphi^{-n} C$ is continuous; we can hence extend f by continuity to all of W.

Consider the regularity of the function f in the direction of the λ_1-eigenspace V_1. If $|\lambda_1| < |\lambda_2|$ then f is Lipschitz on all of W in the direction V_1, because this is true on the dense set $\cup_n \varphi^{-n} C$ ($|\lambda_2|^{-1}$ contracts more quickly than $|\lambda_1|^{-1}$). If $|\lambda_1| = |\lambda_2|$ then f will not in general be Lipschitz, because of the shear. However V_1 contains an invariant direction v_1, which is mapped to the invariant direction v_2 in V_2. v_i may be a line or a plane, depending on whether or not λ_i is real or complex.

We show that f is Lipschitz in the direction v_1. Let S be the set of values $f(x + w) - f(x)$, for w a vector of length ≤ 1 in v_1 and x ranging over W. By quasiperiodicity, S is compact; let $K = \max\{|y|, y \in S\}$. Then for any $w \in v_1$ of length $|w| > 1$, letting $m_w = \lfloor |w| \rfloor$,

$$|f(x + w) - f(x)| \leq |f(x + w) - f(x + m_w \frac{w}{|w|})| +$$

$$\sum_{k=1}^{m_w} \left| f(x - k\frac{w}{|w|}) - f(x - (k-1)\frac{w}{|w|}) \right| \leq 2K|w|$$

and if $|w| \leq 1$, the invariance gives

$$|\lambda_2^n(f(x + w) - f(x))| = |f(\varphi^n x + \lambda_1^n w) - f(\varphi^n x)| \leq 2K|\lambda_1|^n|w|,$$

and dividing by $|\lambda_2^n|$, we have $|f(x + w) - f(x)| \leq 2K|w|$.

Now a Lipschitz function is differentiable almost everywhere; let $z \in W$ be a point at which the partial derivatives in the direction v_1 exists. In a sufficiently small v_1-neighborhood of z, the function f is approximately linear along v_1; by invariance, as n increases, at $\varphi^n(z)$ f looks more and more linear along v_1.

By quasiperiodicity, there must be a point not too far from the origin at which f looks like it does at $\varphi^n(z)$ (up to a translation); taking a limit we find that the function f at some point must be exactly linear along the entire space $x + v_1$. Now quasiperiodicity again implies that on a *dense* set of x, f is linear along $x + v_1$. By continuity f is linear along v_1 for every $x \in W$.

Now $f(\lambda_1 w) = \lambda_2 f(w)$ for $w \in v_1$, so if λ_1 is real, we have $\lambda_2 = \lambda_1$, and if λ_1 is complex, we must have $\lambda_2 = \lambda_1$ or $\overline{\lambda}_1$.

This completes the proof of the lemma.

Remark. We pause here to note that we have not yet used the fact that there is only one tile in the tiling; the above Lemma (8) holds for general SSTs and is in fact the condition for the existence of an SST with expansion φ: In order for $\varphi \in M_n(\mathbb{R})$ to be the expansion for an SST, the blocks $C^r(\lambda)$ in the Jordan form can be grouped into blocks of the same size having Galois-conjugate eigenvalues, so that in each group, the conjugates which appear are strictly larger in modulus than those which do not appear. The construction of an SST with an expansion satisfying this condition is quite similar to that found in [8].

Continuation of Proof of Theorem 7:

Back in W we have:

$$\prod_{\lambda \text{ ev of } \varphi} |\lambda| = b$$

(taken with the correct multiplicities) which is an integer by Lemma 6; since each eigenvalue occurs with with the same multiplicity as its complex conjugate, this implies

$$\prod_{\lambda \text{ ev of } \varphi} \lambda = \pm b.$$

If we apply a Galois automorphism $g \in Gal(D_i)$ acting on the space D_i only, we find

$$\prod_{\lambda \text{ ev of } \varphi} g(\lambda) = \pm b$$

but by the lemma, if $|g(\lambda)| < |\lambda|$ and yet the multiplicities satisfy $m(\lambda) > m(g(\lambda))$ then the product decreases in modulus, a contradiction. Thus we see that $m(\lambda) = m(g(\lambda))$ for all $g \in Gal(D_i)$, so by irreducibility all the k_i above are equal.

Hence we have $\varphi_s \in M_{|W_s|}(\mathbb{Z})$ completing the claim, and so

$$\varphi = \varphi_1 \times \varphi_2 \times \ldots \times \varphi_m \in M_n(\mathbb{Z}).$$

We now prove the second statement of the theorem. We already know that $V \subset \Gamma$, and Γ is a discrete rank-N lattice. The above proof shows that either $W_i = W_{D_i}$ or $W_i = \{0\}$, and we know that W_{D_i} is a rational subspace of $\mathbf{R}^N$. Thus $W = W_1 \oplus \ldots \oplus W_m$ is itself a rational subspace of $\mathbf{R}^N$, and $C \subset \Gamma \cap W$. Since W has dimension n we are done. □

The remaining problem of expansions for *non-quasiperiodic* SRTs in $\mathbf{R}^n$, $n > 2$ is still open. For the line every SRT is periodic, and for the plane we have:

Theorem 9 *An expanding map $\varphi \in M_2(\mathbf{R})$ is the expansion for a non-quasiperiodic SRT of the plane if and only if the eigenvalues of φ are integers, at least one of which has modulus larger than 2.*

Proof: From Theorem 3, a non-quasiperiodic SRT of the plane is half-periodic; if φ does not preserve the direction of periodicity the tiling is periodic in two independent directions and hence quasiperiodic. Thus the eigenvalues of φ are real; the eigenvalue, a, in the periodic direction must preserve the periodicity and so must be an integer.

Since the product of the eigenvalues is an integer, we need only show that the other eigenvalue λ is an algebraic integer.

Assume that the fault lines are vertical; consider the set of horizontal distances $\{d_i\}$ between control points. Since there are locally only a finite set of such distances, the expansion gives an integer linear relationship

$$\lambda d_i = \sum \beta_{ij} d_j,$$

where the β_{ij} are integers. So λ is the eigenvalue of the integer matrix (β_{ij}), and hence must be an algebraic integer.

Since $\lambda a = \pm b$, where a and b are integers, λ must also be an integer.

To construct a non-quasiperiodic SRT from

$$\varphi = \begin{pmatrix} n & 0 \\ 0 & m \end{pmatrix} \text{ or } \begin{pmatrix} n & 0 \\ 1 & n \end{pmatrix},$$

choose digits $D = \{(i,j) | 1 \le i \le n, 1 \le j \le m\}$ but shift the last column of digits (n,j) by an irrational amount $(0,\epsilon)$ in the y-direction. (See for example figure 2). This gives a non-quasiperiodic SRT unless n and m both have modulus 2.

If n and m are both ± 2 this will still give a periodic tiling; in section 4.1 we show that every choice of digits in this case which gives an SRT gives a periodic SRT of the plane.   □

4. Digits

Let $D \subset \mathbf{R}^n$, D^* be the set of finite strings of elements of D, and for any linear map ψ, define $R_\psi : D^* \to \mathbf{R}^n$ by

$$R_\psi(d_{i_0} d_{i_1} \ldots d_{i_k}) = d_{i_0} + \psi(d_{i_1}) + \ldots + \psi^k(d_{i_k}).$$

Let φ be an expanding element of $M_n(\mathbb{Z}) \subset M_n(\mathbf{R})$. It has an inverse $\varphi^{-1} \in M_n(\mathbf{R})$. Given $D \subset \mathbf{R}^n$, let $T_0 = T_0(D) \stackrel{\text{def}}{=} \text{closure}(\varphi^{-1} R_{\varphi^{-1}}(D^*))$. We also define T_0^k to be $\varphi^{-1} R_{\varphi^{-1}}(D^k)$, where D^k is the set of sequences of length k. T_0 is a compact subset of $\mathbf{R}^n$, and

$$\varphi(T_0) = \bigcup_{d \in D} T_0 + d.$$

Thus T_0 has the self-replicating property; we need to know whether or not T_0 contains open sets.

Example. Suppose we have $n = 1$, $D = \{-1, 0, 4\}$ with $\varphi(x) = 3x$. The graph of the integral of the characteristic function of T_0 is given in figure 3.

The reader may show that for example T_0 contains the interval $[0, 1/6]$.

Theorem 10 *Let D be a set of b points on a discrete rank-n φ-invariant lattice $\Gamma \subset \mathbf{R}^n$, with $0 \in D$, $b = |\det \varphi|$. Then the set $T_0(D)$ contains an open set iff R_φ is injective on D^*.*

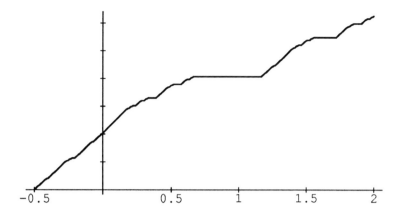

Figure 3: The integral of the characteristic function of T, T a tile for base 3 with digits $\{-1, 0, 4\}$.

Proof of Thm 10: Suppose R_φ is not injective. Let $x = d_{i_0} d_{i_1} \ldots d_{i_k}$ and $y = d_{j_0} d_{j_1} \ldots d_{j_l}$ be two elements of D^* with the same image under R_φ. Let $m = \max\{k, l\}$, then $\varphi^m(T_0)$ can be covered by $b^m - 1$ translates of T_0. Since φ^m expands volume by b^m, this implies that the Lebesgue measure of T_0 is zero. Thus T_0 has no interior.

So suppose R_φ *is* injective. Let λ be the smallest eigenvalue (in modulus) of φ. T_0 has radius at most

$$\frac{|d_*|}{|\lambda|} + \frac{|d_*|}{|\lambda|^2} + \frac{|d_*|}{|\lambda|^3} \cdots \leq \frac{\max_D |d|}{|\lambda| - 1}$$

and so is a bounded, hence compact, set.

The strategy is to show that T_0 has positive Lebesgue measure, and then expand around a Lebesgue point and use self-similarity to give an open set in T_0.

Let m_k denote the measure which assigns mass b^{-k} to each point of T_0^k; then $m_k(T_0^k) = 1$.

Define $\Gamma_k = \varphi^{-k}\Gamma$; then $T_0^k \subset \Gamma_k$. The volume of the fundamental region of Γ_k is $c_1 b^{-k}$, where c_1 is the volume of the fundamental region for Γ.

Let $B = B_r(x)$ be a ball in $\mathbf{R}^n$, and let $N_\epsilon \partial B$ denote the ϵ-neighborhood of ∂B. If ϵ is small with respect to r, say $\epsilon < r/2$, we have for some constant $c_2 = c_2(r)$,

$$\mathrm{Vol}(N_\epsilon \partial B) \leq c_2 \epsilon,$$

so for some constant $c_3 = c_3(r)$ and k sufficiently large with respect to ϵ (for example any k for which $\epsilon > 2|\lambda|^{-k}$) we have

$$\mathrm{card}(\Gamma_k \cap N_\epsilon \partial B) \leq c_3 \epsilon b^k / c_1.$$

Now let $c_4 = \max_D |d|$ so that the masses for m_{k+1} are all within $c_4|\lambda|^{-k}$ of those of m_k; then

$$|m_{k+1}(B) - m_k(B)| \leq b^{-k}\mathrm{card}(\Gamma_k \cap N_{c_4\lambda^{-k}}\partial B) = b^{-k}\frac{c_3 c_4 b^k}{c_1 \lambda^k} = c_5/\lambda^k.$$

Thus the measures m_k converge, to a measure m_∞.

Also note that there is a constant $c_6 > 0$, depending only on Γ so that for all k sufficiently large wrt r,

$$\mathrm{card}(B \cap \Gamma_k) \leq c_6\mathrm{Vol}(B)b^k$$

so that

$$m_k(B) \leq c_6\mathrm{Vol}(B) = c_6\mu(B)$$

μ being Lebesgue measure. Thus $m_\infty(B) \leq c_6\mu(B)$ and $1 = m_\infty(T_0) \leq c_6\mu(T_0)$ so that T_0 has positive Lebesgue measure.

Let u be a Lebesgue point of T_0. Given $0 < \epsilon < 1$, choose a nested sequence of balls U_k around u of density $1 - \epsilon^k$, that is,

$$\frac{\mu(U_k \cap T_0)}{\mu(U_k)} \geq 1 - \epsilon^k.$$

Fix $\delta > 0$. For all $k \geq 1$ choose $l = l(k)$ such that $\varphi^l(U_k)$ contains a ball of radius δ. Now $\varphi^l(U_k \cap T_0)$ is a big ellipsoid of density $1 - \epsilon^k$; we need to find in it a ball of reasonable radius with density also close to 1.

Lemma 11 *There exists an $\eta > 0$ depending only on n such that any ellipsoid in $\mathbb{R}^n$ with semiaxis lengths $\gamma_1 \leq \gamma_2 \leq \cdots \leq \gamma_n$ can be packed to density η with pairwise disjoint balls of radius $\gamma_1/10$.*

Proof of lemma: Use a cubical packing, and the fact that the $\gamma_1/10$-neighborhood of the boundary has only a constant amount of the volume. $\square$

Let $\{B\}$ be the set of balls in such a packing of the ellipsoid $E = \varphi^l(U_k)$, so that

$$\sum \mu(B) = \eta'\mu(E) \geq \eta\mu(E).$$

We have

$$(1 - \epsilon^k)\mu(E) = \mu(E \cap T_0) = \mu(T_0 \cap (\cup B)) + \mu(T_0 \cap (E - \cup B))$$

$$\leq \mu(T_0 \cap (\cup B)) + \mu(E - \cup B) = \mu(T_0 \cap (\cup B)) + (1 - \eta')\mu E.$$

Hence

$$\mu(T_0 \cap (\cup B)) \geq (\eta' - \epsilon^k)\mu E = \frac{\eta' - \epsilon^k}{\eta'}\mu(\cup B)$$

and so

$$\frac{\mu(T_0 \cap (\cup B))}{\mu(\cup B)} \geq 1 - \frac{\epsilon^k}{\eta} \geq 1 - \frac{\epsilon^k}{\eta}$$

so that some ball has density at least $1 - \epsilon^k/\eta$.

Thus we have a sequence of balls $B_k \subset \varphi^{l(k)}(U_k \cap T_0)$ of radii between $\delta/10$ and $|\lambda|\delta/10$, and densities increasing to 1 as $k \to \infty$. Each such ball is covered by copies of T_0 since each is a subset of some $\varphi^N T_0$. All the copies covering a given ball are disjoint in measure, so there is an upper bound to the number of copies occurring in any ball. Choose a subsequence of balls such that the relative translates of the subtiles T_0 occurring are convergent. In the limit the translates of these finite number of T_0s cover all of a ball, and so one of them (and hence all of them) must contain an open set.   □

Theorem 12 *Given $\varphi \in M_n(\mathbb{Z})$ with eigenvalues $|\lambda_i| > 1$. Let $D \subset \mathbb{Z}^n$, with $|D| = b = |\det \varphi|$ and $0 \in D$ such that (1) $d_i - d_j \notin \varphi \mathbb{Z}^n$ and (2) D is contained in no proper φ-invariant subgroup of $\mathbb{Z}^n$. Then there is a periodic SRT with digits D and expansion φ.*

In particular if φ is irreducible we can always take $D = \{0, 1, \dots, b - 1\}$ in $\mathbb{Z}[\varphi] \cong \mathbb{Z}^n$.

Proof: For any set $D \subset \mathbb{Z}^n \subset \mathbb{R}^n$ which contains 0, define a measure μ_D^k which has the point masses $1/b^k$ at each distinct point of $\varphi^k T_0^{2k}(D)$ (recall that $T_0^{2k}(D)$ is the set of points with finite expansions of length $2k$; each such point sits on the lattice $\varphi^{-2k} \mathbb{Z}^n$.) As in Theorem 10, there is a constant c independent of k such that for any ball B_r of radius r not too small, say $r \geq 2\lambda^{-k}$, we have $\mu_D^k(B_r) \leq c \cdot \mu(B_r)$, where μ is Lebesgue measure.

The sets $\varphi^k T_0^k$ are nested subsets of $\mathbb{Z}^n$ since $0 \in D$; thus the sets $\varphi^k T_0^{2k}$ are nested in $\mathbb{R}^n$. As in Theorem 10, the measures converge to a limit measure $\mu_D = \lim_{k \to \infty} \mu_D^k$.

Lemma 13 *Let φ be as before. Suppose $D \subset \mathbb{Z}^n$ is a finite set, $0 \in D$ and $\mathbb{R}^n = \cup_{l \geq 0} \varphi^l T_0(D)$. Then $\mu_D = c \cdot \mu$ for some constant $c > 0$, where μ is Lebesgue measure.*

Proof of Lemma: By hypothesis, $\mathbb{R}^n$ is covered with translates of the compact set T_0. Each translate "sits" at a point of $\mathbb{Z}^n$. Thus there are only a finite number of local arrangements of translates possible. As in Lemma 2, the self-replicating property implies that the *covering* of $\mathbb{R}^n$ is quasiperiodic.

Fix $\delta > 0$, and choose k such that $\text{diam}(\varphi^{-k} T_0) < \delta$. Then for any ball B_δ of radius δ in $\mathbb{R}^n$, there is a copy of $\varphi^{-k} T_0$ sitting completely inside B_δ. Thus $\mu_D(B_\delta) \geq b^{-k}$.

By the homogeneity of the covering, we can find x such that the ball $B_\delta(x)$ has smallest μ_D-mass among all balls of radius δ in $\mathbb{R}^n$. Using the same argument as that after Lemma 11, we blow up this ball by φ^K for some sufficiently large K and find for any $y \in \mathbb{R}^n$ an translated copy of $B_\delta(y)$ sitting inside $\varphi^K(B_\delta(x))$; if $\mu_D(B_\delta(y)) \geq \mu_D(B_\delta(x))$ then since the average mass of the blown-up ball is $\mu_D(B_\delta(x))$ we can find another ball in $\varphi^K(B_\delta(x))$ which has mass smaller than $\mu_D(B_\delta(x))$, contradicting the choice of x. Thus every ball of radius δ must have the same (finite, nonzero) μ_D-mass.

Since this holds for any $\delta > 0$, μ_D is translation-invariant and so must be a multiple of Lebesgue measure.   □

Continuation of proof of Theorem 12 For a set A define $\Delta A = A - A = \{\alpha - \beta | \alpha, \beta \in A\}$, and define $\Delta^{m+1}A = \Delta(\Delta^m A)$.

Now let D satisfy the hypothesis of the theorem. Condition (1) on D and Theorem 10 show that $T_0 = T_0(D)$ contains open sets. Hence $\Delta T_0(D) = T_0(\Delta D)$ contains an open set around the origin, and

$$(4) \qquad \bigcup_{l \geq 0} \varphi^l T_0(\Delta D) = \mathbf{R}^n.$$

A subset of $\mathbf{Z}^n$ which is closed under taking differences is a subgroup. Thus $\cup_{m \geq 0} \Delta^m D$ is a subgroup A (the smallest subgroup containing D), and $\cup_{k \geq 0} \varphi^k A = \mathbf{Z}^n$ by hypothesis (2) on D.

Fix a large neighborhood U of the origin in $\mathbf{Z}^n$, so that $\mathbf{Z}^n = R_\varphi(U^*)$. For some $m \geq 0$, $S^m \overset{\text{def}}{=} \cup_{k \geq 0} \varphi^k \Delta^m D$ contains U. Then $\mathbf{Z}^n = R_\varphi((\Delta^m D)^*)$. Let m be the smallest integer with this property.

If $m = 0$ then the tiling

$$\bigcup_{z \in \mathbf{Z}^n} T_0 + z$$

is an SRT and we are done (see the last paragraph of this proof).

Otherwise, consider S^{m-1}. **We claim:** there are arbitrarily large balls in $\mathbf{Z}^n$ (not necessarily at the origin) such that every point in them has an expansion in S^{m-1}.

To prove this claim, let $U = \{u_1, u_2, \ldots, u_K\}$ be all the points in a large ball around the origin in $\mathbf{Z}^n$. We will construct a point a with an expansion in S^{m-1} such that $\forall u \in U$, $a + u$ also has an expansion in S^{m-1}.

We construct a by defining successively pieces of its expansion. Take first some finite expansion for u_1 in S^m:

$$u_1 = \sum_{j=0}^{J_1-1} \varphi^j(r_j - s_j), \quad r_j, s_j \in \Delta^{m-1}D.$$

We set the first J_1 digits of a to be the digits s_j, and the first J_1 digits of $a + u_1$ to be the digits r_j. The first J_1 digits of the $a + u_i$, for $i > 1$, are chosen to be any digits $r_j^{(i)} \in \Delta^{m-1}D$ so that for all $0 \leq k \leq J_1 - 1$,

$$(5) \qquad u_i \equiv \sum_{j=0}^{k} \varphi^j r_j^{(i)} \quad \text{mod } \varphi^{k+1}.$$

We can choose such digits by condition (i) on D: $\forall x \in \mathbb{Z}^n$ $\exists d_i \in D$ such that $x - d_i \in \varphi\mathbb{Z}^n$.

At this point both the a and $a + u_1$ that we have constructed so far have finite expansions, of length J_1. Now let $u_2' = \varphi^{-J_1}(u_2 - \sum_{j=0}^{k} \varphi^j r_j^{(2)})$ so that u_2' is again in $\mathbb{Z}^n$, and take an expansion of u_2' in S^m,

$$u_2' = \sum_{j=J_1}^{J_2-1} \varphi^{j-J_1}(r_j^{(2)} - s_j), \quad r_j^{(2)}, s_j \in \Delta^{m-1}D.$$

We extend the expansions of a and the $a + u_i$ as follows: take new digits with indices j for $J_1 \leq j \leq J_2 - 1$ in both a and $a + u_1$ to be the digits s_j, and the digits in the same interval of $a + u_2$ to be the $r_j^{(2)}$; the digits in the expansions of $a + u_i$, $i > 2$ are again chosen to satisfy condition (5) above, in the new range $J_1 \leq j \leq J_2 - 1$.

At this point, using the a constructed so far, a, $a + u_1$ and $a + u_2$ all have finite expansions of the same length J_2. We continue, doing this for each $u_i \in U$, until we have constructed an a such that a and each $a + u_i$ has a finite expansion. This completes the proof of the claim.

By Lemma 13 and (4), $\mu_{\Delta^{m'}D} = c_{m'} \cdot \mu$ for all $m' > 0$. But this implies that if every lattice point in a ball of radius $\text{diam}(T_0(\Delta^{m-1}D))$ in $\mathbb{R}^n$ is covered by a translated copy of $T_0(\Delta^{m-1}D)$ then *every* point of $\mathbb{Z}^n$ must be so covered; the measure of each ball is the same. Thus by the above claim every point of $\mathbb{Z}^n$ has a finite expansion with digits $\Delta^{m-1}D$, contradicting minimality of m; so that we must have $m = 1$.

In conclusion, S^1 gives finite expansions to every point of $\mathbb{Z}^n$. Thus for any $\alpha \in \mathbb{Z}^n$, there are two tiles in S^0 which differ by the translation α. Moreover, these tiles have disjoint interiors. (Recall that all tiles in S^0 have disjoint interiors.)

Now we can define our SRT to be

$$\bigcup_{z \in \mathbb{Z}^n} T_0(D) + z,$$

which covers all of $\mathbb{R}^n$ with tiles which have disjoint interiors, and has the self-replicating property

$$\varphi(T_0 + z) = \bigcup_{d \in D} T_0 + \varphi z + d.$$

$\square$

Example. Let

$$\varphi = \begin{pmatrix} 0 & -4 \\ 1 & -2 \end{pmatrix}.$$

Then $\varphi^2 = -2\varphi - 4$, and we embed $\mathbb{Z}[\varphi]$ in the complex plane, sending I to 1 and φ to $\lambda = -1 + \sqrt{-3}$. Then φ acts on $\mathbf{C}$ as multiplication by λ so we can think of our digits as complex numbers, and our tile becomes a set of expansions in base λ with these as digits. We choose $D = \{0, 1, 2, -1 - \lambda\}$ (no two are congruent modulo φ, and $GCD(D) = 1$).

The points of T_0^n are polynomials of degree n in $1/\lambda$ with coefficients $0, 1, 2$ or $-1 - \lambda$. See figures 4 and 5.

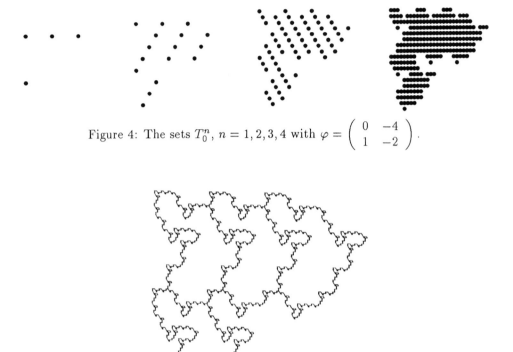

Figure 4: The sets T_0^n, $n = 1, 2, 3, 4$ with $\varphi = \begin{pmatrix} 0 & -4 \\ 1 & -2 \end{pmatrix}$.

Figure 5: ...and part of the resulting self-replicating tiling.

4.1 Example. Base 2 in the plane. Let $\varphi(x, y) = (2x, 2y)$ in $\mathbb{R}^2$. What SRTs can we construct with expansion φ? After an affine change of coordinates, we can assume our digits are

$$D = \{(0, 0), (1, 0), (0, 1), (x, y)\}$$

(They cannot be all collinear.)

Suppose first that the tiling is not quasiperiodic; then it is half-periodic. Consider a strip bounded by two fault lines l, l'; the tiles which are adjacent to l but outside the strip have two or more sons which lie at the same distance from l. From the same argument at l', we see that our digits must line up in two parallel rows of two digits each. We can thus assume without loss of generality that $(x, y) = (1, \alpha)$, with the fault lines vertical.

Each tile is situated at a point

$$\sum_{i=1}^{N} r_i 2^i, \quad r_i \in D$$

so the translation between two tiles is given by

$$\sum_{i=1}^{N} (r_i - s_i) 2^i,$$

where $r_i, s_i \in D$. The x-coordinates of two tiles always differ by an integer amount, so two tiles with different x-coordinates cannot be close together. If two tiles have the same x-coordinate, we can write the difference in their y-coordinates as $m\alpha + n$, where m and n are integers such that (m, n) has an expansion in base 2 using digits $D_y = \{(1, 0), (-1, 0), (0, 0), (0, 1), (0, -1)\}$ (because $r_i - s_i \in D - D$, whose y-coordinates are $1, -1, 0, \alpha$ or $-\alpha$).

Conversely, for any (m, n) which has an expansion in base 2 with these digits, there is a pair of tiles T_1, T_2 in our tiling with translation distance $m\alpha + n$.

But for every pair $(m, n) \in \mathbb{Z}^2$ such that exactly one of m, n is odd, there is such an expansion. For if $(m, n) \equiv (1, 0) \bmod 2$ then either $((m, n) + (1, 0))/2$ or $((m, n) + (-1, 0))/2$ has again exactly one odd component, and similarly if $(m, n) \equiv (0, 1) \bmod 2$ then either $((m, n) + (0, 1))/2$ or $((m, n) + (0, -1))/2$ has exactly one odd component. By induction, using these contracting maps we eventually arrive at $(0, 0)$, this implies that we can expand (m, n) in base 2 using D_y above as digits.

If α is irrational, then the distance of the lattice of points $(1, 0) + (1, 1)\mathbb{Z} + (1, -1)\mathbb{Z}$ to the line of slope α through the origin form a dense set, so that $m\alpha - n$ takes on an infinite number of values in any interval. This implies that the tiling is not periodic in the vertical direction, a contradiction. So α is rational, and the tiling is quasiperiodic.

We can now ask for which rational (x, y) will work to complete our set of digits. We can apply the same argument, letting $\alpha = x + iy$; we have for the translation distance of two tiles an equation of the form $m\alpha + n$, where m is an integer and n is in $\mathbb{Z}[i]$, such that (m, n) can be expanded in base 2, this time with the digits

$$D'' = \{(0, 0), \pm(1, 0), \pm(1, -1), \pm(0, i), \pm(1, -i), \pm(0, 1 - i), \pm(0, 1)\}.$$

If we let $n = n_1 + n_2 i$ then we can show that as long as not all of m, n_1, n_2 are odd (or all even), then (m, n) has such an expansion. The 6 cases using 'o' for odd and 'e' for even are

$$(o, o + ei), (o, e + oi), (o, e + ei), (e, o + oi), (e, e + oi), (e, o + ei).$$

The reader can show that for each case there are two digits $\pm d_i$ such that $((m, n) \pm d_i)/2$ is again of the above form, and any such number eventually leads to $(0, 0)$.

Thus we see that in order for the tiling to exist the line through the point (x, y) cannot pass through any integer lattice points of one of the above forms; thus (x and y being rational), we see that in lowest terms, $x = p_1/q_1$, $y = p_2/q_2$ where the p_i and q_i are all odd.

Conversely, given such an (x, y), then there is a tiling, since multiplying all the digits by $q_1 q_2$ gives digits in the four residue classes of $\mathbb{Z}[i]$ modulo 2.

Theorem 14 *For base 2 in $\mathbf{R}^2$, the digit set D containing $(0, 0)$ gives an SRT if and only if after a linear coordinate change, $D = \{(0, 0), (1, 0), (0, 1), (x, y)\}$, where x, y are both rational and the quotient of two odd integers.*

If we next consider the expansion

$$\varphi = \begin{pmatrix} 2 & 1 \\ 0 & 2 \end{pmatrix}$$

then since φ has a preferred direction we cannot take in general our digits to be of the same form as D, above; we can only conjugate with an affine map which leaves horizontal lines horizontal.

If we have fault lines, however, then they must be horizontal since these are the only invariant directions. Thus in this case we can arrange our digits to be $D = \{(0, 0), (1, 0), (0, 1), (\alpha, 1)\}$. Using the same argument as above we can conclude that α is rational and a quotient of two odd integers.

Thus any SRT with this expansion is quasiperiodic. It is more difficult in this case to determine the exact sets of digits which work since in general there is four-parameter family of possibilities.

The three remaining cases

$$\varphi = \begin{pmatrix} -2 & 0 \\ 0 & -2 \end{pmatrix}, \quad \begin{pmatrix} -2 & 1 \\ 0 & -2 \end{pmatrix}, \quad \text{or} \quad \begin{pmatrix} -2 & 0 \\ 0 & 2 \end{pmatrix},$$

are similar.

4.2. Digits for SRTs in the line. One nice way of working out which digits can be used is to use a bit of Fourier analysis.

Theorem 15 *Let $\varphi(x) = bx$ in the line, $b \in \mathbb{Z}$, $b > 1$. Let D be a set of b non-negative integers, with $0 \in D$. Then T_0, the associated tile, has open sets* *if and only if for all divisors $k > 1$ of b, either $\phi_k(x)|p_D(x)$ or there exists an $l = l(k) > 0$ such that*

$$\phi_b(x^{b^l/k})|p_D(x),$$

where

$$p_D(x) = \sum_D x^d,$$

and $\phi_k(x)$ is the minimal polynomial for the primitive kth roots of unity.

Proof: Consider the image of T_0 under the projection $\pi : \mathbf{R} \to \mathbb{R}/\mathbb{Z}$. If T_0 contains open sets, then for k large, $b^k T_0$ contains an interval of length at least 1. Thus $\pi(b^k T_0)$ covers all of the circle $\mathbf{R}/\mathbb{Z}$. On the other hand, since $b^k T_0$ is the union of translates of T_0, in which any two copies differ by an integral translation, each copy has the same image under π, so that a *single* copy of T_0 covers the circle.

What is more, the measure on T_0 projects to a constant multiple of Lebesgue measure on the circle. As in Theorem 10, this measure μ_D is an infinite convolution of δ-measures.

We have

(6) $$\mu_D = h_0 * h_1 * h_2 * \ldots$$

where

$$h_k = b^{-1} \sum_{d \in D} \delta(db^{-k} \bmod 1).$$

The coefficients in the Fourier series of h_k are

$$\hat{h}_k(n) = b^{-1} \sum_D x^{nd}, \quad x = e^{2\pi i b^{-k}}.$$

We see that the convolution (6) converges, with coefficients

(7) $$\hat{\mu}_D(n) = \prod_{k=1}^{\infty} b^{-1}\left(\sum_{d \in D} e^{2\pi i dn b^{-k}}\right)$$

Since μ_D is a constant multiple of Lebesgue measure, we must have that $\hat{\mu}_D(n) = 0$ for $n \neq 0$. This implies that for all $n \neq 0$ some term in the product (7) must be zero.

If we define $p_D(x) = \sum_D x^d$, then this means

(8) $$\forall n \exists k \quad p_D(e^{2\pi i n b^{-k}}) = 0.$$

This in turn means that, setting $n/b^k = r/s$ in lowest terms, that $\phi_s(x)|p_D(x)$.

If n and b^k are relatively prime, then this relation (8) implies $\phi_{b^k}(x)|p_D(x)$, and since

$$\phi_{b^k}(x) = \phi_b(x^{b^{k-1}}),$$

we have
$$\phi_b(x^{b^{k-1}})|p_D(x).$$
If $GCD(n,b) = m > 1$ then $n/b^k = (n/m)/(b^k/m)$ in lowest terms, so that (8) implies
$$\phi_{b^k/m}(x)|p_D(x).$$
Now it is easy to see that if $k > 1$, $\phi_{b^k/m}(x) = \phi_b(x^{b^{k-1}/m})$, so we see that the condition stated in the theorem is necessary. It is also sufficient: for any n, write $n = n'b^r$, where $n' < b$, and let $k = GCD(n', b)$. Then from
$$\phi_{b^l/k}(x)|p_D(x)$$
we see that
$$0 = p_D(e^{2\pi i k b^{-l}}) = p_D(e^{2\pi i n' b^{-l}}) = p_D(e^{2\pi i n b^{-l-r}}),$$
so that condition (8) is fulfilled.   □

Example (i) Let b be a prime (this question is asked in [16]). then
$$\phi_b(x^{b^l})|p_D(x)$$
implies (taking $l = 0$, which is the same as assuming not all digits have b as a factor)
$$1 + x + \ldots + x^{b-1}|p_D(x),$$
so that
$$p_D(x) = \sum_{k=0}^{b-1} x^{k+l_k b}, \quad l_k \in \mathbb{Z},$$
and hence D has a complete set of residues modulo b. If $l > 0$, then all the digits have a common factor b.

(ii) Let $b = 4$. Then
$$x^4 - 1 = (x-1)(x+1)(x^2+1),$$
and so there exists k_1, k_2 such that
$$(1 + x^{4^{k_1}})|p_D(x) \quad \text{and} \quad (1 + x^{2\cdot 4^{k_2}})|p_D(x),$$
these imply
$$p_D(x) = (1 + x^{4^{k_1}+2l_1 4^{k_1}})(1 + x^{2\cdot 4^{k_2}+l_2 4^{k_2+1}}), \quad l_i \in \mathbb{Z},$$
and if 4 does not divide $GCD(D)$, we can write $D = \{0, \alpha, \beta, \gamma\}$, where
$$\alpha \equiv 1 \bmod 2, \quad \beta \equiv 2 \cdot 4^k \bmod 4^{k+1}, \quad \gamma \equiv \alpha + \beta \bmod 4^{k+1}.$$
Thus for example $D = \{0, 1, 8, 9\}$ gives open sets, yet D has only two distinct residues modulo 4. If we shift each of these digits by -1, then $D' = \{-1, 0, 7, 8\}$ form a **simple** SRT, in the sense that every real has an expansion with these digits.

References

[1]   C. Bandt, Self-Similar Sets 5. Integer Matrices and fractal tilings of $\mathbf{R}^n$, Proc. AMS. **112** number 2 (1991), 549-562.

[2]   R. Bowen, *Equilibrium states and the ergodic theory of Anosov diffeomorphisms*, Lecture Notes in Mathematics **470**, Springer-Verlag, 1975.

[3]   F.M. Dekking, Recurrent Sets, Adv. in Math **44** (1982), 78-104.

[4]   F.M. Dekking, Replicating superfigures and endomorphisms of free groups, J. Combin. Th. Ser. A **32** (1982), 315-320.

[5]   K.J. Falconer, *The Geometry of Fractal Sets*, Cambridge University Press, 1985.

[6]   W. Gilbert, Radix Representations of Quadratic Fields, J. Math. Anal. and Appl. **83** (1981).

[7]   B. Grunbaum and G.C. Shephard, *Tilings and patterns*, W.H. Freeman and Company, 1987.

[8]   R. Kenyon, Self-Similar Tilings, thesis, Princeton University, 1990.

[9]   R. Kenyon, Rigidity of planar tilings, to appear, Invent. Math.

[10]   D.E. Knuth, *The Art of Computer Programming, vol 2.: Seminumerical Algorithms,* Addison Wesley, 1981.

[11]   D. W. Matula, Basic digit sets for radix representations, J. ACM. **29** number 1 (1982) 1131-1143.

[12]   A. M. Odlyzko, Non-negative digit sets in positional number systems, Proc. London Math. Soc. **37** (1978), 213-229.

[13]   R. Penrose, The role of aesthetics in pure and applied mathematical research, Bull. Inst. Math. Appl. **10** (1974), 266-271.

[14]   R. Robinson, Undecidability and non-periodicity of tilings in the plane, *Inventiones Math.* **12** (1971), 177-209.

[15]   Y. Sinai, Constructions of Markov partitions, *Func. Anal. and its Appl.* **2** (1968) no. 2., 70-80.

[16]   R. S. Strichartz, Wavelets and Self-Affine Tilings, preprint, Cornell Univ.

[17]   W.P. Thurston, Groups, Tilings, and Finite State Automata, *AMS Colloquium lectures*, 1990.

INSTITUT FOURIER, B.P 74,
38402 SAINT-MARTIN-D'HERES, FRANCE.

Contemporary Mathematics
Volume **135**, 1992

MARKOV SUBGROUPS OF $(\mathbb{Z}/2\mathbb{Z})^{\mathbb{Z}^2}$

BRUCE KITCHENS AND KLAUS SCHMIDT

ABSTRACT. We examine the closed translation invariant subgroups of $(\mathbb{Z}/2\mathbb{Z})^{\mathbb{Z}^2}$. There is a one-to-one correspondence between these subgroups and the ideals in the ring of Laurent polynomials $\mathbb{Z}/2\mathbb{Z}[x^{\pm 1}, y^{\pm 1}]$, and we relate many of the dynamical properties of the subgroup, acted on by translation, to algebraic properties of the associated ideal.

0. INTRODUCTION

A finite set $F \subseteq \mathbb{Z}^2$ defines a closed translation invariant subgroup, X_F, of the compact 0-dimensional group $(\mathbb{Z}/2\mathbb{Z})^{\mathbb{Z}^2}$ by letting $x \in X_F$ if the sum of the coordinates of x over every translate of F is zero. If σ_H is the horizontal shift on $(\mathbb{Z}/2\mathbb{Z})^{\mathbb{Z}^2}$ and σ_V is the vertical shift then $X_F = \{x \in (\mathbb{Z}/2\mathbb{Z})^{\mathbb{Z}^2} : \Sigma(\sigma_H^n \sigma_V^m x)_{(i,j)} = 0$ for all $(n,m) \in \mathbb{Z}^2\}$, where the summation extends over all $(i,j) \in F$. The shifts σ_H and σ_V are commuting automorphisms of X_F. We call $(X_F, \sigma_H, \sigma_V)$ or X_F a *Markov subgroup*. Theorem 2.1 states that every closed translation invariant subgroup of $(\mathbb{Z}/2\mathbb{Z})^{\mathbb{Z}^2}$ is defined by a finite number of such shapes (= finite subsets of $\mathbb{Z}^2$). The character group of $(\mathbb{Z}/2\mathbb{Z})^{\mathbb{Z}^2}$ is isomorphic to $\{F : F \subseteq \mathbb{Z}^2, F \text{ finite}\}$ and hence to the ring of Laurent polynomials $\mathbb{Z}/2\mathbb{Z}[x^{\pm 1}, y^{\pm 1}]$, and the automorphisms σ_H and σ_V dualize to multiplication by x and y, respectively. We prove in theorem 2.1 that the Markov subgroups of $(\mathbb{Z}/2\mathbb{Z})^{\mathbb{Z}^2}$ are in one-to-one correspondence with the ideals, their annihilators, in $\mathbb{Z}/2\mathbb{Z}[x^{\pm 1}, y^{\pm 1}]$. In view of this we have to relate the algebra of the ideal to the dynamics of the Markov subgroup. An example of this is proposition 2.12: X is ergodic if and only if its associated ideal is principal.

These Markov subgroups are an important special case of general expansive $\mathbb{Z}^d$ actions on compact topological groups. For a discussion of the general problem see [KS1]. We have singled out this special case, $(\mathbb{Z}/2\mathbb{Z})^{\mathbb{Z}^2}$, because it is the simplest case of an expansive $\mathbb{Z}^2$-action on a compact 0-dimensional group, it is very concrete, it displays much of the underlying algebraic complexity, there are many simple examples, and most of the basic questions are unanswered.

In section 1 we discuss the basics, Haar measure, directional entropy, two dimensional entropy, periodic points, etc.. In section 2 we set up the algebraic framework and prove some things about transitivity, ergodicity and algebraic factor maps. In section 3 we discuss the problem of higher order mixing. These examples were first introduced by F. Ledrappier [L] to get an example of a $\mathbb{Z}^2$ action that is mixing but not mixing of order three. We see that the mixing

1980 *Mathematics Subject Classification* (1985 *Revision*). 58F11 54H20.
Key words and phrases. Markov shift, $\mathbb{Z}^2$ action, automorphisms.
This paper is in final form and no version of it will be submitted for publication elsewhere.

properties are related to factoring polynomials in $\mathbb{Z}/2\mathbb{Z}[x^{\pm 1}, y^{\pm 1}]$ over $\overline{\mathbb{F}_2}$, the algebraic closure of $\mathbb{Z}/2\mathbb{Z}$. Here, the problem of determining the maximal order of mixing is unsolved. In section 4 we discuss the problems of topological conjugacy, measurable isomorphism and topological and measurable factors. Once again, the basic questions are unanswered. Section 5 is concerned with subshifts of Markov subgroups and consists mainly of examples and questions.

We would like to thank J. Ashley, D. Berend, F. Blanchard, D. Coppersmith, M. Keane, F. Ledrappier, and J. P. Thouvenot for helpful discussions.

1. Basics

Begin with the two point set $\{0, 1\}$ with the discrete topology. Let $\{0, 1\}^{\mathbb{Z}^2}$ be the space of all infinite arrays of 0's and 1's with the product topology. It is a compact, totally disconnected, metric space. The horizontal shift, defined by $(\sigma_H(x))_{(i,j)} = x_{(i+1,j)}$, and the vertical shift defined by $(\sigma_V(x))_{(i,j)} = x_{(i,j+1)}$ are commuting homeomorphisms of $\{0, 1\}^{\mathbb{Z}^2}$. For a finite set $F \subseteq \mathbb{Z}^2$ we define the projection map $\pi_F : \{0, 1\}^{\mathbb{Z}^2} \to \{0, 1\}^F$ by setting $(\pi_F(x))_{(i,j)} = x_{(i,j)}$ for $(i, j) \in F$. For a finite subset $L \subseteq \{0, 1\}^F$ define the two dimensional subshift of finite type (or Markov shift) X_L as $\{x \in \{0, 1\}^{\mathbb{Z}^2} : \pi_F(\sigma_H^n \sigma_V^m(x)) \in L$ for all $(n, m) \in \mathbb{Z}^2\}$. This is a closed, translation invariant subspace of $\{0, 1\}^{\mathbb{Z}^2}$, and we furnish it with the subspace topology.

Example 1.1. Let F be a 3×3 square and L consist of the 3×3 blocks with exactly four 1's. This seemingly has no ergodic measures with full support.

Example 1.2. (two dimensional golden mean): Let F be a 2×2 square and L be the 2×2 squares with no horizontally or vertically adjacent 1's. Here it seems that no formula is known for the two dimensional entropy.

There are a number of difficulties that come up in dealing with these two dimensional subshifts of finite type (see for example [MP]).

Theorem 1.3. (Berger 1966 [B])

(i) There exist nonempty two-dimensional subshifts of finite type that contain no periodic points (points with finite orbits under σ_H and σ_V).

(ii) Given a finite list L the problem of determining whether X_L is empty or not is undecidable.

We are interested in a special class of these two dimensional subshifts of finite type. To define these we think of $\{0, 1\}$ as $\mathbb{Z}/2\mathbb{Z}$. Then $(\mathbb{Z}/2\mathbb{Z})^{\mathbb{Z}^2}$ under coordinate by coordinate addition is a compact 0-dimensional topological group, and the transformations σ_H and σ_V are automorphisms. A closed shift invariant subgroup of $(\mathbb{Z}/2\mathbb{Z})^{\mathbb{Z}^2}$ will be called a *Markov subgroup*. This is the class of subshifts we will examine. Let $F \subseteq \mathbb{Z}^2$ be a finite set, define a homomorphism (group character) $\chi_F : (\mathbb{Z}/2\mathbb{Z})^{\mathbb{Z}^2} \to \mathbb{Z}/2\mathbb{Z}$ by $\chi_F(x) = \sum_{(i,j) \in F} x_{(i,j)}$. Let $X_F = \bigcap_{(n,m) \in \mathbb{Z}^2} \{x \in (\mathbb{Z}/2\mathbb{Z})^{\mathbb{Z}^2} : \chi_F(\sigma_H^n \sigma_H^m x) = 0\}$. If we think of F as a shape, then the Markov subgroup X_F consists of the points where the sum over the entries of F is zero (modulo 2) no matter where you place F on x. Markov subgroups were first defined by F. Ledrappier [L].

Examples 1.4. F is:

Let X be a Markov subgroup. Then it has a unique Haar measure, μ, which is invariant under σ_H and σ_V. Suppose that $E \subseteq \mathbb{Z}^2$ is a finite set and $\pi_E : X \to (\mathbb{Z}/2\mathbb{Z})^E$ is the projection map. Then $\pi_E(X) \subseteq (\mathbb{Z}/2\mathbb{Z})^E$ is a subgroup. Let $k_E = {}^\#\pi_E(X)$, it will be a power of two. If $A \in \pi_E(X)$, then $A_E = \{x \in X : \pi_E(x) = A\}$, and $\mu(A_E) = 1/k_E$. Let $F \subseteq \mathbb{Z}^2$ be another finite set, denote by $0 \in \pi_E(X)$ the element of all zeros, and note that $\pi_F(0_E) \subseteq \pi_F(X)$ is a subgroup. For $A \in \pi_E(X)$, $\pi_F(A_E)$ is a coset of $\pi_F(0_E)$ in $\pi_F(X)$, and ${}^\#\pi_{E \cup F}(X) = k_E \cdot {}^\#\pi_F(0_E)$. If $A \in \pi_E(X)$, $B \in \pi_F(X)$, and $A_E \cap B_F \neq \emptyset$, then $\mu(A_E \cap B_F) = \mu(A_E)\big({}^\#\pi_F(0_E)\big)^{-1} = \big(k_E \cdot {}^\#\pi_F(0_E)\big)^{-1}$.

Definition 1.5. Finite sets $E, F \subseteq \mathbb{Z}^2$ are *independent* for X if $\pi_F(0_E) = \pi_F(X)$ or, equivalently, if $\pi_E(0_F) = \pi_E(X)$. A finite collection of finite sets $E_1, \cdots, E_k \subseteq \mathbb{Z}^2$ is independent if, for each partition $\{i_1, \cdots, i_m\}, \{j_1, \cdots, j_n\}$ of $\{1, \cdots, k\}$, the sets $E_{i_1} \cup \ldots \cup E_{i_m}$ and $E_{j_1} \cup \ldots \cup E_{j_n}$ are independent.

The next observation is a consequence of the previous discussion.

Observation 1.6. *For finite* $E, F \subseteq \mathbb{Z}^2$, *$E$ and F are independent if and only if, for any* $A \in \pi_E(x)$, $B \in \pi_F(x)$, $\mu(A_E \cap B_F) = \mu(A_E) \cdot \mu(B_F)$. *If E and F are not independent, then either* $\mu(A_E \cap B_F) = 0$ *or* $\mu(A_E \cap B_F) > \mu(A_E) \cdot \mu(B_F)$.

Let X_F be a Markov subgroup defined by a finite set $F \subseteq \mathbb{Z}^2$ and let $(\alpha, \beta) \in \mathbb{Z}^2$ with $(\alpha, \beta) \neq (0,0)$, $gcd(\alpha, \beta) = 1$. The map $\sigma_H^\alpha \sigma_V^\beta : X_F \to X_F$ is an automorphism of a compact totally disconnected group. We recall three results on dynamical properties of such automorphisms.

Theorem 1.7 [H]. *(X, T, μ) is ergodic if and only if $(X^\wedge, T^\wedge)$, the character group and induced automorphism, has no nontrivial periodic orbits.*

Theorem 1.8 [Li], [MT]. *(X, T, μ) is ergodic if and only if it is Bernoulli.*

Theorem 1.9 [K]. *If X is zero-dimensional and T is expansive then (X, T) is topologically conjugate to $(G \times \Lambda, \tau \times \sigma)$ where (Λ, σ) is a full shift, (G, τ) is an automorphism of a finite group and $\tau \times \sigma$ is the product automorphism.*

Given a shape (i.e. a finite set) $F \subseteq \mathbb{Z}^2$, we define the convex hull, $C(F)$, as a subset of $\mathbb{R}^2$ in the usual way. It is a solid polygon. To each face of the polygon we associate a vector in $\mathbb{Z}^2$ by going clockwise around the polygon and making each face a vector that we are traversing from tail to head. For any shape, the sum of these vectors is zero. We assign to each face f its *size* $s = s(f)$ by taking the vector $v_f = (\alpha, \beta)$ corresponding to the face f and letting $s = gcd\{\alpha, \beta\} + 1$. Geometrically, the size of a face is the number of lattice points that lie on the face in the convex hull. For $(\alpha, \beta) \neq 0$ let $\ell_{(\alpha, \beta)}$ be the line through $(0, 0)$ and (α, β) We say that F has *width* w in the (α, β) direction if

$$w = {}^\#\{\ell' \in \ell_{(\alpha, \beta)} + \mathbb{Z}^2 : \ell' \cap C(F) \neq \emptyset\}.$$

Example 1.10. For 1.4a there are three faces each of size 2 in the $(-1, 0), (0, 1)$, and $(1, -1)$ directions. It has width 2 in the $(0, 1)$ direction and width 3 in the $(1, 1)$ direction.

Proposition 1.11. *If F has width w in direction (α, β), where $(\alpha, \beta) \neq (0, 0)$ and $gcd(|\alpha|, |\beta|) = 1$, then either:*

(i) *F doesn't have a face in the (α, β) or $(-\alpha, -\beta)$ directions in which case $\sigma_H^\alpha \sigma_V^\beta$ is expansive on X_F and $(X_F, \sigma_H^\alpha \sigma_V^\beta)$ is topologically conjugate to a full 2^{w-1} shift; or*

(ii) *F has a face in one of the directions (α, β) or $(-\alpha, -\beta)$, in which case $\sigma_H^\alpha \sigma_V^\beta$ is not expansive on X_F, and $(X_F, \sigma_H^\alpha \sigma_V^\beta)$ has entropy $\log 2^{w-1}$.*

Proof. By applying an element of $GL(2, \mathbb{Z})$ we may assume that $(\alpha, \beta) = (1, 0)$. For each $k, \ell \geq 0$, let $\pi_{-k}^\ell : X_F \to (\mathbb{Z}/2\mathbb{Z})^{\mathbb{Z} \times \{-k, \dots, \ell\}}$ be the projection map $\left(\pi_{-k}^\ell(x) \right)_{(i,j)} = x_{(i,j)}$ for $(i, j) \in \mathbb{Z} \times \{-k, \dots, \ell\}$. For all k, ℓ, σ_H acts expansively on $\pi_{-k}^\ell(X_F)$, so by theorem 1.9 $\left(\pi_{-k}^\ell(X_F), \sigma_H \right)$ is topologically conjugate to an automorphism of a finite group cross a full shift. This means (X_F, σ_H) is topologically conjugate to $\varprojlim(G_k \times \Lambda_k, \tau_k \times \sigma_k)$, an inverse limit of such systems. We also see that $\pi_0^{w-2}(X_F)$ consists of all possible "strips" of 0's and 1's of width $w - 1$, i.e. that $\pi_0^{w-2}(X_F)$ is the full 2^{w-1} shift.

For case (i), notice that if we place F in a strip of width $w - 1$ as well as possible, one element, a "corner" of F, will stick out of either the bottom or the top. In $\mathbb{Z}^2$ put down an arbitrary strip of width $w - 1$ in the subset $\mathbb{Z} \times \{0, \dots, w - 2\}$. Place F so that a corner sticks out above the strip. Because all but one of the coordinates of F are specified there is a unique solution for the entry above the strip. Sliding F along shows that a specified strip on $\mathbb{Z} \times \{0, \dots, w - 2\}$ determines a strip on $\mathbb{Z} \times \{0, \dots, w - 1\}$. We may do the same working downwards. We see that each strip on $\mathbb{Z} \times \{0, \dots, w - 2\}$ determines a point in X_F. This means for all $\ell \geq w - 2, k \geq 0$, $\left(\pi_{-k}^\ell(X_F), \sigma_H \right)$ is topologically conjugate to a full 2^{w-1} shift. An examination of example 1.4a in the $(1,1)$ direction will make this argument clear: it is a 2^2 shift.

For case (ii), suppose F has a face on the bottom but not on the top, as in example 1.4a. When we place F in a strip of width $w - 1$ as well as possible either a corner will stick out of the top or a face will stick out the bottom. Since a corner sticks out the top we see that the projection map $\pi_0^\ell(X_F) \to \pi_0^{w-2}(X_F)$ for $\ell \geq w - 2$ is a topological conjugacy. In fact for any k, ℓ sufficiently large $\pi_{-k}^\ell(x_F)$ is a full 2^{w-1} shift. But, notice that the map $\pi_{-1}^{w-2}(X_F) \to \pi_0^{w-2}(X_F)$ is not a conjugacy, it is a 2^{s-1} to 1 factor map, where s is the size of the bottom face. This follows because when we fix the coordinates $\mathbb{Z} \times \{0, \dots, w - 2\}$ and go to extend this to the $\mathbb{Z} \times \{-1, \dots, w - 2\}$ coordinates, we have some choice. We are free to choose $s - 1$ consecutive entries and then the entire row is determined. In this case we see that $\varprojlim(\pi_{-k}^k(X_F), \sigma_H)$ is an inverse limit of 2^{w-1} shifts where each bonding map is of degree 2^{s-1}. This means (X_F, σ_H) is not expansive but has entropy $\log 2^{w-1}$. It is clearly not expansive because this choice we have extending downward means we can find two different points in x_F that agree on arbitrarily large horizontal strips.

The other case of (ii) is where F has horizontal faces on the top and bottom, as in example 1.4e. Now when extending up or down we have a choice and $\pi^{\ell}_{-k}(X_F)$ need not be conjugate to a full shift. By theorem 1.9 it must be a finite group automorphism cross a full shift. The map $\pi^{w-1}_{-1}(X_F) \to \pi^{w-2}_0(X_F)$ is $2^{s-1} \cdot 2^{s'-1}$ to one, where s is the size of the bottom face and s' the size of the top face. This means $\pi^{\ell}_{-k}(X_F)$ is always a finite group automorphism cross a 2^{w-1} shift. Because of the inverse limit, (X_F, σ_H) will have entropy $\log 2^{w-1}$ and is not expansive. $\square$

Corollary 1.12. *If F is not linear, every directional entropy is positive.*

In the previous construction it is useful to examine the image of Haar measure under the projection maps, π^{ℓ}_{-k}. If F has one or no faces in the $(\pm 1, 0)$ direction then $\pi^{\ell}_{-k}(X_F)$ is a conjugate to a full shift and the image of Haar measure is the equidistributed Bernoulli measure. In this case, (X_F, σ_H) is an inverse limit of Bernoulli shifts and by [O] we know that an inverse limit of Bernoulli's is Bernoulli. If F has two faces in the $(\pm 1, 0)$ direction then $\pi^{\ell}_k(X_F)$ is conjugate to $(G \times \Lambda, \tau \times \sigma)$. The image of Haar measure on the finite group G is the equidistributed measure, the image of Haar measure on the full shift Λ is the equidistributed Bernoulli measure, and the image on $G \times \Lambda$ is the product of these. If the G is trivial for all, π^{ℓ}_k, then again (X_F, σ_H) is Bernoulli, but if G is not trivial for some π^{ℓ}_{-k} then (X_F, σ_H) is not ergodic. In section 2 we will identify exactly the directions where $\sigma^{\alpha}_H \sigma^{\beta}_V$ is not ergodic. Notice there can be at most a finite number of these directions. We will see that example 1.4b is Bernoulli in every direction but example 1.4e is not ergodic in the $(1,0)$ or $(0,1)$ directions.

Proposition 1.13. *If $F \neq \emptyset$, then the two dimensional topological entropy of X_F is equal to zero.*

Proof. We want to count the number of elements in $\pi_{\overline{N}}(X_F)$ where $\overline{N}$ is an $N \times N$ square. For large N the projection of the point $x \in X_F$ to the boundary

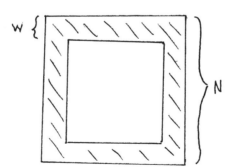

of width w of the square $\overline{N}$ determines x completely in $\overline{N}$. Therefore, $^{\#}\pi_{\overline{N}}(X_F)$ is bounded by 2^{4wN}, and the entropy, $h(X_F)$, is equal to

$$\lim_N \frac{1}{N^2} \log {}^{\#}\pi_{\overline{N}}(X_F) \leq \lim_N \frac{1}{N^2} \log 2^{4wN} = 0. \quad \square$$

Next we examine the periodic points of a Markov subgroup. We say that a point $x \in X_F$ is *periodic in the* (α, β) *direction* if $(\sigma_H^\alpha \sigma_V^\beta)^p(x) = x$ for some $p \neq 0$. We say that $x \in X$ is *periodic* if the orbit of x under σ_H and σ_V is finite.

Proposition 1.14. *For any finite* $F \subseteq \mathbb{Z}^2$ *the periodic points of* X_F *are dense.*

Proof. If $F = \emptyset$ this is obvious. If $F \neq \emptyset$ then by proposition 1.11 we may choose $(\alpha, \beta) \in \mathbb{Z}^2$ so that $\sigma_H^\alpha \sigma_V^\beta$ is expansive on X_F, and $(X_F, \sigma_H^\alpha \sigma_V^\beta)$ is topologically conjugate to a full 2^{w-1} shift. This means that the periodic points for $\sigma_H^\alpha \sigma_V^\beta$ are dense in X_F and that there are a finite number of points of each period. The automorphism $\sigma_H^\alpha \sigma_V^{\beta+1}$ permutes the points of period p for $\sigma_H^\alpha \sigma_V^\beta$. Each will have a finite orbit for $\sigma_H^\alpha \sigma_V^{\beta+1}$. Every periodic point for $\sigma_H^\alpha \sigma_V^\beta$ will be periodic for $\sigma_H^\alpha \sigma_V^{\beta+1}$, which means it will have a finite orbit under σ_H and σ_V. $\square$

This proposition is a special case of a theorem about expansive $\mathbb{Z}^d$ actions on compact topological groups [KS1]. In particular, every Markov subgroup $X \subseteq (\mathbb{Z}/2\mathbb{Z})^{\mathbb{Z}^2}$ has a dense set of periodic points.

2. Algebra

Here we study the connection between Markov subgroups and polynomial rings. We start with the group $(\mathbb{Z}/2\mathbb{Z})^{\mathbb{Z}^2}$ and observe that every group character χ is defined by a finite subset $S(\chi) \subseteq \mathbb{Z}^2$ so that for $x \in (\mathbb{Z}/2\mathbb{Z})^{\mathbb{Z}^2}$, $\chi(x) = \Sigma x_{(i,j)}$, where the sum is over $(i,j) \in S(\chi)$. The character group $((\mathbb{Z}/2\mathbb{Z})^{\mathbb{Z}^2})^\wedge$ of $(\mathbb{Z}/2\mathbb{Z})^{\mathbb{Z}^2}$, is isomorphic to $\oplus_{\mathbb{Z}^2} \mathbb{Z}/2\mathbb{Z}$, which can be identified with the finite subsets of $\mathbb{Z}^2$ and hence with the Laurent polynomial ring $\mathbb{Z}/2\mathbb{Z}[x^{\pm 1}, y^{\pm 1}]$, where each finite subset $S \subseteq \mathbb{Z}^2$ corresponds to the polynomial with support S. We will think of $((\mathbb{Z}/2\mathbb{Z})^{\mathbb{Z}^2})^\wedge$ as $\mathbb{Z}/2\mathbb{Z}[x^{\pm 1}, y^{\pm 1}]$ and write $S(p) \subseteq \mathbb{Z}^2$ for the support of a polynomial $p \in \mathbb{Z}/2\mathbb{Z}[x^{\pm 1}, y^{\pm 1}]$. The automorphisms σ_H and σ_V on $(\mathbb{Z}/2\mathbb{Z})^{\mathbb{Z}^2}$ induce automorphisms of $\mathbb{Z}/2\mathbb{Z}[x^{\pm 1}, y^{\pm 1}]$ consisting of multiplication by x and y, respectively. In the terminology we have been using, a shape $F \subseteq \mathbb{Z}^2$ corresponds to a polynomial, p_F, and

$$X_F = \{z \in (\mathbb{Z}/2\mathbb{Z})^{\mathbb{Z}^2} : p_F(\sigma_H^i \sigma_V^j z) = 0 \text{ for all } (i,j) \in \mathbb{Z}^2\}$$
$$= \{z \in (\mathbb{Z}/2\mathbb{Z})^{\mathbb{Z}^2} : (x^i y^j p_F)(z) = 0 \text{ for all } (i,j) \in \mathbb{Z}^2\}.$$

If we let $\langle p_F \rangle$ denote the ideal in $\mathbb{Z}/2\mathbb{Z}[x^{\pm 1}, y^{\pm 1}]$ generated by p_F we see that $X_F = \langle p_F \rangle^\perp$, the annihilator of the ideal $\langle p_F \rangle$. More generally, if $I \subseteq \mathbb{Z}/2\mathbb{Z}[x^{\pm 1}, y^{\pm 1}]$ is an ideal, then $I^\perp \subseteq (\mathbb{Z}/2\mathbb{Z})^{\mathbb{Z}^2}$ is a closed, translation invariant, subgroup. Conversely, if $X \subseteq (\mathbb{Z}/2\mathbb{Z})^{\mathbb{Z}^2}$ is a closed, translation invariant, subgroup then its annihilator $X^\perp \subseteq \mathbb{Z}/2\mathbb{Z}[x^{\pm 1}, y^{\pm 1}]$ is an ideal. Since every ideal in $\mathbb{Z}/2\mathbb{Z}[x^{\pm 1}, y^{\pm 1}]$ is finitely generated, every Markov subgroup X is of the form $X = X_{F_1} \cap \ldots \cap X_{F_n}$ for a finite collection of shapes $\{F_1, \ldots, F_n\}$.

Theorem 2.1. *The Markov subgroups of* $(\mathbb{Z}/2\mathbb{Z})^{\mathbb{Z}^2}$ *are in one-to-one correspondence with the ideals in* $\mathbb{Z}/2\mathbb{Z}[x^{\pm 1}, y^{\pm 1}]$. *Any Markov subgroup is the intersection of finitely many subgroups defined by shapes.*

From now on we will be dealing with ideals $I \subseteq \mathbb{Z}/2\mathbb{Z}[x^{\pm 1}, y^{\pm 1}]$. The two operations on ideals we will use are addition, where $I + J$ is the ideal generated by the sums of elements in I and in J, and intersection $I \cap J$. For basic properties of ideals we refer to [AM], and for duality theory to [HR]. We have that $\left((\mathbb{Z}/2\mathbb{Z})^{\mathbb{Z}^2}\right)^\wedge \simeq \mathbb{Z}/2\mathbb{Z}[x^{\pm 1}, y^{\pm 1}]$ and $\mathbb{Z}/2\mathbb{Z}[x^{\pm 1}, y^{\pm 1}]^\wedge \simeq (\mathbb{Z}/2\mathbb{Z})^{\mathbb{Z}^2}$. For I, an ideal, $I^\perp$ is a Markov subgroup and $(I^\perp)^\wedge \simeq \mathbb{Z}/2\mathbb{Z}[x^{\pm 1}, y^{\pm 1}]/I$. Furthermore, $(I + J)^\perp = I^\perp \cap J^\perp$ and $(I \cap J)^\perp = I^\perp + J^\perp$. Our aim now is to make a correspondence between algebraic properties of ideals $I \subseteq \mathbb{Z}/2\mathbb{Z}[x^{\pm 1}, y^{\pm 1}]$ and dynamical properties of the associated Markov subgroups.

Let I be an ideal, and $I^\perp \simeq (\mathbb{Z}/2\mathbb{Z}[x^{\pm 1}, y^{\pm 1}]/I)^\wedge$ the associated Markov subgroup.

Proposition 2.2. *A continuous shift commuting homomorphism, φ, from $I^\perp$ into*
$(\mathbb{Z}/2\mathbb{Z})^{\mathbb{Z}^2}$ *is defined by a polynomial q with $\varphi(z)_{(i,j)} = q\bigl(\sigma_H^i \sigma_V^j(z)\bigr)$.*

Proof. The map $\varphi : I^\perp \to \mathbb{Z}/2\mathbb{Z}$ defined by $z \mapsto \varphi(z)_{(0,0)}$ is a group character, so is defined by a polynomial q. The rest follows because φ is shift commuting. $\square$

Our notation will be to use q interchangeably as a polynomial, a group character, and a map on Markov subgroups.

Proposition 2.3. *Let I be an ideal and $q \in \mathbb{Z}/2\mathbb{Z}[x^{\pm 1}, y^{\pm 1}]$. We have an exact sequence*
$$0 \to (I + \langle q \rangle)^\perp \to I^\perp \overset{q}{\to} J^\perp \to 0,$$
where $J = \{r : rq \in I\}$.

Proof. $(I + \langle q \rangle)^\perp = I^\perp \cap \langle q \rangle^\perp$ which is clearly the kernel of q. Next observe that
$$q(I^\perp)^\perp = \{r : rq(x) = 0 \ \forall x \in I^\perp\} = \{r : rq \in I\}. \quad \square$$

Lemma 2.4. $0 \to \langle q \rangle^\perp \cap \langle p \rangle^\perp \to \langle p \rangle^\perp \overset{q}{\to} \langle p \rangle^\perp \to 0$ *is exact if and only if $\langle q \rangle^\perp \cap \langle p \rangle^\perp = \langle p, q \rangle^\perp$ is finite.*

Proof. By proposition 1.11 we may pick $\alpha, \beta \in \mathbb{Z}$, with $gcd(\alpha, \beta) = 1$, and so that $(\langle p \rangle^\perp, \sigma_H^\alpha \sigma_V^\beta)$ is topologically conjugate to a full shift. The polynomial q defines a shift commuting map from $(\langle p \rangle^\perp, \sigma_H^\alpha \sigma_V^\beta)$ to itself. A well known theorem of Hedlund [He] asserts that this is onto if and only if it is finite-to-one. $\square$

Lemma 2.5. $\langle p \rangle^\perp \cap \langle q \rangle^\perp = \langle p, q \rangle^\perp$ *is finite if and only if $gcd(p, q) = 1$.*

Proof. Consider the exact sequence $0 \to \langle q \rangle^\perp \cap \langle p \rangle^\perp \to \langle p \rangle^\perp \to J^\perp \to 0$. The ideal $J = \{r : rq \in \langle p \rangle\}$ is equal to $\langle p \rangle$ if and only if $gcd(p, q) = 1$. Now apply lemma 2.4. $\square$

For an ideal $I \subseteq \mathbb{Z}/2\mathbb{Z}[x^{\pm 1}, y^{\pm 1}]$ we define $gcd(I)$, as expected, to be the highest common factor of all elements of I.

Lemma 2.6. *If $q = gcd(I)$ then $\langle q \rangle/I$ and $I^\perp/\langle q \rangle^\perp$ are finite.*

Proof. Since $(I^\perp/\langle q \rangle^\perp)^\wedge \simeq \langle q \rangle/I$, either both are finite or both are infinite. We use induction on the number of generators for I. If $I = \langle f_1, f_2 \rangle$, $gcd(I) = q, f_1 = qp_1, f_2 = qp_2$, then $\langle f_1, f_2 \rangle = \langle q \rangle \langle p_1, p_2 \rangle \subseteq \langle q \rangle$ and
$$0 \to \langle q \rangle^\perp \to \langle f_1, f_2 \rangle^\perp \overset{q}{\to} \langle p_1, p_2 \rangle^\perp \to 0.$$

By lemma 2.5, $\langle p_1, p_2 \rangle^\perp \simeq \langle f_1, f_2 \rangle^\perp / \langle q \rangle^\perp$ is finite.

Next suppose the lemma is true when the number of generators is n, and $I = \langle f_1, \ldots, f_{n+1} \rangle$. Let $q' = gcd(\langle f_1, \ldots, f_n \rangle)$ and $q = gcd(q', f_{n+1}) = gcd(I)$. By the induction hypothesis $\langle q' \rangle / \langle f_1, \ldots, f_n \rangle$ and $\langle q \rangle / \langle q', f_{n+1} \rangle$ are finite. Note that

$$\langle q', f_{n+1} \rangle / \langle f_1, \ldots, f_{n+1} \rangle \simeq \langle q' \rangle + \langle f_{n+1} \rangle / \langle f_1, \ldots, f_n \rangle + \langle f_{n+1} \rangle$$

has cardinality less than or equal to the cardinality of $\langle q' \rangle / \langle f_1, \ldots, f_n \rangle$. As a set, $\langle q \rangle / I \simeq \langle q \rangle / \langle q', f_{n+1} \rangle \times \langle q', f_{n+1} \rangle / I$ and so is finite. $\square$

Every Markov subgroup $X \subseteq (\mathbb{Z}/2\mathbb{Z})^{\mathbb{Z}^2}$ has a unique minimal open shift invariant subgroup, which we call the *irreducible component of the identity*. It is equal to the maximal subgroup of X with a dense orbit. If X is the irreducible component of the identity then X is *transitive* or *irreducible*.

Proposition 2.7. *If $q = gcd(I)$ then $\langle q \rangle^\perp$ is the irreducible component of the identity.*

Proof. $\langle q \rangle^\perp$ is an invariant subgroup of $I^\perp$. It is irreducible because there is an $(\alpha, \beta) \in \mathbb{Z}^2$ so that $(\langle q \rangle^\perp, \sigma_H^\alpha \sigma_V^\beta)$ is conjugate to a full shift (proposition 1.11). The group $I^\perp / \langle q \rangle^\perp$ is a finite homomorphic image of $I^\perp$ with an induced $\mathbb{Z}^2$ action. $\square$

Proposition 2.8. *$I^\perp$ is transitive if and only if I is principal.*

Proof. This follows immediately from proposition 2.7. $\square$

Proposition 2.9. *$I^\perp$ is infinite if and only if $gcd(I) \neq 1$.*

Proof. If $q = gcd(I)$ then $I^\perp / \langle q \rangle^\perp$ is finite by lemma 2.6. If $q = 1, I^\perp \simeq I^\perp / \langle q \rangle^\perp$. If $q \neq 1$ then $\langle q \rangle^\perp \subseteq I^\perp$ is infinite. $\square$

Remark 2.10. With reference to Lemma 2.4 we note that it is not true in general that

$$0 \to \langle q \rangle^\perp \cap I^\perp \to I^\perp \overset{q}{\to} I^\perp \to 0$$

if and only if $\langle q \rangle^\perp \cap I^\perp$ is finite. To see this let p_1, p_2, q be non-trivial polynomials with $gcd(p_1, p_2 q) = 1$. Then

$$0 \to \langle p_1 \rangle^\perp \cap \langle p_1 q, p_2 q \rangle^\perp \to \langle p_1 q, p_2 q \rangle^\perp \overset{p_1}{\to} \langle q \rangle^\perp \to 0$$

and $\langle p_1 \rangle^\perp \cap \langle p_1 q, p_2 q \rangle^\perp \subseteq \langle p_1, p_2 \rangle^\perp$ is finite by Proposition 2.9.

Proposition 2.11. *Let $p = p_1 \cdots p_l$ with p_i irreducible over $\mathbb{Z}/2\mathbb{Z}$. Then $\langle p \rangle^\perp$ is not ergodic in the (n, m) direction if and only if some p_i is a polynomial in $x^n y^m$.*

Proof. By theorem 1.7, $\langle p \rangle^\perp$ is nonergodic in the (n, m) direction if and only if there is a character with a finite orbit under $\sigma_H^n \sigma_V^m$. This means there exists a $q \in \mathbb{Z}/2\mathbb{Z}[x^{\pm 1}, y^{\pm 1}]$, $q \notin \langle p \rangle$ and $t > 0$ such that $q(1 + (x^n y^m)^t) \in \langle p \rangle$, so that $\sigma_H^n \sigma_V^m$ is nonergodic if and only if $1 + (x^n y^m)^t$ is either in $\langle p \rangle$ or a zero divisor for $\langle p \rangle$, i.e. if and only if one of the p_i's divides $1 + (x^n y^m)^t$. If a polynomial p_i divides $1 + (x^n y^m)^t$ then it is a polynomial in $\mathbb{Z}/2\mathbb{Z}[x^n y^m]$. Conversely, if p_i is in $\mathbb{Z}/2\mathbb{Z}[x^n y^m]$ then it divides $1 + (x^n y^m)^t$ for some t. $\square$

Proposition 2.12. $I^\perp$ *is ergodic if and only if I is principal.*

Proof. If I is principal then by proposition 2.11 there are ergodic directions. If I is not principal we set $q = gcd(I)$ and note that $I^\perp/\langle q \rangle^\perp$ is a non-trivial, non-ergodic, finite factor. Hence $I^\perp$ is not ergodic. $\square$

Proposition 2.13. $I^\perp$ *is infinite and contains no proper infinite Markov subgroups if and only if $I = \langle p \rangle$ is principal for some irreducible polynomial $p \neq 1$.*

Proof. If I is not principal, let $q = gcd(I)$. If $q = 1$, $I^\perp$ is finite by proposition 2.9. If $q \neq 1$, then $\langle q \rangle^\perp$ is a proper, infinite, Markov subgroup of $I^\perp$.

If $I = \langle p \rangle$ and p factors as $p = p_1 \cdots p_n$ then $\langle p \rangle \subseteq \langle p_i \rangle$ and $\langle p_i \rangle^\perp$ is a proper infinite Markov subgroup of $I^\perp$. Conversely, suppose $I^\perp$ contains a proper infinite Markov subgroup $J^\perp$. If $q' = gcd(J)$, then $\langle q' \rangle^\perp \subseteq J^\perp \subseteq I^\perp$ is infinite. This means $I \subseteq \langle q' \rangle$ and q' divides $gcd(I)$. $\square$

Proposition 2.14. $\langle p \rangle^\perp$ *is an ergodic subgroup of $I^\perp$ if and only if p divides $gcd(I)$.*

Proof. Let $q = gcd(I)$. If p divides q then $\langle p \rangle^\perp \subseteq \langle q \rangle^\perp \subseteq I^\perp$. If $\langle p \rangle^\perp$ is an ergodic subgroup of $I^\perp$ then $I \subseteq \langle p \rangle$ and p divides $gcd(I)$. $\square$

Remark. There may be non-ergodic infinite subgroups. For example, let $I = \langle q_1^2 q_2 \rangle$ and $J = \langle q_1^2, q_1 q_2 \rangle$. Then I is a proper subideal of J, $J^\perp \subseteq I^\perp$, and $gcd(J)$ divides $gcd(I)$.

Proposition 2.15. $I^\perp \cap J^\perp = \{0\}$ *if and only if I and J are relatively prime.*

Proof. $I^\perp \cap J^\perp = (I + J)^\perp = \{0\}$ if and only if $I + J = \mathbb{Z}/2\mathbb{Z}[x^{\pm 1}, y^{\pm 1}]$ which is the definition of relatively prime. $\square$

Proposition 2.16. $I^\perp \cap J^\perp$ *is infinite if and only if $gcd(I, J) \neq 1$.*

Proof. Let $q = gcd(I, J) = gcd(I + J)$. Then $I^\perp \cap J^\perp = (I + J)^\perp$ is infinite if and only if $q \neq 1$ by proposition 2.9. $\square$

3. MIXING

This section is concerned with the mixing properties of ergodic subgroups, $\langle p \rangle^\perp$, which motivated F. Ledrappier's original example [L]. The subgroup $\langle p \rangle^\perp$ is *mixing of order n* (or *n-mixing*) if, for any n measurable sets $U_1, \ldots, U_n$, $\mu\left(\sigma_H^{r_1} \sigma_V^{s_1}(U_1) \cap \ldots \cap \sigma_H^{r_n} \sigma_V^{s_n}(U_n)\right)$ goes to $\Pi_i \mu(U_i)$ as $d\left((r_i, s_i), (r_j, s_j)\right)$ goes to infinity.

The subgroup $\langle p \rangle^\perp$ is *mixing on a shape $E = \{(r_1, s_1), \ldots, (r_n, s_n)\} \subseteq \mathbb{Z}^2$* if for any n measurable sets $U_1, \ldots, U_n$ in $\langle p \rangle^\perp$, $\mu\left((\sigma_H^{r_1} \sigma_V^{s_1})^t(U_1) \cap \ldots \cap (\sigma_H^{r_n} \sigma_V^{s_n})^t(U_n)\right)$ goes to $\Pi \mu(U_i)$ as t goes to infinity.

The subgroup $\langle p \rangle^\perp$ is *mixing on $E = \{(r_1, s_1), \ldots, (r_n, s_n)\}$ at times $a2^{lt}$* if, for any n measurable sets $U_1, \ldots, U_n$, $\mu\left((\sigma_H^{r_1} \sigma_V^{s_1})^{a2^{lt}}(U_1) \cap \ldots \cap (\sigma_H^{r_n} \sigma_V^{s_n})^{a2^{lt}}(U_n)\right)$ goes to $\Pi \mu(U_i)$ as t goes to infinity. We sometimes say *mixing on a polynomial q* instead of on a set, when $S(q) = E$.

Mixing properties with respect to Haar measure are tied to the independence properties of subsets of $\mathbb{Z}^2$ as discussed in definition 1.5 and observation 1.6. Let $\overline{N} \subseteq \mathbb{Z}^2$ be the $N \times N$ square, $E = \{(r_1, s_1), \ldots, (r_n, s_n)\} \subseteq \mathbb{Z}^2$, $A_1, \ldots, A_n \in \pi_{\overline{N}}(\langle p \rangle^\perp)$, and consider, for each $t \in \mathbb{N}$, $\mu\left((\sigma_H^{r_1} \sigma_V^{s_1})^t A_1 \cap \ldots \cap \right.$

$(\sigma_H^{r_n} \sigma_V^{s_n})^t A_n)$. Observation 1.6 and the properties of Haar measure imply that $\mu((\sigma_H^{r_1} \sigma_V^{s_1})^t A_1 \cap \ldots \cap (\sigma_H^{r_n} \sigma_V^{s_n})^t A_n)$ is either zero or larger than $c \cdot \Pi \mu(A_i)$, where $c > 1$ is independent of t, when $(\sigma_H^{r_1} \sigma_V^{s_1})^{-t} \overline{N}, \ldots, (\sigma_H^{r_n} \sigma_V^{s_n})^{-t} \overline{N}$ are not independent. This means that $\langle p \rangle^\perp$ is mixing on E if and only if, for every $N \in \mathbb{N}$, there exists a t_0 so that $(\sigma_H^{r_1} \sigma_V^{s_1})^{-t} \overline{N}, \ldots, (\sigma_H^{r_k} \sigma_V^{s_k})^{-t} \overline{N}$ are independent for all $t \geq t_0$. Next we examine independence of subsets in terms of the algebra of the ideal $\langle p \rangle$. The shape $S(q) \subseteq \mathbb{Z}^2$ of a polynomial $q \in \mathbb{Z}/2\mathbb{Z}[x^{\pm 1}, y^{\pm 1}]$ is defined as in section 2.

Observation 3.1. *A finite collection of disjoint finite sets $E_1, \ldots, E_n \subseteq \mathbb{Z}^2$ is independent for $\langle p \rangle^\perp$ if and only if there do not exist polynomials $q_1, \ldots, q_n \in \mathbb{Z}/2\mathbb{Z}[x^{\pm 1}, y^{\pm 1}]$ with $S(q_i) \subseteq E_i$ for all $i = 1, \ldots, n$, $q_i \notin \langle p \rangle$ for some i, and $q_i + \ldots + q_n \in \langle p \rangle$.*

This observation, which is clear from the definition of $\langle p \rangle^\perp$ and independence, illustrates the relationship between the mixing properties of $\langle p \rangle^\perp$ and the algebraic properties of $\langle p \rangle$.

We will examine the problem of mixing on shapes. The proof of theorem 3.6 requires fairly complicated algebraic arguments, and will appear in [KS3]. We do prove some partial results which indicate why these conditions arise, and illustrate the theorem with a number of examples.

We restrict our attention to ergodic subgroups. By proposition 2.8 these are the ones defined by a principal ideal. Proposition 2.11 answers the question of directional mixing. A subgroup $\langle p \rangle^\perp$ is two-mixing if and only if no irreducible factor of p, over $\mathbb{Z}/2\mathbb{Z}$, is a polynomial in $x^r y^s$ any $(r, s) \in \mathbb{Z}^2$ (cf. theorem 2.4 in [KS1]). To discuss the higher order mixing problem we need some facts about finite fields. For more information see [LN]. Let $\mathbb{F}_2 = \mathbb{Z}/2\mathbb{Z}$, the finite field with two elements. For each $l \geq 1$, let $\mathbb{F}_{2^l}$ be the field with 2^l elements, and let $\overline{\mathbb{F}_2}$ be the algebraic closure of $\mathbb{F}_2$. We are interested in factoring polynomials over these fields. For all $l \geq 1$, the rings $\mathbb{F}_{2^l}[x, y]$ and $\overline{\mathbb{F}_2}[x, y]$ are unique factorization domains and $\mathbb{F}_2[x, y] \subseteq \mathbb{F}_{2^l}[x, y] \subseteq \overline{\mathbb{F}_2}[x, y]$. For a polynomial $p = a_0 x^{r_0} y^{s_0} + \ldots + a_n x^{r_n} y^{s_n} \in \overline{\mathbb{F}_2}[x, y]$, $a_i \neq 0$, we say the shape of p, $S(p) \subseteq \mathbb{Z}^2$, is $\{(r_0, s_0), \ldots, (r_n, s_n)\}$. We may assume our shapes are contained in the positive quadrant of $\mathbb{Z}^2$, i.e. that their corresponding polynomials have no negative exponents.

We begin with the following observation due to F. Ledrappier.

Observation 3.2. *If $q = q_1 q_2 \in \langle p \rangle, q_1 \notin \langle p \rangle$, then $\langle p \rangle^\perp$ is not mixing on q_2 at times 2^t.*

Proof. Let $S(q_2) = \{(r_1, s_1), \ldots, (r_k, s_k)\}$, observe that $q_1 q_2^{2^t} = \Sigma q_1 (x^{r_i} y^{s_i})^{2^t} \in \langle p \rangle$, and apply observation 3.1. □

Example 3.3. (Ledrappier): Example 1.4a is ergodic in every direction and so by theorem 2.4 in [KS1] it is mixing of order two. It is not mixing on $1 + x + y$ because it is $\langle 1 + x + y \rangle^\perp$. This means it is not mixing of order three.

The next lemmas generalize this observation. Their proofs are sketched.

Lemma 3.4. *Let $p \in \mathbb{F}_2[x, y]$ be an irreducible polynomial. If $f = a_0 + a_1(x^{r_1} y^{s_1}) + \ldots + a_n(x^{r_n} y^{s_n}) \in \mathbb{F}_{2^l}[x, y]$ divides p in $\mathbb{F}_{2^l}[x, y]$ then there are*

$c_0, \ldots, c_n \in \mathbb{F}_2[x, y]$, not all in $\langle p \rangle$, so that $c_0 + c_1 (x^{r_1} y^{s_1})^{2^{lt}} + \ldots + c_n (x^{r_n} y^{s_n})^{2^{lt}} \in \langle p \rangle$ for all t. This means $\langle p \rangle^{\perp}$ is not mixing on $S(f)$ at times 2^{lt}.

Proof. Let f be defined as in the statement. Consider the two rings $\mathbb{F}_2[x, y]/\langle p \rangle \subseteq \mathbb{F}_{2^l}[x, y]/\langle f \rangle$. Define the matrix

$$F = \begin{bmatrix} 1 & (x^{r_1} y^{s_1}) & \cdots & (x^{r_n} y^{s_n}) \\ 1 & (x^{r_1} y^{s_1})^{2^l} & \cdots & (x^{r_n} y^{s_n})^{2^l} \\ \vdots & \vdots & & \vdots \\ 1 & (x^{r_1} y^{s_1})^{2^{ln}} & \cdots & (x^{r_n} y^{s_n})^{2^{ln}} \end{bmatrix}$$

and the vector

$$\bar{a} = \begin{bmatrix} a_0 \\ a_1 \\ \vdots \\ a_n \end{bmatrix}.$$

Observe that $F\bar{a} = 0$ in $(\mathbb{F}_{2^l}[x, y]/\langle f \rangle)^{n+1}$ and hence that $\det F \in \langle p \rangle$. This produces the desired $c_0, \ldots, c_n \in \mathbb{F}_2[x, y]$. $\square$

For a polynomial $g = b_0 + b_1(x^{r_1} y^{s_1}) + \ldots + b_m(x^{r_m} y^{s_m}) \in \mathbb{F}_2[x, y]$ and $a \in \mathbb{N}$ we set $g^{(a)} = b_0 + b_1(x^{r_1} y^{s_1})^a + \ldots + b_m(x^{r_m} y^{s_m})^a$.

Lemma 3.5. *Suppose that $f = a_0 + a_1(x^{r_1} y^{s_1}) + \ldots + a_n(x^{r_n} y^{s_n}) \in \mathbb{F}_{2^l}[x, y]$ and that $f^{(a)}$ divides $p^{(b)}$ in $\overline{\mathbb{F}_2}[x, y]$. Then there are $c_0, \ldots, c_n \in \mathbb{F}_2[x, y]$ and a sequence of t going to infinity so that $c_0 + c_1(x^{r_1} y^{s_1})^t + \ldots + c_n(x^{r_n} y^{s_n})^t \in \langle p \rangle$ for all t in in this sequence. This means that $\langle p \rangle^{\perp}$ is not mixing on $S(f)$.*

Proof. Let $f^{(a)}, p^{(b)}$ be as stated and in $\mathbb{F}_{2^l}[x, y]$. Consider the isomorphism

$$\mathbb{F}_2[x, y]/\langle p \rangle \xrightarrow{\varphi_b} \mathbb{F}_2[x^b, y^b]/\langle p^{(b)} \rangle \subseteq \mathbb{F}_2[x, y]/\langle p^{(b)} \rangle \subseteq \mathbb{F}_{2^l}[x, y]/\langle f^{(a)} \rangle,$$

where $\varphi_b(g) = g^{(b)}$ for every $g \in \mathbb{F}_2[x, y]$. From lemma 3.4 we know that there are $d_0, \ldots, d_n \in \mathbb{F}_2[x, y]$ so that $d_0 + d_1(x^{r_1} y^{s_1})^{a2^{lt}} + \ldots + d_n(x^{r_n} y^{s_n})^{a2^{lt}} \in \langle p^{(b)} \rangle$ for all t. For an infinite sequence of $k \in \mathbb{N}$ we have $a2^{lk} = c \pmod{b}$ for some c. Define the matrix F and the vector $\bar{d}$ by

$$F = \begin{bmatrix} 1 & (x^{r_1} y^{s_1})^{a2^{lk_0} - c} & \cdots & (x^{r_n} y^{s_n})^{a2^{lk_0} - c} \\ 1 & (x^{r_1} y^{s_1})^{a2^{lk_1} - c} & \cdots & (x^{r_n} y^{s_n})^{a2^{lk_1} - c} \\ r\vdots & \vdots & & \vdots \\ 1 & (x^{r_1} y^{s_1})^{a2^{lk_n} - c} & \cdots & (x^{r_n} y^{s_n})^{a2^{lk_n} - c} \end{bmatrix}$$

$$\bar{d} = \begin{bmatrix} d_0 \\ d_1(x^{r_1} y^{s_1})^c \\ \vdots \\ d_n(x^{r_n} y^{s_n})^c \end{bmatrix}.$$

For any $k_0, \ldots, k_n$ in our sequence, $F\bar{d} = 0$ in $\mathbb{F}_2[x, y]/\langle p^{(b)} \rangle$, and $\det F \in \mathbb{F}_2[x^b, y^b]$. This means $\det F$ is a zero divisor in $\mathbb{F}_2[x^b, y^b]/\langle p^{(b)} \rangle$ and implies that we can find $c_0^{(b)}, \ldots, c_n^{(b)} \in \mathbb{F}_2[x^b, y^b]$ with $c_0^{(b)} + c_1^{(b)}(x^{r_1} y^{s_1})^{2^{lk} - c} + \cdots +$

$c_n^{(b)}(x^{r_n}y^{s_n})^{2^{lk}-c} \in \langle p^{(b)} \rangle$. By taking $t = (2^{lk} - c)/b$ for each k in our sequence we obtain a sequence of t going to infinity for which $c_0^{(b)} + c_1^{(b)}(x^{r_1}y^{s_1})^t + \cdots + c_n^{(b)}(x^{r_n}y^{s_n})^t \in \langle p \rangle$. $\square$

These lemmas give an indication of why the following conditions are sufficient. The proof of the necessity of the conditions relies on an unpublished theorem due to D.W. Masser [M]. The full proof can be found in [KS3].

Theorem 3.6. *The Markov subgroup* $\langle p \rangle^\perp$ *fails to be mixing on a shape* $S \subseteq \mathbb{Z}^2$ *if and only if there exist* a, $b \in \mathbb{N}$ *and* $f \in \mathbb{F}_{2^l}[x, y]$ *with* $S(f) \subseteq S$ *so that* $f^{(a)}$ *and* $p^{(b)}$ *have a common factor. Moreover,* a, b, l *can be explicitly bounded as a function of* p.

Example 3.7. Let $p = 1 + x + y + x^2 + y^2 + xy$. This is irreducible in $\mathbb{F}_2[x, y]$ but factors in $\mathbb{F}_{2^2}[x, y]$ into $(1 + \alpha x + \alpha^2 y)(1 + \alpha^2 x + \alpha y)$, where $\mathbb{F}_{2^2} = \{0, 1, \alpha, \alpha^2\}$ with $\alpha^3 = 1$ and $\alpha + \alpha^2 = 1$. We can solve for c_0, c_1, c_2 and obtain that $(x + y) + (1 + y)x^{2^{2t}} + (1 + x)y^{2^{2t}} \in \langle p \rangle$ for all t. We see that $\langle p \rangle^\perp$ fails to be mixing on $S = \{(0, 0), (0, 1), (1, 0)\}$.

Example 3.8. The polynomial $p = 1 + x + y + x^2 + y^2 + xy + x^3 + x^2 y + xy^2 + y^3$ is absolutely irreducible, but $p^{(3)}$ is divisible by $f = 1 + x + y$. So $\langle p \rangle^\perp$ fails to be mixing mixing on $S = \{(0, 0), (0, 1), (1, 0)\}$.

Example 3.9. Let $p = 1 + x^3 + y^3$. It is absolutely irreducible but for $f = 1 + x + y$, $f^{(3)} = p$. So $\langle p \rangle^\perp$ fails to be mixing on $S = \{(0, 0), (1, 0), (0, 1)\}$.

Example 3.10. Let p be as in example 3.7. Let $f = (1 + \alpha x + \alpha^2 y)(1 + \alpha x) = 1 + \alpha^2 + \alpha^2 x^2 + xy$, so $S(f) = \{(0, 0), (0, 1), (2, 0), (1, 1)\}$. Then $\langle p \rangle^\perp$ fails to be mixing on $S(f)$ even though $f^{(a)}$ never divides $p^{(b)}$ and $\langle p \rangle^\perp$ is mixing on every subset of $S(f)$.

Although we can relate the factors of $p^{(b)}$ in $\overline{\mathbb{F}}_2[x, y]$ to mixing on a shape, we are not able to answer the most basic question about mixing: given p, what is $\langle p \rangle^\perp$'s maximal order of mixing? We end this section with the following conjecture.

Conjecture 3.11. *Let* $p \in \mathbb{F}_2[x, y]$ *and let* k *be the least number of nonzero terms that appear in a polynomial in* $\langle p^{(b)} \rangle$ *for* $b \in \mathbb{N}$. *Then* $\langle p \rangle^\perp$ *is mixing of order* $k - 1$ *but not of order* k.

4. MAPPINGS

We would like to know when two subgroups are topologically conjugate or measurably isomorphic. We would also like to know when one subgroup is a factor, either measurable or topological, of another. We cannot answer these questions in any generality, but shall look at some special cases, define some invariants, and state some conjectures.

First we observe that no subgroup, $I^\perp$, $I \neq \{0\}$ and $I \neq \mathbb{Z}/2\mathbb{Z}[x^{\pm 1}, y^{\pm 1}]$, can be a measurable factor of $(\mathbb{Z}/2\mathbb{Z})^{\mathbb{Z}^2}$. If $I \neq \{0\}$ then there is a $p \in I, p \neq 0$ and $I^\perp$ fails to be mixing on p at times 2^t by observation 3.2. $(\mathbb{Z}/2\mathbb{Z})^{\mathbb{Z}^2}$ is mixing of all orders so $I^\perp$ cannot be a factor of $(\mathbb{Z}/2\mathbb{Z})^{\mathbb{Z}^2}$.

Similar reasoning leads to the following observation.

Observation 4.1. *Suppose that $q = 1 + x + y$, $p \in \mathbb{Z}/2\mathbb{Z}[x^{\pm 1}, y^{\pm 1}]$, $\langle p \rangle^{\perp}$ is ergodic in the x and y directions, and $\varphi : \langle p \rangle^{\perp} \to \langle q \rangle^{\perp}$ a continuous factor map. Then φ is a group homomorphism.*

Proof. Let φ be an $(2\ell + 1)^2$ block map and let $L \subseteq \mathbb{Z}^2$ be the $(2\ell + 1)^2$ square centered at the origin. Let $A, B, O \in \pi_L(\langle p \rangle^{\perp})$, where O is the square of all 0's. Let $A_{(0,0)} = \{x \in \langle p \rangle^{\perp} : \pi_L(x) = A\}$ and define $B_{(0,0)}$ and $O_{(0,0)}$ similarly. Since $\langle p \rangle^{\perp}$ is ergodic in the x and y directions, the sets L and $L + (2^t, 0)$, as well as the sets L and $L + (0, 2^t)$, are independent in $\langle p \rangle^{\perp}$ for all sufficiently large t. Fix such a t and choose $x \in A_{(0,0)} \cap \sigma_H^{-2^t}(O_{(0,0)})$ and $y \in B_{(0,0)} \cap \sigma_V^{-2^t}(O_{(0,0)})$. Then $\varphi(x)_{(2^t, 0)} = \varphi(y)_{(0, 2^t)} = 0$ because φ must map the fixed point of all zeros in $\langle p \rangle^{\perp}$ to the only fixed point in $\langle q \rangle^{\perp}$.

Note that $\varphi(x)_{(0,0)} = \varphi(A)$, $\varphi(y)_{(0,0)} = \varphi(B)$, where we are using φ as a map on both blocks and points. Since $q^{2^t} = 1 + x^{2^t} + y^{2^t} \in \langle q \rangle$, $\varphi(x)_{(0,0)} + \varphi(x)_{(2^t, 0)} + \varphi(x)_{(0, 2^t)} = \varphi(A) + \varphi(x)_{(0, 2^t)} = 0$ and $\varphi(y)_{(0,0)} + \varphi(y)_{(2^t, 0)} + \varphi(y)_{(0, 2^t)} = \varphi(B) + \varphi(y)_{(2^t, 0)} = 0$. Next consider the point $x + y \in \langle p \rangle^{\perp}$, whose image satisfies that $\varphi(x+y)_{(0,0)} + \varphi(x+y)_{(2^t, 0)} + \varphi(x+y)_{(0, 2^t)} = \varphi(A + B) + \varphi(y)_{(2^t, 0)} + \varphi(x)_{(0, 2^t)} = 0$. But comparing the three equations we see that $\varphi(A + B) = \varphi(A) + \varphi(B)$. This means φ is a homomorphism. $\square$

We would like to apply this type of argument to arbitrary p and q but because we do not understand the mixing properties of these subgroups well enough, we are unable to do it.

Conjecture 4.2. *If $p, q \in \mathbb{Z}/2\mathbb{Z}[x^{\pm 1}, y^{\pm 1}], p, q \neq 0$ and $\varphi : \langle p \rangle^{\perp} \to \langle q \rangle^{\perp}$ is a measurable factor map then either φ is a homomorphism or $\tau \circ \varphi$ is a homomorphism, where $\tau : \langle q \rangle^{\perp} \to \langle q \rangle^{\perp}$ is the map that interchanges 0 and 1. This can happen only when q has an even number of non-zero coefficients.*

Because of the results in Section 2, the structure of shift commuting group homomorphisms is easy to understand. In particular we would obtain the following.

Corollary 4.3 to Conjecture 4.2. *If $\langle p \rangle^{\perp}$ and $\langle q \rangle^{\perp}$ are, measurably isomorphic, then $p = q$.*

Next we examine some special sigma algebras which are invariants for measurable isomorphism and which supply some evidence for the conjecture. For $p \in \mathbb{Z}/2\mathbb{Z}[x^{\pm 1}, y^{\pm 1}]$ we define the convex hull of p in the obvious way by $C(p) = C(S(p))$. Faces, sizes and widths are defined as in section 1. We multiply convex hulls in the obvious way: $C(p) \cdot C(q) = C(pq)$.

The partition $\mathcal{P}_0 = \{[0]_{(0,0)}, [1]_{(0,0)}\}$, where $[i]_{(0,0)} = \{x \in X_F : x_{(0,0)} = i\}$, generates $\mathcal{B}$ under the $\mathbb{Z}^2$ action of σ_H and σ_V. For the rest of this discussion everything is done measurably with respect to this sigma algebra. Suppose $\mathcal{P}$ is a finite partition that generates $\mathcal{B}$ under the $\mathbb{Z}^2$ action. Let $\alpha_N(\mathcal{P}) = \vee \sigma_V^n(\mathcal{P})$, for $|n| \leq N$, $\alpha(\mathcal{P}) = \vee \sigma_V^n(\mathcal{P})$ for all n, $\mathcal{A}_N(\mathcal{P})$ be the sigma algebra generated by $\vee \sigma_H^m \alpha_N(\mathcal{P})$, for $m \geq 0$, and $\mathcal{A}(\mathcal{P})$ the sigma algebra generated by $\vee \sigma_H^m \alpha(\mathcal{P})$, for $m \geq 0$. An atom of $\mathcal{A}(\mathcal{P})$ is determined by the $\mathcal{P}$-coordinates (m, n) with $m \leq 0, n \in \mathbb{Z}$. An atom of $\mathcal{A}(\mathcal{P}_0)$ consists of all of the points in $\langle p \rangle^{\perp}$ with fixed coordinates in the left half plane, i.e. x and y are in the same atom if and only if $x_{(i,j)} = y_{(i,j)}$ for all $i \leq 0$ and $j \in \mathbb{Z}$. From the definition of $\mathcal{A}(\mathcal{P})$ the following is clear.

Lemma 4.4.

 (i) $\sigma_H \mathcal{A}(\mathcal{P}) \subseteq \mathcal{A}(\mathcal{P})$
 (ii) $\sigma_V \mathcal{A}(\mathcal{P}) = \mathcal{A}(\mathcal{P})$
 (iii) $\bigvee_{m \leq 0} \sigma_H^m \mathcal{A}(\mathcal{P}) = \mathcal{B}$

Proposition 4.5. *If p doesn't have a face with direction* $(0, -1)$ *then* $\mathcal{A}(\mathcal{P}) = \mathcal{B}$.

Proof. We take $\mathcal{P}_0$ as before and conclude from lemma 4.4 (iii) that there exists, for every $\varepsilon > 0$, an $N \geq 0$ such that $\sigma_H^N \mathcal{P}_0 \underset{\varepsilon}{\subseteq} \mathcal{A}(\mathcal{P})$. Then, by lemma 4.4 (i) and (ii), for all $n \geq N$, $m \in \mathbb{Z}$, $\sigma_H^n \sigma_V^m \mathcal{P}_0 \underset{\varepsilon}{\subseteq} \mathcal{A}(\mathcal{P})$. We can find $\ell \geq 1$ so that a translate of the set (=polynomial) p^{2^ℓ} contains (i, j) and has all other points with

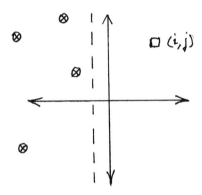

first coordinate less than or equal to $-N$. The sets $[0]$ and $[1]$ at each coordinate of p^{2^l}, except (i, j), are ε contained in $\mathcal{A}(\mathcal{P})$. This means the (i, j) coordinate is $(k - 1)\varepsilon$ contained in $\mathcal{A}(\mathcal{P})$. Since (i, j) and ε are arbitrary $\mathcal{A}(\mathcal{P}) = \mathcal{B}$. $\square$

Given a partition α and a sigma algebra $\mathcal{C}$ we define the information function $I(\alpha|\mathcal{C})$ in the usual manner. Then the conditional entropy is defined by the equation $H(\alpha|\mathcal{C}) = \int I(\alpha|\mathcal{C})d\mu$.

Let $\mathcal{A}^\infty(\mathcal{P}) = \cap \sigma_H^n \mathcal{A}(\mathcal{P})$, for $n \geq 0$. If p doesn't have a face in the $(0, -1)$ direction we have just seen that $\mathcal{A}^\infty(\mathcal{P}) = \mathcal{B}$. If p doesn't have a face in the $(0, -1)$ direction and is irreducible then $\mathcal{A}^\infty(\mathcal{P})$ is the trivial sigma algebra. If p has a face in the $(0, -1)$ direction but is not ergodic in the $(0, -1)$ direction then $\mathcal{A}^\infty(\mathcal{P})$ will be an intermediate sigma algebra. In any case, $\mathcal{A}^\infty(\mathcal{P}_0)$ will be a subsigma algebra of every $\mathcal{A}^\infty(\mathcal{P})$. Proofs of these assertions are contained in [KS3].

Proposition 4.6. *If p has a face with direction* $(0, -1)$ *then*

 (i) $H(\alpha(\mathcal{P}_0)|\sigma_H \mathcal{A}(\mathcal{P}_0)) = (s - 1)\log 2$
 (ii) *For any finite partition (not necessarily generating),*

$$H(\alpha(\mathcal{P})|\sigma_H \mathcal{A}(\mathcal{P})) = \overline{\lim_t} \frac{1}{t} H(\alpha(\mathcal{P})|\sigma_H^t \mathcal{A}(\mathcal{P})) \leq (s - 1)\log 2.$$

Proof. To see (i) consider the picture below and observe that for a fixed half

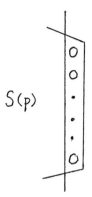

$$S(p)$$

plane there are exactly $2^{(s-1)}$ extensions by vertical strips to the right. This means the function $I(\alpha(\mathcal{P}_0)|\sigma_H \mathcal{A}(\mathcal{P}_0))$ is constant and equal to $(s-1)\log 2$. To prove the inequality in (ii) first observe that $H(\alpha(\mathcal{P})|\sigma_H \mathcal{A}(\mathcal{P}))$ is finite. We have that

$$H(\alpha_N(\mathcal{P})|\sigma_H \mathcal{A}(\mathcal{P})) \le H(\alpha_N(\mathcal{P})|\sigma_H \mathcal{A}_N(\mathcal{P})) \le h(\sigma_H).$$

Since

$$H(\alpha_N(\mathcal{P})|\sigma_H \mathcal{A}(\mathcal{P})) \le H(\alpha_{N+1}(\mathcal{P})|\sigma_H \mathcal{A}(\mathcal{P}))$$

we see that $H(\alpha(\mathcal{P})|\sigma_H \mathcal{A}(\mathcal{P}))$ is well defined and finite.

Secondly, observe that there exists, for every $\varepsilon > 0$, an $N \ge 0$ such that

$$H(\alpha(\mathcal{P})|\sigma_H \mathcal{A}(\mathcal{P})) \ge H(\alpha(\mathcal{P})|\sigma_H \mathcal{A}(\mathcal{P}) \vee \sigma_H^n \mathcal{A}(\mathcal{P}_0)) \ge H(\alpha(\mathcal{P})|\sigma_H \mathcal{A}(\mathcal{P})) - \varepsilon$$

for all $n \ge N$. The left inequality is immediate. The right hand one follows from the martingale theorem because $\sigma_H^n \mathcal{A}(\mathcal{P}_0)$ decreases to $\mathcal{A}^\infty(\mathcal{P}_0)$ as $n \to \infty$ and $\mathcal{A}^\infty(\mathcal{P}_0) \subseteq \sigma_H^t \mathcal{A}(\mathcal{P})$ for every $\mathcal{P}$ and t. It follows that, for the same ε and N, and for all $t \ge 0$ and $n \ge N$,

$$
\begin{aligned}
(*) \qquad tH(\alpha(\mathcal{P})|\sigma_H \mathcal{A}_\mathcal{P}) &= H(\alpha(\mathcal{P})|\sigma_H^t \mathcal{A}_\mathcal{P}) \\
&\ge H(\alpha(\mathcal{P})|\sigma_H^t \mathcal{A}_\mathcal{P} \vee \sigma_H^{n+t} \mathcal{A}_{\mathcal{P}_0}) \\
&\ge tH(\alpha(\mathcal{P})|\sigma_H \mathcal{A}_\mathcal{P}) - t\varepsilon.
\end{aligned}
$$

Thirdly, observe that, since $\sigma_H^{-m} \mathcal{A}_{\mathcal{P}_0}$ increases to $\mathcal{B}$ as m goes to infinity, there exists, for any $\varepsilon > 0$, an $M \ge 0$ with $H(\alpha(\mathcal{P})|\sigma_H \mathcal{A}_\mathcal{P} \vee \sigma_H^{-m} \mathcal{A}(\mathcal{P}_0)) < \varepsilon$ for every $m \ge M$. Hence

$$(**) \qquad H(\alpha(\mathcal{P})|\sigma_H^t \mathcal{A}_\mathcal{P} \vee \sigma_H^{-m} \mathcal{A}_{\mathcal{P}_0}) < t\varepsilon$$

for every $t \geq 0$ and $m \geq M$. Finally, we see that

$$
\begin{aligned}
tH(\alpha(\mathcal{P})|\sigma_H \mathcal{A}_{\mathcal{P}}) & \\
= {} & H(\alpha(\mathcal{P})|\sigma_H^t \mathcal{A}_{\mathcal{P}}) \\
\leq {} & H(\alpha(\mathcal{P})|\sigma_H^t \mathcal{A}_{\mathcal{P}} \vee \sigma_H^{n+t} \mathcal{A}_{\mathcal{P}_0}) + t\varepsilon \quad \text{(by *)} \\
\leq {} & H(\alpha(\mathcal{P}) \vee \sigma_H^{-m} \alpha(\mathcal{P}_0)|\sigma_H^t \mathcal{A}_{\mathcal{P}} \vee \sigma_H^{n+t} \mathcal{A}_{\mathcal{P}_0}) + t\varepsilon \\
= {} & H(\alpha(\mathcal{P})|\sigma_H^t \mathcal{A}_{\mathcal{P}} \vee \sigma_H^{-m} \mathcal{A}_{\mathcal{P}_0}) + H(\sigma_H^{-m} \alpha(\mathcal{P}_0)|\sigma_H^t \mathcal{A}_{\mathcal{P}} \vee \sigma_H^{n+t} \mathcal{A}_{\mathcal{P}_0}) + t\varepsilon \\
\leq {} & 2t\varepsilon + H(\sigma_H^{-m} \alpha(\mathcal{P}_0)|\sigma_H^{n+t} \mathcal{A}_{\mathcal{P}_0}) \quad \text{(by **)} \\
= {} & 2t\varepsilon + (m+n+t)H(\alpha(\mathcal{P}_0)|\sigma_H \mathcal{A}_{\mathcal{P}_0})
\end{aligned}
$$

so that $H(\alpha(\mathcal{P})|\sigma_H \mathcal{A}_{\mathcal{P}}) \leq H(\alpha(\mathcal{P}_0)|\sigma_H \mathcal{A}_{\mathcal{P}_0})$. $\quad\square$

Corollary 4.7. *The convex hull of p is an invariant of measurable conjugacy for $\langle p \rangle^{\perp}$.*

Proof. Propositions 4.5 and 4.6 show that the measurable dynamics of $\langle p \rangle^{\perp}$ determine the direction and length of each face of $C(p)$. $\quad\square$

Theorem 4.8. *Suppose $p, q \in \mathbb{Z}/2\mathbb{Z}[x^{\pm 1}, y^{\pm 1}]$, $p, q \neq 0$ and $\varphi : \langle p \rangle^{\perp} \to \langle q \rangle^{\perp}$ is a measurable factor map. Then $C(p) = C(q)C(k)$ for some polynomial $k \in \mathbb{Z}/2\mathbb{Z}[x^{\pm 1}, y^{\pm 1}]$.*

Proof. Let $F(p) \subseteq \mathbb{Z}^2$ and $F(q) \subseteq \mathbb{Z}^2$ denote the sets of faces of p and q, respectively, and recall that the sum of all elements of $F(p)$ (resp. $F(q)$) is equal to zero. From proposition 4.6 it is clear that there exists, for every $f \in F(q)$, an $f' \in F(p)$ with the same direction, but possibly with greater length. Put, for every $g \in F(p)$, $g' = f' - f$ if $g = f'$ for some $f \in F(q)$, and $g' = g$ otherwise. Then $\sum_{g \in F(p)} g' = 0$, and the set $\{g' : g \in F(p)\}$ is the set of faces of a finite, convex subset $K \subseteq \mathbb{Z}^2$. The proof is completed by setting k equal to the polynomial corresponding to K. $\quad\square$

Putting together propositions 4.5 and 4.6 and corollary 4.7 gives the following result.

Theorem 4.9. *If $p, q \in \mathbb{Z}/2\mathbb{Z}[x^{\pm 1}, y^{\pm 1}]$ and $\langle p \rangle^{\perp}$, $\langle q \rangle^{\perp}$ are measurably isomorphic then*

    (i) *$C(p) = C(q)$;*
    (ii) *the factors of p and q in $\overline{\mathbb{F}}_2[x^{\pm 1}, y^{\pm 1}]$ have the same shapes.*

Jon Ashley previously proved directly that if $\langle p \rangle^{\perp}$ and $\langle q \rangle^{\perp}$ are isomorphic then $C(p)$ and $C(q)$ have the same corners. Our proof involving invariant σ-algebras is essentially a more abstract version of his proof concerning corners.

5. SUBSHIFTS

Given $p \in \mathbb{Z}/2\mathbb{Z}[x^{\pm 1}, y^{\pm 1}]$ we would like to describe the subshifts of $\langle p \rangle^{\perp}$, i.e. the closed, $\mathbb{Z}^2$-invariant subsets. More ambitiously, we would like to describe the invariant measures. A similar problem, but in a different setting, is to take the squaring and cubing maps of the circle as a $\mathbb{Z}^2$ action on the circle and ask for closed invariant sets and invariant measures. In [F2], H. Furstenberg showed that the only closed invariant sets are finite or the entire circle, and in [R], D. Rudolph showed that if a measure has positive entropy under either map, it is Lebesgue measure. The answer when the measure has zero entropy under both maps is still unknown.

We shall discuss the polynomial $p = 1 + x + y$, which seems - at least intuitively - related to the problem discussed by Furstenberg and Rudolph. Although we cannot classify the shift invariant measures in this case, we construct examples of such subshifts which may through analogy shed some light on the possible construction of measures on the circle which are invariant both under squaring and cubing. Let $\tau : (\mathbb{Z}/2\mathbb{Z})^{\mathbb{Z}} \to (\mathbb{Z}/2\mathbb{Z})^{\mathbb{Z}}$ be defined by $\tau(x)_i = x_i + x_{i+1}$. This is a 2-to-1 group homomorphism that commutes with the shift. We may think of $\langle p \rangle^{\perp}$ as the inverse limit $\varprojlim\left((\mathbb{Z}/2\mathbb{Z})^{\mathbb{Z}}, \tau\right)$ with σ_H and σ_V on $\langle p \rangle^{\perp}$ corresponding to σ and τ^{-1} on $\varprojlim\left((\mathbb{Z}/2\mathbb{Z})^{\mathbb{Z}}, \tau\right)$. A subshift of $\langle p \rangle^{\perp}$ corresponds to a subshift, $\Lambda \subseteq (\mathbb{Z}/2\mathbb{Z})^{\mathbb{Z}}$ with $\tau(\Lambda) = \Lambda$. This change of view will be useful, and also exhibits the formal similarity with the problem on the circle mentioned above: the shift σ corresponds to squaring on the circle (i.e. multiplication by 2 in dyadic expansion), and τ is similar to the cubing map (in dyadic expansion, τ is just like multiplication by 3, except that digits are not carried).

Observation 5.1. *Suppose that $\Lambda \subseteq (\mathbb{Z}/2\mathbb{Z})^{\mathbb{Z}}$ is a subshift, $\tau(\Lambda) = \Lambda$, and σ is topologically mixing on Λ. Then $\Lambda = (\mathbb{Z}/2\mathbb{Z})^{\mathbb{Z}}$ or Λ is the point of all 0's.*

Proof. Observe that $\tau^{2^t}(x)_i = x_i + x_{i+2^t}$. Let $[i_0, \dots, i_n]_0 = \{x \in \Lambda : x_k = i_k \quad 0 \le k \le n\}$, define $[j_0, \dots, j_n]_0$ similarly, and assume that these sets are nonempty. Since σ is mixing on Λ we can find $x \in \Lambda$ with $x \in [i_0, \dots, i_n]_0$ and $\sigma^{2^t}(x) \in [j_0, \dots, j_n]_0$, for some large t. So, $\tau^{2^t}(x)_k = i_k + j_k$ for $0 \le k \le n$. This means that Λ is closed under addition, i.e. a subgroup. By [K], the only shift invariant subgroups of $(\mathbb{Z}/2\mathbb{Z})^{\mathbb{Z}}$ are itself or finite, and the only finite subgroup that is mixing under σ is the trivial one. $\square$

Next we construct some examples of non-trivial subshifts of $\langle p \rangle^{\perp}$. These are not subgroups, but have an algebraic nature, and they are the only examples we know. We will do this in $(\mathbb{Z}/2\mathbb{Z})^{\mathbb{Z}}$ using τ.

Construction 5.2. *Although every closed, shift invariant subgroup of $(\mathbb{Z}/2\mathbb{Z})^{\mathbb{Z}}$ is either equal to $(\mathbb{Z}/2\mathbb{Z})^{\mathbb{Z}}$ or finite (cf. [K]), there are nontrivial, infinite, closed subgroups of $Y \subseteq (\mathbb{Z}/2\mathbb{Z})^{\mathbb{Z}}$ which are invariant under σ^n for $n > 1$: indeed, every nontrivial subgroup $H \subseteq (\mathbb{Z}/2\mathbb{Z})^n$ gives rise to such a Y through free concatenation of elements of H with the zero coordinate as the first coordinate of one of the elements in H. Such a subgroup Y gives rise to a subshift $\overline{Y} \subseteq (\mathbb{Z}/2\mathbb{Z})^{\mathbb{Z}}$ of the form $\overline{Y} = \bigcup_{k=0}^{n-1} \sigma^k(Y)$. Clearly, $\overline{Y}$ will in general not be τ-invariant; however, if n is a power of 2, then the τ-orbit of $\overline{Y}$ will be small, and $\Lambda = \bigcup_{k=0}^{n-1} \tau^k(\overline{Y})$ is a subshift of $(\mathbb{Z}/2\mathbb{Z})^{\mathbb{Z}}$ which is invariant under σ and τ. If Λ is obtained from a subgroup $H \subseteq (\mathbb{Z}/2\mathbb{Z})^n$ in the manner just described then the entropy of σ and τ is positive, and is given by $\frac{1}{n}\log(\text{cardinality of } H)$. As an explicit example one could take $n = 2$ and $H = \{(0,0), (1,1)\} \subseteq (\mathbb{Z}/2\mathbb{Z})^2$. Then Y is the set of all sequences (x_n) in $(\mathbb{Z}/2\mathbb{Z})^{\mathbb{Z}}$ with $x_{2n} = x_{2n+1}$ for all $n \in \mathbb{Z}$, and $\overline{Y}$ and Λ can be calculated explicitly.*

This procedure can be iterated as follows. Take a sequence of $n_i \to \infty$. Choose a proper subgroup H_1 of $(\mathbb{Z}/2\mathbb{Z})^{2^{n_1}}$ and form Λ_1. Take a proper subgroup H_2

of $(H_1)^{2^{n_2}}$ and form Λ_2, and so on. Then let $\Lambda = \bigcap \Lambda_i$. It is clear that the restrictions of σ and τ to Λ have zero entropy.

As an example of this, consider the Morse sequence

$$M = 0110100110010110\ldots$$

obtained by letting $b_0 = 0, b_1 = \overline{b_0} = 1, b_2 = \overline{b_0 b_1} = 10$, $b_n = \overline{b_0 \cdots b_{n-1}} = b_{n-1}\overline{b_{n-1}}$, for $n > 1$. Then

$$M = b_0 b_1 b_2 \cdots$$

For each k let $H_k = \{0^{2^{k-1}}, b_k, \overline{b_k}, 1^{2^{k-1}}\} \subseteq (\mathbb{Z}/2\mathbb{Z})^{2^{k-1}}$, and note that $H_{k+1} \subseteq H_k \times H_k$. Construct Λ_k from H_k for each k and let $\Lambda = \bigcap_k \Lambda_k$.

For $k \geq 2$, $h(\Lambda_k, \sigma) = h(\Lambda, \tau) = \log 4/2^{k-1}$ and $h(\Lambda, \sigma) = h(\Lambda, \tau) = 0$. Λ is a subshift of $(\mathbb{Z}/2\mathbb{Z})^{\mathbb{Z}}$ invariant under τ with zero entropy that contains the Morse minimal set. It contains two fixed points, the point of all 0's and the point of all 1's, but no other periodic points. This is because $(\Lambda_k)_0^0$ contains no points of period less than 2^{k-1} except the two fixed points.

REFERENCES

[AM]   M. F. Atiyah and I. G. MacDonald, *Introduction to commutative algebra*, Addison Wesley, Reading, Mass., 1969.

[B]   R. Berger, *The Undecidability of the Domino Problem*, Mem. AMS no. 66 (1966).

[F1]   H. Furstenberg, *Recurrence in Ergodic Theory and Combinatorial Number Theory*, Princeton University Press, 1981.

[F2]   H. Furstenberg, *Disjointness in Ergodic theory, Minimal sets, and a problem in Diophantine approximation*, Math. Sys. Th. 1 (1967), 1–49.

[H]   P. Halmos, *On automorphisms of compact groups*, Bull. Am. Math. Soc. 49 (1943), 619–624.

[He]   G. A. Hedlund, *Endomorphisms and automorphisms of the shiftdynamical system*, Math. Sys. Th. 3 (1969), 320–375.

[HR]   E. Hewitt and K. Ross, *Abstract Harmonic Analysis*, Academic Press, Inc. and Springer Verlag, 1963.

[K]   B. Kitchens, *Expansive dynamics on zero-dimensional groups*, Ergodic Th. and Dynam. Sys. 7 (1987), 249–261.

[KS1] B. Kitchens and K. Schmidt, *Automorphisms of compact groups*, Ergodic Th. and Dynam. Sys. 9 (1989), 691–735.

[KS2] B. Kitchens and K. Schmidt, *Periodic points, decidability and Markov subgroups*, Dynamical Systems, Proceeding of the special Year (j. c. Alexander, ed.), Springer-Verlag, 1988, pp. 440–454.

[KS3] B. Kitchens and K. Schmidt, *Expansive $\mathbb{Z}^d$-actions on zero dimensional groups*, preprint.

[L]   F. Ledrappier, *Un champ markovian peut être d'entropie nulle and mélangeant*, C. R. Acad. Sc. Paris, Ser. A 287, 1978, pp. 561–562.

[LN]   R. Lidl and H. Niederreiter, *Finite Fields*, Addison-Wesley, 1983.

[L]   D. Lind, *The structure of skew products with ergodic group automorphisms*, Israel J. Math. 28 (1977), 205–248.

[LSW] D. Lind, K. Schmidt and T. Ward, *Mahler measure and entropy for commuting automorphisms of compact groups*, Invent. Math. 101 (1990), 593–629.

[LW]   D. Lind and T. Ward, *Automorphisms of solenoids and p-adic entropy*, Ergodic Th. and Dynam. Sys. 8 (1988), 411–419.

[M]   W. Masser, letters to D. Berend.

[MP]   N. Markley and M. Paul, *Matrix subshifts for $\mathbb{Z}^\nu$ symbolic dynamics*, Proc. London Math. Soc. 43 (1981), 251–272.

[MT] G. Miles and R. K. Thomas, *The breakdown of automorphisms of compact topological groups*, Studies in Probability and Ergodic Theory, Advances in Mathematics Supplementary Studies, vol. 2, Academic Press, New York-London, 1978, pp. 207–218.

[O] D. Ornstein, *Ergodic Theory, Randomness, and Dynamical Systems*, Yale University Press, 1974.

[R] D. Rudolph, ×2 *and* ×3 *invariant measures and entropy*, preprint.

[S1] K. Schmidt, *Automorphisms of compact abelian groups and affine varieties*, to appear, Proc. London Math. Soc..

[S2] K. Schmidt, *Mixing automorphisms of compact groups and a theorem by Kurt Mahler*, Pacific J. of Math. **137** (1989), 371-385.

MATHEMATICAL SCIENCES DEPARTMENT, IBM T. J. WATSON RESEARCH CENTER, YORKTOWN HEIGHTS, NY 10598, USA

MATHEMATICS INSTITUTE, UNIVERSITY OF WARWICK, COVENTRY CV4 7AL, UK

E-mail: kitch@watson.ibm.com and ks@maths.warwick.ac.uk

Contemporary Mathematics
Volume **135**, 1992

ON THE DIMENSION OF SOME GRAPHS

FRANÇOIS LEDRAPPIER

ABSTRACT. We compute the Hausdorff dimension of some subsets of $\mathbb{R}^2$. The typical example where our method applies is the following set T:

$$T = \{(x,y), \quad y = \sum_{n=0}^{\infty} 2^{-\frac{n}{2}} \phi_0(2^n x)\},$$

where $\phi_0(z)$ is the distance of the real z to the nearest integer.

We present in this paper the calculation of the Hausdorff dimension of examples of subsets Γ of R^2 which can be presented as graphs of a function defined by a series á la Weierstrass:

$$\Gamma(\{b_n\}, \phi, s) = \{(x,y)y = \sum_{n=0}^{\infty} b_n^{s-2} \phi(b_n x)\}$$

where $\{b_n, \ n \geq 0\}$ is a sequence of numbers satisfying $b_{n+1}/b_n \geq \lambda > 1$, ϕ is a $\mathbb{Z}$ periodic Lipschitz function and s is a real number, $1 \leq s < 2$.

Besicovitch and Ursell [BU₁] showed in 1937 that every number between 1 and s can be obtained as the dimension of such a graph. They assume for this that $b_{n+1}/b_n \to \infty$. Then there is less and less interaction between the terms of the series and the choice of ϕ is not important. In [BU₁] they consider the function ϕ_0, the distance to the nearest integer. Of course we are interested in the case when b_{n+1}/b converges to some finite b and in particular in the classical case $b_n = b^n$, $\phi_1 = \cos 2\pi$ when it is widely believed that the Hausdorff dimension of the corresponding graph Γ is s (see e.g. [BL]). Known are estimate from below of $\dim \Gamma$ ([MW], [PU]). Here we are able to give at least one explicit example where

$$\dim \Gamma(\{b^n\}, \phi, s) = s$$

namely $b = 2$, $\phi = \phi_0$, $s = 1.5$, and this is the content of the paper.

Our approach is dynamical, because when $b_n = b^n$, the set Γ can be considered as the repeller for some expanding self-mapping of $[0,1] \times \mathbb{R}$. The fact that b is an integer helps us because there is a natural Markov partition for the mapping

---

1991 Mathematics Subject Classifications. Primary 28A78, 58F11.

This paper is in final form and no version of it will be submitted for publication elsewhere.

$x \rightarrow bx(\bmod 1)$. The dynamical framework developed in section II would also make sense with a general b; we would have to use a countable Markov partition (as constructed e.g. in [H], [T]) and the local product structure of the absolutely continuous invariant measure ([L$_1$]) but the replacement of Proposition 1 below would be more complicated to state and to use. Now for $b_n = 2^n$, $n \geq o$, the result clearly depends on the function ϕ and on s: in particular if there is a Lipschitz ψ such that $\phi(x) = \psi(x) - 2^{s-2}\psi(2x)$, we obtain a Lipschitz graph Γ of dimension 1. In Proposition 1, we shall state a property of a piecewise $C^{1+\epsilon}$ function ϕ which ensures that $\dim \Gamma(\{2^n\}, \phi, s) = s$. Verifying this property is a priori simpler than the original problem, but we have been able to perform this verification only with the function ϕ_0, where it reduces to a well-known problem in harmonic analysis. The verification is trivial for $s = 1 + 1/n$, $n \geq 1$, wrong for $s_0 = \log_2(1 + \sqrt{5})$, done by Erdös [E] for a.e. s in an interval $(1, 1 + \delta)$. The fact that our approach fails for $s_0 = \log_2(1 + \sqrt{5})$ does not mean that the corresponding dimension is smaller than s_0. In fact, when our approach works, it yields a much stronger property, less likely to hold for all s, namely that any subgraph above a set of positive Lebesgue measure has already dimension s. This is because we use the ergodic theory of the measure of maximal entropy $\log 2$ on the repeller, which projects onto the Lebesgue measure on the first coordinate.

In this introduction, I had to use the words "Markov partition", "absolutely continuous invariant measure" and "maximal entropy", as one almost inevitably does when one tries to apply dynamics and ergodic theory to other problems. But there is a further reason why I want to dedicate this paper to Roy Adler: the core of the argument comes from [LY$_1$], where we observed that in smooth ergodic theory, when suitably defined, dimensions usually add. For the convenience of the reader I reproduce the relevant part of the proof here and I hope that it will be clear that this argument, for some part, roots in [A] and [AR].

I. SETTING AND RESULTS

Let $\phi : \mathbb{R} \rightarrow \mathbb{R}$ be a C^0, piecewise $C^{1+\epsilon}$, $\mathbb{Z}$ periodic function and define for $1 \leq s < 2$ the function $\psi_{s,\phi}(x) = \sum_{n=0}^{\infty} 2^{n(s-2)}\phi(2^n x)$, the graph $\Gamma_{s,\phi} = \{(x, \psi_{s,\phi}(x)), \; x \in [0,1]\}$ and the measure $m_{s,\phi}$ on $\Gamma_{s,\phi}$ which projects onto the Lebesgue measure on $[0, 1]$ (we shall omit s, ϕ or both when there is no ambiguity). The function ϕ_0 will be the distance to the closest integer. We say that a measure μ on a metric space X has dimension d if the following limit exists and is equal to d for μ a.e x in X:

$$\lim_{\eta \to 0} \frac{\ln \mu B(x, \eta)}{\ln \eta}$$

where $B(x, \eta)$ is the ball of radius η about x. When a measure has dimension d, every set of positive measure has Hausdorff dimension at least d (see e.g. [Y]). We call θ, $0 < \theta < 1$, an Erdös number if the distribution law of $\sum_{n=0}^{\infty} \theta^n \epsilon_n$ has dimension 1, where ϵ_n are Bernoulli (i.i.d. $\{1/2, 1/2\}$) variables with values $+1$ or -1. Recall that $2^{-1/p}$, $p \geq 1$ is an Erdös number, Pisot - Vijayaraghavan numbers are not ([PU] Theorem 8) and that there is a $\gamma < 1$ such that a.e number in $(\gamma, 1)$ is Erdös ([E]). Also from the entropy / dimension formula [L$_2$], (see the remark at the end of the paper) follows that the set of Erdös numbers is a G_δ subset of $[1/2, 1)$. That it is dense in $[1/2, 1]$ is the weakest of the conjectures about prevalence of Erdös numbers.

With the above notations our result is the following:

Theorem. *Let 2^{1-s} be an Erdös number, then m_{s,ϕ_0} has dimension s.*

Corollary. *Let 2^{1-s} be an Erdös number, then the Hausdorff dimension of $\Gamma_{s,\phi}$ is s.*

In fact $\dim \Gamma_{s,\phi}$ is not smaller than s by the theorem, and not bigger by [BU₁]. The corollary applies in particular to $s = 1 + \frac{1}{p}$, $p \geq 1$ or to a.e. s in the interval $(1, \log_2 2/\gamma)$ or conjecturally to a generic s in $(1, 2)$. The theorem follows easily from the definition of Erdös number and from the following criterion, where ϕ' denotes the derivative of ϕ.

Proposition 1. *Let $\underline{\xi} = \{\xi_i, i = 1, 2, \ldots\}$ be a sequence of Bernoulli variables with values 0 or 1. For each $(x, \underline{\xi})$ define*

$$x_n(\underline{\xi}) = \frac{x}{2^n} + \frac{\xi_1}{2^n} + \frac{\xi_2}{2^{n-1}} + \cdots + \frac{\xi_n}{2}.$$

Assume that for Lebesgue a.e. x, the distribution law of the variable $Y_x(\underline{\xi})$ defined by $Y_x(\underline{\xi}) = \sum_{n=1}^{\infty} 2^{(1-s)n} \phi'(x_n(\underline{\xi}))$ has dimension 1. Then, $m_{s,\phi}$ has dimension s.

One has only to observe that $\phi_0'(x_n(\underline{\xi})) = (-1)^{\xi_n}$.

The rest of the paper is devoted to the proof of Proposition 1.

II. Proof of Proposition 1

We shall first construct a mapping which inverts the natural repeller associated with $\Gamma_{s,\phi}$. Let $\Omega = \{0, 1\}^{\mathbf{N}^*}$ be the space of sequences $\underline{\xi} = \{\xi_1, \xi_2, \ldots\}$ and define on $[0, 1] \times \mathbb{R} \times \Omega$ the mapping $F_{s,\phi}$ by

$$F_{s,\phi}(x, y, \underline{\xi}) = \left(\frac{x}{2} + \frac{\xi_1}{2}, 2^{s-2}y + \phi\left(\frac{x}{2} + \frac{\xi_1}{2} \right), \sigma\underline{\xi} \right)$$

where σ is the shift transformation on Ω.

Clearly, the mapping $F_{s,\phi}$ leaves invariant the set $X_{s,\phi} = \Gamma_{s,\phi} \times \Omega$ and the measure $\mu_{s,\phi} = m_{s,\phi} \times d\underline{\xi}$, where $d\underline{\xi}$ is the Bernoulli measure on Ω. For fixed $\underline{\xi}$, the set $\Gamma \times \{\underline{\xi}\}$ is a local stable manifold of F. The measure m therefore can be considered as the conditional measure of the invariant measure μ associated with a partition in local stable manifolds. In our example, the results of smooth ergodic theory pertaining to such objects can be established directly, as we explain now. For each fixed $\underline{\xi}$, we apply to $\Gamma_{s,\phi}$ a sequence of mappings F_0 or F_1, chosen according to $\underline{\xi}$, where

$$F_0(x, y) = \left(\frac{x}{2}, 2^{s-2}y + \phi\left(\frac{x}{2} \right) \right)$$

$$F_1(x, y) = \left(\frac{x}{2} + \frac{1}{2}, 2^{s-2}y + \phi\left(\frac{x}{2} + \frac{1}{2} \right) \right)$$

For $\xi = 0, 1$ the Jacobian $D_{x,y}F_\xi$ of F_ξ is given in natural coordinates by

$$D_{(x,y)}F_\xi = \begin{bmatrix} 2^{-1} & 0 \\ 1/2\phi'(x/2 + \xi/2) & 2^{s-2} \end{bmatrix}.$$

For such a product of matrices, Osseledets theory is elementary: there is one direction $\mathcal{J}(x, y, \underline{\xi}) = \binom{1}{v(x,y,\underline{\xi})}$ such that

$$D_{(x,y)}F_{\xi_1}\mathcal{J} = 1/2\mathcal{J}(F(x, y, \underline{\xi})),$$

it is given by

$$v(x, y, \underline{\xi}) = -\sum_{n=1}^{\infty} 2^{(1-s)n}\phi'(F_{\xi_n}\dots F_{\xi_1}(x,));$$

all other directions $\mathcal{J}$ satisfy $\lim_n 1/n \ln \left\| D_{(x,y)}(F_{\xi_n}\dots F_{\xi_1})\mathcal{J} \right\| = (s-2)$. For each fixed $\underline{\xi}$, the field $\mathcal{J}(x, y, \underline{\xi})$ of *strong stable directions* is locally ϵ-Hölder continuous. Moreover it depends only on x. It defines locally a $C^{1+\epsilon}$ foliation by parallel $C^{1+\epsilon}$ curves, the *local strong stable manifolds*. Smooth ergodic theory gives us the dynamical description of this foliation (see [R]): two points (x, y, ξ) and (x', y', ξ) in the same local stable manifold belong to the same local strong stable manifold if and only if the corresponding $x_n(\underline{\xi})$ and $(x'_n)(\underline{\xi})$ always belong to the same interval of continuity of ϕ' and we have:

$$\limsup_n 1/n \ln d\left((F_{\xi_n}\dots F_{\xi_1})(x, y), (F_{\xi_n}\dots F_{\xi_1})(x', y')\right) \le -\ln 2.$$

Observe that in the case when $\phi = \phi_0$, the direction $\mathcal{J}$ is *constant* on each local stable manifold $\Gamma \times \{\xi\}$, so that the local strong stable manifolds are straight segments of slope $v(\underline{\xi})$ crossing $[0, 1] \times \mathbb{R}$. Also, the above dynamical description is easy to check directly.

Following [LY$_1$], we consider the local strong stable foliation as a measurable partition and index the quotient space by some transversal, e.g. a vertical line $x = cst$. We consider conditional measures $m_{s,\phi,x,y,\xi}$ of the measure $m_{s,\phi}$ associated with the partition of local strong stable manifolds, and the transverse measures $m_{s,\phi,\underline{\xi},x}$ which are the projection of the measure $m_{s,\phi}$ on the x vertical line along the strong stable foliation defined by the sequence $\underline{\xi}$. Observe that since the foliation is made of parallel curves, the measures $m_{s,\phi,\underline{\xi},x}$ depend on the choice of x only by a translation.

Proposition 2. [LY$_1$] *There exists numbers δ_1, γ_2 and δ_2 such that:*

    i) $\delta_2 = \delta_1 + \gamma_2$,

    ii) $1 = \delta_1 + \gamma_2(2 - s)$,

    iii) *for μ a.e. $(x, y, \underline{\xi})$ the measure $m_{s,\phi,x,y,\xi}$ has dimension δ_1,*

    iv) *for a.e. ξ, the measures $m_{x,\phi,\underline{\xi},x}$ have dimension γ_2,*

    v) *the measure $m_{s,\phi}$ has dimension δ_2.*

We shall return to Proposition 2 in Section III. Let us now discuss why Proposition 2 implies Proposition 1. We read from properties i), ii) and v) that the measure $m_{s,\phi}$ has the dimension s if and only if $\gamma_2 = 1$.

Consider first the case when $\phi \equiv \phi_0$. We have a measure m_{s,ϕ_0} for which we know that the dimension is δ_2 (property v). We know that $\delta_2 \geq 1$ since m_{s,ϕ_0} projects on Lebesgue on the horizontal axis. We project on the vertical direction along a random direction with slope $v(\xi)$. We know that for m_{s,ϕ_0} a.e. point (x_0, y_0) in Γ_{s,ϕ_0}, a.e. direction $\mathcal{J}_0$ the projection of m_{s,ϕ_0} onto the vertical $(x = x_0)$ along the direction $\mathcal{J}_0$ has dimension γ_2 (property iv). That $\gamma_2 = 1$ and thus Proposition 1 follow from the above discussion and Lemma 1:

Lemma 1. [M] *Let m be a finite measure on $\mathbb{R}^2$ such that*

$$\liminf_{\eta \to \infty} \frac{\ln mB((x,y),\eta)}{\ln \eta} \geq 1 \text{ for } m \text{ a.e. } (x,y),$$

ν a measure on directions with dimension one. For m a.e. (x,y), ν a.e. θ, we have

$$\limsup_{r \to 0} \frac{\ln m_{\theta,x}(y-r,y+r)}{\ln r} = 1$$

where $m_{\theta,x}$ is the projection of m on the vertical line going through x and along the direction θ.

Proof of Lemma 1. For any fixed θ, the lim sup cannot be bigger than 1 a.e, since we are considering a measure on a real line. The proof of the other inequality follows the proof of an analogous result of Marstrand concerning sets (and not merely measures). Fix $\delta > 0$ and consider $C(x, y, \delta)$ such that, for all $\eta > 0$

$$mB((x,y),\eta) \leq C(x,y,\delta)\eta^{1-\delta}.$$

We have $C(x, y, \delta) < +\infty$ m a.e.. By replacing ν by a smaller positive measure, we may assume that θ varies in some interval A away from the vertical direction and that ν satisfies

$$\nu B(\theta, \eta) \leq C\eta^{1-\delta}$$

for all θ in A. Define for (x, y) in $\mathbb{R}$, θ in A, $r > 0$

$$a(x, y, \theta, r) = m_{\theta,x}(y-r, y+r).$$

Marstrand's claim is that for all $r > 0$, and some constant C':

(*) $$\int a(x,y,\theta,r)\, d\nu(\theta) \leq C'C(x,y,\delta)r^{1-2\delta}.$$

From (*) and Fatou's Lemma follows that for m.a.e (x, y), ν.a.e. θ:

$$\liminf_{r \to 0} \frac{a(x,y,\theta,r)}{r^{1-2\delta}} < +\infty$$

and Lemma 1 follows from the arbitrariness of δ.

To prove (*), write (following [M]):

$$\int a(x,y,\theta,r)\,d\nu(\theta) = \int_A m\{(x',y') : |\operatorname{proj}_{\theta,x}(x',y') - y| \le r\}\,d\nu(\theta)$$

$$= \int_X \nu\{\theta : |\operatorname{proj}_{\theta,x}(x',y') - y| \le r\}\,dm(x',y')$$

For a fixed (x',y') the set of directions such that $|\operatorname{proj}_{\theta,x}(x',y') - y| \le r$ is an interval of size smaller than $\frac{C''r}{|(x,y)-(x',y')|}$, so that the last integrand can be estimated from above by a function of the distance to (x,y): $\min\Big((A),$ $\frac{CC''r^{1-\delta}}{|(x,y)-(x',y')|^{1-\delta}}\Big)$, and (*) follows, e.g. by summing on the dyadic annuli about (x,y) (see [M]). This achieves the proof of the lemma and thus of Proposition 1 in the case $\phi = \phi_0$.

In the general case of a $C^{1+\epsilon}$ function ϕ, the strong stable leaves might curve and the above argument has to be modified. One can estimate in a similar way as in the lemma the following quantities $b(x,y,\underline{\xi},r,t)$: $b(x,y,\underline{\xi},r,t) = m\{(x',y') : |\operatorname{proj}_{\underline{\xi},x}(x',y') - y| < r$ and $|(x,y)-(x',y')| < r^t\}$ for some $t < 1$ close enough to 1. Using property iii) in Proposition 2, this suffices to conclude that $\delta_2 - \delta_1 = 1$ and therefore to prove Proposition 1. Details for the argument in a very similar setting are given in [LY$_2$] section 4.

III. PROOF OF PROPOSITION 2

For the sake of notational simplicity, we shall assume that the discontinuities of the derivative ϕ' may only occur at 0 and 1/2. Let P be the partition of $[0,1] \times \mathbb{R} \times \Omega$ defined by $P = \{P_0, P_1\}$ where

$$P_0 = \{0 \le x < 1/2\}, \quad P_0 = \{1/2 \le x < 1\}.$$

The partition P is Bernoulli (i.i.d. $\{1/2, 1/2\}$) and generates under F. The entropy $h(P,F)$ is $\ln 2$. We denote $P(x)$ or $P(x,y,\underline{\xi})$ the element of P which contains $(x,y,\underline{\xi})$.

Let ζ be the partition defined by the local strong stable manifold. The partition ζ is decreasing, $F\zeta = \zeta \vee P$ and we denote h_1 its entropy

$$h_1 = H(F\zeta/\zeta) = -\int \ln m_{x,y,\underline{\xi}}(P(x,y,\underline{\xi}))\,d\mu.$$

We also have (see [LY$_1$] Lemma 9.3.1) for μ a.e. $x,y,\underline{\xi}$:

$$h_1 = \lim_{n\to\infty} -\frac{1}{n}\ln m_{x,y,\underline{\xi}}\Big(\bigcap_{j=0}^{n-1} F^j P(F^{-j}(x,y,\underline{\xi}))\Big).$$

Since $\bigcap\limits_{j=0}^{n-1} F^j P(F^{-j}(x,y,\underline{\xi}))$ is the rectangle based upon the dyadic interval of order n containing x it follows easily from this limit that δ_1 exists μ a.e. (property iii of Proposition 2) and that $h_1 = \delta_1 \ln 2$.

In general, the partition ζ does not catch the whole entropy of the system, and the next step is to consider an analogous "vertical" entropy and prove analogous properties, namely that γ_2 exists (property iv) and that

$$\ln 2 - h_1 = (2 - s)\gamma_2 \ln 2$$

(which is property ii given that $h_1 = \delta_1 \ln 2$). We follow [LY$_1$] section 11: Let $B^T(x, y, \underline{\xi}, \delta)$ be the set of local strong stable manifolds defined by $\underline{\xi}$ and at a distance less than δ of the local strong stable manifold of (x, y). Here the distances between leaves is easy to define since they are parallel curves. Define $g(x, y, \underline{\xi}) = m_{x,y,\underline{\xi}}(P(x, y, \underline{\xi}))$ and

$$g_\delta(x, y, \underline{\xi}) = \frac{m(B^T(x, y, \underline{\xi}, \delta) \cap P(x, y, \underline{\xi}))}{mB^T(x, y, \underline{\xi}, \delta)}.$$

As $\delta \to 0$, $g_\delta \to g$ μ a.e. and in L^1 by the Lebesgue density theorem. We shall use the fact that for μ a.e. $(x, y, \underline{\xi})$

$$(\ast\ast) \qquad \lim_{n \to \infty} \frac{1}{n} \sum_{i=0}^{n-1} \ln g_{\delta_{n,k}}(F^{-k}(x, y, \underline{\xi})) = \int \ln g \, d\mu = -h_1$$

where $\delta_{n,k} = 2^{(n-k)(s-2)}$. The proof of $(\ast\ast)$ parallels the proof of the Mc Millan Breiman theorem as e.g. in [B]. We can write $mB^T(x, y, \underline{\xi}, 2^{n(s-2)})$ as:

$$mB^T(F^{-n}(x, y, \underline{\xi}), 1) \times \prod_{k=0}^{n-1} \frac{mB^T(F^{-k}(x, y, \underline{\xi}), 2^{(n-k)(s-2)})}{mB^T(F^{-k-1}(x, y, \underline{\xi}), 2^{(n-k-1)(s-2)})}$$

By applying F we see that

$$mB^T(F^{-k-1}(x, y, \underline{\xi}), 2^{(n-k-1)(s-2)}) =$$

$$= \frac{m(B^T(F^{-k}(x, y, \underline{\xi}), 2^{(n-k)(s-2)}) \cap P(F^{-k}(x, y, \underline{\xi})))}{mP(F^{-k}(x, y, \underline{\xi}))}$$

$$= 2g_{\delta_{n,k}}(F^{-k}(x, y, \underline{\xi}))mB^T(F^{-k}(x, y, \underline{\xi}), 2^{(n-k)(s-2)}).$$

We obtain:

$$-\frac{1}{n} \ln mB^T(x, y, \underline{\xi}, 2^{n(s-2)}) = \ln 2 + \frac{1}{n} \sum_{i=0}^{n-1} \ln g_{\delta_{n,k}} \cdot F^{-k} - \frac{1}{n} \ln mB^T(F^{-n}(\cdot), 1).$$

By $(\ast\ast)$ we get the existence of γ_2 and the entropy/dimension formula ii. The final step is to conclude that δ_2 exists and equals to $\delta_1 + \gamma_2$ (properties v) and i)). Let us denote $\underline{\delta}_2$ and $\bar{\delta}_2$ the corresponding a.e. value of lim inf and lim sup of $\frac{\ln mB(.,n)}{\ln \eta}$ as $\eta \to 0$. Firstly, given that the space is C^1 equivalent to a product of two intervals the inequality $\delta_1 + \gamma_2 \leq \underline{\delta}_2$ follows from general dimension theory ([LY$_1$] Lemma 11.3.1). The proof that

$$(2 - s)\bar{\delta}_2 \ln 2 \leq \ln 2 - h_1 + (2 - s)\delta_1 \ln 2$$

is the content of [LY$_1$] section 10.2. The idea is that a typical ball of radius $2^{n(s-2)}$ has measure less than $2^{n(s-2)\bar{\delta}_2+n\epsilon}$ and contains some number of rectangles of measure 2^{-n}. A lower estimate of that number of rectangles is obtained by intersecting with a typical strong stable leaf and by considering the conditional measure on this leaf. The conditional measure of the ball is at least $2^{(s-2)n\delta_1-n\epsilon}$ and good rectangles are of conditional measure smaller than $e^{-n(h_1-\epsilon)}$. Handling the precise meaning of "typical" and "good" in the above argument is routine in measure theory, see [LY$_1$] section 10.2 for a more precise treatment, and this achieves the proof of Proposition 2.

Observe that Proposition 2 and its proof are valid when the unstable foliation do not depend on the past either because ϕ' is constant a.e. like in the Rademacher function [PU]

$$\begin{aligned} \phi_R(x) &= 1 \quad \text{for } (x) \leq 1/2 \\ &= 0 \quad \text{for } (x) > 1/2 \end{aligned}$$

or the twisted Rademacher function $\phi_T(x) = 2x(\bmod 1)$ or for other examples of graphs like self-affine graphs [PU], [BU$_2$], [K], [U]. Proposition 2 subsumes the results in these papers saying that if the projection of m on the vertical axis has dimension 1, then the whole set has dimension s. In some sense, Proposition 1 is the generalization of this idea to Weierstrass graphs.

The last remark is that when applied to the Rademacher function ϕ_R, Proposition 2 iv says that the law of $\sum_{n \geq 0} \theta^n \epsilon_n$, ϵ_n Bernoulli variables, has always some dimension. When a measure on the interval has a dimension, then all notions of dimension coincide [Y]. This justifies the use of entropy dimension formula in section I, although this formula is only established in [L$_2$] with a weaker notion of dimension.

REFERENCES

[A]   R. L. Adler, *Entropy of Skew Product Transformations*, Proc. Amer. Math. Soc. **14** (1963), 655–669.

[AR]   L. M. Abramov and V. A. Rohlin, *Entropy of a Skew Product Transformation with invariant measure*, Vestnik Leningrad University **17** (1962), 5–13; A.M.S. Transl. Series 2 **48** (1966), 255-265.

[B]   P. Billingsley, *Ergodic thoery and information*, John Wiley and Sons, New York, 1965.

[BL]   M. V. Berry and Z. V. Lewis, *On the Weierstrass Mandelbrot fractal function*, Proceedings Royal Soc. London A370 (1980), 459–486.

[BU$_1$] A.S. Besicovitch and H.D. Ursell, *Sets of fractional dimensions (V). On dimensional numbers of some continuous curves*, J. London Math. Soc. **12** (1937), 18–25.

[BU$_2$] T. Bedford and M. Urbanski, *The box and Hausdorff dimension of self-affine sets*, Ergod. Th. &. Dyn. Sys. **10** (1990), 627–644.

[E]   P. Erdös, *On the smoothness properties of a family of Bernoulli convolutions*, Amer. J. Math **62** (1940), 180–186.

[H]   F. Hofbauer, *β shifts have unique maximal measure*, Mh. Math **85** (1978), 189–198.

[K]   N. Kôno, *On self-affine functions*, Japan J. Appl. Maths **3** (1986), 259–269.

[L₁]   F. Ledrappier, *Some properties of absolutely continuous invariant measures on an interval*, Ergod. Th. & Dyn. Sys **1** (1981), 77–93.

[L₂]   F. Ledrappier, *Une relation entre entropie, dimension et exposant pour certaines marches aléatoires*, C. R. Acad. Sc. Paris **296** (1983), 369–372.

[LY₁]   F. Ledrappier and L-S Young, *The metric entropy of diffeomorphisms. Part II: Relations between entropy, exponents and dimension*, Ann Maths **122** (1985), 540–574.

[LY₂]   F. Ledrappier and L-S Young, *Dimension Formula for Random Transformations*, Commun. Math. Phys. **177** (1988), 529–548.

[M]   J. M. Marstrand, *Some fundamental geometrical properties of plane sets of fractional dimensions*, Proc. London Math. Soc. **4** (1954), 257–302.

[MW]   R. D. Mauldin and S. C. Williams, *On the Hausdorff dimension of some graphs*, TAMS **298** (1986), 793–803.

[PU]   F. Przytycki and M. Urbański, *On the Hausdorff dimension of some fractal sets.*, Studia Mathematica **93** (1989), 155–186.

[R]   D. Ruelle, *Ergodic theory of differentiable dynamical systems*, Publ. Math. I. H. E. S. **50** (1979), 27–58.

[T]   Y. Takahashi, *Isomorphisms of β. automorphisms to Markov automorphisms.*, Osaka J. Math **10** (1973), 175–184.

[U]   M. Urbanski, *The probability distribution and Hausdorff Dimension of Self affine functions*, Prob. Th. and Rel. Fields **84** (1990), 377–391.

[Y]   L-S Young, *Dimension, entropy and Lyapunov exponents*, Ergod. Th. & Dyn. Sys. **2** (1982), 109–124.

Laboratoire de Probabilités, Université Paris VI, 4, Place Jussieu - T. 46–56, F - 75252 PARIS Cédex 05

Contemporary Mathematics
Volume **135**, 1992

SYNCHRONIZING PREFIX CODES AND AUTOMATA
AND THE ROAD COLORING PROBLEM

Dominique Perrin, Marcel-Paul Schützenberger

LITP

Institut Blaise Pascal

Paris

Abstract

We prove two new results concerning the existence of synchronizing words for prefix codes. Both results assert that any finite aperiodic maximal prefix code is equivalent to a synchronizing one under two equivalence relations to be defined more precisely below. One of these equivalence relations is that of tree isomorphism and is the subject of a conjecture, known as the road coloring conjecture, that is settled in the case corresponding to our hypotheses.

1. INTRODUCTION

The notion of a *synchronizing word* is a basic and elementary notion in automata theory. Given a finite deterministic automaton, a word x is called synchronizing if the state reached after processing the word x is independent of the initial state in which the automaton was started. This notion has been studied since the beginning of automata theory and appeared with E.F. Moore's "gedanken experiments". It also appears in many recent developments concerning automata (see e.g. Aho, 1988 or Eppstein, 1990). The term "synchronizing word" is however not universally in use and one may find instead *resolving block* (Adler, Marcus, 1979) or *reset sequence* (Eppstein, 1990).

From the abstract point of view, synchronizing words correspond, in the semigroup of transitions of the automaton, to elements of minimal possible rank. This algebraic formulation allows a generalization to non-deterministic automata (see Berstel, Perrin, 1985). From another viewpoint, the existence of synchronizing words guarantees an almost everywhere one-to-one correspondance between paths and their labels in an

---

1991 Mathematics Subject Classification. Primary 68Q70; Secondary 20M05.

This paper is in final form and no version will be submitted for publication elsewhere.

appropriate measure space. It is this property which is of interest in the applications to coding since it guarantees stability against errors.

It is curious that such a simple notion gives rise to several unsolved problems. We mention two of them in this introduction : The Cerny-Pin conjecture and the road coloring conjecture.

The Cerny conjecture asserts that any n-state synchronizing automaton has a synchronizing word of length at most $(n-1)^2$. It is easy to prove the existence of a synchronizing word of length bounded by a cubic polynomial in n but no quadratic bound has yet been obtained. The simple example of the automaton of Figure 1.1 shows that the bound $(n-1)^2$ cannot be improved.

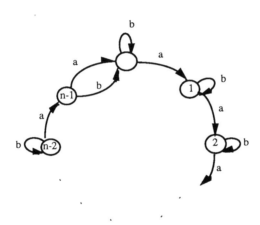

Figure 1.1. A worst case for Cerny's problem

The conjecture has been put in a more general form by Pin : if there is a word of rank d (as a mapping from the state set into itself) then there is one of length at most $(n-d)^2$. A bibliography on this problem can be found in (Berstel, Perrin, 1985). A recent result by A. Carpi (1988)shows that a cubic bound also holds in the case of unambiguous automata.

The road coloring problem is encountered in the study of isomorphism of symbolic dynamical systems (Adler, Goodwin, Weiss, 1977). It is conjectured that, except for the trivial case of periodicity, it is always possible to modify the labeling of the graph underlying a deterministic automaton to make it synchronizing. The name comes from the analogy where the states of the automaton are cities and the edges roads connecting them. An appropriate coloring would allow a traveller to find his way to some city by following a fixed rule specifying the appropriate succession of colors, irrespective of his starting point. The conjecture is presently unsettled.

In this paper, we prove two new results on synchronizing words. Our results start with a prefix code instead of an automaton. Both notions are strongly related since, for any deterministic automaton, the set of first returns to a given state is a prefix code. However, our hypothesis are more easily formulated in terms of prefix codes.

We introduce an equivalence on prefix codes, called *the flipping equivalence*. It corresponds to isomorphism of the associated unlabeled trees.

Our first result is that any finite aperiodic maximal prefix code is flipping equivalent to a synchronizing one. The proof uses in a crucial way a theorem of Reutenauer (1985) giving a non-commutative factorization of the polynomial associated with a prefix code.

Our second result is a modification of the first one for another equivalence relation : the commutative equivalence which identifies words differing only in the relative ordering of their letters. The proof is quite similar to that of the previous result.

The first result settles, under our hypotheses, the road coloring problem. In terms of the original formulation, it settles it in the case of graphs satisfying the additional assumption that all vertices except one have exactly one entering edge. Such graphs are sometimes refered to as "renewal systems" in symbolic dynamics.

Our paper is organized as follows. In Section 2, we recall the definitions and results to be used later, especially the factorization theorem of Reutenauer. In Section 3, we discuss the case of an equivalence relation which is a common refinement of the two equivalence relations considered above. We reproduce a result of (Schützenberger, 1967) with part of its proof with the intention both of updating the statement and to prepare the study of the road coloring problem given in Section 6. In Section 4, we prove our main result concerning flipping equivalence. The corresponding result for commutative equivalence is proved in Section 5. Finally, in Section 6 we discuss the exact relationship of our results with the road coloring problem.

2. PREFIX CODES

In all that follows we use the notation and terminology of (Berstel, Perrin, 1985). For the sake of readability we recall most of the definitions.

Let A be an alphabet. We denote by A^* the free monoid on the set A, which is the set of all finite sequences on A equipped with the concatenation as a product, the neutral element being the empty sequence, called the empty word. We denote by 1 the empty word and by $A^+ = A^* - 1$ the free semigroup on A. In general, we recall that a monoid is a set with a binary associative operation and a neutral element whereas a semigroup is the same but without the necessity of a neutral element.

A *prefix code* on A is a subset X of A^+ which contains no proper prefix of any of its elements. A prefix code can be identified with a labeled tree. Thus the prefix code $X = \{aa, ab, baa, bab\}$ on $A = \{a, b\}$ corresponds to the binary tree represented on Figure 2.1 with an obvious convention for the labeling using

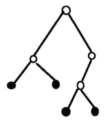

Figure 2.1. A prefix code

a for left and b for right. The words of X are in 1-1 correspondance with the leaves of the associated tree.

A *prefix* of X is a proper prefix of some word of X. The set of prefixes thus corresponds bijectively to the internal nodes of the associated tree.

For a subset X of A^*, we denote by X^* the submonoid generated by X. When X is a prefix code, X^* is free with basis X. This is the origin of the term "code" which refers in general to the uniqueness of parsing or deciphering.

A prefix code is said to be *maximal* when it is maximal under inclusion among the prefix codes on the alphabet A. It is easy to verify that a prefix code X is maximal iff it is *right complete*, that is to say that for each word w in A^* one has

$$wA^* \cap XA^* \neq \emptyset \tag{2.1}$$

Equation (2.1) means that every word is comparable to a codeword for the prefix ordering. It is not difficult to prove that it is equivalent to the fact that each word w in A^* is a prefix of some word in X^*, i.e.

$$wA^* \cap X^* \neq \emptyset \tag{2.2}$$

In terms of trees, a prefix code is maximal iff the associated tree is a complete k-ary tree, where $k = Card(A)$.

We shall mainly discuss here *finite* prefix codes. We shall however occasionnally consider a much weaker condition defined as follows. A prefix code X on the alphabet A is called *thin* if there exists a word w in A^* that does not appear inside words of X, i.e. such that

$$A^* w A^* \cap X = \emptyset \tag{2.3}$$

A finite prefix code X is thin since only words of bounded length may appear inside words of X.

We now come to the definition of the objects of central interest to us. A word x is said to be *synchronizing* for a prefix code X if wx is in X^* for all words w in A^*. Hence x is synchronizing iff

$$A^* x \subset X^* \tag{2.4}$$

A prefix code X is called synchronizing if there exists a synchronizing word for X. A synchronizing prefix code is obviously maximal since Formula (2.4) is a uniformisation of Formula (2.2) It is also thin since no element of X contains x properly.

For instance, the prefix code $X = \{aa, ab, baa, bab, bb\}$ represented on Figure 2.2. (i) admits $x = baa$ as a synchronizing word as the reader may check by a little reasoning. On the contrary, the code $X = \{aa, ab, ba, bb\}$ of Figure 2.2 (ii) is not synchronizing and the same is true of any code in which all words have the same length not equal to one.

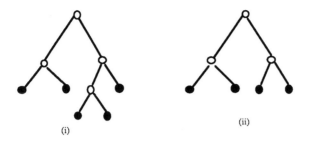

(i)   (ii)

Figure 2.2. (i) A synchronizing prefix code and (ii) a non synchronizing one

We shall see below how to systematically look for synchronizing words.

We also need some terminology from automata theory. Let Q be a set. An *automaton* on Q is given by a function

$$\delta : Q \times A \to Q$$

This function defines a right action of A^* on Q. We denote this action by a dot, writing $q.a$ instead of $\delta(q, a)$.

Given an element $i \in Q$, the *stabilizer* of i is the set

$$Stab(i) = \{x \in A^* \mid i.x = i\}$$

It can be verified that $Stab(i)$ has the form $Stab(i) = X^*$ with X a prefix code, sometimes called the set of *first returns*. Conversely any prefix code can be obtained in this way. One may further assume that all elements q of Q play a role in the sense that there exist u, v in A^* such that $i.u = q$ and $q.v = i$. We say in this case that the automaton is *trim* or *irreducible*.

We define the *rank* $r(w)$ of a word w as the number of elements of Q reachable through w, i.e.

$$r(w) = Card\{q.w \mid q \in Q\}$$

A word $x \in X^*$ is clearly synchronizing iff $r(x) = 1$. In general, the *degree* of X denoted $d(X)$ is the minimal non-zero value of the ranks of the words of A^*. It can proved that it does not depend on the automaton used to obtain X (provided it is trim). Hence X is synchronizing iff $d(X) = 1$.

A finite prefix code can be obtained from a finite automaton. It is synchronizing iff its degree is equal to one. In this case, the automaton itself is also called synchronizing. For example, the prefix code of Figure 2.2 (i) corresponds the first return at node 1 in the automaton given on Figure 2.3.

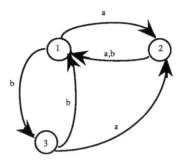

Figure 2.3. A finite automaton

The search for a synchronizing word is easily done with a finite automaton. It reduces to a search in the graph obtained by considering the action of the letters on the subsets of the state set. A synchronizing word is one that is the label of a path from the set of all states to a singleton set. The graph corresponding to the automaton of Figure 2.3 is represented on Figure 2.4 with only part of the edges represented. It allows one to find easily the synchronizing word $x = baa$.

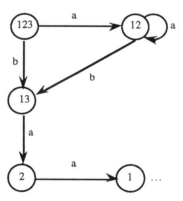

Figure 2.4. The action on subsets(partial drawing)

It is of course not true that, conversely a prefix code obtained from a finite automaton is itself finite, since there may be cycles in the graph of the automaton that do not use the special state i. One may show however that the code is thin when the automaton is finite (see Berstel, Perrin, 1984).

The *period* of a prefix code X is the gcd of the lengths of its elements. It is known that, *for a maximal prefix code of finite degree, the period is a divisor of the degree* (see Berstel, Perrin, 1984 p. 242). This implies that a prefix code can be synchronizing only when it is of period 1. The study of synchronizing automata or codes deals with the problem of finding additional conditions ensuring that the converse implication holds.

Several other properties relating the degree to other parameters are also known. A useful one is the following : For a finite maximal prefix code X, *the degree is a divisor of each of the integers n such that $a^n \in X$ for $a \in A$* (see Berstel, Perrin, 1984 p. 117).

We will use on several occasions non-commutative polynomials and series. We recall here the basic notions on this subject. A systematic exposition can be found in (Cohn, 1985) or (Berstel, Reutenauer, 1988).

We denote by $\mathbb{Z} << A >>$ the ring of series with coefficents in $\mathbb{Z}$ and non-commutative variables in A and by $\mathbb{Z} < A >$ the corresponding ring of polynomials. For a serie S, we denote by (S, w) the value of S on the word w, also called the coefficient of w in S. We shall write

$$S = \sum_{w \in A^*} (S, w) w$$

The *support* of a series S, denoted $supp(S)$ is the set of words w such that $(S, w) \neq 0$. A serie is a polynomial iff its support is finite.

We shall not distinguish between a subset X of A^* and its characteristic series, writing therefore

$$X = \sum_{x \in X} x$$

We denote by $|P|$ the *degree* of a polynomial P, which is the maximum of the lengths $|w|$ of the elements w in its support. We also denote by $\widehat{P}$ the homogeneous component of P of maximum degree. Therefore

$$(\widehat{P}, w) = \begin{cases} (P, w) & \text{if } |w| = |P| \\ 0 & \text{otherwise} \end{cases}$$

For u in A^* and S in $\mathbb{Z} << A >>$ we denote $u^{-1}S$ the series defined by

$$(u^{-1}S, w) = (S, uw)$$

with the symmetric definition for Su^{-1}.

We shall use several times the fact that the set of homogeneous polynomials is a *free* subsemigroup of $\mathbb{Z} < A >$.

We now show the interplay between codes and polynomials.

Let X be a maximal prefix code on A and let P be the set of its proper prefixes including the empty prefix. We have the equality

$$X - 1 = P(A - 1) \tag{2.5}$$

in which 1 denotes the empty word.

Formula (2.5) expresses a factorisation property. It is easy to derive from the equality between sets

$$PA + 1 = X + P$$

expressing the fact that a prefix followed by a letter is either still a prefix or is a word of X.

A much deeper factorisation property was given by Reutenauer(1985). We state it below in its simplified version concerning prefix codes although his result is more general and holds for general codes.

THEOREM 2.1 (REUTENAUER). — *Let X be a finite maximal prefix code on the alphabet A. There exists two polynomials $L, D \in \mathbb{Z} < A >$ such that*

$$X - 1 = L(d + (A - 1)D)(A - 1) \tag{2.6}$$

where d denotes the degree of X.

A proof of the result is presented in the book of (Berstel, Reutenauer, 1988). It is important to see that when X is not synchronizing, i.e. when $d > 1$, the central factor in the right handside of (2.6) is non trivial. In fact, assuming that the constant term of L is 1, the constant term of D must be $d - 1$, which implies $D \neq 0$.

Also comparing (2.5) and (2.6), we obtain the equality

$$P = L(d + (A - 1)D) \tag{2.7}$$

which expresses a factorisation of the polynomial of prefixes of X.

Equality (2.6) can be rewritten

$$X - 1 = L(A - 1)(d + D(A - 1)) \tag{2.8}$$

By inverting both sides and using the identity $X^* = (1 - X)^{-1}$ we obtain

$$A^* = (d + D(A - 1))X^*L \qquad (2.9)$$

We conjecture that for any finite maximal prefix code X of degree d there exist a finite collection of d disjoint maximal prefix codes $T_i(1 \leq i \leq d)$ and a set L such that

$$A^* = (\sum_{i=1}^{d} T_i)X^*L \qquad (2.10)$$

Such an equality implies the existence of a factorization like (2.9) since, letting $T_i - 1 = U_i(A - 1)$ we have

$$A^* = (d + (\Sigma U_i)(A - 1))X^*L \qquad (2.11)$$

It implies the stronger property that the polynomials L, D in (2.6) can be chosen to have positive coefficients. It also implies that the degree of X is at least equal to d according to the following observation.

PROPOSITION 2.3.. — *Let X be a finite maximal prefix code on the alphabet A such that*

$$X - 1 = L(A - 1)R \qquad (2.12)$$

with L, R two subsets of A^*. If R the disjoint union of d maximal prefix codes, then X is of degree at least equal to d.

Proof : We first show that each element of R is a suffix of an element of X. Let indeed r be in R an let $l \in L$ be chosen of length $|L|$. Then, for any letter a in A, lar has coefficient at least one in $X + LR$. Since l is of maximal length, this implies that either r is a suffix of X or it is a suffix of some other element of R. This proves the property by ascending induction on $|r|$.

We now consider an automaton on Q such that X is the set of first returns to a state i. We will show that any word in A^* has at least d states in its range. Let $w \in A^*$ be longer than $|X|$. Then w has d prefixes $t_1, \ldots, t_d$ in R. Since each t_k is a suffix of an element of X, there is a state q_k such that $q_k.t_k = i$.

For each $k = 1, \ldots, d$, let r_k be the state defined by

$$r_k = q_k.w$$

We will verify that all r_k are distinct and this will prove the claim. Let indeed k, ℓ be such that $r_k = r_\ell$. Since we may concatenate w on the right by any word we may suppose that $r_k = r_\ell = i$. Let $w = t_k x_k = t_\ell x_\ell$. Then x_k, x_ℓ are in X^* since they stabilize i. But then the word w has two distinct factorizations in the product RX^*L namely (t_k, x_k, ε) and $(t_\ell, x_\ell, \varepsilon)$, (see Figure 2.5). This is a contradiction since (2.12) is equivalent to the equation

$$A^* = RX^*L$$

□

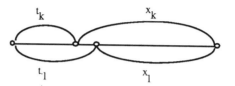

Figure 2.5. Two parsings of w

It is known that a factorization like Eq. (2.10) holds for biprefix codes with $L = 1$ (see Berstel, Perrin, 1985). In the general case, the answer is not known. It is a particular case of a more general conjecture on codes known as the factorization conjecture (ibid. p. 423).

3. LENGTH DISTRIBUTIONS

For a subset X of A^*, the sequence of numbers $\alpha = (\alpha_n)_{n \geq 0}$ given by

$$\alpha_n = Card(X \cap A^n)$$

is called the *length distribution* of X. We also denote

$$f_X(t) = \sum_{n \geq 0} \alpha_n t^n$$

the corresponding generating series. We denote by ρ_X or ρ_α the radius of convergence of the series $f_X(t)$. Let $q = Card(A)$. Since $\alpha_n \leq q^n$, we have $\rho_X \geq 1/q$.

When X is thin, we have $\rho_X > 1/q$ (see Eilenberg, 1974 p. 230 or Berstel, Perrin 1985 p. 67).

A sequence $(\alpha_n)_{n \geq 0}$ is the length distribution of a prefix code on a q-letter alphabet iff it satisfies the inequality

$$\sum_{n \geq 1} \alpha_n q^{-n} \leq 1 \tag{3.1}$$

Inequality (3.1) goes back to C. Shannon and it is at times referred to as *Kraft Inequality*. When X is a thin maximal prefix code, we have

$$\sum_{n \geq 1} \alpha_n q^{-n} = 1 \tag{3.2}$$

Indeed, by Equality (2.5) we have

$$f_X - 1 = f_P(qt - 1) \tag{3.3}$$

Since X is thin, P is thin and therefore $\rho_P > 1/q$. Evaluating both sides of (3.3) at $t = 1/q$ gives the desired equality. Conversely, we have the following

THEOREM 3.1 (SCHÜTZENBERGER, 1967). — *If α is a sequence satisfying Equality (3.2) and $\rho_\alpha > 1/q$, it is the enumerating sequence of some thin maximal prefix code. Moreover, the code can also be chosen synchronizing provided the sequence $(\alpha_n)_{n \geq 0}$ satisfies the additional requirement that the integers α_n are relatively prime.*

We shall reproduce here the part of the proof of this result needed for the purpose of a discussion presented in Section 6.

It is not difficult to see that the first part is true. Indeed, if $\rho_\alpha > 1/q$, we may build a prefix code X with length distribution α such that some word w does not appear within any word of X. Then X is thin and maximal. We may always choose $w = a^k$ for some letter a in A. Then X satisfies the following inclusion

$$A^* w \subset X^* a^*$$

and to choose X synchronizing, we only need a word $x \in X^*$ such that

$$a^* x \subset X^*$$

One may show that except for a trivial case where the sequence α_n is ultimately equal to one, we may rearrange the words of X in a length-preserving way so that for some $b \in A$ and some integer $n \geq 0$ the prefix code $Y = X \cap (a^* \cup a^* b a^*)$ satisfies

$$Y = a^n \cup \{y_0, y_1, \cdots, y_{n-1}\} \tag{3.6}$$

where each $y_i = a^i b a^{\lambda_i - i - 1}$ is a word of length λ_i satisfying

$$i + 1 \leq \lambda_i \leq n \tag{3.7}$$

and there is an integer t with $1 \leq t \leq n$ such that $\lambda_i = n$ iff $i \geq n - t$ and finally the number λ_i are relatively prime.

The above conditions are satisfied in particular when $\lambda_0 \leq \lambda_1 \leq \cdots \leq \lambda_{n-1}$ and the λ_i are relatively prime.

The following lemma therefore completes the proof of Theorem 3.1.

LEMMA 3.2. — *If Y satisfies the above conditions, there exists a word y in Y^* such that $a^* y \subset Y^*$*

Proof : We denote $Q = \{0, 1, \cdots, n - 1\}$ and we define an action on Q by

$$i.a = (i + 1) \quad \mod.n$$
$$i.b = (i - \lambda_i + 1) \quad \mod.n$$

The corresponding automaton is such that $Y^* = Stab(0)$. Let M be the transition monoid of the automaton, which is the monoid of all mappings from Q into Q obtained by the action of all words. For each d with $1 \leq d \leq n$, let

$$I_d = \{n - d, \cdots, n - 2, n - 1\}$$

and let M_d be the monoid

$M_d = \{m \in M \mid Q.m = I_d$ and $i.m = i$ for all $i \in I_d\}$.

We want to prove that M_1 is not empty. This implies our conclusion since a word z defining an element of M_1 satisfies

$$a^*(za) \subset Y^*$$

In the sequel we do not distinguish between a word and the element of M that it defines. We first observe that for all $i \in Q$ we have

$$i.ba^{n-1} \geq i$$

with equality iff $i \in I_t$. Thus ba^{n-1} has a power which belongs to M_t. This proves that M_t is not empty. We shall now prove that if M_s with $1 < s \leq t$ is not empty, then some M_r with $1 \leq r < s$ is not empty. For q in Q, we denote $[q]_s$ the integer in $\{1, 2, \cdots, s\}$ congruent to q mod. s. Let m be an element of M_s.

Case 1. There exists a p with $1 \leq p \leq n$ such that $(n - p).m \neq n - [p]_s$. We choose the smallest p satisfying this inequality. Then $p > s$ by the definition of M_s. Let

$$m' = ma^{n+s-p}m$$
$$J_s = I_s - (n - [p]_s)$$

Since $Q.m = I_s.m$, we have $Q.m' = I_s.m'$. For all s' with $1 \leq s' < s$ we have

$$
\begin{aligned}
(n - s').m' &= (n - s').a^{n+s-p}m \\
&= (n + (s - s') - p).m \\
&= n - [p + s']_s
\end{aligned}
$$

and

$$
\begin{aligned}
(n - s).m' &= (n - s).a^{n+s-p}m \\
&= (n - p).m \\
&\neq n - [p]_s
\end{aligned}
$$

Hence $Q.m' = J_s$. If $[p]_s = 1$, the element $m_1 = m'a$ belongs to M_{s-1}. Otherwise, let $[p]_s = k > 1$, and let

$$m_2 = m'(a^{n-1}m)^{s-k}$$

We have, for $1 \leq s' < s$,

$$(n - s').m' = n - [k + s']_s$$

and for $1 \leq s' \leq s$

$$(n - s').(a^{n-1}m) = (n - s' - 1).m = n - s' - 1$$

whence

$$(n - s').(a^{n-1}m)^{s-k} = n - s' - k$$

and finally

$$(n - s').m_2 = n - s'.$$

Hence m_2 belongs to M_{s-1} whence the desired conclusion in this case also.

Case 2. For all p with $1 \leq p \leq n$ we have $(n - p).m = n - [p]_s$. We first suppose that s does not divide n. Let then $n = n's + d$ with $1 \leq d \leq s$. For all d' with $1 \leq d' \leq d$ we have

$$(n - d').a^d m = (d - d').m$$
$$= n - d'$$

Hence $a^d m$ fixes pointwise the set I_d. Also for $d < d' \leq s$ we have

$$(n - d').a^d m = (n + d - d').m$$
$$= n + d - d'$$

Hence some power of $a^d m$ belongs to M_d.

We are finally left with the case where s divides n. It is easy to see that this implies that $p.m \equiv p \mod.s$ for all $p \in I$ and $m \in M$. Since the λ_i are relatively prime, this implies $s = 1$, a contradiction. $\square$

An additional problem concerning length distributions is the following. When a prefix code X is the stabilizer of a state in a finite automaton, then the series $f_X(t)$ is rational (see Eilenberg, 1974). It is not completely known under which conditions the converse holds, i.e. under which additional assumptions Theorem 3.1 holds with the additional conclusion that the prefix code is a stabilizer in a finite automaton. See (Perrin, 1989) for a partial answer.

4. FLIPPING EQUIVALENCE

We introduce a transformation on prefix codes called *flipping*. It is defined as follows. Let X be a prefix code on the alphabet A. Let $a, b \in A$ be two letters and let u be a proper prefix of X. Let

$$X = X' + uaR + ubS$$

with X', R, S prefix codes. One has in fact $R = (ua)^{-1}X, S = (ub)^{-1}X$.

The prefix code

$$Y = X' + uaS + ubR$$

is said to be the image of X under an *elementary flip*. A flip is a composition of elementary flips. The flipping transformation defines an equivalence called the *flipping equivalence*. We denote

$$X \sim Y$$

two prefix codes X, Y which are flipping equivalent.

The flipping transformation is of course a very natural and simple one on the trees associated with prefix codes. Indeed, an elementary flip is just an exchange of two subtrees rooted at the sons of some vertex.

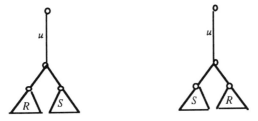

Figure 4.1. An elementary flip

We have represented on Figure 4.2 an equivalence class of the flipping equivalence. Actually two unlabeled complete binary trees correspond to flipping equivalent maximal prefix codes iff they are isomorphic, as one may easily verify.

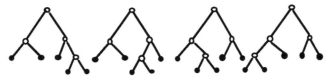

Figure 4.2. A flipping equivalence class

The flipping equivalence preserves some of the properties of prefix codes. First of all, two flipping equivalent pregix codes have the same length distribution (the converse implication is however not true). As a consequence, equivalent prefix codes have the same period. Also, two equivalent prefix codes are simultaneously maximal or not maximal.

We are going to prove the following result. It is, in the case of finite prefix codes, a reinforcement of Theorem 3.1

THEOREM 4.1. — *The flipping equivalence class of any finite maximal prefix code of period 1 contains at least one synchronizing prefix code.*

The proof relies on two lemmas. In the first lemma we start with a Reutenauer's factorization (2.8)

$$X - 1 = L(A - 1)(d + (A - 1)D)$$

and consider $R = (d + (A - 1)D)$. The homogeneous component of highest degree is, when $D \neq 0$

$$\widehat{R} = A\widehat{D}$$

The lemma shows that, except for the periodic case, the homogeneous polynomial $\widehat{R}$ is not equal to A^n. The condition given is the lemma is of course also sufficient.

LEMMA 4.2. — *If X is a finite maximal prefix code of period p such that*

$$X - 1 = L(A - 1)R$$

where $\widehat{R} = A^n$ for some $n \geq 1$, then R is a polynomial in A dividing $1 + A + \ldots + A^{p-1}$.

Proof.Let $E = (A - 1)R$. We first show that E is a polynomial in A. Let us suppose by induction on $m < n$ that

$$E = E' + \sum_{i=m+1}^{n} s_i A^i \tag{4.1}$$

with $\mid E' \mid \leq m$. Let g be in the support of $\widehat{L}$ and let h be a word of length m. For all words k of length $n - m$ we have $ghk \in supp(\widehat{L}\widehat{E}) \subset supp(X)$ and thus $ghk \in X$. Since X is prefix, we have $(LE, gh) = 0$.
But, by Formula (4.1) we have

$$(LE, gh) = (L, g)(E', h) + \sum_{i=m+1}^{n} r_{t+m-i} s_i \tag{4.2}$$

where r_i is the coefficient in L of the prefix of length i of g and $t = \mid g \mid$. Since $(LE, gh) = 0$, we deduce from (4.2) the Formula

$$(E', h) = -(1/(L, g)) \sum_{i=m+1}^{n} r_{t+m-i} s_i$$

It shows that (E', h) does not depend on the word h but only on its length m and proves that Equality (4.1) is true for $m - 1$. Thus we have proved by induction that E is a polynomial in A, i.e.

$$E = \sum_{i=0}^{n} s_i A^i$$

Let x be a word of X of length q. Let r, s be the polynomials in the variable z

$$r(z) = \sum_{i=0}^{q} r_i z^i \quad s(z) = \sum_{i=0}^{n} s_i z^i$$

where r_i is the coefficient in L of the prefix of length i of x. We have for each integer m such that $0 < m < q$ the equality similar to (4.2)

$$\sum_{i+j=m} r_i s_j = 0$$

since, X being prefix, the coefficient of the prefix of length m of x in LE is zero. We therefore have

$$z^q - 1 = r(z)s(z)$$

and the lemma is proved.   □

We now prove a second lemma. It shows that, in the non periodic case, we may use the flipping transformation to destroy the possibility of a non trivial factorization of the polynomial $X - 1$. For a finite maximal prefix code X, we denote by $e(X)$ the integer defined by

$$e(X) = \max\{e \geq 0 \mid X - 1 = L(A - 1)R, e = \mid R \mid\}$$

Thus, $e(X) > 0$ iff X has a non-trivial factorization. Consequently, $e(X) = 0$ implies that X is synchronizing.

LEMMA 4.3. — *Let X be a finite maximal prefix code such that*

$$X - 1 = L(A - 1)R \qquad (4.3)$$

with $\mid R \mid = n \geq 1$ and $\widehat{R} \neq A^n$ Then there exists a prefix code X' flipping equivalent to X such that

$$e(X') < e(X)$$

Proof. Let $E = (A-1)R$. We first note that Eq. (4.3) implies that $\widehat{X} = \widehat{L}A\widehat{R} = \widehat{L}\widehat{E}$. Therefore the homogeneous polynomials $\widehat{L}, \widehat{E}$ are unambiguous, i.e. have 0-1 coefficients. Let $g \in \widehat{L}$ and let Y be the finite maximal prefix code $Y = g^{-1}X$. We have $\widehat{Y} = \widehat{E}$. Since $\widehat{Y} \neq A^{n+1}$, there exists a prefix code Y' flipping equivalent to Y such that $\widehat{Y} \neq \widehat{Y}'$. Let X' be the maximal prefix code defined by the equality

$$X' - gY' = X - gY \qquad (4.4)$$

We have $X' \sim Y'$. Consider a non trivial factorization

$$X' - 1 = L'E' \qquad (4.5)$$

and suppose by contradiction that $\mid E \mid \leq \mid E' \mid < \mid X \mid$. Since $g\widehat{Y}' \subset \widehat{X}' = \widehat{L}'\widehat{E}'$, the set $\widehat{L}'$ contains a prefix g' of g. Let $g = g'h$ (see Figure 4.3).

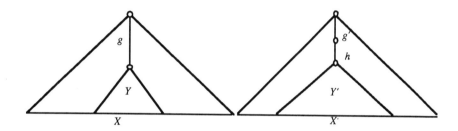

Figure 4.3. The codes X and X'

Let $F = h^{-1}E'$, $H = g'^{-1}L - h$ and let L_1, E_1', L_1' be defined by the following equalities

$$L = L_1 + g'H + g'h$$
$$E' = E_1' + hF$$
$$L' = L_1' + g'$$

Substituting into (4.3) and (4.5), we obtain

$$\widehat{X} = (\widehat{L}_1 + g'\widehat{H} + g'h)\widehat{E} \qquad (4.6)$$

$$\widehat{X}' = \widehat{L}'(\widehat{E}_1' + h\widehat{F}) \qquad (4.7)$$

By restricting Equality (4.4) on both sides to the words of maximal length beginning with g', we derive

$$\widehat{E'_1} = \widehat{H}\,\widehat{E} \qquad (4.8)$$

And by restricting (4.4) to the words of maximal length that do not begin by g' we obtain

$$\widehat{L'_1}(\widehat{E'_1} + h\widehat{F}) = \widehat{L_1}\widehat{E} \qquad (4.9)$$

Substituting in (4.9) the value of $\widehat{E'_1}$ given by (4.8) we have

$$\widehat{L'_1}h\widehat{F} = (\widehat{L_1} - \widehat{L'_1}\widehat{H})\widehat{E} \qquad (4.10)$$

Since $|\,\widehat{F}\,| = |\,\widehat{E}\,|$, we deduce from (4.10) that $\widehat{F} = \widehat{E}$.

This contredicts the hypothesis $\widehat{Y} \neq \widehat{Y'}$ since on one hand $\widehat{Y} = \widehat{E}$ and on the other hand $\widehat{F} = \widehat{Y'}$.□

We can now complete the proof of Theorem 4.1. We use an induction on the integer $e(X)$. The property is true when $e(X) = 0$ since then X itself is synchronizing. When $e(X) \geq 1$, we have $X - 1 = L(A - 1)R$ with $|\,R\,| = n \geq 1$. If $\widehat{R} = A^n$, then by Lemma 4.2, R divides $1 + A + \ldots + A^{p-1}$ with p the period of X. Hence, $p \geq n + 1 \geq 2$ in contradiction with the hypothesis $p = 1$. Therefore, $\widehat{R} \neq A^n$ and by Lemma 4.3, there exists an X' flipping equivalent to X such that $e(X') < e(X)$ whence the property by induction.

5. COMMUTATIVE EQUIVALENCE

There is another equivalence on prefix codes which is also a refinement of the length distxribution equivalence. This equivalence, called *commutative equivalence* is of more general interest since it applies to all subsets of the free monoid. We first recall its definition.

Two words u, v in A^* are said to be commutatively equivalent if for all a in A the number of occurrences of a in u is equal to the number of occurrences of a in v. We denote this equivalence by the symbol $\equiv$. Two subsets X, Y of A^* are said to be commutatively equivalent if there is a one-to-one mapping f from X onto Y such that for all x in X, one has $f(x) \equiv x$. We again denote $X \equiv Y$. Figure 5.1 represents two commutatively equivalent maximal prefix codes. Actually their equivalence class does not contain other prefix codes (compare with Figure 4.2)

Figure 5.1. Two commutatively equivalent prefix codes

We will basically use the same arguments as in the preceding section to prove the following result.

THEOREM 5.1. — *The commutative equivalence class of any finite maximal prefix code of period 1 contains at least one synchronizing prefix code.*

The proof goes along the same lines as the proof of Theorem 4.1. We choose X such that the integer $e(X)$ is minimal in its commutative equivalence class. Suppose, by contradiction, that X is not synchronizing. Then we have

$$X - 1 = L(A - 1)(d + D(A - 1))$$

with $\mid D \mid \geq 2$. By Lemma 4.2 we have $\widehat{D} \neq A^n$ since otherwise X would be of period $p \geq 2$. Consequently, there exists a word h such that for some pair of letters a, b in A we have

$$(ha)^{-1}\widehat{D} \neq (hb)^{-1}\widehat{D}$$

Let $U = (ha)^{-1}\widehat{D}, V = (hb)^{-1}\widehat{D}$. Let $g \in \widehat{G}$ and $Y = g^{-1}X$. We have $\widehat{Y} = A\widehat{D}$ and therefore

$$Y = W + ahbU + bhaV$$

Let

$$Y' = W + ahbV + bhaU$$

Since $ahb \equiv bha$, we have $Y \equiv Y'$. Let X' be the prefix code commutatively equivalent to X defined by

$$X' - gY' = X - gY$$

Then, one may use the same proof as in Lemma 4.3 to show that $e(X') < e(X)$, a contradiction. This proves Theorem 5.1

To close this section, we mention the fact that the commutative equivalence is the object of an important open problem about codes. It is indeed conjectured that any finite maximal code is commutatively equivalent to a prefix code (see Berstel, Perrin, 1985).

6. THE ROAD COLORING PROBLEM

We finally discuss the road coloring problem mentioned in the introduction and we relate it to the results of the previous sections.

Let $\mathcal{A}$ be a finite automaton given by a function

$$\delta : Q \times A \to Q$$

The underlying graph of $\mathcal{A}$ is the directed graph having Q as set of vertices and an edge (p, q) iff there is an $a \in A$ such that $\delta(p, a) = q$. It is therefore the graph obtained from the familiar diagram associated with the automaton after removing the labels of the edges. It has the property that all its vertices have the same outdegree, in fact equal to the number of symbols in A.

A graph is said to be *road colorable* if it is the underlying graph of some synchronizing automaton.

Recall that a graph is called aperiodic if there is an integer n such that the n-th power of its adjacency matrix has all its elements strictly positive. This is of course equivalent to the graph being strongly connected and the g.c.d of the cycle lengths being equal to one.

The conjecture formulated in (Adler, Goodwin, Weiss, 1977) is the following : *any aperiodic graph with all vertices of the same outdegree is road colorable.*

We reformulate Theorem 4.1 as follows to obtain a solution of this conjecture in a particular case.

THEOREM 6.1. — *Any aperiodic graph such that*
(i) all vertices have the same outdegree
(ii) all vertices except one have indegree one
is road colorable.

Proof. We define the *renewal automaton* of a finite maximal prefix code X to be the automaton having the set P of prefixes of X as set of states and the transition function defined by $\delta(p, a) = pa$ if $pa \in P$ and $\delta(p, a) = 1$ otherwise. The graph underlying the renewal automaton of X is therefore obtained from the unlabeled tree associated with X by merging all leaves with the root. Clearly, an elementary flip does not affect this graph. Hence, when X and X' are flipping equivalent, the underlying graphs of their renewal automata are the same and the result follows from Theorem 4.1.□

The road coloring conjecture is known to be true in some other particular cases. One of them (O'Brien, 1981) is that of graphs satisfying the additional assumptions

(i) there are no multiple edges

(ii) there is a simple cycle of prime length.

Another case, proved by Friedman (1990) is that of graphs containing a simple cycle of length prime to the weight of the graph. The weight of a graph is defined to be the sum of the components of an integer Perron left eigenvector chosen with its components relatively prime.

Some further particular cases have been investigated by A. Mahieux (1986).

In the paper of (Adler et al., 1977) the following result is proved : let G be an aperiodic graph with constant outdegree. Let M be the adjacency matrix of G and let n be an integer such that M^n has all its coefficients positive. For $k > 0$, let $G^{(k)}$ denote the graph having as vertices the paths of length h in G and edges the pairs (s, t) with $s = (s_1, \ldots, s_k), t = (s_2, \ldots, s_k, s_{k+1})$. Then $G^{(2n)}$ is road colorable. In terms of symbolic dynamics, this means that the system of finite type associated with G is conjugate to one that is road colorable. This result can actually also be proved using the construction of (Schützenberger, 1967) reproduced in Theorem 3.1. Indeed a splitting of the states of the graph will allow to label the cycles in such a way as to obtain a set of first returns containing the words described by Equations (3.3.6-7)

REFERENCES

[1] Adler R.L., Goodwin L.W., Weiss, B.,1977, Equivalence of topological Markov shifts, *Israel J. Math.* t.**27**, p. 49-63.

[2] Adler R.L., Marcus, B., 1979, Topological entropy and equivalence of dynamical systems, *Memoirs AMS*, **219**.

[3] Aho A., Dahbura, A., Lee, D., Uyar, M., 1988, An optimization technique for protocol conformance test generation based on UIO sequences and rural chinese postman tours, in *Protocol Specification, Testing and Verification VIII*, S. Aggarwal and K. Sabnani eds., North Holland.

[4] Berstel J., Perrin D., 1985,*Theory of Codes*, Academic Press.

[5] Berstel J., Reuteneauer, C., 1988, *Rational Series and their Languages*, Springer.

[6] Carpi, A., 1988, On synchronizing unambiguous automata, *Theoret. Comput. Sci.*,**60**, p.285-296.

[7] Cohn, P.M., 1985, *Free Rings and their Relations*, Academic Press (2nd edition).

[8] Eilenberg, S., 1974, *Automata, Languages and Machines*, Vol. A, Academic Press.

[9] Eppstein, D., 1990, Reset sequences for monotonic automata, *SIAM J. Comput.*, **19**, p. 500-510.

[10] Friedman, J., 1990, On the road coloring problem, *Proc. Amer. Math. Soc.* **110**, 1133-35.

[11] Mahieux, A., 1986, unpublished manuscript.

[12] O'Brien, G.L., 1981, The road coloring problem, *Israel J. Math*, t. **39**, p. 145-154.

[13] Perrin, D., 1989, Arbres et séries rationnelles, *C.R. Acad. Sci. Paris*, **309**, 713-716.

[14] Reutenauer, C.,1985, Noncommutative factorisation of variable-length codes *J. Pure applied Algebra* t.*36*, p.167-186.

[15] Schützenberger, M.P.,1967, On synchronizing prefix codes, *Information and Control*, t. **11**, p.396-401.

Contemporary Mathematics
Volume **135**, 1992

A zero entropy, mixing of all orders tiling system

SHAHAR MOZES

The purpose of this note is to give an explicit example of a two-dimensional subshift of finite type ("tiling") $(\Omega, \mathcal{B}, \mu, \mathbb{Z}^2)$ having zero two-dimensional entropy and such that the action of $\mathbb{Z}^2$ on it is mixing of all orders. This example is a specific instance of a large family of systems originating from the study of actions of Cartan subgroups of p-adic semisimple Chevalley groups, G, on homogeneous spaces, $\Gamma\backslash G$ (see [**M1**]). In [**M1**] we show how, via the use of the affine Bruhat-Tits building associated with G, one can obtain a subshift of finite type such that the action of the Cartan subgroup on $\Gamma\backslash G$ is a compact group extension of the subshift. The system described here is the subshift of finite type obtained for a specific group $G = PGL(2, \mathbb{Q}_p) \times PGL(2, \mathbb{Q}_l)$ modulo a certain quaternion lattice. In [**M2**] we prove a much more general result concerning mixing of all orders which implies that the system described here is mixing of all orders. However the method used there is completely different. Moreover for the case considered here we get a stronger quantitative assertion concerning the rate of mixing.

Let p, l be two distinct primes s.t. $p \equiv l \equiv 1(\mod 4)$ and $p, l \geq 13$. Let A, B be the following two sets of quaternions:

$$A = \left\{ a = a_0 + a_1 i + a_2 j + a_3 k \middle| \begin{array}{l} a_0, a_1, a_2, a_3 \in \mathbb{Z}, \ a_0 > 0, \ |a|^2 = p \\ a_0 \equiv 1, a_1 \equiv a_2 \equiv a_3 \equiv 0(\mod 2) \end{array} \right\}$$

$$B = \left\{ b = b_0 + b_1 i + b_2 j + b_3 k \middle| \begin{array}{l} b_0, b_1, b_2, b_3 \in \mathbb{Z}, \ b_0 > 0, \ |b|^2 = l \\ b_0 \equiv 1, b_1 \equiv b_2 \equiv b_3 \equiv 0(\mod 2) \end{array} \right\}$$

1991 *Mathematics Subject Classification*. 28D15, 28D20, 54H20.

Sponsored in part by the Edmund Landau Center for research in Mathematical Analysis supported by the Minerva Foundation (Federal Republic of Germany).

This paper is in final form and no version will be submitted for publication elsewhere.

PROPOSITION 1. (Jacobi, see [**L**])
A contains $p + 1$ elements. B contains $l + 1$ elements.   □

Let H_0 be the set of integral quaternions, congruent to 1 modulo 2. We will say that two quaternions are equivalent if one is a multiple of the other by some rational number. Using this equivalence relation H_0 is a group (w.r.t. multiplication). Let Γ be the subgroup of H_0 generated by $A \cup B$.

The symbols ("tiles") of the subshift will be unit squares whose edges are labeled (colored) by the elements of A and B in the following way:

    i. Vertical edges are labeled by elements of A.

    ii. Horizontal edges are labeled by elements of B.

    iii. $a, a' \in A$ and $b, b' \in B$ label a square as in figure 1. if and only if $ab = b'a'$. (Where equality is up to sign).

FIGURE 1

There are two adjacency rules:

    i. ("Domino rule") Two neighbouring squares have to have the same label on the common edge.

    ii. Two consecutive edges (vertical or horizontal) are not labeled by conjugate quaternions.

Let Ω be the set of all legal tilings. Ω inherits a topology from the product topology. Let $\mathcal{B}$ be the Borel σ-algebra of Ω.

$\mathbb{Z}^2$ acts on Ω by translations. We define an invariant measure μ on Ω by setting $\mu(R_{m,n}) = 1/((p+1)p^{m-1}(l+1)l^{n-1})$, where $R_{m,n}$ any cylindrical set determined by a legal labeling of a rectangle of height m and width n. One has to verify that μ actually extends to a countably additive measure on Ω. This can be deduced from the following proposition about factorization of quaternions. (In [**M1**] we show that this is the measure induced on Ω from the Haar measure on $\Gamma \backslash G$.)

PROPOSITION 2. (see [**M1**] or [**P**])
Let $a \in A$, $b \in B$ then there are unique elements $a' \in A$, $b' \in B$ s.t. $ab = b'a'$ (equality up to sign).   □

COROLLARY 1. $(\Omega, \mathbb{Z}^2)$ *has 0 two-dimensional topological entropy.*

PROOF:. It follows from the proposition that a legal labeling of a rectangle is determined by the labeling of, say, its bottom and right sides. Hence the number of legal rectangles of size $m \times n$ is $(p+1)p^{m-1}(l+1)l^{n-1}$. It follows that the topological entropy is 0.   □

We define now certain sub σ-algebras of $\mathcal{B}$. Enumerate the horizontal lines by $\mathbb{Z}$ and let $\mathcal{H}_j$ ($\mathcal{L}_j$, respectively) be the σ-algebra generated by the cylindrical sets determined by specifying the labelings of edges above (below) the j'th line.

The following theorem, together with its obvious "vertical" analogue, is the quantitative assertion alluded to in the introduction.

THEOREM 1. *Given any $\epsilon > 0$ there exists $n \in \mathbb{N}$ such that if $X \in \mathcal{L}_j$ and $Y \in \mathcal{H}_{j+n}$ then*

$$|\mu(X \cap Y) - \mu(X)\mu(Y)| < \epsilon$$

Moreover one can give an explicit bound for n of the form: $n \leq -C\log(\epsilon)$ for some constant $C > 0$.

Define a family of finite graphs $F_k = (V_k, E_k)$. The vertices V_k of F_k are the legal labelings of k horizontal (tiling's) edges by elements of B. Two vertices are adjacent if there is a legal labeling of a $1 \times k$ rectangle s.t. the two vertices are the labelings of its horizontal boundaries. Note that from proposition 2 it follows that these graphs are $p + 1$ regular.

Observe the following:

1. A legal labeling of a $m \times k$ rectangle corresponds to a path of length m in the graph F_k with no "backtracking" (i.e. we don't return immediately on the edge we came from).
2. The measure μ induces a probability distribution on the space of paths with no backtracking in F_k which correspond to the following random walk: Choose the starting vertex and an edge to be taken from it uniformly, at each consecutive step choose one of the p edges at the current vertex, different from the one just taken, with equal probability $1/p$.

Let M_k be the Markovian matrix defining this random walk on the directed edges of the graph F_k. Let P_k be the Markovian matrix corresponding to a uniform random walk on the directed edges allowing backtracking, i.e. from the edge (a, b) we go with probability $1/(p + 1)$ to any of the edges (b, c). Let L_k be the matrix corresponding to the uniform random walk on the vertices of F_k. Let $\vec{E}_k$ denote the set of directed edges of F_k. Define an inner product $< \cdot, \cdot >$ on $\mathbb{R}^{\vec{E}_k}$ by: $< f, h >= 1/|\vec{E}_k| \sum_{\alpha \in \vec{E}_k} f(\alpha)h(\alpha)$. Let $\bar{1}_k \in \vec{E}_k$ be the constant 1 vector. Notice that $M_k \bar{1}_k = \bar{1}_k$. Let π_0 be the orthogonal projection on the subspace spanned by $\bar{1}_k$ and $\pi_1 = \text{id} - \pi_0$. For a vector $h \in \mathbb{R}^{\vec{E}_k}$ let $\hat{h} \in \mathbb{R}^{\vec{E}_k}$ be the vector obtained by reversing the edges' orientations, i.e. $\hat{h}(a, b) = h(b, a)$. It is easy to see that

$$(*) \qquad M_k f = \frac{p+1}{p}(P_k f - \frac{1}{p+1}\hat{f}) = \frac{p+1}{p}P_k f - \frac{1}{p}\hat{f}$$

Define two linear transformations, $T : \mathbb{R}^{V_k} \to \mathbb{R}^{\vec{E}_k}$, $S : \mathbb{R}^{\vec{E}_k} \to \mathbb{R}^{V_k}$ by:

$$(Th)(a, b) = \frac{1}{p+1}h(a)$$

$$(Sf)(a) = \sum_{(b,a)\in\vec{E}_k} f(b,a)$$

It is easy to verify that $TS = P_k$, $ST = L_k$. Notice that the transformations T, S, P_k and L_k preserve the sum of coordinates (in the respective spaces $\mathbb{R}^{\vec{E}_k}$, $\mathbb{R}^{V_k}$).

In order to use these graphs to prove Theorem 1 we need a bound on the second eigenvalues of their adjacency matrices.

PROPOSITION 3. *The graphs F_k are connected.*

PROOF. Γ can be embedded as an irreducible uniform lattice in $G = PGL(2,\mathbb{Q}_p) \times PGL(2,\mathbb{Q}_l)$. (Here and in the sequel the embeddings referred to are via the familiar representation of quaternions as 2×2 matrices where the "$\sqrt{-1}$" is taken in the corresponding field $\mathbb{Q}_p$ or $\mathbb{Q}_l$.) In [M1] it is shown that $(\Omega, \mu, \mathbb{Z}^2)$ is a factor of $(\Gamma\backslash G, \nu, \mathbb{Z}^2)$ where ν is the Haar measure on $\Gamma\backslash G$ and $\mathbb{Z}^2$ acts via a certain homomorphism into the Cartan subgroup of G. By the Howe-Moore theorem (see [Z]) it follows that that the action of $\mathbb{Z}^2$ on $(\Gamma\backslash G, \nu)$ is mixing. Hence $(\Omega, \mu, \mathbb{Z}^2)$ is also mixing. This implies that some vertical translate of the cylindrical set corresponding to any given labeling of k horizontal edges intersects the cylindrical set corresponding to another such labeling. This implies that the graph F_k is connected. $\square$

Fix some vertex in F_k. Let β be the product of the k quaternions associated with it. Look at any loop in the graph F_k. This is a $m \times k$ rectangle whose bottom and top boundaries are labeled by β (we use here a slight abuse of notation which is justified by the fact that β can be written uniquely as a product of k quaternions from B (see [L])). Let α be the product of the quaternions from A labeling the left side of this rectangle and α' be the product of the quaternions labeling the right side of it. By the adjacency rules of the subshift we have:

$$(**) \qquad\qquad \alpha\beta = \beta\alpha'$$

where equality is up to sign. Let $\Gamma_p < \Gamma$ be the subgroup generated by A. It is a free group with A a symmetric set of generators (see [L]). Let $\Gamma_p(k) < \Gamma_p$ be the subgroup of the elements satisfying $(**)$. Look at the universal cover of F_k. This is a $p + 1$-regular tree. By labeling each of its edges by the element of A labeling the left side of the $1 \times k$ rectangle corresponding to the edge it covers in F_k, and associating the vertex above the vertex labeled β in F_k with the identity element, we can identify this covering tree with the Cayley graph of Γ_p. The group $\Gamma_p(k)$ may be viewed as the covering group of this cover, and the graph F_k is the quotient of this tree by $\Gamma_p(k)$. Alternatively one may think of this tree as the affine Bruhat-Tits building of $PGL(2,\mathbb{Q}_p)$ and F_k as its quotient by the group $\Gamma_p(k) < \Gamma_p$ embedded as a lattice in $PGL(2,\mathbb{Q}_p)$. The group $\Gamma_p(k)$ is the stabilizer of a certain segment of length k in the tree of $PGL(2,\mathbb{Q}_l)$ when we embed Γ_p in $PGL(2,\mathbb{Q}_l)$. It follows that $\Gamma_p(k)$ is a congruence subgroup of Γ_p. This allows us to use the results of Lubotzky, Phillips and Sarnak:

THEOREM. (see [**LPS**], [**L**])

The graphs F_k are Ramanujan graphs. I.e. the second eigenvalues of their adjacencies matrices are smaller or equal in absolute value to $2\sqrt{p}$. □

PROPOSITION 4. *Let $\mathbb{R}^{V_k}$ have the usual inner product. Then:*

   a. $||S|| = \sqrt{p+1}|\vec{E}_k|$

   b. $||T|| = \frac{1}{\sqrt{p+1}|\vec{E}_k|}$

   c. *The norm of the restriction of ST to the subspace H_k of the 0 sum vectors in $\mathbb{R}^{V_k}$ is $\leq \frac{2\sqrt{p}}{p+1}$.*

   d. *The restriction of $P_k = TS$ to the 0 sum vectors in $\mathbb{R}^{\vec{E}_k}$ has norm 1.*

PROOF. a., b., d. follow by direct computation. c. follows from Lubotzky-Phillips-Sarnak theorem. □

PROPOSITION 5. *The restriction of P_k to 0 sum vectors satisfies:*

$$\left|\left| \left(\frac{p+1}{p}\right)^2 P_k^2 \right|\right| \leq \frac{2(p+1)}{p\sqrt{p}}$$

PROOF.

$$\left|\left| \left(\frac{p+1}{p}\right)^2 P_k^2 \right|\right| = \left(\frac{p+1}{p}\right)^2 ||TSTS|| \leq \left(\frac{p+1}{p}\right)^2 ||T||||ST||||S|| =$$

$$= \left(\frac{p+1}{p}\right)^2 \frac{1}{\sqrt{p+1}|\vec{E}_k|} \frac{2\sqrt{p}}{p+1} \sqrt{p+1}|\vec{E}_k| = \frac{2(p+1)}{p\sqrt{p}}$$

Where all the norms of the various transformations are with respect to the subspaces of 0-sum vectors (these transformations map 0-sum vectors to 0-sum vectors). □

COROLLARY 2. *The restriction of M_k^2 to the subspace of 0-sum vectors in $\mathbb{R}^{\vec{E}_k}$ satisfies*

$$||M_k^2|| \leq \frac{2(p+1)(1+\sqrt{p})+1}{p^2}$$

In particular for $p \geq 13$ it follows that $||M_k^2|| < \theta < 1$ (θ is independent of k).

PROOF. Using (*) we obtain:

$$M_k^2 f = \frac{p+1}{p} P_k M_k f - \frac{1}{p}\widehat{M_k f} =$$

$$= \frac{p+1}{p} P_k \left(\frac{p+1}{p} P_k f - \frac{1}{p}\hat{f}\right) - \frac{1}{p}\left(\frac{p+1}{p}\widehat{P_k f} - \frac{1}{p}\hat{f}\right) =$$

$$= \left(\frac{p+1}{p}\right)^2 P_k^2 f - \frac{p+1}{p^2} P_k \hat{f} - \frac{p+1}{p^2}\widehat{P_k f} - \frac{1}{p^2}f$$

Note that the transformation $h \to \hat{h}$ preserves norm, and deduce that:

$$||M_k^2|| \leq || \left(\frac{p+1}{p}\right)^2 P_k^2|| + ||\frac{p+1}{p^2}P_k|| + ||\frac{p+1}{p^2}P_k|| + \frac{1}{p^2} \leq$$

$$\leq \frac{2(p+1)}{p\sqrt{p}} + \frac{2(p+1)}{p^2} + \frac{1}{p^2} = \frac{2(p+1)(1+\sqrt{p})+1}{p^2}$$

$\square$

COROLLARY 3. *There exist constants $\lambda < 1$ and $C > 0$ such that for any* $k \in \mathbb{Z}$, $f \in \mathbb{R}^{\vec{E}_k}$, $n \in \mathbb{Z}$:

$$||\pi_0(M_k^n f)|| \leq C\lambda^n ||\pi_0(f)||$$

$\square$

We can prove now theorem 1: approximating X, Y by sets which are measurable with respect to some finite "windows", it is enough to prove the theorem for sets X and Y which are measurable with respect to a rectangle of width k and height m, where X is measurable with respect to the part of the rectangle below the j'th line and Y is measurable with respect to the part of the rectangle above the $j + N$'th line. Define two vectors $f_X, f_Y \in \mathbb{R}^{\vec{E}_k}$:

$f_X(a, b)$ = the probability that a random walk without backtracking in the graph $\vec{F}_k$ (i.e. the directed graph) will end at the edge (a, b), and the corresponding cylindrical set (placing the last vertex at the j'th horizontal line) will be contained in X.

$f_Y(a, b)$ = the probability that a random walk with no backtracking in $\vec{F}_k$ starting at the edge (a, b) will correspond to a cylindrical set contained in Y (when the starting vertex is placed at the $j + N$'th horizontal line).

It follows from the definitions that

$$\mu(X) =< f_X, \bar{1}_k >_k$$

$$\mu(Y) =< f_Y, \bar{1}_k >_k$$

$$\mu(X \cap Y) =< M_k^{N+1} f_X, f_Y >_k$$

Hence:

$$\mu(X \cap Y) =< M_k^{N+1} f_X, f_Y >_k =< M_k^{N+1}(\pi_0(f_X) + \pi_1(f_X)), \pi_0(f_Y) + \pi_1(f_Y) >_k =$$

$$=< M_k^{N+1}(\pi_0(f_X)), f_Y >_k + < \pi_1(f_X), \pi_1(f_Y) >_k =$$

$$=< M_k^{N+1}(\pi_0(f_X)), f_Y >_k + < f_X, \bar{1}_k >_k < f_Y, \bar{1}_k >_k$$

It follows using corollary 3 that for N such that $C\lambda^N < \epsilon$, we have

$$|\mu(X \cap Y) - \mu(X)\mu(Y)| < \epsilon$$

$\square$

COROLLARY 4.

The system $(\Omega, \mathcal{B}, \mu, \mathbb{Z}^2)$ is mixing of all orders.

PROOF. It is enough to prove the claim for sets which are measurable with respect to finite windows. We proceed by induction on k - the number of sets. We have already observed (see proof of proposition 3) that $(\Omega, \mathcal{B}, \mu, \mathbb{Z}^2)$ is 2-mixing. Assume $k > 2$ and that the assertion was proved for less then k sets. $\mathbb{Z}^2$ acts by pushing the windows determining the sets away from each other. As the k sets are far enough from each other there is a partition of them into two groups which belong to disjoint half planes (horizontal or vertical) which are at distance N from one another for any chosen N. Using theorem 1 or its "vertical" analogue, we conclude that the intersection of the sets in one group is ϵ independent of the intersection of the sets in the other group. Use the induction hypothesis to complete the proof. $\square$

REFERENCES

[L]   A. Lubotzky, *Discrete Groups, Expanding Graphs and Invariant Measures.*, Notes of CBMS Conference University of Oklahoma, May 1989.

[LPS]   A. Lubotzky, R. Phillips and P. Sarnak, *Ramanujan Graphs.*, Combinatorica **8** (1988), 261 – 277.

[M1]   S. Mozes, *Actions of Cartan Subgroups*, Ph.D Dissertation Hebrew University, Jerusalem (1991).

[M2]   S. Mozes, *Mixing of All Orders Of Lie Groups Actions*, Inventiones Math. **107** (1992), 235 – 241.

[P]   G. Pal, *On the factorization of generalized quaternions*, Duke Math. J. **4** (1938), 696 – 709.

[S]   J. P. Serre, *Trees.*, Springer-Verlag, 1980.

[Z]   R. J. Zimmer, *Ergodic Theory and Semisimple Groups.*, Birkhauser, 1984.

MATHEMATICAL SCIENCES RESEARCH INSTITUTE
1000 CENTENNIAL DR., BERKELEY, CA 94720 U.S.A.

Contemporary Mathematics
Volume **135**, 1992

A Cocycle Equation for Shifts

For Roy Adler on his 60th birthday

WILLIAM PARRY

We shall be concerned with a cocycle equation which arises in problems associated with lifting dynamical properties and with periodic orbits.

Let T be an invertible measure-preserving transformation of the Lebesgue space $(X, \mathcal{B}, m)$ and let $U : X \to G$, $V : X \to G$ be two measurable maps to the compact Lie group G. Then $T_1 : X \times G \to X \times G$, $T_2 : X \times G \to X \times G$ defined by

$$T_1(x,g) = (Tx, gU(x)) \text{ and}$$
$$T_2(x,g) = (Tx, gV(x))$$

are two measure-preserving transformations of $X \times G$ (endowed with the measure $m \times dg$) commonly known as *skew-products*. T_1, T_2 commute with the left G action given by $(x,h) \to (x,gh)$, $g \in G$ and the quotient of T_1, T_2 with respect to this action is, of course, T. A necessary and sufficient condition for T_1, T_2 to be isomorphic by an isomorphism which commutes with G and whose quotient is the identity is that there should exist a measurable map $F : X \to G$ such that

$$(0.1) \qquad\qquad U \cdot F = F \circ T \cdot V$$

and this is the cocycle equation we shall study.

A closely related equation takes the form

$$U \cdot F = F \circ T$$

where $F : X \to \mathbb{C}^d$ and U is unitary.

1991 Mathematics Subject Classification. Primary 28D05; Secondary 58F11.

This paper is in final form and no version will be submitted for publication elsewhere.

§1. The cocyle equation when restricted

In this section we shall suppose there is an *exhaustive* σ-algebra $\mathcal{A} \subset \mathcal{B}$ (i.e. $T^{-1}\mathcal{A} \subset \mathcal{A}$ and $T^n\mathcal{A} \uparrow \mathcal{B}$) and U,V are $\mathcal{A}$ measurable. Then we conclude:

Proposition 1. If F is a $\mathcal{B}$ measurable solution to (0.1) then F is $\mathcal{A}$ measurable.

Proof. Without loss in generality we may suppose $G = U(d)$ the group of d dimensional unitary matrices.

Applying the conditional expectation operator $E(\cdot|\mathcal{A})$ to (0.1) (for each term in the equation) we get

$$U \cdot E(F|\mathcal{A}) = E(F \circ T|\mathcal{A}) \cdot V$$

or

$$U \cdot E(F|\mathcal{A}) = E(F \mid T\mathcal{A}) \circ T \cdot V$$

and multiplying this with (0.1) inverted we have

$$F(x)^{-1}E(F|\mathcal{A})(x) = V(x)^{-1}F(Tx)^{-1}E(F|T\mathcal{A})(Tx)V(x) \text{ a.e.}$$

so that

$$\text{Trace } F^{-1} \cdot E(F|\mathcal{A}) = \text{Trace } F^{-1} \circ T \cdot E(F|T\mathcal{A}) \circ T \text{ a.e.}$$

Integration then gives

$$\int \text{Trace } F^{-1} \cdot E(F|\mathcal{A})dm = \int \text{Trace } F^{-1} \cdot E(F|T\mathcal{A})dm.$$

As F is U(d) valued $F(x)^{-1} = \overline{F(x)}'$

and we have

$$\int \sum_{i,j} \bar{F}_{ji} E(F_{ji}|\mathcal{A})dm = \int \sum_{i,j} \bar{F}_{ji} E(F_{ji}|T\mathcal{A})dm.$$

By exactly the same argument, using $T^n\mathcal{A}$ in place of $\mathcal{A}$, we have

$$\int \sum_{i,j} \bar{F}_{ji} E(F_{ji}|T^n\mathcal{A})dm = \int \sum_{i,j} \bar{F}_{ji} E(F_{ji}|T^{n+1}\mathcal{A})dm$$

so that

$$\int \sum_{i,j} \bar{F}_{ji} E(F_{ji}|\mathcal{A})dm = \int \sum_{i,j} \bar{F}_{ji} E(F_{ji}|T^n\mathcal{A})dm$$

and taking limits, using the Martingale theorem, we get

$$\int \sum_{i,j} \bar{F}_{ji} E(F_{ji}|\mathcal{A})dm = \int \sum_{i,j} |F_{ji}|^2 dm .$$

It is now clear that

$$E(F_{ji}|\mathcal{A}) = F_{ji} \text{ a.e.}$$

in other words each F_{ji} is $\mathcal{A}$ measurable which completes the proof.

As before we define the skew-product $T_1(x,g) = (Tx, gU(x))$ on $X \times G$. A well known criterion for the ergodicity of T_1, given that T is ergodic, is that for every non-trivial irreducible unitary representation R the equation

$$(1.1) \qquad\qquad R(U(x))F(x) = F(Tx) \text{ a.e.}$$

should have no measurable solution $F : X \to \mathbb{C}^d$ where d is the degree of R (cf. [3]). Of course, in general, this criterion is very difficult to test. However, when U is $\mathcal{A}$ measurable where $\mathcal{A}$ is an exhaustice σ-algebra the criterion is simplified to some extent by

Proposition 2. When U is $\mathcal{A}$ measurable and if the equation (1.1) has a $\mathcal{B}$ measurable solution F then F is $\mathcal{A}$ measurable.

The proof is very similar to the proof of Proposition 1.

§2. Markov cocyles

Here we shall specialise to the context of a two-sided shift of finite type σ defined on the shift space $X = X_A$ where A is an aperiodic 0-1 matrix. We shall suppose that σ preserves the Markov probability m defined by a stochastic matrix compatible with A.

When $U,V : X \to G$ are Hölder continuous, Livsic [4] has provided the basic criteria for the equation

$$(2.1) \qquad\qquad U \cdot F = F \circ \sigma \cdot V$$

to have a measurable solution (a.e) $F : X \to G$. In fact there is a measurable solution if and only if there is a Hölder continuous solution and such a solution exists if and only if

$$F(x)^{-1}U(\sigma^{n-1})\cdots U(x)F(x) = V(\sigma^{n-1})\cdots V(x)$$

for all periodic points x of period n and all $n = 1,2,3,\dots$.

When U,V depend only on 2 coordinates (i.e. $U(x) = U(x_0,x_1)$, $V(x) = V(x_0,x_1)$, $x = \{x_n\}$) the equation (2.1) can be decided in a straightforward way in view of

Theorem 3. If $U,V : X \to G$ depend only on the coordinates x_0,x_1 and if the equation has a measurable solution F then F depends only on the coordinate x_0.

Proof. Let $\mathcal{A} = \displaystyle\bigvee_{i=0}^{\infty} \sigma^{-i}\alpha$ where α is the state partition $\alpha =(A_1,\dots A_k)$ $(A_i = \{x : x_0 = i\}$, $1,2,\dots k$ being the states) then $\mathcal{A}$ is an exhaustive σ-algebra and U,V are $\mathcal{A}$ measurable. Thus by Proposition 1 F is $\mathcal{A}$ measurable. Now consider the inverse transformation σ^{-1} and the exhaustive σ-algebra $\mathcal{A} = \displaystyle\bigvee_{i=0}^{\infty} \sigma^{i}\alpha$. The equation (2.1) may be written

$$U(\sigma^{-1})F(\sigma^{-1}) = FV(\sigma^{-1})$$

and $U(\sigma^{-1}), V(\sigma^{-1})$ depend on coordinates x_{-1}, x_0 so that $U(\sigma^{-1}), V(\sigma^{-1})$ are $\mathcal{A}'$ measurable. Thus F (or F^{-1}) is $\mathcal{A}'$ measurable. However, it is not difficult to show, when σ is equipped with a Markov measure, that $\mathcal{A}' \cap \mathcal{A} = \alpha$ i.e. F is α measurable. In other words F depends only on the coordinate x_0 .

Remark. Theorem 3 was proved by Adler, Kitchens and Marcus [1] in the special case where $G = S_n$, the symmetric group on n symbols. The analogue for the case $G = \mathbb{R}$ was proved in [5], so one wonders for what locally compact (non–compact) groups G the theorem is true.

Remark. If we know that the equation

$$U{\cdot}F = F \circ \sigma{\cdot}V$$

has a *continuous* solution (when U,V depend on x_0, x_1) the Markov measure can always be provided (and is not really part of the hypothesis) and F is a function of x_0 alone.

Remark. When σ is a full shift and is endowed with a Bernoulli probability and if

U,V depend only on x_0, then F depends on *no coordinates* i.e. F is constant and $U = FVF^{-1}$.

In the same way if we wish to test the ergodicity of T_1, the skew-product over σ given by U, where $U(x) = U(x_0, x_1)$, we have:

Proposition 4. If R is a unitary representation of G and if

$$F \circ \sigma = R(U)F \text{ a.e.}$$

where $F : X \to \mathbb{C}^d$ is measurable, $U(x) = U(x_0, x_1)$, then F is a function of x_0.

This proposition renders the problem of deciding when T_1 is ergodic (given that σ is endowed with a Markov probability) relatively easy. If we wish to decide when T_1 is weak-mixing (and therefore a K-automorphism [10], and therefore Bernouilli [7]), once we have shown that T_1 is ergodic, we need only consider 1-dimensional representations χ of G and then a necessary and sufficient condition is that the equation

$$F(x_1) = e^{ia}\chi(U(x_0,x_1)) F(x_0)$$

should have only the trivial solution F a constant, $e^{ia} = 1, \chi \equiv 1$.

The same problem arises when we consider a 'suspension flow' σ_t^f i.e. a flow with base σ and ceiling the positive function $f(x) = f(x_0,x_1)$. The flow σ_t^f is weak-mixing precisely when the equation

$$F(x_1) = e^{iaf(x_0,x_1)} F(x_0)$$

has only the solution F constant.

§3. Future cocycles

For a number of problems, especially those concerned with periodic orbits, it is convenient to make the assumption that a cocycle depends only on future coordinates, for then one may replace the two-sided shift by a one-sided shift and benefit from the Ruelle (Perron-Frobenius) operator. This technique was employed by Sinai [9] (but see also [2], [8]).

For example if $f : X \to \mathbb{R}$ is Hölder continuous then there exists a Hölder continuous $f' : X \to \mathbb{R}$ which is cohomologous to f where f' depends only on the future, i.e. $f'(x) = f'(x_0,x_1,...)$. More precisely

$$f' = f + h \circ \sigma - h$$

where, also, h is Hölder continuous. (The Hölder exponent for f',h is usually a half of that for f.)

A similar result holds for a continuous unitary valued cocycle $U : X \to U(d)$:

Theorem 4. If U is a Hölder continuous $U(d)$ valued cocycle then there exists a Hölder continuous $U(d)$ valued cocycle U' where $U'(x) = U'(x_0,x_1,...)$ and

$$U' \cdot F = F \circ \sigma \cdot U$$

for some Hölder continuous $U(d)$ valued function.

The proof is very similar to the proof of the $\mathbb{R}$ valued cocycle analogue and just as this latter case is useful for the study of zeta functions Theorem 4 can be employed in the study of L functions associated with dynamical systems. For details see Appendix II of [6].

Of course, it would not be difficult to show that Theorem 4 holds for G valued cocycles where G is a compact group and in view of Sinai's theorem it holds for $\mathbb{R}^n$ valued cocycles so the problem arises as to how general G may be.

References

1.   R. Adler, B. Kitchens & B. Marcus, Almost topological classification of finite-to-one factor maps between shifts of finite type, Ergodic Th. and Dynam. Sys. **5** (1985), 485-500.

2.   R. Bowen, Equilibrium states and the ergodic theory of Anosov diffeomorphisms, S.L.N. 470, Springer, N.Y., 1975.

3.   H.B. Keynes & D. Newton, Ergodicity in (G,σ) extensions, S.L.N. 819, Springer, N.Y.

4.   A. Livsic, Cohomology properties of dynamical systems, Math. USSR, Izv. **6** (1972), 1278-1301.

5.   W. Parry, Endomorphisms of a Lebesgue space III, Israel J. Math. **21** (1975), 167-172.

6.   W. Parry & M. Pollicott, Zeta functions and the periodic orbit structure of hyperbolic dynamics, Astérisque **187-188**, Société Mathématique de France, 1990.

7.   D. Rudolph, Classifying the isometric extensions of a Bernoulli shift, J. d'Analyse Mathématique, **34** (1978), 36-60.

8.   D. Ruelle, Thermodynamic formalism, Addison-Wesley, 1978.

9.   Y.G. Sinai, Gibbs measures in ergodic theory, Russ. Math. Surv. **27** (1972), 21-70.

10. R.K. Thomas, Metric properties of transformations of G-spaces, T.A.M.S. **160** (1971), 103-117.

Mathematics Institute, University of Warwick, Coventry CV4 7AL

Contemporary Mathematics
Volume **135**, 1992

In General A Degree Two Map Is An Automorphism

For Roy Adler on his 60^(th) birthday

WILLIAM PARRY

We are all familiar with the category theorems of ergodic theory which, in some cases, assert striking and unexpected results. For example, one can show that in the weak topology, for the set of endomorphisms of a Lebesgue space, automorphisms comprise a dense G_δ . Our surprise, however, is frequently tempered by a recognition that our knowledge of the topology employed to formulate such theorems is hazy. In this note I wish to prove a category theorem which, to my mind, gains surprise from the fact that the topology involved, vis a vis the uniform topology, is as unmysterious as one could wish.

Let K denote the circle with unit circumference and let $S(K) \subset C(K,K)$ denote the space of all orientation preserving, Lebesgue measure–preserving, degree 2, maps of K onto itself. We endow $C(K,K)$ with the uniform topology.

Theorem

$S(K)$ is a closed subset of $C(K,K)$ in the uniform topology and the set of all ergodic and zero-entropy maps in $S(K)$ is a dense G_δ in $S(K)$.

The entropy referred to here, of course, is computed with respect to Lebesgue measure. In other words, most of the endomorphisms are *measure-theoretically automorphisms*. This observation is a by-product of a simpler observation answering a question raised by Shereshevsky who asked for the existence of ergodic, zero-entropy continuous measures preserved by the one–sided two–shift.

We begin with the one–sided two shift σ and the set M of probabilities which are preserved by σ . We denote the set of ergodic probabilities by E , the set of zero entropy probabilities by Z , the set of probabilities μ in M such that $\mu(C) > 0$ (C a cylinder) by S_C . Clearly $S = \bigcap_C S_C$ is the set of probabilities μ in M which are supported. Since

---

1991 Mathematics Subject Classification. Primary 28D05; Secondary 58F11.

This paper is in final form and no version will be submitted for publication elsewhere.

$$E = \bigcap_{k,m} \bigcap_{N} \left\{ \mu : \left| \frac{1}{N} \sum_{n=0}^{N-1} \int f_k \circ \sigma^n f_k d\mu - \left(\int f_k d\mu \right)^2 \right| < \frac{1}{m} \right\}$$

where $\{f_k\}$ is a countable dense sequence in the space of continuous functions, E is a dense G_δ. Also

$$Z = \bigcap_{k} \bigcup_{N} \left\{ \mu : \frac{1}{N} H_\mu(\alpha v \sigma^{-1} \alpha v \ldots v \sigma^{-N} \alpha) < \frac{1}{k} \right\}$$

where $\alpha = \{[0], [1]\}$ is the basic partition, so, with respect to the weak-topology, Z is a dense G_δ. Obviously S is a dense G_δ. We therefore conclude:

Lemma 1.

The set of ergodic, zero entropy, supported (and non-atomic) $\mu \in M$ is a dense G_δ.

Let $\pi(x_0, x_1, \ldots) = \frac{x_0}{2} + \frac{x_1}{4} + \ldots$ and $Tx = 2x \bmod 1$. Then $\pi\sigma = T\pi$ and π maps the invariant probabilities M onto the set of invariant probabilities for T in a one-one fashion, so we use M to denote the latter set as well.

Clearly we have

Lemma 2.

The set M' of T invariant probabilities which are continuous, supported, ergodic, with zero entropy is a dense G_δ in the weak topology.

Let $\mu \in M'$ and define $\varphi_\mu(x) = \mu[o, x)$ so that $\varphi_\mu(y) - \varphi_\mu(x) = \mu[x, y) = \ell\varphi_\mu[x, y)$, where ℓ is Lebesgue measure (i.e. $\mu = \ell\varphi_\mu$) and $\ell\varphi_\mu T^{-1}\varphi_\mu^{-1} = \ell$. Thus $T_\mu = \varphi_\mu T \varphi_\mu^{-1}$ is Lebesgue measure preserving and φ_μ is an orientation preserving (i.e. strictly increasing) homeomorphism of the unit interval.

Lemma 3.

Each T_μ ($\mu \in M$) is ergodic, orientation preserving, Lebesgue measure-preserving, has zero entropy and interpreted as a map of K has degree 2.

Now note that every orientation preserving, Lebesgue measure-preserving, degree 2 map S of K is T_μ for some μ (if we assume, as we may, that $S(0) = 0$). This is the same as saying there is an orientation preserving homeomorphism φ such that $S = \varphi T \varphi^{-1}$. One constructs φ as follows:

Since S is Lebesgue measure-preserving $S^{-1}[x, y) = [a, b) \cup [c, d)$ where $Sa = Sc = x$, $Sb = Sd = y$ and $\ell[x, y) = \ell[a, b) + \ell[c, d) > \ell[a, b)$. Thus $\ell S[a, b) > \ell[a, b)$ where $[a, b) \subset I_0$ or I_1 ($[0, 1) = I_0 \cup I_1$, I_0, I_1 disjoint intervals on each of which S is one-one). Hence $S(b) - S(a) > b - a$. Hence if $x \neq y$ then

$x_0, x_1, \ldots \neq y_0, y_1, \ldots \quad S^n x \in I_{x_n} , \quad S^n y \in I_{y_n} .$

Thus sequences generated by S coincide with sequences generated by T and we can map x to z by φ when the S sequence of x equals the T sequence of z.

Proof of Theorem

First we note that $S(K)$ is closed in $C(K,K)$ in the uniform topology. This is clear since, if f is continuous and $\int f(S_n) d\ell = \int f d\ell$ and $S_n \to S$ uniformly then $\int f(S) d\ell = \int f d\ell$,

To prove that the set of ergodic zero entropy members of $S(K)$ is a dense G_δ in $S(K)$ we let S be a degree 2 , ergodic, orientation preserving, Lebesgue measure preserving map and write

$$S = \varphi_\mu T \varphi_\mu^{-1} = T_\mu$$

where $\varphi_\mu(x) = \mu[o,x)$. T preserves μ so we approximate μ by another T invariant ergodic, continuous, zero entropy probability m so closely that $\sum_I |mI - \mu I| < \varepsilon$ where $\beta = \{I\}$ is a fine partition into intervals and $\varepsilon > 0$ is small.

Then $T_m = \varphi_m T \varphi_m^{-1}$ is uniformly close to $S = T_\mu$.

Thus we see that density is proved. To prove that the set in question is a G_δ we go through the usual procedure: The set of ergodic maps and the set of zero entropy maps form G_δ's .

Constructions

To conclude we give a construction of a degree two, orientation-preserving, Lebesgue measure-preserving map of the circle with zero entropy. The construction of such a map which is ergodic would require extra work. It will be clear that the construction we give is quite general.

Let A be a Borel subset of K (interpreted as the unit interval with $0, 1$ identified) with the property that $\ell(A \cap I) > 0$ and $\ell(A^c \cap I) > 0$ for all non-empty intervals I .

For $0 \leq x < \ell(A)$ define S by

$$\ell A \cap [0, Sx) = x$$

and for $\ell(A) \leq x < 1$ define S by

$$\ell A + \ell A^c \cap [0, Sx) = x .$$

In other words, on the interval $[0, \ell(A))$, S is defined as the inverse of the map $x \to \ell A \cap [0,x)$ and on the interval $[\ell(A), 1)$ it is defined as the inverse of the map

$$x \to \ell(A) + \ell A^c \cap [0,x) \,.$$

It is not difficult to show that S has the desired properties and that every such map arises in this way (assuming $S(0) = 0$).

Mathematics Institute, Univrsity of Warwick, Coventry CV4 7AL, UK

Contemporary Mathematics
Volume **135**, 1992

$\mathbf{Z}^n$ versus $\mathbf{Z}$ Actions for Systems of Finite Type

CHARLES RADIN

ABSTRACT. We consider dynamical systems of finite type with $\mathbf{Z}^n$ actions, and discuss the differences between the cases $n = 1$ and $n \geq 2$. For the latter we examine the degree of "order" which is possible when the system is uniquely ergodic.

Systems of finite type

We begin with some notation. Let A be a finite alphabet, and consider the "infinite arrays" $A^{\mathbf{Z}^n}$ as functions on $\mathbf{Z}^n$. We say that the dynamical system which consists of the natural action of $\mathbf{Z}^n$ on the compact $X \subseteq A^{\mathbf{Z}^n}$ is "of finite type" if there is some finite $C \subset \mathbf{Z}^n$ and finite set F of finite arrays from A^C (thought of as restrictions $x|_C$ to C of functions x in $A^{\mathbf{Z}^n}$) such that

$$(1) \qquad X = X_F \equiv \left\{ x \in A^{\mathbf{Z}^n} : \text{for all } t \in \mathbf{Z}^n, \ x_t|_C \notin F \right\}$$

where $x_t(j) \equiv x(j - t)$ for $t, j \in \mathbf{Z}^n$.

It is easy to see that certain choices of F will lead to an empty X_F. For this and other reasons it is useful to "redefine" X_F, whereby instead of **forbidding** restrictions in F from appearing in the arrays we just **minimize their appearance**. We do this using the "energy function" $E : A^C \to \mathbf{R}$ defined to be the characteristic function of F. (That is, $E(f) = 1$ for $f \in F$, $E(f) = 0$ for $f \notin F$.) We then define:

$$(2) \qquad \tilde{X}_E \equiv \big\{ x \in A^{\mathbf{Z}^n} : \text{for every finite } B \subset \mathbf{Z}^n,$$
$$E^B(x) = \inf\{E^B(y) : y = x \text{ outside } B\} \big\}$$

where

$$E^B(z) \equiv \sum_{t : B_{-t} \cap C \neq \emptyset} E(z_t|_C).$$

1991 *Mathematics Subject Classification*. 58F11, 52C20.
Research supported in part by NSF Grant No. DMS-9001475.
This paper is in final form and no version of it will be submitted for publication elsewhere.

X_F is the set of arrays in $A^{\mathbf{Z}^n}$ in which no translate of an array in F appears, while $\tilde{X}_E$ just minimizes the appearance of such arrays. Clearly $X_F \subseteq \tilde{X}_E$. Furthermore it is easy to prove using the compactness of $A^{\mathbf{Z}^n}$ that $\tilde{X}_E$ is always nonempty, in fact for an arbitrary function $E : A^C \to \mathbf{R}$, not just characteristic functions. We therefore define a "zero temperature" dynamical system as one defined by 2) for any fixed function E. (See [8] for other motivation, from physics.)

Unique ergodicity

We say the compact $X \subset A^{\mathbf{Z}^n}$ is "uniquely ergodic" if there is one and only one Borel probability measure on X invariant under the natural action of $\mathbf{Z}^n$. Consider the following three classes of uniquely ergodic systems.

    i) All uniquely ergodic $X \subset A^{\mathbf{Z}^n}$
    ii) All uniquely ergodic zero temperature $X \subset A^{\mathbf{Z}^n}$
    iii) All uniquely ergodic $X \subset A^{\mathbf{Z}^n}$ of finite type

It is clear by construction that there is containment as one descends the list, but proper containment is not obvious. Our first result along this line is the following (known as the Third Law of Thermodynamics).

THEOREM. (J. Miękisz and C. R. [7]). All uniquely ergodic zero temperature $X \subset A^{\mathbf{Z}^n}$ have zero topological entropy.

(It can be shown that if $n = 1$ then the unique invariant measure is supported by a finite set, so the result is obvious in that case. For a discussion of the general case see [8].) This together with the theorem of Jewett-Kreiger-Weiss [12] shows that the first containment is proper. We do not know a proof that the second containment is proper, but there is a preprint by Miękisz [3] going part way.

$\mathbf{Z}^n$ Versus $\mathbf{Z}$ Actions

To say that a symbolic system $X \subset A^{\mathbf{Z}^n}$ is of finite type implies that each variable, with values in A, corresponding to a point of $\mathbf{Z}^n$ can only directly affect nearby variables. (A convenient generalization of systems of finite type to actions of $\mathbf{R}^n$ is discussed in [9].) This is reminiscent of the differential equations of the natural sciences. One of the main points we wish to make follows by analyzing such nonmathematical applications of $\mathbf{Z}^n$ and $\mathbf{Z}$ actions. We envision $\mathbf{Z}$ actions as typically modeling **evolution** problems; that is, $\mathbf{Z}$ represents time. We reformulated the condition of finite type above as an optimization condition, for reasons we will soon discuss. With this in mind, we note that evolution problems can also sometimes be reformulated as optimization problems – think of the least action principle for Hamiltonian systems for example. However in such a reformulation it is typical that one seeks as solutions critical points, not global optima, and that such critical points can represent a wide variety of curves. On the other hand, the n translation variables of $\mathbf{Z}^n$ actions often represent spatial translations, and, as in the crystal problem of condensed matter physics, or the sphere packing problem, or the problems of space tiling, what one seeks is typically (generically [5,6,7]) a **unique** solution (a well defined "structure", so to speak) to a global optimization problem of the general form of our zero temperature condition [8]. When properly formulated, the solution is sought in a space of invariant probability measures, and the uniqueness of the solution translates into the property of unique ergodicity.

Thus **Z** actions and **Z**^{n} actions naturally represent very different situations, the former accomomodating a very flexible class of arrays (in particular they are highly nonunique), and the latter representing some unique structure such as a crystal or quasicrystal. This is "why" unique ergodicity is not as natural for **Z** actions as it is for **Z**^{n} actions.

Our interest is primarily with **Z**^{n} actions, and more specifically in the degree to which uniquely ergodic zero temperature systems (or systems of finite type) tend to be "ordered". (Why does low temperature matter tend to be crystalline, why do there always seem to be periodic examples among the densest sphere packings in any dimension, why is it hard to find tiles which can only tile space nonperiodically? See [8].) Consider the following extreme cases of "order" for a zero temperature system X with unique invariant measure ρ.

  a) Periodic; (corresponding to a finite set X, the orbit of a periodic array)
  b) Quasiperiodic; (corresponding to X with purely discrete dense spectrum)
  c) Weakly mixing ρ; (corresponding to purely singular continuous spectrum)
  d) Strongly mixing ρ; (roughly corresponding to purely absolutely continuous spectrum)

(Note that as in probability theory we are using measure theoretic – chiefly spectral – properties to analyze the order of the dynamical system.)

Examples of a) are easy to obtain. Examples of b) were first obtained by R. Berger in 1966 [1], with nicer examples by R. Robinson [10] and others. Examples of c) are due to S. Mozes in 1989 [4,7]. It is unknown if there are examples of d), and this is an important open problem.

There are several reasons for our introduction of the class of zero temperature systems. They constitute a natural generalization of systems of finite type with the conditions defining the system still strictly local, and there are real parameters for the class so that one can address questions of genericity. Furthermore, ever since the work of G. Toulouse [11,2] it has been commonly felt by condensed matter theorists that energy functions E as above which are "frustrated" (that is, cannot be reformulated as a characteristic function as is the case for systems of finite type), are more likely to lead to "disorder" – or smooth spectrum, as in models of spin glasses. In other words, physical intuition suggests that the class of zero temperature systems is broader than, and should contain more disordered examples than (perhaps of type d) above), the class of systems of finite type. Needless to say, it would be most interesting if this could be proved true.

References

1. R. Berger, *The undecidability of the domino problem*, Mem. Amer. Math. Soc. **66** (1966).
2. J. Miękisz, *Frustration without competing interactions*, J. Stat. Phys. **55** (1989), 351–355.
3. J. Miękisz, *The global minimum of energy is not always a sum of local minima – a note on frustration*, preprint, University of Louvain-la-Neuve, June, 1991.
4. S. Mozes, *Tilings, substitution systems and dynamical systems generated by them*, J. d'Analyse Math. **53** (1989), 139–186.
5. C. Radin, *Correlations in classical ground states*, J. Stat. Phys. **43** (1986), 707–712.
6. C. Radin, *Low temperature and the origin of crystalline symmetry*, Int. J. Mod. Phys. **1** (1987), 1157–1191.
7. C. Radin, *Disordered ground states of classical lattice models*, Revs. Math. Phys. **3** (1991), 125–135.
8. C. Radin, *Global order from local sources*, Bull. Amer. Math. Soc. **25** (1991), 335–364.
9. C. Radin and M. Wolff, *Space tilings and local isomorphism*, Geometriae Dedicata (to appear).
10. R.M. Robinson, *Undecidability and nonperiodicity for tilings of the plane*, Invent. Math. **12** (1971), 177–209.
11. G. Toulouse, *Theory of frustration effect in spin glasses, I*, Commun. Phys. (G.B) **2** (1977), 115–119.
12. B. Weiss, *Strictly ergodic models for dynamical systems*, Bull. Amer. Math. Soc. **13** (1985), 143–146.

Department of Mathematics, The University of Texas at Austin, Austin, Texas 78712-1082

E-mail address: radin@math.utexas.edu

Contemporary Mathematics
Volume **135**, 1992

PRINCIPAL VECTORS OF COMMUTING BLOCK MAPS

FRANK RHODES

ABSTRACT. Endomorphisms of the full 2-shift which are linear in the first but not in all variables are studied via the principal vectors of their block maps. It is shown that the principal vector of a block map g which commutes with a block map f is determined by the principal vectors of f and of f^2.

1. Introduction

Endomorphisms of the full 2-shift can be studied via the polynomial presentations of the corresponding block maps [2]. Considerable progress has been made in this way in understanding commutativity and factorization for the important class of endomorphisms which are linear in the first variable but are not linear in all variables [1][5]. It can now be seen that all the techniques rely on the fact that if suitable conditions are placed on the principal vector of a map f then there is a particularly close relationship between the principal vectors of f, of the powers of f and of maps g which commute with f. In this note I show that similar close relationships always occur.

The set of block maps which are linear in all their variables is denoted by $\mathcal{L}$, while the set of block maps which depend linearly on the first variable is denoted by $\mathcal{L}_1$. The principal vector $\mathcal{V}f$ and the principal auxiliary vector $\mathcal{A}f$ of a block map f are defined in [5]. The key result of this note is that whenever f is in $\mathcal{L}_1 \backslash \mathcal{L}$ and $h = f \circ g = g \circ f$ the principal vectors of f, g and h take the forms

$$\mathcal{V}Rf = (r_1, r_2, \ldots, r_p)(r_1 + k, r_2, \ldots, r_p)^{a-1}(r_{ap+1}, r_{ap+2}, \ldots, r_m)$$

$$\mathcal{V}Rg = (r_1, r_2, \ldots, r_p)(r_1 + k, r_2, \ldots, r_p)^{b-1}(r_{ap+1}, r_{ap+2}, \ldots, r_m)$$

$$\mathcal{V}Rh = (r_1, r_2, \ldots, r_p)(r_1 + k, r_2, \ldots, r_p)^{c-1}(r_{ap+1}, r_{ap+2}, \ldots, r_m)$$

where $c = a + b$. This is a basis for extending the results on commutativity and factorization of block maps.

1991 Mathematics Subject Classification. Primary 54 H 20

This paper is in final form and no version of it will be submitted for publication elsewhere.

2. Principal Vectors of Compositions

The following lexicographic ordering of vectors was introduced in [4]:
$(r_1, r_2, \ldots, r_m) < (s_1, s_2, \ldots, s_n)$ if

(i) $m < n$ and $r_i = s_i, 1 \le i \le m$, or

(ii) there exists $j \le \min(m, n)$ such that $r_j > s_j$ while $r_i = s_i$ whenever $i < j$.

Note that if $m = n = 1$ then $(r_1) < (s_1)$ as vectors precisely when $r_1 > s_1$ as numbers. Thus sequences will be enclosed in parentheses to identify vectors. Additional parentheses may be introduced to emphasize the form of a vector which is built up by concatenation, and a power notation may be used to show repetitions in concatenations.

Note also that $(r_1, r_2, \ldots, r_m) < (s_1, s_2, \ldots, s_n)$ is equivalent to $(k, r_1, r_2, \ldots, r_m) < (k, s_1, s_2, \ldots, s_n)$ and to $(r_1 + k, r_2, \ldots, r_m) < (s_1 + k, s_2, \ldots, s_m)$. If $i \le \min(m, n)$ then these imply $(r_1, r_2, \ldots, r_i) \le (s_1, s_2, \ldots, s_i)$.

The operator notations introduced in [1] and extended in [3] will be used in the proof of the following proposition.

2.1 Proposition. *Suppose that $g \in \mathcal{L}_1 \backslash \mathcal{L}$ and that ψ is non-trivial with $\mathcal{V}Rg = (s_1, s_2, \ldots, s_n), \mathcal{V}\psi = (r_1, r_2, \ldots, r_m)$ and $\mathcal{A}\psi = (k_1, k_2, \ldots, k_{m+1})$. Then*

$$\mathcal{V}(\psi \circ g) = \underset{0 \le i \le m}{\text{Max}} \, (r_1, r_2, \ldots r_i, s_1 + k_{i+1}, s_2, \ldots, s_n)$$

the maximum being taken over those values of i for which $k_{i+1} \ne r_{i+1}$. This maximum is attained for precisely one value of i. If

$$\mathcal{V}(\psi \circ g) = (r_1, r_2, \ldots, r_p)(s_1 + k_{p+1}, s_2, \ldots, s_n)$$

then for $1 \le j \le p$

$$\mathcal{V}(Q_{r_j} \ldots Q_{r_1} \psi \circ g) = (r_{j+1}, \ldots, r_p)(s_1 + k_{p+1}, s_2, \ldots, s_n)$$

Proof. The first part of the proposition will be proved by induction on the length of the principal vector of ψ. The result is valid when ψ is linear. Suppose that it is valid for all block maps with principal vector of length less than m, and let ψ be a block map whose principal vector has length m. If $k_1 < r_1$ then, similar to the first equation in the proof of Theorem 4.8 of [4], we have

$$\psi \circ g = T^{k_1 - 1}(I + T\Lambda Rg + T^{s_1}I.T^{s_1+1}Q_{s_1}Rg + T^{s_1+1}R_{s_1}Rg)$$

$$+ T^{k_1}\theta \circ g + T^{r_1 - 1}g.T^{r_1}Q_{r_1}\psi \circ g + T^{r_1}R_{r_1}\psi \circ g$$

for some linear block map θ. The principal vector of $\psi \circ g$ is that of $T^{k_1-1}(T^{s_1}I.T^{s_1+1}Q_{s_1}Rg)$ or that of $T^{r_1-1}g.T^{r_1}Q_{r_1}\psi \circ g$, since the principal vectors of these two expressions are different from each other. Thus if $k_1 < r_1$

$$\mathcal{V}(\psi \circ g) = \text{Max}\,\{(s_1 + k_1, s_2, \ldots s_n), (r_1, \mathcal{V}Q_{r_1}\psi \circ g)\}$$

$$= \underset{0 \le i \le m}{\text{Max}} \, (r_1, r_2, \ldots r_i)(s_1 + k_{i+1}, s_2, \ldots s_n),$$

the maximum being taken over those values of i for which $k_{i+1} \neq r_{i+1}$. If $k_1 = r_1$ then the first two summands in the expansion for $\psi \circ g$ given above are omitted and the same final result holds. The definition of the ordering of vectors guarantees that the maximum is attained for precisely one value of i. The last part of the proposition follows from the observation that inequalities between vectors are preserved when common initial terms are dropped.   □

When a block map is linear in the first variable the first term in its principal auxiliary vector is 1. In the next proposition it is convenient to set $k_1 = 0$.

2.2. Proposition. *Suppose that $f, g \in \mathcal{L}_1 \backslash \mathcal{L}$ with*

$$\mathcal{V}Rg = (s_1, s_2, \ldots, s_n), \quad \mathcal{V}Rf = (r_1, r_2, \ldots, r_m)$$

and $\mathcal{A}f = (1, k_2, k_3, \ldots, k_{m+1})$. Set $k_1 = 0$. Then

$$\mathcal{V}R(f \circ g) = \underset{0 \leq i \leq m}{\text{Max}} \; (r_1, r_2, \ldots, r_i)(s_1 + k_{i+1}, s_2, \ldots, s_n),$$

the maximum being taken over the values of i for which $k_{i+1} \neq r_{i+1}$. This value is attained for precisely one value of i. Moreover, $\mathcal{V}R(f \circ g) \neq (r_1, r_2, \ldots, r_m)$.

Proof. The principal vector of $f \circ g$ is either $\mathcal{V}(g) = (s_1 + 1, s_2, \ldots s_n)$ or $\mathcal{V}(T^{r_1} g . T^{r_1+1} Q_{r_1} f \circ g) = (r_1 + 1, \mathcal{V}Q_{r_1} f \circ g)$. Thus

$$\mathcal{V}R(f \circ g) = \text{Max} \begin{cases} (s_1, s_2, \ldots, s_n) \\[2mm] \underset{0 \leq i \leq m}{\text{Max}} \; (r_1, r_2, \ldots r_i)(s_1 + k_{i+1}, s_2, \ldots, s_n) \end{cases}$$

$$= \underset{0 \leq i \leq m}{\text{Max}} \; (r_1, r_2, \ldots, r_i)(s_1 + k_{i+1}, s_2, \ldots, s_n),$$

the maximum being taken over those values of i for which $k_{i+1} \neq r_{i+1}$. That the principal vector of $R(f \circ g)$ can not be $(r_1, r_2, \ldots, r_m)$ can be proved by contradiction using the formula for $Q_{r_m} \ldots Q_{r_1} R(f \circ g)$ in the proof of Theorem 4.8 of [4].   □

2.3 Corollary. *Suppose that $f \in \mathcal{L}_1 \backslash \mathcal{L}$ with $\mathcal{V}Rf = (r_1, r_2, \ldots, r_m)$ and $\mathcal{A}f = (1, k_2, k_3, \ldots, k_{m+1})$. Then*

$$\mathcal{V}Rf^2 = \underset{1 \leq i \leq m}{\text{Max}} \; (r_1, r_2, \ldots, r_i)(r_1 + k_{i+1}, r_2, \ldots, r_m)$$

the maximum being taken over values of i for which $k_{i+1} \neq r_{i+1}$. The maximum is attained for precisely one value of i.

An *exact period* of a vector $(r_1, r_2, \ldots r_m)$ is a number p which divides m and for which $r_i = r_{i+p}$ whenever $1 \leq i \leq m\text{-}p$. Suppose that $f \in \mathcal{L}_1 \backslash \mathcal{L}$ with $\mathcal{V}Rf = (r_1, r_2, \ldots, r_m)$ and $\mathcal{V}Rf^2 = (r_1, r_2, \ldots, r_\pi)(r_1 + k, r_2, \ldots, r_m)$. Let the least exact period of $(r_1 + k, r_2, \ldots, r_\pi)$ be p with $\pi = ap$. Let $(r_{ap+1}, r_{ap+2}, \ldots, r_m) = (r_1 + k, r_2, \ldots, r_p)^e (r_{ap+ep+1}, \ldots, r_m)$ with e maximal. Then $\mathcal{V}T^k Rf = (r_1 + k, r_2, \ldots, r_p)^{a+e} (r_{ap+ep+1}, \ldots, r_m)$ will be called the *canonical presentation* of the principal vector of f.

3. Principal Vectors of Commuting Block Maps

It is shown in [1] that to study the non-constant block maps which commute with a given block map f it is sufficient to study the set $C^*(f)$ of those block maps which commute with f and which depend on the first variable. The relationship between the canonical presentations of the principal vectors of commuting block maps is given in the next theorem.

3.1 Theorem. *Suppose that $f \in \mathcal{L}_1 \backslash \mathcal{L}$ with $\mathcal{V}Rf = (r_1, r_2, \ldots, r_m)$. Suppose also that $g \in C^*(f)$ is non-trivial and $h = f \circ g = g \circ f$. Then $g, h \in \mathcal{L}_1 \backslash \mathcal{L}$ and the canonical presentations of the principal vectors of f, g and h take the forms*

$$\mathcal{V}T^k Rf = (r_1 + k, r_2, \ldots, r_p)^{a+e}(r_{ap+ep+1}, r_{ap+ep+2}, \ldots, r_m)$$

$$\mathcal{V}T^k Rg = (r_1 + k, r_2, \ldots, r_p)^{b+e}(r_{ap+ep+1}, r_{ap+ep+2}, \ldots, r_m)$$

$$\mathcal{V}T^k Rh = (r_1 + k, r_2, \ldots, r_p)^{c+e}(r_{ap+ep+1}, r_{ap+ep+2}, \ldots, r_m)$$

where $a, b \geq 1$ and $c = a + b$.

Proof. That $g \in \mathcal{L}_1 \backslash \mathcal{L}$ follows from Theorem 2.12 and Lemma 3.3 of [1]. Suppose that $\mathcal{V}Rf = (r_1, r_2, \ldots, r_m), \mathcal{V}Rg = (s_1, s_2, \ldots, s_n), Af = (1, k_2, k_3, \ldots, k_{m+1})$ and $Ag = (1, \ell_2, \ell_3, \ldots, \ell_{n+1})$. Then by Proposition 2.2 there exist u, v such that

$$\mathcal{V}R(f \circ g) = (r_1, r_2, \ldots, r_u)(s_1 + k_{u+1}, s_2, \ldots, s_n),$$

$$\mathcal{V}R(g \circ f) = (s_1, s_2, \ldots, s_v)(r_1 + \ell_{v+1}, r_2, \ldots, r_m)$$

with $u \geq 1, v \geq 1, k_{u+1} \neq r_{u+1}$ and $\ell_{v+1} \neq s_{v+1}$. The first $u + v$ terms of these vectors are equal term by term. Thus by an argument similar to that of the first part of the proof of Proposition 3.5 of [5] $k_{u+1} = \ell_{v+1} = k$, say, and

$$(r_1 + k, r_2, \ldots, r_u) = (r_1 + k, r_2, \ldots, r_p)^a$$

$$(s_1 + k, s_2, \ldots, s_v) = (r_1 + k, r_2, \ldots, r_p)^b$$

where p divides $[u, v], u = ap$ and $v = bp$ and p is the least exact period of $(r_1 + k, r_2, \ldots, r_p)$. Moreover, $(r_{u+1}, r_{u+2}, \ldots, r_m) = (s_{v+1}, s_{v+2}, \ldots, s_n) = (r_1 + k, r_2, \ldots, r_p)^e(r_{ap+ep+1}, \ldots, r_m)$, say, with e maximal. In the remainder of the proof we set $(r_1 + k, r_2, \ldots, r_p) = P$ and $(r_{ap+ep+1}, \ldots, r_m) = E$.

Now suppose that

$$\mathcal{V}T^k Rf^2 = (r_1 + k, r_2, \ldots, r_q)(r_1 + k_{q+1}, r_2, \ldots, r_m).$$

Let j be the maximal integer such that $(r_1, r_2, \ldots, r_q) = P^j(r_{jp+1}, \ldots, r_q)$. Then Corollary 2.3 applied to $\mathcal{V}Rf^2$ with $i = q$ and $i = ap$ gives

$$\mathcal{V}T^k Rf^2 = P^j(r_{jp+1}, \ldots, r_q)(r_1 + k_{q+1}, r_2, \ldots, r_p)P^{a+e-1}E > P^{2a+e}E.$$

Proposition 2.2 applied to $\mathcal{V}R(f \circ g)$ with $i = ap$ and $i = q$ gives

$$\mathcal{V}T^k R(f \circ g) = P^{a+b+e}E > P^j(r_{jp+1}, \ldots, r_q)(r_1 + k_{q+1}, \ldots, r_q)P^{b+e-1}E.$$

If $q - jp \geq p$ then $j = a + e$ and by comparing the first $(a + e + 1)p$ terms on each side of each of the two inequalities one obtains

$$P^{a+e}(r_{ap+ep+1}, \ldots, r_{ap+ep+p}) = P^{a+e+1}$$

so that e is not maximal. Thus $q - jp < p$. If $q - jp > 0$ then by comparing the first $q + p$ terms on each side of the two inequalities one obtains

$$P^j(r_{jp+1}, \ldots r_q)(r_1 + k_{q+1}, r_2, \ldots, r_p) = P^{j+1}(r_1 + k, r_2, \ldots, r_{q-jp}).$$

Then $k_{q+1} = k$ and $[p, q - jp]$ is an exact period of P. Since this contradicts the assumption that p is the least exact period of P it follows that $q = jp$ for some integer j. In this case also $k_{q+1} = k$, and the two inequalities take the form $P^{j+a+e}E > P^{2a+e}E$ and $P^{a+b+e}E > P^{j+b+e}E$ from which it follows that $j = a$. Thus the presentation of the principal vector of f is canonical. Similarly, the presentations of the principal vectors of g and h are canonical.

3.2 Corollary. *If the presentation*

$$VT^k Rf = (r_1 + k, r_2, \ldots, r_p)^{a+e}(r_{ap+ep+1}, \ldots, r_m)$$

is canonical then for each positive integer n the presentation

$$VT^k Rf^n = (r_1 + k, r_2, \ldots, r_p)^{na+e}(r_{ap+ep+1}, \ldots, r_m)$$

is canonical.

3.3 Corollary. *Suppose that in the canonical presentation*

$$VT^k Rf = (r_1 + k, r_2, \ldots, r_p)^{a+e}(r_{ap+ep+1}, \ldots, r_m)$$

we have $a = 1$. Then f is not a power (greater than one) of a block map g, and f is not a sum of powers Σg^i where the maximal power is greater than one.

REFERENCES

1. E. M. Coven, G. A. Hedlund and F. Rhodes, *The commuting block maps problem*, Trans. Amer. Math. Soc. **249** (1979), 113 – 138.
2. G. A. Hedlund, *Endomorphisms and automorphisms of the shift dynamical system*, Math. Systems Theory **3** (1969), 320 – 375.
3. F. Rhodes, *The sums of powers theorem for commuting block maps*, Trans. Amer. Math. Soc. **271** (1982), 225 – 236.
4. F. Rhodes, *The principal part of a block map*, J. Combinatorial Theory Ser. A **33** (1982), 48 – 64.
5. F. Rhodes, *The role of the principal part in factorizing block maps*, Math. Proc. Cambridge Phil. Soc. **96** (1984), 223 – 235.

Contemporary Mathematics
Volume **135**, 1992

On The Recurrence of Countable Topological Markov Chains

IBRAHIM A. SALAMA

ABSTRACT. We consider a countable graph Γ (and the shift dynamical system or chain that it determines) and we use a pictorial approach to classify Γ as transient, null recurrent, or positive recurrent. This approach avoids the computational and combinatorial problems usually encountered. We consider joining two graphs Γ_1 and Γ_2 at some state, creating a new graph Γ. We study the recurrence property of Γ based on the recurrence properties of Γ_1 and Γ_2.

1. Introduction

Chains on a countably infinite set are used to analyze problems in various fields (see [3], [7], [6], [2], [1]). One of the important questions in studying such chains is to decide whether the chain is transient, null recurrent, or positive recurrent. This classification was introduced by Vere–Jones [9], and Gurevic [4] showed that the chain has a Borel probability measure with maximal entropy if and only if the chain is positive recurrent. If we consider the usual approach for the classification problem, mentioned in [9], we encounter computational and combinatorial problems. In [8] we introduced

1991 Mathematics Subject Classification: Primary 54H20.
This paper is in final form and no version of it will be submitted for publication elsewhere.

a geometric or pictorial approach to deal with this problem. In this paper, we clarify some of the issues raised in [8] and extend some of the results. The main theorem for classification is presented in Section 1, followed by some applications and examples in Section 2. We start by fixing notations and giving some background.

Let Γ be a strongly connected directed graph on a countable set of vertices $S = \{s_1, s_2, ...\}$, and let

$$X(\Gamma) = \{X \in S^Z \mid \text{for all } i,$$

$$\text{there is an edge in } \Gamma \text{ from } x_i \text{ to } x_{i+1}\}.$$

If S has the discrete topology and S^Z the product topology, then in the induced topology $X(\Gamma)$ (or simply X), together with the shift transformation σ defined by $(\sigma x)_i = x_{i+1}$ for all i, is a (non–compact) dynamical system, called the *chain* determined by the directed graph Γ. The topological entropy of X may be determined by using Gurevic's definition (Gurevic, [4]), $h_G(X) = h_G(\Gamma) = \sup_{\Gamma' < \Gamma} h(\Gamma')$, where the sup is taken over all (connected) finite subgraphs Γ' of Γ (and $h(\Gamma')$ is the usual topological entropy of the subshift of finite determined by Γ', the logarithm of the maximal eigenvalue of the transition matrix).

Fix a vertex in Γ and define

$B_s^{(n)}$ = number of paths of length n in Γ from s to s;

$f_s^{(n)}$ = number of paths of length n in Γ from s to s with no other occurrence of s in between;

$F(\Gamma, s, z) = \Sigma f_s^{(n)} z^n;$

$$B(\Gamma,s,z) = \Sigma \ B_s^{(n)} \ z^n;$$

$$F'(\Gamma,s,z) = \Sigma \ n \ f_s^{(n)} \ z^{n-1}.$$

Let ℓ_Γ and R_Γ be the radii of convergence of $F(\Gamma,s,z)$ and $B(\Gamma,s,z)$, respectively. Note that R_Γ is independent of s while ℓ_Γ may depend on s. We will abbreviate ℓ_Γ by ℓ_s and R_Γ by R (generally, ℓ_{Γ_i} by ℓ_{is} and R_{Γ_i} by R_i). We also let $L_i = \inf_s \ell_{is}$.

Finally, with these notations, Gurevic ([4],[5]) showed that if X is a connected graph with $h_G(\Gamma) < \infty$, then $h_G(\Gamma) = -\log R$. Also, the Vere–Jones ([9],[10]) results on classification of a graph Γ as transient, null recurrent, or positive recurrent, may be summarized by the following table.

	transient	null recurrent	positive recurrent
$F(\Gamma,s,R)$	< 1	$= 1$	$= 1$
$F'(\Gamma,s,R)$		$= \infty$	$< \infty$
$B(\Gamma,s,R)$	$< \infty$	$= \infty$	$= \infty$
$\lim_{n\to\infty} B_s^{(n)} R^n$	$= 0$	$= 0$	> 0

The missing entry in this table is about the possible values of $F'(\Gamma,s,R)$, where Γ is transient. Sheldon Newhouse raised a question about the possibility of $F'(\Gamma,s,R)$ being finite when Γ is transient. We give an affirmative answer for this question; thus for a transient Γ, $F'(\Gamma,s,R)$ may be finite or infinite.

2. Recurrence of Chains

In this section we present a criterion enabling us to decide if a connected graph Γ is transient, null recurrent, or positive recurrent. Our approach is geometric or pictorial in nature, avoiding the computational and combinatorial problems usually encountered. The idea, which we presented first in [8], is to relate the value of $h_G(\Gamma)$ to the values of $h_G(\Gamma')$, where Γ' is either a subgraph of Γ or a graph containing Γ. The main result is given in Theorem 1.

THEOREM (1): (i) Γ_0 is transient if and only if there exists $\Gamma_2 > \Gamma_0$ such that $h_G(\Gamma_0) = h_G(\Gamma_2)$. Moreover, if Γ_0 is transient, then there exists $\Gamma_1 < \Gamma_0$ such that $h_G(\Gamma_1) = h_G(\Gamma_0)$.

(ii) Γ_0 is recurrent if and only if for all $\Gamma_2 > \Gamma_0$, $h_G(\Gamma_2) > h_G(\Gamma_0)$. Moreover, if Γ_0 is null recurrent or Γ_0 is positive recurrent with $R_0 = L_0$, then there exists $\Gamma_1 < \Gamma_0$ such that $h_G(\Gamma_1) = h_G(\Gamma_0)$.

(iii) Γ_0 is positive recurrent with $R_0 < L_0$ if and only if for all $\Gamma_1 < \Gamma_0 < \Gamma_2$, $h_G(\Gamma_1) < h_G(\Gamma_0) < h_G(\Gamma_2)$.

In proving Theorem (1), we first prove the following.

THEOREM (2). If Γ_0 is transient or null recurrent, then $R_0 = L_0$. Hence, if $R_0 < L_0$, then Γ_0 is positive recurrent.

PROOF. If Γ_0 is null recurrent, then $F'(\Gamma_0,s,R_0) = \infty$, $R_0 = \ell_{os}$, thus $R_0 = L_0$. Let Γ_0 be transient; then $F(\Gamma_0,s,R_0) < 1$. Let Γ_1 be the loop graph corresponding to Γ_0, that is, the graph with $f_s^{(n)}$ loops of length n based at a state s_1 of Γ_1, $n = 1,2,\dots$. Consider the loop series $F(\Gamma_0,s,R) = \Sigma\ f_s^{(n)}\ R^n$

and its truncations $F_k(\Gamma_0,s,R) = \Sigma_{n=1}^k f_s^{(n)} R^n$.   Let Γ_k be the subgraph of Γ_1 containing all loops, based at s_1, of length k or less.   Let R_k be the solution for the equation $F_k(\Gamma_0,s,R) = 1$.   By Petersen ([7]), $h_G(\Gamma_k) = -\log R_k$, and $h_G(\Gamma_1) = \sup_k h_G(\Gamma_k)$; thus $R_k \searrow R_1 = R_0$.   Now, if $R_0 < \ell_{os}$, then we can find large k such that $R_0 < R_k < \ell_{os}$, which implies $F_k(\Gamma_0,s,R_k) > F(\Gamma_0,s,R_k)$, contradicting $F_k(\Gamma_0,s,R) \leq F(\Gamma_0,s,R)$ for all R.   Thus $R_0 = \ell_{os}$, hence $R_0 = L_0$.   ∎

COROLLARY (3):   *If $R_0 = L_0$, then $\ell_{os} = L_0$ for all s in Γ_0.   In particular, if Γ_0 is transient or null recurrent, then ℓ_{os} is independent of s.*

THEOREM (4):   *There exists $\Gamma_1 < \Gamma_0$ such that $h_G(\Gamma_1) = h_G(\Gamma_0)$ if and only if $R_0 = L_0$.*

PROOF: (i)   For a state s in Γ_0, let $Fol(s) = \{s' \mid$ there is an arrow from s to s' in $\Gamma_0\}$.   Let s_0 be a state in Γ_0 such that $\#Fol(s_0) \geq 2$.   Let $s^* \in Fol(s_0)$, let Γ_1 be the largest connected subgraph obtained by removing the edge $s_0 s^*$ from Γ_0, and let Γ_2 be the largest connected subgraph obtained by removing the edges $s_0 s'$, $s' \in Fol(s_0) - s^*$.   Let i be such that $\ell_{is_0} = \min(\ell_{1s_0}, \ell_{2s_0})$.   Then $R_0 \leq R_i \leq \ell_{is_0} = \ell_{os_0} = L_0$.   If $R_{os} = L_0$, then we have $R_0 = R_i$.

(ii)   If $\Gamma_1 < \Gamma_0$ with $h_G(\Gamma_1) = h_G(\Gamma_0)$, then $R_0 = R_1$.   But Γ_1 is transient, that is $L_1 = R_1$, and $L_0 \leq L_1$ (since $\Gamma_1 < \Gamma_0$).   Thus $R_0 = L_0$.   ∎

PROOF OF THEOREM (1):   (i)   If $\Gamma_0 < \Gamma_2$, is a state in

both Γ_0 and Γ_2, and $R_0 = R_2$, then $F(\Gamma_0, s, R_0) < F(\Gamma_2, s, R_2)$
≤ 1 and Γ_0 is transient. On the other hand, assume Γ_0 is
transient. Then we can find an m such that $\Sigma_{n \neq m} f_s^{(n)} R_0^n +$
$(f_s^{(m)}+1)R_0^m < 1$. Let Γ_2 be the graph obtained by adding to
Γ_0 a loop of length m based at s. Then $\Gamma_0 < \Gamma_2$ and
$F(\Gamma_2, s, R_0) < 1$. Since $R_2 \leq R_0$, $F(\Gamma_2 s, R_2) \leq F(\Gamma_2, s, R_0) < 1$,
and Γ_2 is transient. Note that $R_0 = L_0 = L_2$, and by
Theorem (4) $R_2 = L_2$; hence $R_0 = R_2$. The second part of
Theorem (1): (i) follows from Theorem (4); (ii) and (iii),
follow from (i) above and Theorem (4). ∎

The following lemma can be useful for classification of Γ_0
using $F(\Gamma_0, s, L_0)$.

LEMMA (5): (i) $R_0 < L_0$ *if and only if* $F(\Gamma_0, s, \ell_{os}) > 1$.
(ii) *If* $F(\Gamma_0, s, \ell_{os}) > 1$ *for some* s, *then* Γ_0 *is positive
recurrent.*
(iii) Γ_0 *is transient if and only if* $F(\Gamma_0, s, \ell_{os}) < 1$ *for some* s.

PROOF: (i) If $R_0 < L_0$, then Γ_0 is positive recurrent and
$F(\Gamma_0, s, R_0) = 1$. Thus $F(\Gamma_0, s, \ell_{os}) \geq F(\Gamma_0, s, L_0) > 1$. If
$F(\Gamma_0, s, \ell_{os}) > 1$, then $R_0 < L_0$, since $F(\Gamma_0, s, R_0) \leq 1$.

(ii) Follows from (i) and Theorem (2).

(iii) If Γ_0 is transient, then $R_0 = L_0 = \ell_{os}$ for all s in Γ_0,
and $F(\Gamma_0, s, \ell_{os}) < 1$. If $F(\Gamma_0, s, \ell_{os}) < 1$, then $F(\Gamma_0, s, R_0) < 1$,
since $R_0 = \ell_{os}$.

3. Applications and Examples

We consider a natural operation to join two graphs Γ_1 and
Γ_2, producing a new graph Γ. This may be described

informally as follows: let Γ_1 and Γ_2 be two graphs with state space S_1 and S_2, respectively. Let $s_1 \in S_1$ and $s_2 \in S_2$. Join s_1 and s_2 by an edge and shrink this edge until its length becomes zero, thus joining s_1 and s_2 and creating a new state, s_0, in place of both of them, and a new graph Γ. Note that the state space, S, of Γ is $S = (S_1 - \{s_1\}) \cup (S_2 - \{s_2\}) \cup \{s_0\}$. If we let $\text{Fol}(s_0) = \{s \in S_i |$ there is an arrow from s_i to s in $\Gamma_i\}$, and $\text{Pre}(s_i) = \{s \in S_i |$ there is an arrow from s to s_i in $\Gamma_i\}$, then $\text{Fol}(s_0) = \text{Fol}(s_1) \cup \text{Fol}(s_2)$, and $\text{Pre}(s_0) = \text{Pre}(s_1) \cup \text{Pre}(s_2)$. The same notion may be used to join two states in the same graph.

A question of interest is: if we know the recurrence properties of Γ_1 and Γ_2, and if we join Γ_1 and Γ_2 at some state (as described above) creating a new graph Γ, then what can we say about the recurrence property of Γ? The same question holds if we join two states in the same graph.

THEOREM (6): *Let Γ_0 be a graph obtained by joining a recurrent graph Γ_1 with another graph Γ_2 at some state. If $R_1 \leq R_2$, then Γ_0 is positive recurrent with $R_0 < L_0$.*

PROOF: Since Γ_1 is recurrent,

$$R_0 < R_1 \leq \min(L_1, L_2) = L_0.$$

Thus $R_0 < L_0$, and Γ_0 is positive recurrent.

COROLLARY (7): *Let Γ_0 be a graph obtained by joining two recurrent graphs Γ_1 and Γ_2 at some state. Then Γ_0 is positive recurrent with $R_0 < L_0$.*

THEOREM (8): *Let Γ_0 be the graph obtained by joining two states s_1 and s_2 of Γ_1. Then $R_0 \leq R_1$ and $L_1 \leq L_0$.*

Moreover, if Γ_1 is recurrent, then Γ_0 is positive recurrent with $R_0 < L_0$.

PROOF: Let $f_{s_i}^{(n)}$ be the number of loops in Γ_1 based at s_i, $i = 1, 2$. For any three states a, b, and c in Γ_1, let $_c f_{ab}^{(n)}$ be the number of paths starting at and arriving at b after n steps for the first time without visiting c in between. Let s_0 be the state in Γ_0 created by joining s_1 and s_2, and let $f_{s_0}^{(n)}$ be the number of loops in Γ_0 based at s_0. Then

$$f_{s_0}^{(n)} = f_{s_1}^{(n)} + {}_{s_1}f_{s_1 s_2}^{(n)} - \sum_{k=1}^{n-1} {}_{s_1}f_{s_1 s_2}^{(n-k)} \; {}_{s_1}f_{s_2 s_1}^{(k)} +$$

$$f_{s_2}^{(n)} + {}_{s_2}f_{s_2 s_1}^{(n)} - \sum_{k=1}^{n-1} {}_{s_2}f_{s_2 s_1}^{(n-k)} \; {}_{s_2}f_{s_1 s_2}^{(k)} \; .$$

What this equation is saying is: to count the number of loops of length n in Γ_0 based at s_0, we consider all loops in Γ_1 based at s_1. Then, we take away from them all loops of the form $s_1 \, x_1 \, \cdots \, x_{n-1} \, s_1$, where $x_i = s_2$ for some i, then we add to them all paths of the form $s_1 \, x_1 \, \cdots \, x_{n-1} \, s_2$, $x_i \neq s_1$ or s_2. (We do the same for loops based at s_2 in Γ_1.) Now, let k_1 (k_2) be the length of the shortest path from s_2 to s_1 (s_1 to s_2) in Γ_1. Then ${}_{s_1}f_{s_1 s_2}^{(n)} \leq f_{s_1}^{(n+k_1)}$ and ${}_{s_2}f_{s_2 s_1}^{(n)} \leq f_{s_2}^{(n+k_2)}$; thus we have

$$f_{s_0}^{(n)} \leq f_{s_1}^{(n)} + f_{s_1}^{(n+k_1)} + f_{s_2}^{(n)} + f_{s_2}^{(n+k_2)},$$

and $\ell_{s_i} \leq \ell_{s_0}$ for i = 1,2. Hence $L_1 \leq \ell_{s_0}$ and $L_1 \leq L_0$ (since s_1 and s_2 were arbitrary). Since $\Gamma_1 < \Gamma_0$, $R_0 \leq R_1$. If Γ_1 is recurrent, then $R_0 < R_1$ and Γ_0 is positive recurrent with $R_0 < L_0$. ∎

We conclude by giving some examples and remarks.

1.   There exists a transient graph Γ_1 with $F'(\Gamma_1, s, R_1) < \infty$. We give an example for such graph.   This also shows the existence of a positive recurrent graph Γ_0 with $R_0 = L_0$.   To see this: Let Γ_1 be a loop graph based at s with

$$f_s^{(n)} = \left\lfloor \frac{2^{n+2}}{n(n+1)(n+2)} \right\rfloor, \text{ where } \lfloor x \rfloor \text{ is the largest integer} \leq x.$$

Then $\ell_{1s} = \frac{1}{2}$, $\Sigma f_s^{(n)} (\frac{1}{2})^n < 1$, and Γ_1 is transient (since $R_1 \leq \ell_{1s}$); thus $\ell_{1s} = L_1 = R_1 = \frac{1}{2}$ with $\Sigma n f_s^{(n)}(\frac{1}{2})^n < \infty$. Now, let $\Sigma f_s^{(n)}(\frac{1}{2})^n = \delta$, and $1 - \delta = \Sigma_{n=1}^{\infty} \frac{x_i}{2^n}$, when each $x_i = 0$ or 1.   Let Γ_0 be a loop graph based at s with number of loops given by $(f_s^{(n)} + x_n)$.   Note that $\ell_{0s} = \frac{1}{2}$ and by a similar argument to what we have in the proof of Theorem (2) we get $R_0 = \frac{1}{2}$.   Thus $\ell_{0s} = L_0 = R_0 = \frac{1}{2}$, and Γ_0 is recurrent.   Noting $\Sigma n(f_s^{(n)} + x_n)(\frac{1}{2})^n < \infty$, we have that Γ_0 is positive recurrent.

2.   Let Γ_0 be the graph obtained by joining positive recurrent Γ_1 and transient Γ_2 with $R_2 < R_1$.   The graph Γ_0 can be transient (join ⬭ with ⬭⬭ ..... to obtain

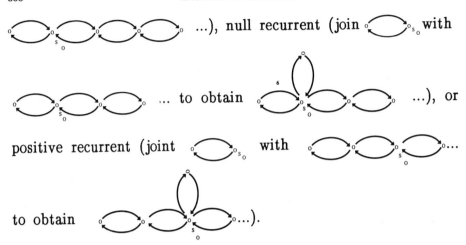

...), null recurrent (join ⬭ with ... to obtain ⬭ ...), or positive recurrent (joint ⬭ with ⬭... to obtain ⬭...).

3. As in (2) above, let Γ_1 be transient and Γ_2 null recurrent with $R_2 < R_1$. Then Γ_0 can be transient; this is seen by joining Γ_1 and Γ_2 at s_0

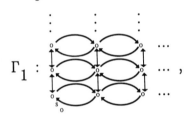

$$\Gamma_1 :$$

$$\Gamma_2 : \cdots$$

4. If Γ_1 and Γ_2 are both transient, then Γ_0 can be transient (let $\Gamma_2 = \Gamma_1$, Γ_1 as given in Remark 3 above), or null recurrent ($\Gamma_1 = \Gamma_2$: ...).

5. In reference to Theorem (8), consider the graph

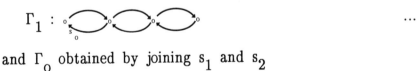

$$\Gamma_1 : \qquad \cdots$$

and Γ_0 obtained by joining s_1 and s_2

$$\Gamma_0 :$$

Now, Γ_0 may be seen as the join of Γ_1' and Γ_1'',

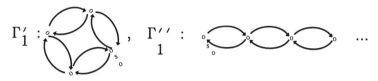

and both have topological entropy $\log 2$. Since Γ_1' is recurrent, then by Theorem (6) Γ_0 is positive recurrent with $R_0 < L_0$. That shows we can have a positive recurrent Γ_0 without Γ_1 being recurrent.

Acknowledgment. The author would like to thank Professors Karl Petersen and Sheldon Newhouse for their helpful discussions.

REFERENCES

[1] F. Blanchard, and G. Hansel, *Systems codes*, Theoretical Computer Science, **44** (1986), 17–49.

[2] F. Blanchard, *Systemes dynamiques topologiques associes a des automates recurrents*, Z. Wahr. Verw. Geb., **58** (1981), 549–564.

[3] E.E. Dinaburg, *An example of the computation of topological entropy*, Uspehi Mat. Nauk, **23** (1968), 249–250.

[4] B.M. Gurevic, *Topological entropy of Enumerable Markov chains*, Soviet Math. Dokl., no. 4, **10** (1969), 911–915.

[5] B.M. Gurevic, *Shift entropy and Morkov measures in the path space of a denumerable graph*, Soviet Math. Dokl., No. 3, **11** (1970), 744–747.

[6] W. Krieger, *On the uniqueness of the equilibrium state*, Math. Systems Theory, **8** (1974), 97–104.

[7] K. Petersen, *Chains, entropy, coding*, Ergodic Theory Dynamical Systems, **6** (1986), 415–448.

[8]   I.A. Salama, *Topological entropy and recurrence of countable chains*, Pacific J. Math. **134** (1988), 325–341; enrata: Pacific J. Math. **140** (1989), 397–398.

[9]   D. Vere–Jones, *Geometric erogodicity in denumerable Markov chains*, Quart. J. Math. Oxford, (2), **13** (1962), 7–28.

[10] D. Vere–Jones, *Ergodic properties of nonnegative matrices – I*, Pacific J. Math., No. 2, **22** (1967), 361–386.

School of Business
North Carolina Central University
Durham, NC 27707

Contemporary Mathematics
Volume **135**, 1992

SUBSTITUTIONS, ADIC TRANSFORMATIONS, AND
BETA-EXPANSIONS

BORIS SOLOMYAK

ABSTRACT. It is shown that dynamical systems corresponding to a class of substitutions of non-constant length have purely discrete spectrum. The proof relies on properties of digit expansions with non-integral bases.

1. INTRODUCTION

Our purpose is to investigate spectral properties of a class of measure-preserving transformations. A transformation of this class has three isomorphic representations: as a substitution dynamical system, as a stationary adic transformation, and as an analog of "+1" transformation on the set of "integral" β-expansions.

1.1. Substitutions. Let ζ be the following substitution on the alphabet $\mathcal{A} = \{1, \dots, m\}, m \geq 2$:

$$(1) \quad \zeta(1) = \underbrace{1 \dots 1}_{a_1} 2, \; \zeta(2) = \underbrace{1 \dots 1}_{a_2} 3, \dots, \zeta(m-1) = \underbrace{1 \dots 1}_{a_{m-1}} m, \; \zeta(m) = \underbrace{1 \dots 1}_{a_m}.$$

Let u be the infinite word $\lim_{k \to \infty} \zeta^k(1)$, which is a fixed point of the substitution. Further, consider the shift transformation $(\sigma v)_n = v_{n+1}$ on $\mathcal{A}^{\mathbf{N}}$. Let $\mathcal{X}_\zeta$ denote the closure of the orbit of u under the shift. The resulting "substitution dynamical system" $(\mathcal{X}_\zeta, \sigma)$ is uniquely ergodic [M]. The spectral theory of such systems is far from being complete, especially in the "nonconstant length case" (see [Q]), to which (1) in general belongs.

1.1.1. **Main Theorem.** *If $a_1 \geq a_2 \geq \dots \geq a_m \geq 1$, then the dynamical system $(\mathcal{X}_\zeta, \sigma)$ corresponding to the substitution (1), has purely discrete spectrum (=complete set of eigenvectors).*

By the theorem of Halmos and von Neumann (see [W, Ch.3]), a measure-preserving transformation with discrete spectrum is metrically isomorphic to a translation on a compact group with the Haar measure. In some cases this group can be identified, using the description of the set of eigenvalues [H, L, So3].

1991 *Mathematics Subject Classification.* 28D05, 58F19; Secondary 11A67, 68Q68.

Key words and phrases. Measure-preserving transformation, discrete spectrum, substitution, beta-expansion.

This paper is in final form and no version of it will be submitted for publication elsewhere

1.1.2. Special case. Suppose that $a_1 \geq a_2 \geq \ldots \geq a_m = 1$. Then our substitution dynamical system is isomorphic to a translation on the torus $\mathbf{T}^{m-1}$. For $m = 3$, $a_1 = a_2 = a_3 = 1$, this was proved by Rauzy [R1]. For $a_1 = a_2 = \ldots = a_m = 1$ the result was obtained in [So1] in the framework of adic transformations.

1.1.3. Remark. In a recent paper [R2] there is an interesting example of a substitution dynamical system isomorphic to a translation on $\mathbf{T} \times \mathbf{Z}_2$ ($\mathbf{Z}_2$ = group of 2-adic integers). It is also mentioned (a precise formulation is not given) that if the matrix of the substitution ($M_\zeta = (M_{ij})$, M_{ij} = number of "i"s in $\zeta(j)$) has only one eigenvalue outside the unit disk $|z| < 1$, and determinant equal to ± 1, then the corresponding dynamical system is isomorphic to a torus translation. This includes the special case above, by a result due to Brauer [Br, proof of Th.2].

1.2. Adic Transformations.

The adic transformation on the space of infinite paths of a Bratteli diagram was introduced by Vershik [V1, V2] for the purposes of the theory of operator algebras. He proved that any ergodic automorphism on a Lebesgue space is metrically isomorphic to an adic transformation. For definitions and precise formulations the reader is referred to [VL].

We shall be concerned with the stationary case only, when the adic transformation acts on the space of realizations of a one-sided topological Markov chain. Let $M = (M_{ij})$ be a non-negative integral $m \times m$ matrix, and let $\Gamma(M)$ be the directed graph with the set of vertices $\mathcal{A} = \{1, 2, \ldots, m\}$ and adjacency matrix M. Consider the space $\mathcal{X}_M$ of all one-sided infinite paths in this graph, endowed with the product topology (a path is regarded as a sequence of edges). To define the adic transformation, a partial order is introduced on $\mathcal{X}_M$. Let $\mathcal{E}_i$ be the set of edges with the terminal vertex i. Fix a total order on each of the sets $\mathcal{E}_i$, $i = 1, 2, \ldots, m$, agreeing with the natural order on the initial vertices. The order on $\mathcal{X}_M$ is lexicographical from the right: a path x is said to precede a path y if they differ in finitely many terms (edges), and in the last such term the edge of x precedes the edge of y. The adic transformation of x is defined as the minimum of y in $\mathcal{X}_M$ preceded by x. (In other words, this is the "immediate successor transformation".) It is easy to see that this transformation is well-defined on the set of non-maximal paths. If the matrix M is aperiodic, deleting a countable set from $\mathcal{X}_M$, we obtain a dynamical system which turns out to be uniquely ergodic (see [VL]).

The connection between stationary adic transformations and substitutions was discovered by Livshitz [L]. Consider the substitution ζ on the alphabet $\{1, 2, \ldots, m\}$ defined by

$$\zeta(i) = \underbrace{1 \ldots 1}_{M_{1i}} \underbrace{2 \ldots 2}_{M_{2i}} \ldots \underbrace{m \ldots m}_{M_{mi}}.$$

It is proved in [L] that if ζ satisfies the property of unique admissible decoding, then the systems $(\mathcal{X}_M, T)$ and $(\mathcal{X}_\zeta, \sigma)$ are metrically isomorphic. It is indicated in Section 2 that this isomorphism can be obtained from the prefix automaton of the substitution.

Let M be the companion matrix of the polynomial $p(X) = X^m - a_1 X^{m-1} - \cdots - a_m$, $a_i \geq 0$. Then the result of Livshitz applies, and the adic transformation is isomorphic to the dynamical system corresponding to the substitution (1).

1.3. Beta-expansions. The third description of our transformation is based on the notion of β-expansion. Let $\beta > 1$ be a real number. Every $x \geq 0$ can be written as

$$x = \sum_{i=-k}^{\infty} \varepsilon_i \beta^{-i},$$

where the "digits" ε_i are computed by the "greedy algorithm":

Denote by $[y]$ and $\{y\}$ the integer part and the fractional part of a number y. There exists $k \in \mathbf{Z}$ such that $\beta^k \leq x < \beta^{k+1}$. Let $\varepsilon_{-k} = [x/\beta^k]$, and $r_{-k} = \{x/\beta^k\}$. Then for $-k < i < \infty$, put $\varepsilon_i = [\beta r_{i-1}]$, and $r_i = \{\beta r_{i-1}\}$. We shall use the notation

$$x \overset{\beta}{\sim} \varepsilon_{-k}\varepsilon_{-k+1}\ldots\varepsilon_0.\varepsilon_1\varepsilon_2\ldots$$

Note that the β-expansion of Rényi [Re] coincides with this expansion for $0 \leq x < 1$, but is different for $x = 1$. We have $1 \overset{\beta}{\sim} 1.00\ldots$, while the Rényi expansion is

$$1 = \sum_{i=1}^{\infty} d_i \beta^{-i}, \quad \text{where} \quad d_1 = [\beta], \quad \beta - [\beta] \overset{\beta}{\sim} .d_2 d_3 \ldots$$

The sequence $d_1 d_2 d_3 \ldots$ will be denoted by $d(1, \beta)$. Let $\mathcal{N}$ be the set of all "integral" β-expansions:

$$\mathcal{N} = \{x \overset{\beta}{\sim} \varepsilon_{-k}\varepsilon_{-k+1}\ldots\varepsilon_0.0^\omega\}$$

(0^ω stands for $000\ldots$). Let τ be the "immediate successor" transformation on $\mathcal{N}$ with respect to the induced order from $\mathbf{R}$. Consider $(\overline{\mathcal{N}}, \overline{\tau})$, where $\overline{\mathcal{N}}$ is the completion of the set $\mathcal{N}$ with respect to the product topology, and $\overline{\tau}$ is the extension of τ. If the Rényi expansion of 1 in base β is finite: $d(1, \beta) = a_1 a_2 \ldots a_m$ (such β are called "simple β-numbers" [P]), then $(\overline{\mathcal{N}}, \overline{\tau})$ is topologically conjugate to the stationary adic transformation corresponding to the companion matrix of $p(X) = X^m - a_1 X^{m-1} - \cdots - a_m$.

A few words about the structure of the paper. In Section 2 we recall some basic facts about substitutions. Then we establish a condition for discreteness of the spectrum of a substitution dynamical system, generalizing a result due to Host. Our proof is based on the "operator theoretic" approach suggested in [So1, So2]. Section 2 is concluded with the discussion of a "system of numeration related to substitution" developed by Rauzy and his associates, which reveals the connections with the adic transformation.

In Section 3 we restrict ourselves to the case of companion matrices. Such a matrix naturally determines a renewal system which is a shift of finite type if the companion matrix comes from a β-expansion of a simple β-number.

In Section 4 we prove the Main Theorem using recent results of Frougny and the author [FS] on finite β-expansions. In fact, a "conditional statement" is proved: if β is a Pisot number of degree m, such that $d(1, \beta) = a_1 a_2 \ldots a_m$, and every element of the set $\mathbf{Z}_+[\beta^{-1}]$ has a finite β-expansion, then the substitution (1) defines a dynamical system with purely discrete spectrum. In [FS] it is shown that this is the case when $a_1 \geq a_2 \geq \ldots \geq a_m$.

In this paper we work mostly in the framework of substitutions. However, it was the adic transformation viewpoint which provided the original motivation and intuition, and we believe that it deserves to be publicized.

A preliminary version of this paper is contained in [So3].

I am grateful to V. De Angelis, C. Kraaikamp, D. Lind, and B. Praggastis for many helpful discussions.

2. SUBSTITUTIONS

First we recall the basic definitions and results, using the "one-sided" approach in [Q].

The set $\mathcal{A} = \{1, \ldots, m\}$ will be our *alphabet*. Its elements are called *letters*. The set of finite sequences (b_i), $b_i \in \mathcal{A}$, will be denoted by $\mathcal{A}^*$, and its elements will be called the *words* of the language. The "empty word" e is also included into $\mathcal{A}^*$. The length of a word B is denoted by $|B|$. Consider the set $\mathcal{A}^{\mathbf{N}}$ of "infinite words" with the product topology, which makes it compact. If v_1 and v_2 are two finite words, they can be concatenated to form a word $v_1 v_2$. If $b = b_1 b_2 \ldots$ is a word (finite or infinite), $b[k, l] = b_k b_{k+1} \ldots b_l$ is said to be a factor of b.

A *substitution* is a mapping $\zeta : \mathcal{A} \to \mathcal{A}^*$. The substitution will be assumed one-to-one. Using the concatenation, the mapping ζ is naturally extended to $\mathcal{A}^*$ and to $\mathcal{A}^{\mathbf{N}}$. We shall assume that the word $\zeta(1)$ starts with 1 and $|\zeta(1)| > 1$. Then $\zeta^2(1)$ starts with $\zeta(1)$, $\zeta^3(1)$ starts with $\zeta^2(1)$ etc. As a result, we get an infinite word

$$u = u_1 u_2 \ldots u_m \ldots = \lim_{i \to \infty} \zeta^i(1),$$

which is a fixed point of the substitution: $\zeta(u) = u$. The dynamical system is formed by the *shift transformation*:

$$(\sigma v)_n = v_{n+1}, \; v \in \mathcal{A}^{\mathbf{N}},$$

restricted to the closure of the orbit

$$\mathcal{X}_\zeta = \mathrm{clos}\{\sigma^n u : n \geq 0\}.$$

The word u is said to be *aperiodic* if its orbit under the shift is infinite. With a substitution ζ one can associate a matrix $M_\zeta = (M_{ij})$ such that M_{ij} is the number of letters i in the word $\zeta(j)$.

We shall assume that the substitution ζ is *primitive*, that is, there is a $k \in \mathbf{Z}_+$, such that for any $\alpha \in \mathcal{A}$, the word $\zeta^k(\alpha)$ contains all letters. This implies that the system $(\mathcal{X}_\zeta, \sigma)$ is minimal (= all orbits are dense) and uniquely ergodic [M]. Let μ be the σ-invariant measure.

If ζ is a primitive substitution and u is aperiodic, then there exists a countable set Y such that $\sigma : \mathcal{X}_\zeta \setminus Y \to \mathcal{X}_\zeta \setminus Y$ is a bijection [Q]. The σ-invariant measure μ on $\mathcal{X}_\zeta$ is continuous, $\mu(Y) = 0$, and we can confine ourselves to the measure-preserving transformation $(\mathcal{X}_\zeta \setminus Y, \sigma, \mu)$ [Q]. By abuse of notation, we will let below $\mathcal{X}_\zeta := \mathcal{X}_\zeta \setminus Y$, keeping in mind that we have to discard the countable set Y.

The following result will be used to prove the discreteness of the spectrum. For a set $\mathcal{L} \subset \mathbf{N}$ let

$$\overline{\mathrm{dens}}(\mathcal{L}) = \limsup(K^{-1} \mathrm{card}\,(\mathcal{L} \cap [1, K])), \quad K \to \infty,$$

be its upper density.

Theorem 2.1. *Let ζ be a one-to-one primitive substitution on $\mathcal{A} = \{1, \ldots, m\}$, such that $\zeta(1) = 1w$, $|w| > 0$, and $u = (u_i)_{i=1}^{\infty} = \lim_{k \to \infty} \zeta^k(1)$ is aperiodic. Set $p_n = |\zeta^n(1)|$ and $D_n = \{i : u_i \neq u_{i+p_n}\}$. If*

$$(2) \qquad \sum_n \overline{dens}(D_n) < \infty,$$

then the substitution dynamical system $(\mathcal{X}_\zeta, \sigma, \mu)$ has purely discrete spectrum.

This form of the theorem is suggested by a result due to Host (see [Q, Lemma VI.27], which is concerned with substitutions on $\{0, 1\}$. Our proof is based on the "operator theoretic approach" of [So1, So2].

Lemma 2.2. *Let U be a unitary operator on a Hilbert space $\mathcal{H}$. Suppose that there exists a total set $\mathcal{F} \subset \mathcal{H}$, and a sequence of integers $q_n \to \infty$, satisfying a linear integral recurrence relation, such that*

$$\forall f \in \mathcal{F}, \quad \sum_{n \geq 1} \| U^{q_n} f - f \|^2 < \infty.$$

Then U has purely discrete spectrum.

Proof. The proof is essentially contained in [So2] but we provide it for the reader's convenience.

Let $E(\cdot)$ be the projection spectral measure of the unitary operator U on the circle $\mathbf{T}$, furnished by the Spectral Theorem. Put $\mu_f = (E(\cdot)f, f)$. We have

$$\int_{\mathbf{T}} |\lambda^{q_n} - 1|^2 d\mu_f(\lambda) = \| U^{q_n} f - f \|^2 .$$

The assumption of the lemma implies that $\sum_n \mu_f(\{\lambda : |\lambda^{q_n} - 1| > 2^{-l}\}) < \infty$, for all $l > 0$. By the Borel-Cantelli Lemma, $\mu_f(\mathbf{T} \setminus \{\lambda : \lambda^{q_n} \to 1\}) = 0$. Since this is true for all f from a total set $\mathcal{F}$, we conclude that $E(\mathbf{T} \setminus \{\lambda : \lambda^{q_n} \to 1\}) = 0$. It is a well-known fact, going back to Pisot (see also [So2, 3.3]), that the set $\{\lambda : \lambda^{q_n} \to 1\}$ is countable, for q_n satisfying a linear integral recurrence relation. Thus, the spectral measure is a countable sum of "point masses", and hence U has purely discrete spectrum. $\square$

Corollary 2.3. *Let T be a measure-preserving transformation on a Lebesgue space $(\mathcal{X}, \mathcal{B}, \nu)$, and let $\mathcal{G}$ be a generator partition, that is, $\bigvee_{n=-\infty}^{\infty} T^n \mathcal{G}$ generates $\mathcal{B}$. Suppose that there exists a sequence of integers $q_n \to \infty$ satisfying a linear integral recurrence relation, such that*

$$(3) \qquad \forall G \in \mathcal{G}, \quad \sum_{n \geq 1} \nu(T^{q_n} G \triangle G) < \infty.$$

Then T has purely discrete spectrum.

Proof. We apply Lemma 2.2 to the unitary operator $U : f(x) \mapsto f(Tx)$ on $\mathcal{H} = L^2(\mathcal{X}, \nu)$. The total set $\mathcal{F}$ is the collection of characteristic functions χ_G of the elements of the partitions $\bigvee_{-k}^{k} T^n \mathcal{G}$, $k > 0$. We have

$$\| U^{q_n} \chi_G - \chi_G \|^2 = \nu(T^{q_n} G \triangle G),$$

and it remains to note that (3) implies that $\sum_{n\geq 1} \nu(T^{q_n}G \triangle G) < \infty$ for all $G \in \bigvee_{-k}^{k} T^n \mathcal{G}$. $\square$

Now we are in a position to prove Theorem 2.1. We are going to apply Corollary 2.3 to the system $(\mathcal{X}_\zeta, \sigma, \mu)$. Put $q_n = p_n = |\zeta^n(1)|$. This sequence satisfies the linear recurrence relation $p_{n+m} = k_1 p_{n+m-1} + k_2 p_{n+m-2} + \cdots + k_m p_n$, where $X^m - k_1 X^{m-1} - k_2 X^{m-2} - \cdots - k_m$ is the characteristic polynomial of the matrix M_ζ. The partition $\mathcal{G}$ will be the *state partition* $\mathcal{G} = \{C_\alpha\}_{\alpha \in \mathcal{A}}$:

$$C_\alpha = \{x \in \mathcal{X}_\zeta : x_1 = \alpha\}, \ \alpha \in \mathcal{A}.$$

It is known (see [Q, V.4]) that the measure of any cylinder $C_v = \{x \in \mathcal{X}_\zeta : x[1, |v|] = v\}$, $v \in \mathcal{A}^*$ is determined by the frequency of the orbit $\{\sigma^n u\}$ hitting C_v :

$$(4) \qquad \mu(C_v) = \lim_{K \to \infty} K^{-1} \mathrm{card} \{i \leq K : u[i, i + |v| - 1] = v\}$$

We have

$$\sigma^{p_n} C_\alpha \triangle C_\alpha \subset B_n \overset{def}{=} \{x \in \mathcal{X}_\zeta : x_1 \neq x_{p_n+1}\}.$$

The set B_n is a disjoint union of cylinders corresponding to words of length $p_n + 1$. Thus we have from (4)

$$\mu(B_n) = \lim_{K \to \infty} K^{-1} \mathrm{card} \{i \leq K : u_i \neq u_{i+p_n}\} = \overline{\mathrm{dens}}(D_n)$$

(in fact, it follows that for the set D_n exists density, not just upper density). Now (2.1) implies that $\sum_{n\geq 1} \mu(\sigma^{p_n} C_\alpha \triangle C_\alpha) < \infty$, and Corollary 2.3 can be applied. $\square$

2.1. Systems of Numeration.

We are going to use the system of numeration related to substitution developed by Rauzy and his associates (see [R3, DT]). We shall say that a word B is a (proper) prefix of the word C, and write $B \in \mathrm{pref}(C)$, if C starts with B and $C \neq B$. The empty word e is a prefix of every word. We shall denote by $\mathrm{pref}^*(C)$ the set of improper prefixes of the word C: $\mathrm{pref}^*(C) = \{C\} \cup \mathrm{pref}(C)$, and set $\mathrm{Pref}(\zeta) = \cup_{1 \leq i \leq m} \mathrm{pref}(\zeta(i))$.

Consider the *prefix automaton of the substitution*. By definition, it is a directed labeled graph. The vertices are indexed by the letters $\{1, 2, \ldots, m\}$, and the edges are labeled by the set $\mathrm{Pref}(\zeta)$. An edge labeled by a word A goes from vertex i to vertex j if $Aj \in \mathrm{pref}^*(\zeta(i))$.

Definition 2.4. A sequence $(X_i, \alpha_i)_{i=-n}^{0}$, $X_i \in \mathrm{Pref}(\zeta)$, $\alpha_i \in \mathcal{A}$, is called *admissible* if it corresponds to a path in the prefix automaton, starting from the vertex 1. That is to say, $X_{i+1}\alpha_{i+1}$ is a prefix (maybe improper) of the word $\zeta(\alpha_i)$, $i \leq -1$, and $X_{-n}\alpha_{-n} \in \mathrm{pref}^*(\zeta(1))$.

Theorem 2.5. [R3, p.5], [DT, p.158]. *For any $N \geq 1$ there is a unique $n = n(N)$ and unique admissible sequence $(X_i, \alpha_i)_{i=-n}^{0}$ such that $X_{-n} \neq e$, and*

$$(5) \qquad u[1, N] = \zeta^n(X_{-n})\zeta^{n-1}(X_{-n+1}) \ldots \zeta(X_{-1})X_0 .$$

Conversely, to any finite path in the prefix automaton starting at the vertex 1, there corresponds a prefix of the word u satisfying (5).

Since factors $u[1, N]$ correspond to the orbit of the word u under the substitution dynamical system, the theorem says that this orbit can be coded by finite paths in the prefix automaton starting from the vertex 1. Under some additional conditions, this can be extended to the coding of the space $\mathcal{X}_\zeta$ by infinite paths of the automaton. In this representation the dynamical system $(\mathcal{X}_\zeta, \sigma)$ becomes a sort of "odometer", or "immediate successor" transformation. This can be used to prove the isomorphism of the substitution dynamical system and the adic transformation.

3. CASE OF COMPANION MATRICES

Now we are going to restrict ourselves to a class of examples. Namely, we shall consider the stationary adic transformation on the Markov compactum $\mathcal{X}_M$, where M is a companion matrix of a polynomial $p(X) = X^m - a_1 X^{m-1} - \cdots - a_m$. We assume that $m \geq 2$, a_i are non-negative integers, and $a_1, a_m > 0$. The corresponding substitution has the form

$$
(6) \qquad \zeta(l) = \underbrace{1 \ldots 1}_{a_l}(l+1), \quad l = 1, 2 \ldots, m-1, \quad \zeta(m) = \underbrace{1 \ldots 1}_{a_m},
$$

It is easily checked that ζ is primitive (for any $\alpha \in \mathcal{A}$, the word $\zeta^m(\alpha)$ contains all letters), and $u = \lim_{k \to \infty} \zeta^k(1)$ is aperiodic.

For the substitution (6) we obtain a system of numeration with usual integral digits, rather than words. Let $(X_i, \alpha_i)_{i=-n}^0$ be an admissible sequence. Then X_i is a proper prefix of $\zeta(\alpha_{i-1})$, so X_i consists only of letters 1 (or may be empty). Put $b_i = |X_i|$, $p_n = |\zeta^n(1)|$. It follows from (5) that

$$
(7) \qquad N = \sum_{i=-n}^{0} b_i p_{-i} .
$$

Sequences (b_i) arising in such a way will be called *allowed*. By Theorem 2.5 an allowed sequence $(b_i)_{i=-n}^0$ uniquely determines the admissible sequence (X_i, α_i). Thus the representation (7) with an allowed sequence of coefficients is unique. Note that for $a_1 \geq a_2 \geq \ldots a_m \geq 1$, the system of numeration (7) was considered in [PT]. This is a generalization of the well-known Fibonacci system of numeration.

Let us give a description of allowed sequences. Consider the prefix automaton of (6) with edges labeled by $|A|$ instead of a prefix A. It has a single edge $i \to i+1$ labeled by a_i, for $i = 1, 2, \ldots, m-1$, and a_i edges $i \to 1$, $i = 1, 2, \ldots, m$, labeled by $0, 1, \ldots, a_i - 1$. Clearly, every path starting from vertex 1, is a concatenation of cycles from the set

$$
\{1 \to 1, \ 1 \to 2 \to 1, \ldots, 1 \to 2 \to \ldots \to m \to 1\}
$$

(the last cycle may be incomplete). To a cycle $1 \to 2 \to \ldots \to l \to 1$ there corresponds a block of the allowed sequence having the form $a_1 a_2 \ldots a_{l-1} q$, for some q, $0 \leq q \leq a_l - 1$. Infinite concatenations of words from a finite "vocabulary" $W = \{w_1, \ldots, w_k\}$ constitute a subshift which is called a *renewal system* [GLS]. The following is now obvious.

Lemma 3.1. *The companion matrix M determines a renewal system on the alphabet $\{1, 2, \ldots, \max(a_i)\}$ with the vocabulary*

$$W = \begin{cases} q, & \text{for } 0 \le q \le a_1 - 1; \\ a_1 q, & \text{for } 0 \le q \le a_2 - 1; \\ \cdots\cdots & \cdots\cdots\cdots \\ a_1 a_2 \ldots a_{m-1} q, & \text{for } 0 \le q \le a_m - 1. \end{cases}$$

Allowed sequences b correspond to words of the renewal system with finitely many nonzero elements, such that $b_i = 0$ for $i > 0$.

Recall that a set of infinite words in a finite alphabet is a *shift of finite type* if it can be defined by forbidding a finite number of words. The renewal system in Lemma 3.1 is not always a shift of finite type. However, there is a nice special case when it is. Let $<_{lex}$ denote the lexicographical order on the set of words (finite or infinite) with integral digits; to compare the words of different length we just append the shorter one with zeros. The shift σ acts on words by deleting the first letter.

Lemma 3.2. *Suppose that the word $d = a_1 a_2 \ldots a_m$ has the property*

$$(8) \qquad\qquad \sigma^i d <_{lex} d, \quad i = 1, \ldots, m - 1.$$

A sequence $(y_i)_{-\infty}^{\infty}$ is a word of the renewal system from Lemma 3.1 if and only if

$$(9) \qquad\qquad \forall k, \; y[k, k + m - 1] <_{lex} d.$$

The renewal system is then a shift of finite type.

Proof. Straightforward. $\square$

The last lemma naturally brings us to β-expansions. Recall that $d(1, \beta)$ denotes the Rényi expansion of 1 in base β (see Section 1).

Theorem 3.3. [P] *(see also* [B]*). Let $d = a_1 a_2 \ldots a_m$ be a sequence of non-negative integers. There exists a $\beta > 0$, such that $d(1, \beta) = d$, if and only if condition (8) is satisfied. If this is the case, then condition (9) is necessary and sufficient for a sequence $y_{-n} y_{-n+1} \ldots y_{-1}.y_0 y_1 \ldots y_l 0^{\omega}$ to be a β-expansion of some real number.*

Corollary 3.4. *Under the condition (8) allowed sequences $(b_i)_{i=-n}^{0}$ are precisely those for which $b_{-n} b_{-n+1} \ldots b_{-1} b_0.0^{\omega}$ is a β-expansion.*

We have proved that if (8) holds, then allowed sequences $(b_i)_{i=-n}^{\infty}$ are in one-to-one correspondence with the set $\mathcal{N}$ of "integral" β-expansions. It is not hard to see that the shift on the orbit of u becomes the "immediate successor" transformation on $\mathcal{N}$ under this correspondence. This provides the third representation of our dynamical system described in Section 1.

If (8) holds, we can fix β such that $d(1, \beta) = a_1 a_2 \ldots a_m$. The corresponding shift of finite type is called the β-*shift*.

4. Proof of the Main Theorem

Recall that a *Pisot number* is an algebraic integer whose Galois conjugates are all less than one in modulus. By $\mathbf{Z}_+[\lambda]$ will be denoted the set of polynomials in λ with non-negative integral coefficients. If λ is an algebraic integer, then $\mathbf{Z}_+[\lambda] \subset \mathbf{Z}_+[\lambda^{-1}]$.

Theorem 4.1. *Let β be a Pisot number of degree m, such that $d(1, \beta) = a_1 a_2 \ldots a_m$ is the Rényi expansion of 1 in base β. Suppose also that all numbers of the set $\mathbf{Z}_+[\beta^{-1}]$ have finite β-expansions. Then the measure-preserving system $(\mathcal{X}_\zeta, \sigma)$ corresponding to the substitution (6) has purely discrete spectrum.*

The Main Theorem will follow since all the hypotheses of Theorem 4.1 hold if $a_1 \geq a_2 \geq \ldots \geq a_m$ [FS, Theorem 2].

For the proof of Theorem 4.1 we shall also need the following result from [FS]. Let $\mathcal{F}_m(\beta)$ denote the set of numbers $x > 0$, whose β-expansions terminate at or before β^{-m} :

$$x \overset{\beta}{\sim} \varepsilon_{-k} \varepsilon_{-k+1} \ldots \varepsilon_m 0^\omega$$

Lemma 4.2. [FS, Prop.2]. *Under the hypotheses of Theorem 4.1, there exists $L = L(\beta) \in \mathbf{Z}_+$, such that*

$$(x \in \mathcal{F}_m(\beta), \ y \in \mathcal{F}_m(\beta)) \Rightarrow x + y \in \mathcal{F}_{m+L}(\beta).$$

The key point of the proof of Theorem 4.1 is the following lemma.

Lemma 4.3. *Under the hypotheses of Theorem 4.1, there exists a positive integer R with the following property. Let $X_i = X_i(N)$ be the words of the admissible sequence in Theorem 2.5. If the sequence $(X_i)_{i=-n}^0$ has a "gap" of length R: for some k, $2R < k < n$,*

$$(10) \qquad X_{-k+1} = X_{-k+2} = \ldots = X_{-k+R} = e, \ X_{-k+R+1} = 1,$$

then

$$u_N = u_{N+p_l}$$

for all l such that $k \leq l \leq n$.

Proof. There will be two conditions which determine R, but it is more convenient to state them in the process of the proof.

Fix $l \in [k, n]$. It follows from (10) that the allowed sequence $b = (b_i)_{-n}^0$ corresponding to N has the form

$$b_{-n} b_{-n+1} \ldots b_l \ldots b_{-k} \underbrace{00 \ldots 0}_{R} 1 b_{-k+R+2} \ldots b_0.$$

By Corollary 3.4, b is a β-expansion. Let

$$x = \sum_{i=-n}^{-k} b_i \beta^{-i} \overset{\beta}{\sim} b_{-n} b_{-n+1} \ldots b_{-k} 00 \ldots 0.$$

Consider $x' = x + \beta^l$. Clearly, both x and β^l belong to $\mathcal{F}_{-k}(\beta)$. By Lemma 4.2, $x' \in \mathcal{F}_{-k+L}(\beta)$. The first condition to be imposed on R is that $R > L$. Then, since $x' < \beta^{n+1} + \beta^l < \beta^{n+2}$, we have

$$x' \overset{\beta}{\sim} \varepsilon_{-n-1}\varepsilon_{-n}\ldots\varepsilon_{-k+R-1}0^\omega,$$

where ε_{-n-1} and several digits before 0^ω can be also zeros. Put

$$b'_i = \begin{cases} \varepsilon_i, & -n \le i \le -k+R-1; \\ 0, & i = -k+R; \\ b_i, & i \ge -k+R+1. \end{cases}$$

By Lemma 3.1, $b' = (b'_i)$ is an allowed sequence. We claim that

(11)
$$N + p_l = \sum_{-n-1}^{0} b'_i p_{-i}.$$

This will prove the lemma because the representation (7) with allowed sequence of coefficients is unique, and thereby we shall have $X_i(N) = X_i(N + p_l)$, $i \ge -k + R + 1$. Then by (5) u_N is the last letter of the word

$$\zeta^{k-R-1}(X_{-k+R+1})\zeta^{k-R-2}(X_{-k+R+2})\ldots\zeta(X_{-1})X_0,$$

which is not empty, as $X_{-k+R+1} = 1$, and u_{N+p_l} is the last letter of the same word. Thus, it remains to check (11).

Recall that $p_i = |\zeta^i(1)|$, so this sequence satisfies the recurrence relation

(12)
$$p_{i+m} = a_1 p_{i+m-1} + \cdots + a_m p_i.$$

The polynomial $p(X) = X^m - a_1 X^{m-1} - \cdots - a_m$ is irreducible since $\deg(\beta) = m$ and $p(\beta) = 0$. By assumption, it has zeros $\beta_1 = \beta, \beta_2, \ldots, \beta_m$, with $|\beta_i| < 1$, $i \ge 2$. It is well known that a sequence satisfying a recurrence relation (12) has the form $p_i = \sum_{j=1}^{m} A_j \beta_j^i$, where $A_1 > 0$, since $p_i \to \infty$. It follows that for some $C_1 > 0$,

(13)
$$|p_i - A_1 \beta^i| \le C_1 r^i,$$

where $r = \max\{|\beta_i|, i \ge 2\} < 1$. By the definition of b' we have

$$D \overset{def}{=} \sum_{i=-n-1}^{0} b'_i p_{-i} - \sum_{i=-n}^{0} b_i p_{-i} = \sum_{i=-n-1}^{-k+R-1} \varepsilon_i p_{-i} - \sum_{i=-n}^{-k} b_i p_{-i}.$$

Recall that $x = \sum_{i=-n}^{-k} b_i \beta^{-i}$, $x' = \sum_{i=-n-1}^{-k+R-1} \varepsilon_i \beta^{-i}$. Using the estimates ε_i, $b_i \le \max(a_\kappa) = a_1$, $k > 2R$, and (13), we conclude that

$$|D - A_1(x - x')| \le 2C_1 a_1 (1-r)^{-1} r^{k-R+1} \le C_2 r^R,$$

where $C_2 = 2C_1 a_1 (1-r)^{-1}$. But $A_1(x-x') = A_1 \beta^l$, and $|A_1 \beta^l - p_l| \le C_1 r^l \le C_1 r^{2R}$, so

$$|D - p_l| \le C_2 r^R + C_1 r^{2R}.$$

The second condition to be imposed on R will be that $C_2 r^R + C_1 r^{2R} < 1$. Since $D - p_l$ is an integer, this will imply that $D = p_l$. Since by (7), $N = \sum_{i=-n}^{0} b_i p_{-i}$, the proof of (11) and of the lemma is now complete. $\square$

Now we can conclude the proof of Theorem 4.1. We are going to apply Theorem 2.1, so we have to check that $\sum_l \overline{\mathrm{dens}}(D_l) < \infty$. The set D_l consists of those N for which $u_N \neq u_{N+p_l}$. It follows from Lemma 4.3 that if $N \in D_l$, then the block $\underbrace{0 \ldots 0}_{R} 1$ does not occur in the segment $b[-l+1, -R]$ of the allowed sequence corresponding to N.

Allowed sequences $(b_i)_{-n}^{0}$ of length not exceeding $(n+1)$ by Theorem 2.5 correspond to $N = 1, 2, \ldots, p_{n+1} - 1$. Recall that allowed sequences come from the β-shift, which is a shift of finite type. A shift of finite type is defined by forbidding a finite number of words, or alternatively, by a non-negative integral matrix. If λ is a spectral radius of this matrix, then the number of k-blocks appearing in the shift of finite type is equivalent to λ^k, when $k \to \infty$. (This corresponds to the fact that the topological entropy of the shift of finite type is equal to $\log(\lambda)$.) Evidently, the number of k-blocks in the β-shift is equivalent to β^k. Consider the shift of finite type obtained from β-shift by forbidding the word $\underbrace{0 \ldots 0}_{R} 1$. By the Perron-Frobenius theory (see [Se, Th. 1.1(e)]), the spectral radius of the corresponding matrix β_1 is strictly less than β. The foregoing considerations and (13) imply that for some positive constants C, C', C'',

$$\mathrm{card}\,(D_l \cap [1, p_{n+1} - 1]) < C\beta^{n-l+R+1}\beta_1^{l-R} < C'\beta^{n+1}(\beta_1/\beta)^l < C'' p_{n+1}(\beta_1/\beta)^l.$$

Since p_{n+1}/p_n is bounded (it tends to β), the last inequality gives that

$$\overline{\mathrm{dens}}(D_l) \leq c(\beta_1/\beta)^l,$$

for some $c > 0$. Thus Theorem 2.1 can be applied, and the proof of Theorem 4.1 is now complete. $\square$

REFERENCES

[B] F. Blanchard, *β-expansions and symbolic dynamics*, Theor. Comp. Sci. 65 (1989), 131–141.

[Br] A. Brauer, *On algebraic equations with all but one root in the interior of the unit circle*, Math. Nachr. 4 (1951), 250–257.

[DT] J. M. Dumont and A. Thomas, *Systèmes de numération et fonctions fractales relatif aux substitutions*, Theor. Comp. Sci. 65 (1989), 153–169.

[FS] C. Frougny and B. Solomyak, *Finite beta-expansions*, Preprint LITP 92.02, Institut Blaise Pascal, 1992.

[GLS] J. Goldberger, D. Lind, and M. Smorodinsky, *The entropies of renewal systems*, Israel J. Math. 33 (1991), 1–23.

[H] B. Host, *Valeurs propres de systèmes dynamiques définis par de substitutions de longuer variable*, Ergodic Theory and Dynamical Systems 6 (1986), 529–540.

[L] A. N. Livshitz, *A sufficient condition for weak mixing of substitutions and stationary adic transformations*, Mat. Zametki 44 (1988), no. 6, 785–793 (Russian), English transl. in Math. Notes 44 (1988), no. 6, 920–925.

[M] P. Michel, *Stricte ergodicité d'ensembles minimaux de substitutions*, C. R. Acad. Sci. Paris 278 (1974), 811–813.

[P] W. Parry, *On the β-expansions of real numbers*, Math. Acad. Sci. Hungar. 11 (1960), 401–416.

[PT] A. Pethö and R. Tichy, *On digit expansions with respect to linear recurrences*, J. Number Theory 33 (1989), 243–256.

[Q] M. Queffélec, *Substitution dynamical systems - spectral analysis*, Lecture Notes in Math. vol. 1294, Springer-Verlag, Berlin and New York, 1987.

[R1] G. Rauzy, *Nombres algébriques et substitutions*, Bull. Soc. math. France **110** (1982), 147–178.

[R2] _____, *Rotations sur les groupes, nombres algébriques, et substitutions*, Séminaire de Théorie des Nombres de Bordeaux, 1987–1988, Exp. 21.

[R3] _____, *Sequences defined by iterated morphisms*, Preprint.

[Re] A. Rényi, *Representations for real numbers and their ergodic properties*, Acta Math. Acad. Sci. Hungar. **8** (1957), 477–493.

[Se] E. Seneta, *Non-negative Matrices and Markov Chains*, Springer-Verlag, New York, 1981.

[So1] B. Solomyak, *On a dynamical system with discrete spectrum*, Uspehi Mat. Nauk **41** (1986), no. 2, 209–210 (Russian), English transl. in Russian Math. Surveys **41** (1986), no. 2, 219–220.

[So2] _____, *On the spectral theory of adic transformations*, Adv. of Soviet Math. **9** (1992).

[So3] _____, *Finite β-expansions and spectra of substitutions*, Preprint.

[V1] A. M. Vershik, *Uniform algebraic approximation of shift and multiplication operators*, Dokl. Akad. Nauk. SSSR **259** (1981), no. 3, 526–529 (Russian), English transl. in Soviet Math. Dokl. **24** (1981), no. 1, 97–100.

[V2] _____, *A theorem on periodic Markov approximation in ergodic theory*, Zapiski Nauchn. Semin. LOMI **115** (1982), 72–82 (Russian), English transl. in J. of Soviet Math. **28** (1985), 667–674.

[VL] A. M. Vershik and A. N. Livshitz, *Adic models of ergodic transformations, spectral theory, substitutions, and related topics*, Adv. in Soviet Math. **9** (1992).

[W] P. Walters, *Ergodic theory*, Springer-Verlag, Berlin, Heidelberg, and New York, 1982.

DEPARTMENT OF MATHEMATICS, UNIVERSITY OF WASHINGTON, SEATTLE, WASHINGTON 98195

E-mail address: solomyak@math.washington.edu

Contemporary Mathematics
Volume **135**, 1992

FINITARY ISOMORPHISM OF
m-DEPENDENT PROCESSES

MEIR SMORODINSKY
Dedicated to my friend Roy Adler

1. INTRODUCTION

The purpose of this paper is to prove the following result.

Theorem. *m-dependent stationary processes of the same entropy are finitarily isomorphic.*

If a finite state process is endowed with the infinite product topology of the discrete topology on the state space than the finitary maps coincide with the almost continuous maps and the finite maps are the continuous ones. Finite factors of i.i.d processes are m-dependent processes.

Corollary. *Finite factors of Bernoulli schemes are finitarily isomorphic to Bernoulli schemes of the same entropy.*

It is not known whether m-dependent processes are finite factors of i.i.d processes. Aaronson and al. 1989 [A-G-K-V] showed that there are 1-dependent processes which are not 2 block factors of an i.i.d process. In a subsequent paper [A-G-K], an example of a 5 state Markov chain was constructed which is 1-dependent but not a 2-block factor of an i.i.d. process. The finitary isomorphism theory of stationary stochastic processes was developed in [K-S 1],[K-S 2] [K-S 3] [A-J-R],[R] and [P]. This theory parallels the measure theoretic theory of Sinai [S] and Ornstein [O-1]. However there is a notable gap in this theory, namely a factor theorem is missing. Ornstein, in 1970 [O-2], proved that factors of Bernoulli shifts are (measure theoretically) isomorphic to Bernoulli shifts. It is reasonable to expect the following analog:

Conjecture. *Finitary factors of Bernoulli schemes (i.i.d processes) are finitarily isomorphic to Bernoulli schemes.*

Our result falls short of that. It implies only that *finite* factors of Bernoulli schemes are finitarily isomorphic to Bernoulli schemes.

The proof of the theorem is based on the the marker method. Specifically, it closely follows the pattern of the proof in [K-S 3].

1991 *Mathematics Subject Classification.* 28D05, 54H20, 60G10.
This paper is in final form and no version will be submitted for publication elsewhere.

I would like to thank Jon Aaronson for a discussion which led me to the results presented in this paper.

Added in proof. It was communicated to me by Bob Burton that he together with M.Goulet and R.Meester constructed an example of 1-dependent process which is not a block factor of any i.i.d process.

2. PRELIMINARIES

Let $X = (X_n)_{n \in Z}$ be a stationary process on a finite state space A.

Definition 2.1. The process X is *m-dependent* if the σ-field generated by $(X_{-n})_{n=1,2,...}$ is independent of the one generated by $(X_n)_{n=m,m+1,...}$

We now recall some of the definitions from [K-S 3].

Definition 2.2. The process $(X_n^{(k)})_{n \in Z}$, called *the k-stringing of X* (k, fixed positive integer), is defined as follows. The state space of $X^{(k)}$ is A^k, and

$$X_n^{(k)} = (X_n, X_{n+1}, ..., X_{N+k-1})(n \in Z).$$

Definition 2.3. Let $A' \subset A$ and let $b \notin A$. Put $X'_n = X_n$ if $X_n \notin A'$ and $X'_n = b$ if $X \in A'$. We say that the process X' obtained from X by *collapsing* A'.

Definition 2.4. Let $B = \{b_1, ..., b_l\}$ be a set of symbols not in A, $(q_1, ..., q_l)$ a probability vector and $a_i \in A$. We say that *the process X' is obtained from X by independently splitting a_i according to $q_1, ..., q_l$* if X' is defined as follows : The states of X' are $A \cup B - \{a_i\}$. Let $c_0, ..., c_r$ be a sequence of such states and $J = \{j\}$ a subset of the indices such that $c_j \in B$. Let J' be the set of the other indices. Put
$$P[X'_0 = c_0, ..., X'_r = c_r] = P[X_j = a_i, j \in J \& X_j = c_j, j \in J'] \cdot \prod_{j \in J} q_{c_j}.$$

The proofs of the following lemmas follow immediately from the definitions above.

Lemma 1. *X and $X^{(k)}$ are continuously isomorphic.*

Lemma 2. *If X is m-dependent process then $X^{(k)}$ is m+k+1-dependent.*

Lemma 3. *m-dependence is stable under collapsing, independent splitting and independent products.*

Definition 2.5. Let $a_i \in A$. The *distribution of the state a_i in X* is defined as the distribution of the process obtained from X by replacing all the occurrences of a_i by "1" and the other states by "0".

Proposition. *Let X and Y be m-dependent processes of the same entropy and such that there exists a state in X with the same distribution as some state in Y. Then X and Y are finitarily isomorphic.*

Proof 2.6. We use the states which share the same distribution to construct markers. As usual, there will be markers of rank r for each $r = 1, 2, 3, ...$. A

marker of rank r, M_r will be a block of states of length N_r which is determined
by the process of the common state. (If a_i is the common state, a typical choice
is $M_k = a_i^{N_k}$, where N_k is well chosen sequence of increasing natural numbers).
We say that *a marker of rank r occurs at position* n if $X_n, ..., X_{n+N_r-1} = M_r$.

Let E denote the event that markers of rank r occur at the $i_j, j = 1, ..., s$.
positions. Let $\sigma_i, i = 1, ...s - 1$ denote the σ-field generated by

$$X_{i_j+N_r+m}, ..., X_{i_{j+1}-m-1}.$$

Since X is m-dependent, $\sigma_1, ..., \sigma_{s-1}$ are mutually independent, given E.

As in [K-S 2], *fillers* will be defined by blocks that can occur between markers.
From the discussion above, suffices to determined fillers by blocks of length l_j,
such that

$$l_j < i_{j+1} - N_k - 2m,$$

to make sure that the filler distributions will be mutually independent. The rest
of the proof proceeds along the the same lines as in [K-S 2], with some necessary
modifications. $\square$

3. MARKERS FOR M-DEPENDENT PROCESSES

To prove our theorem we begin with an m-dependent process X. We construct
Y, a $m+m_1+1$-dependent process, for suitably chosen m_1 and Z an independent
process, such that,

(1) $X^{(m_1)}$ and Y have states with the same distribution.

(2) $Z^{(m_1)}$ and Y have states with the same distribution.

(3) the entropies of X, Y and Z are equal.

Once we have constructed Y and Z the proof of the theorem will follow from
the proposition. The rest of the paper is devoted to the construction of the
processes Y and Z.

Let W be a 2-state Bernoulli scheme. Put $Y' = X \mathbf{x} W$ and let $Y'^{(m_1)}$ be the
m_1 stringing of $Y', m_1 > m$. We partition the states of $Y'^{(m_1)}$ into three disjoint
subsets, M, N and O, defined as follows:

$$M = \{(a_1, x_2, ..., x_{m_1}) \mathbf{x} (0, ..., 0, 1) : (x_2, ..., x_{m_1}) \in A^{m_1-1}\}$$

Let $w = w_1, ..., w_{m_1} \in A^{m_1}$ be a fixed allowable sequence of states, $w_0 = a_2$.
Put

$$N = \{(w) \mathbf{x} (y_1, ..., y_{m_1}) : (y_1, ..., y_m) \in \{0, 1\}^{m_1}\},$$

Denote by O the set of the remaining states.

Let Y'' be the 3-state process which is obtained from $Y'^{(m_1)}$ by collapsing
M, N and O. Y'' is an $m + m_1 + 1$-dependent process and we denote its states
by M, N and O. For m_1 sufficiently large, the entropy if Y'' is smaller than that
of X.

Observe that the distribution of M in Y'' is m_1-dependent. Therefore the
distribution of this 2-state m_1-dependent process is determined by the following
restrictions. A "1" must be followed by m zeroes and the entries thereafter are

independent and have the stationary distribution. Such a distribution of a state will also occur if we m_1-string an independent process and consider a state of the form $(b_0, ..., b_0, b_1)$, provided the stationary probability of it will equal that of M. Choose Z, as such independent process. We can assume (using independent splitting if necessary)that the entropy of Z equals that of X.

The distribution of N is the same as that of w in $X^{(m_1)}$. By independently splitting the state O of Y'' we can get a $m + m_1 + 1$-dependent process Y of entropy equal to that of X. $\square$

REFERENCES

[A-G-K-V]   J. Aaronson, D. Gilat, M. Keane, and V. de Valk,, *An algebraic construction of a class of one-dependent processes*, The Ann. of Probability **17** (1989), 128-143.

[A-G-K]   J. Aaronson, D. Gilat, and M. Keane, *On the structure of 1-dependent Markov chains*, J. of Th. Probability (to appear).

[A-J-R]   M. A. Ackoglu, A. del Junco, and M. Rahe, *Finitary codes between Markov processes*, Z. Wahrscheinlichkeitstheorie und Verw. Gebiete **47** (1979), 305-314.

[K-S 1]   M. Keane and M. Smorodinsky, *A class of finitary codes*, Israel J. Math. **26** (1977), 352-371.

[K-S 2]   _____, *Bernoulli Schemes of the same entropy are finitarily isomorphic*, Ann. of Math. **109** (1979), 397-406.

[K-S 3]   _____, *Finitary isomorphism of irreducible Markov processes*, Israel J. Math **34** (1979), 281-286..

[O 1]   D. Ornstein, *Ergodic Theory, Randomness, and Dynamical Systems*, Yale University Press, 1974.

[O 2]   _____, *Factors of Bernoulli shifts are Bernoulli shifts*, Advances in Math. **5** (1970), 349-364.

[P]   B. Petit, *Deux schemas de Bernoulli d'alphabet denombrable et d'entropie infinie sont finitairement isomorphes*, Z. Wahrscheinlichkeitstheorie und Verw. Gebiete **59** (1982), 161-168.

[R]   D. Rudolph, *A characterization of those processes finitarily isomorphic to a Bernoulli shift*, Proceeding of the special year in Ergodic Theory and Dyn. Systems, University of Maryland, 1980.

[S]   Ya. G. Sinai, *A Weak Isomorphism of Transformations Having an Invariant Measure*, Dokl. Akad. Nauk. SSSR **147** (1962), 797-800.

DEPARTMENT OF STATISTICS, TEL-AVIV UNIVERSITY, TEL AVIV 69978, ISRAEL

Contemporary Mathematics
Volume **135**, 1992

Constant-to-One Factor Maps and Dimension Groups

PAUL TROW

ABSTRACT. We give necessary conditions for the existence
of eventual constant-to-one factor maps, of a specified degree,
between shifts of finite type.

1. Introduction

In this paper we give some necessary conditions for the existence of
eventual constant-to-one factor maps between shifts of finite type. A
factor map is *constant-to-one* if every point in the range has exactly d
inverse images, for some positive integer d. Our results are based on
a structure theorem for constant-to-one factor maps due to Nasu ([N,
Theorem 7.3]).

In Corollary 2.6, we show that if A and B are integral eventually
positive (I.E.P.) matrices and A eventually factors onto B by d-to-1 everywhere maps, for a fixed $d \in \mathbb{N}$, and if every prime dividing d divides
every non-leading coefficient of χ_B, then A is shift equivalent to a matrix
of the form $B \oplus C$, where C is a square integral matrix (possibly the zero matrix). In Theorem 2.4, we give a more general necessary condition,
without the assumption on primes dividing d.

We begin by giving some background on shifts of finite type and factor maps. For further background, see [AM], [BMT] or [KMT].

Partially supported by NSF grant #DMS-9003944
1991 Mathematics Subject Classification. Primary 28D05.
This paper is in final form and no version of it will be
submitted for publication elsewhere.

Let A be an $n \times n$ non-negative, integral matrix. A defines a directed graph $G(A)$ on n vertices as follows: $G(A)$ has A_{ij} labelled edges leading from vertex i to vertex j. We will usually assume that A is *irreducible*, which means that for each i, j, there exists a positive integer $m = m(i, j)$ such that $A_{ij}^m > 0$. Let $\mathcal{E}_A$ denote the set of edges of $G(A)$, and $\mathcal{S}_A$ the set of vertices of $G(A)$. Elements of $\mathcal{E}_A$ are called *symbols* and elements of $\mathcal{S}_A$ are called *states*. If $x_1, x_2 \in \mathcal{E}_A$, we say that $x_1 x_2$ is *allowed* if the terminal vertex of x_1 is the initial vertex of x_2. We define the *shift of finite type* Σ_A to be $\{x \in \mathcal{E}_A^{\mathbb{Z}} : x_i x_{i+1}$ is allowed for all $i\}$. We give $\mathcal{E}_A$ the discrete topology and $\mathcal{E}_A^{\mathbb{Z}}$ the product topology, in which Σ_A is a closed subspace of $\mathcal{E}_A^{\mathbb{Z}}$. Σ_A is *irreducible* if A is an irreducible matrix.

A square integral matrix A is *integral eventually positive*, or I.E.P., if $A^m > 0$ for all sufficiently large $m \in \mathbb{N}$. If A is I.E.P., then Σ_{A^m} is a shift of finite type for all sufficiently large m.

We define the *shift map* $\sigma : \Sigma_A \to \Sigma_A$ by $(\sigma x)_i = x_{i+1}$. It is easy to see that σ is a homeomorphism.

If Σ_A and Σ_B are shifts of finite type, a *factor map* is a continuous, surjective map $f : \Sigma_A \to \Sigma_B$ satisfying $f\sigma = \sigma f$. If f is one-to-one, it is a *topological conjugacy*. A factor map $f : \Sigma_A \to \Sigma_B$ is called a *one-block map* if there exists a map $g : \mathcal{E}_A \to \mathcal{E}_B$ such that $(f(x))_i = g(x_i)$. When the meaning is clear, we let f denote both the map on Σ_A and the map on symbols. It follows from the Curtis-Hedlund-Lyndon Theorem [He, Theorem 3.4] and a standard recoding, that up to topological conjugacy of the domain, any factor map can be represented by a one-block map; i.e. there exists a topological conjugacy $\alpha : \Sigma_A \to \Sigma_{\bar{A}}$, such that the induced map $\bar{f} = f\alpha^{-1} : \Sigma_{\bar{A}} \to \Sigma_B$ is a one-block map (see [KMT, p.86]).

A factor map $f : \Sigma_A \to \Sigma_B$ is *right closing* if for all $u, v \in \Sigma_A$, the conditions $f(u) = f(v)$ and $u_i = v_i$ for all $i \leq N$, for some $N \in \mathbb{Z}$, imply that $u = v$. The definition of *left closing* is similar, the only change being $i \geq N$ in place of $i \leq N$.

A one-block map $f : \Sigma_A \to \Sigma_B$ is called *right resolving* if for $x_1 \in \mathcal{E}_A$, $y_1 \in \mathcal{E}_B$, if $f(x_1) = y_1$ and $y_1 y_2$ is allowed, then there exists a unique $x_2 \in \mathcal{E}_A$ such that $x_1 x_2$ is allowed and $f(x_2) = y_2$. f is *left resolving* if for $x_2 \in \mathcal{E}_A$, $y_2 \in \mathcal{E}_B$, if $f(x_2) = y_2$ and $y_1 y_2$ is allowed, then there exists a unique $x_1 \in \mathcal{E}_A$ such that $x_1 x_2$ is allowed and $f(x_1) = y_1$.

A factor map $f : \Sigma_A \to \Sigma_B$ is *finite-to-one* if there exists $N \in \mathbb{N}$ such that $\#f^{-1}(y) \leq N$ for all $y \in \Sigma_B$. The *degree* of a finite-to-one factor map $f : \Sigma_A \to \Sigma_B$, where Σ_A and Σ_B are irreducible, is the unique

$d \in \mathbb{N}$ such that $\#f^{-1}(y) = d$ for every bilaterally transitive $y \in \Sigma_B$. It has long been known that such a d exists (see [CP, Theorem 6.5]). f is *constant-to-one*, or *d-to-1 everywhere*, if for all $y \in \Sigma_B$, $\#f^{-1}(y) = d$. If A and B are I.E.P. matrices, A *eventually factors onto B by d-to-1 everywhere maps* if for all sufficiently large m, there exists a d-to-1 everywhere factor map $f_m : \Sigma_{A^m} \to \Sigma_{B^m}$.

The following theorem, due to M. Nasu, characterizes constant-to-one factor maps.

THEOREM 1.1 ([N, THEOREM 7.3]). *Let $f : \Sigma_A \to \Sigma_B$ be a factor map between irreducible shifts of finite type. Then f is constant-to-one if and only if there exist topological conjugacies $\alpha : \Sigma_A \to \Sigma_{\bar{A}}$ and $\beta : \Sigma_B \to \Sigma_{\bar{B}}$ and a one-block map $\bar{f} : \Sigma_{\bar{A}} \to \Sigma_{\bar{B}}$, which is right and left resolving, such that $f = \beta^{-1}\bar{f}\alpha$.*

PROOF: We sketch a proof of "only if", which is the direction we will need. (This proof was pointed out to us by B. Kitchens.) First, suppose that f is constant-to-one. By recoding, we may assume that f is a one-block map. Since f is constant-to-one, it follows from [KMT, Theorem 2.6] that any two inverse images $u, v \in \Sigma_A$ of a point in Σ_B are mutually separated (u, v are *mutually separated* if $u_i \neq v_i$ for all $i \in \mathbb{Z}$). This implies that f must be right and left closing. It now follows by a theorem of Kitchens (see [Ki] or [BKM, Proposition 1]) that f can be recoded to a map $\bar{f}$ which is right and left resolving. (In [Ki], it is shown that a right closing map f can be recoded to a right resolving one. If f is also left closing, then a similar argument enables one to recode f to a left resolving map, in a way which does not alter the right resolving property of the first recoding.) ∎

For any one-block map $f : \Sigma_A \to \Sigma_B$, we define the induced map $\hat{f}$ on states of Σ_A as follows: if $i \in \mathcal{S}_A$ and $e \in \mathcal{E}_A$ is any edge beginning at i, then $\hat{f}(i)$ is defined to be the initial state of $f(e)$. The *relation matrix* R of $\hat{f}$ is defined as follows.

$$R_{ij} = \begin{cases} 1 & \text{if } \hat{f}(i) = j \\ 0 & \text{otherwise.} \end{cases}$$

If $f : \Sigma_A \to \Sigma_B$ is a one-block map which is right and left resolving, then it is not hard to see that the following equations hold.

(1.2)

(i) $AR = RB$

(ii) $R^T A = BR^T$

(iii) $R^T R = dI$

where d is the degree of f.

Equations (i) and (ii) follow from the facts that f is right resolving and left resolving respectively. Equation (iii) follows from the fact that $(R^T R)_{ij} = 0$ for $i \neq j$, and $(R^T R)_{ii} = \#\hat{f}^{-1}(i) = d$. To see the last equality, note that for any $i \in S_B$, we can choose a point $y \in \Sigma_B$ in which i occurs. Since f is right and left resolving, it is not hard to see that y will have exactly $\#\hat{f}^{-1}(i)$ inverse images, and this must equal d, the degree of f.

We briefly review dimension groups. The definition of dimension group which we give is a more concrete version of the usual definition in terms of direct limits. For further details, see [BMT]. Let A be an $n \times n$ integral matrix and let

$$V_A = \bigcap_{k=0}^{\infty} (\mathbb{Q}^n) A^k$$

be the *eventual image* of A; we assume that A acts on row vectors. Clearly, A acts invertibly on V_A. If $\mathcal{L}$ is an A-invariant sublattice of $\mathbb{Z}^n$ (i.e. $(\mathcal{L})A \subseteq \mathcal{L}$), we define the *dimension group of $\mathcal{L}$*, denoted $G_{\mathcal{L}}$, by $G_{\mathcal{L}} = \{v \in V_A : vA^r \in \mathcal{L} \text{ for some } r \in \mathbb{N}\}$. It is easy to check that the restriction of the linear map A to $G_{\mathcal{L}}$ is a group automorphism of $G_{\mathcal{L}}$, which we also denote by A. We can make $G_{\mathcal{L}}$ into a module over the ring $\mathbb{Z}[x, x^{-1}]$ of Laurent polynomials in one variable, with coefficients in $\mathbb{Z}$, by defining $p(x)v = vp(A)$, where $p(x) \in \mathbb{Z}[x, x^{-1}]$ and $v \in G_{\mathcal{L}}$. $(G_{\mathcal{L}}, A)$ is called the *dimension module pair* of $G_{\mathcal{L}}$. We denote $G_{\mathbb{Z}^n}$ by G_A. Observe that $G_A = G_{A^m}$ for all $m \in \mathbb{N}$. For if $v \in G_A$, then $vA^r \in \mathbb{Z}^n$ for some $r \in \mathbb{N}$, so $vA^{mk} \in \mathbb{Z}^n$ for $mk \geq r$. Therefore $v \in G_{A^m}$. The reverse inclusion is obvious.

If $\mathcal{L}_1$ is an A-invariant sublattice and $\mathcal{L}_2$ is a B-invariant sublattice and $\theta : (G_{\mathcal{L}_1}, A) \to (G_{\mathcal{L}_2}, B)$ is a module homomorphism, then $\theta A = B\theta$. If θ is surjective, it is called a *quotient map*. If θ is bijective, it is called an *isomorphism* and we write $(G_{\mathcal{L}_1}, A) \cong (G_{\mathcal{L}_2}, B)$.

It is easy to see that an A-invariant subgroup H of G_A is a submodule if $A|_H$ is a bijection of H. The following lemma gives another characterization of submodules of (G_A, A).

LEMMA 1.3. *Let A be an $n \times n$ integral matrix and H a subgroup of G_A. Then H is a submodule of G_A if and only if $H = G_{\mathcal{L}}$ for some A-invariant sublattice $\mathcal{L}$.*

PROOF: Suppose H is a submodule of G_A. Then H is A-invariant. Let $\mathcal{L} = H \cap \mathbb{Z}^n$. Clearly $\mathcal{L}$ is A-invariant and we claim that $H = G_{\mathcal{L}}$. If $v \in H$, then since $v \in G_A$, $vA^r \in H \cap \mathbb{Z}^n$ for some $r \in \mathbb{N}$, so $v \in G_{\mathcal{L}}$. For the reverse inclusion, if $v \in G_{\mathcal{L}}$ then $vA^r \in H$ for some $r \in \mathbb{N}$. Since H is a submodule of G_A, A acts invertibly on H, so $vA^r A^{-r} = v \in H$. Therefore $H = G_{\mathcal{L}}$. The converse is obvious. ∎

REMARKS 1.4. If $\mathcal{L}$ is an A-invariant sublattice of $\mathbb{Z}^n$, then there exists an integral matrix C such that $(G_{\mathcal{L}}, A) \cong (G_C, C)$; namely, let C be the matrix for the linear map A restricted to $\mathcal{L} \cap V_A$ with respect to any integral basis. By Lemma 1.3, any submodule of (G_A, A) is isomorphic to (G_C, C) for some integral matrix C. In particular, if A is not nilpotent, then (G_A, A) is isomorphic to (G_C, C) for a non-singular matrix C.

If H is a subset of $\mathbb{Q}^n$, let $\langle H \rangle$ denotes the rational span of H. A submodule H of G_A is *closed* if $\langle H \rangle \cap G_A = H$.

If A is a square integral matrix, then the *characteristic polynomial modulo x* of A, denoted χ_A^*, is defined by $\chi_A^* = \frac{\chi_A}{x^t}$, where x^t is the highest power of x which divides χ_A.

If B and C are square matrices, then $B \oplus C$ denotes the matrix in block form

$$\begin{bmatrix} B & 0 \\ 0 & C \end{bmatrix}.$$

If B and C are integral, then $(G_{B \oplus C}, B \oplus C) \cong (G_B, B) \oplus (G_C, C)$, the direct sum of (G_B, B) and (G_C, C).

The following definition is due to R.F. Williams ([Wi]). Two square integral matrices A and $\bar{A}$ (not necessarily of the same size) are *shift equivalent over $\mathbb{Z}$* if there exist integral matrices R and S, and a positive integer l such that $AR = R\bar{A}, SA = \bar{A}S, RS = A^l, SR = \bar{A}^l$. If in addition R and S are non-negative, A and $\bar{A}$ are *shift equivalent*.

The connection between shift equivalence and dimension groups is given by the following theorem, due to W. Krieger.

THEOREM 1.5 ([K, THEOREM 4.2]). *Let A and $\bar{A}$ be square integral matrices. The following are equivalent.*

(i) $(G_A, A) \cong (G_{\bar{A}}, \bar{A})$.

(ii)   *A and $\bar{A}$ are shift equivalent over $\mathbb{Z}$.*

The following lemma essentially follows from [KR, Lemma 3.2 and Theorem 3.3].

LEMMA 1.6. *Let A and B be square integral matrices and suppose that $\theta : (G_{A^m}, A^m) \to (G_{B^m}, B^m)$ is a homomorphism, where $m \in \mathbb{N}$ is such that the ratio of no pair of distinct eigenvalues of B is an m'th root of unity. Then θ is a homomorphism $(G_A, A) \to (G_B, B)$.*

PROOF: By the Remarks 1.4, we may assume that A and B are nonsingular. Note that $G_{A^m} = G_A$ and $G_{B^m} = G_B$. If S is the matrix which represents θ (with respect to the standard basis), then it suffices to prove that $AS = SB$, and this follows from [BMT, the claim at the beginning of the proof of Theorem 3.2]. In that proof, in addition to the assumption on m, it is assumed that the eventual kernel of A is contained in the kernel of S, which clearly holds in this case, since A is non-singular.   ∎

2. Main Results

LEMMA 2.1. *Let A be a square integral matrix and $e \in \mathbb{N}$. Suppose that p is a prime which divides every non- leading coefficient of χ_A. Then there exists $N \in \mathbb{N}$ such that p^e divides every entry of A^s for $s \geq N$.*

PROOF: By the Cayley-Hamilton Theorem, $\chi_A(A) = 0$, and since p divides every non-leading coefficient of χ_A, it follows that p divides every entry of A^n, where n is the degree of χ_A. If $s \geq N = ne$, then p^e divides every entry of A^s.   ∎

THEOREM 2.2. *Let A be a square integral matrix and $b \in \mathbb{Z}$, $b \neq 0$. The following are equivalent.*

(i) $G_A = bG_A$

(ii) *For every prime p which divides b, p divides every non-leading coefficient of χ_A.*

(iii) *b divides every entry of A^s for some $s \in \mathbb{N}$.*

(iv) *The map $\gamma : G_A \to G_A$ given by $\gamma(v) = bv$ is an automorphism.*

PROOF: (i) $\Rightarrow$ (iii) Suppose $G_A = bG_A$. Assume that A is $n \times n$. For each standard basis vector e_i, $1 \le i \le n$, $e_i A^n \in G_A = bG_A$. Therefore $e_i A^n = bw_i$ for $w_i \in G_A$. For sufficiently large r, $w_i A^r \in \mathbb{Z}^n$, for $1 \le i \le n$. Since $e_i A^{n+r} = bw_i A^r$, b divides every entry of $e_i A^{n+r}$ and so b divides every entry of A^{n+r}.

(iii) $\Rightarrow$ (ii) Suppose that b divides every entry of A^s for $s \in \mathbb{N}$. Let p be a prime dividing b. By a theorem of M. Boyle (see [T1, p. 188]), if s is greater than or equal to the degree of the minimal polynomial for A, there exists a constant-to-one endomorphism of Σ_{A^s} of degree p. If γ is a non-zero eigenvalue for A, then γ^s is a non-zero eigenvalue for A^s, and by [T2, Corollary 2.8], p is a unit in $\mathbb{Z}[\frac{1}{\gamma^s}]$. Since $\mathbb{Z}[\frac{1}{\gamma^s}] \subseteq \mathbb{Z}[\frac{1}{\gamma}]$, this implies p is a unit in $\mathbb{Z}[\frac{1}{\gamma}]$. By [B, Prop. 2.4], p divides every non-leading coefficient of χ_A.

(ii) $\Rightarrow$ (i) Suppose that for every prime p which divides b, p divides every non-leading coefficient of χ_A. Write $b = \prod_{i=1}^k p_i^{e_i}$. By Lemma 2.1, for $1 \le i \le k$, there exist N_i such that $p_i^{e_i}$ divides every entry of A^s for $s \ge N_i$. Let $M = max\{N_i\}_{i=1}^k$. Then b divides every entry of A^M, so $\frac{1}{b} e_i A^M \in \mathbb{Z}^n$. Now, if $v \in G_A$, then $vA^r \in \mathbb{Z}^n$ for some $r \in \mathbb{N}$. Write $vA^r = \sum_{i=1}^n a_i e_i$, $a_i \in \mathbb{Z}$. Then $\frac{1}{b} vA^{r+M} = \sum_{i=1}^n a_i(\frac{1}{b} e_i A^M)$, which is in $\mathbb{Z}^n$. Therefore $\frac{1}{b} v \in G_A$ and so $v \in bG_A$. Consequently $G_A \subseteq bG_A$. The reverse inclusion is obvious. Therefore $G_A = bG_A$.

The equivalence of (i) and (iv) is an easy exercise.   ∎

DEFINITION 2.3. *If B is a square integral matrix and $d \in \mathbb{N}$, then the factorization of d over B is $d = ab$, where a has the property that if p is a prime which divides a, then p does not divide every non- leading coefficient of χ_B, and b has the property that every prime dividing b divides all the non-leading coefficients of χ_B.*

THEOREM 2.4. *Let A and B be I.E.P. matrices and suppose that A eventually factors onto B by d-to-1 everywhere maps. Let $d = ab$ be the factorization of d over B. Then there exists a quotient map $\pi : (G_A, A) \to (G_B, B)$ and an injective homomorphism $\delta : (G_B, B) \to (G_A, A)$ such that $\pi\delta = aI_{G_B}$ and $Ker(\pi)$ and $Im(\delta)$ are closed submodules of (G_A, A). Furthermore, $aG_A \subseteq Im(\delta) \bigoplus Ker(\pi)$.*

PROOF: We first show that it suffices to prove the theorem assuming that Σ_A and Σ_B are irreducible shifts of finite type and $f : \Sigma_A \to \Sigma_B$ is a d-to-1 everywhere factor map. For if this is proved, then assuming A and B are I.E.P., and A eventually factors onto B by d-to-1 everywhere maps, choose $m \in \mathbb{N}$ such that the ratio of no pair of distinct eigenvalues of B is an m'th root of unity and there exists a d-to-1 everywhere factor map $f : \Sigma_{A^m} \to \Sigma_{B^m}$. We have an injective homomorphism $\delta : (G_{B^m}, B^m) \to (G_{A^m}, A^m)$ and a quotient map $\pi : (G_{A^m}, A^m) \to (G_{B^m}, B^m)$ such that $\pi\delta = aI_{G_{B^m}} = aI_{G_B}$. The theorem then follows from Lemma 1.6.

Now, assume that Σ_A and Σ_B are irreducible shifts of finite type and $f : \Sigma_A \to \Sigma_B$ is a d-to-1 everywhere factor map. By Theorem 1.1, there exist conjugacies $\alpha : \Sigma_A \to \Sigma_{\bar{A}}$ and $\beta : \Sigma_B \to \Sigma_{\bar{B}}$ such that $\bar{f} = \beta f \alpha^{-1}$ is right and left resolving. By [W, Theorem A] and Theorem 1.5, $(G_A, A) \cong (G_{\bar{A}}, \bar{A})$ and $(G_B, B) \cong (G_{\bar{B}}, \bar{B})$. It is easy to see that isomorphisms of dimension modules take closed submodules to closed submodules, so it suffices to prove the theorem assuming that f is right and left resolving. Assume that A is $n \times n$ and B is $k \times k$. By (1.2), we have

$$AR = RB$$
$$R^T A = BR^T$$
$$R^T R = dI$$

where R is the relation matrix of f. Note that R must have exactly one 1 in each row. Define $\pi : G_A \to G_B$ by $\pi(v) = vR$, and $\delta : G_B \to G_A$ by $\delta(v) = \frac{1}{b}vR^T$. It follows from the matrix equations that π is a homomorphism of (G_A, A) into (G_B, B). To see that π is surjective, note that $\pi(e_i) = e_j$ if and only if $\hat{f}(i) = j$, and that $\hat{f}$ is surjective (since f is surjective and B is irreducible). So π is a quotient map. By [BMT, Theorem 5.4], $Ker(\pi)$ is a closed submodule of (G_A, A). By Theorem 2.2, $\frac{1}{b}G_B = G_B$, and it follows from this and the matrix equations that δ is a homomorphism of (G_B, B) into (G_A, A). Injectivity follows from the fact the R is a relation matrix. To see that $Im(\delta)$ is a closed submodule of (G_A, A), note that $b\delta(e_j) = e_jR^T = \alpha_j$, where α_j is the j'th row of R^T. Since R is a relation matrix, R^T has exactly one 1 in each column, and so it is easy to see that if $\mathcal{L}$ is the integral span of $\{\alpha_j\}_{j=1}^{k}$, then $\mathcal{L}$ is an A-invariant sublattice of $\mathbb{Z}^n$, with the property that $\langle \mathcal{L} \rangle \cap \mathbb{Z}^n = \mathcal{L}$). It follows from this property and Lemma 1.3 that $G_{\mathcal{L}}$ is a closed submodule of (G_A, A). We claim that $Im(\delta) = G_{\mathcal{L}}$, which will show that $Im(\delta)$ is closed submodule. To see this, observe that $b\delta$ maps $\mathbb{Z}^k$ bijectively onto $\mathcal{L}$, and therefore defines an isomorphism of G_B onto $G_{\mathcal{L}}$, so that $Im(b\delta) = G_{\mathcal{L}}$. Since, by Theorem 2.2 (iv), multiplication by b is an automorphism of G_B, $Im(b\delta) = Im(\delta)$, which proves the claim. To see that $\pi\delta = aI_{G_B}$, observe that if $v \in G_B$, then $\pi\delta(v) = \frac{1}{b}vR^TR = \frac{1}{b}v(dI) = av$.

To prove the last assertion, let $av \in aG_A$, where $v \in G_A$. Then $av =$

$\delta\pi(v) + (av - \delta\pi(v))$. Clearly, $\delta\pi(v) \in Im(\delta)$. Since $\pi(av - \delta\pi(v)) = a\pi(v) - \pi\delta\pi(v) = a\pi(v) - a\pi(v) = 0$, we have $av - \delta\pi(v) \in Ker(\pi)$. The uniqueness of the representation follows from the fact that $Im(\delta) \cap Ker(\pi) = \{0\}$. ∎

COROLLARY 2.5. *Let A and B be I.E.P. matrices and suppose that A eventually factors onto B by d-to-1 everywhere maps. If every prime which divides d divides every non-leading coefficient of χ_B, then $G_A = Im(\delta) \bigoplus Ker(\pi)$.*

PROOF: By hypothesis, $a = 1$ in the factorization of d over B. The Corollary follows immediately from Theorem 2.4. ∎

Under the conditions of Corollary 2.5, G_A is the direct sum of two closed submodules of (G_A, A). We now express this condition in terms of shift equivalence.

COROLLARY 2.6. *Suppose that A and B are I.E.P. matrices, and suppose that A eventually factors onto B by d-to-1 everywhere maps. If every prime dividing d divides every non-leading coefficient of χ_B, then A is shift equivalent over $\mathbb{Z}$ to a matrix of the form $B \bigoplus C$, where C is a square integral matrix.*

PROOF: Let π and δ be as in Theorem 2.4. Since δ is injective, $(Im(\delta), A) \cong (G_B, B)$. By the Remarks 1.4, $(Ker(\pi), A) \cong (G_C, C)$, where C is a square integral matrix. By Corollary 2.5, $G_A = Im(\delta) \bigoplus Ker(\pi)$, so $(G_A, A) \cong (G_B, B) \bigoplus (G_C, C) \cong (G_{B \bigoplus C}, B \bigoplus C)$. By Theorem 1.5, A is shift equivalent over $\mathbb{Z}$ to $B \bigoplus C$. ∎

We note that if $Ker(\pi) = \{0\}$, then C can be taken to be the zero matrix.

EXAMPLE 2.7. Let

$$A = \begin{bmatrix} 1 & 2 \\ 1 & 0 \end{bmatrix}$$

and

$$B = [2].$$

We will show that there does not exist a constant-to-one factor map $\Sigma_A \rightarrow \Sigma_B$ of degree 2^n, for any n. Note that $\chi_A = (x - 2)(x + 1)$, while $\chi_B = x - 2$. Let $\mathcal{E}_2$ and $\mathcal{E}_{-1}$ denote the A-eigenspaces for 2 and -1 respectively (for row vectors). If there were a constant-to-one factor map $f : \Sigma_A \to \Sigma_B$ of degree 2^n, the since 2 divides the non-leading coefficient of χ_B, $G_A = Im(\delta) \bigoplus Ker(\pi)$, where π and δ are as in Theorem 2.4. By elementary linear algebra, since $Ker(\pi)$ and $Im(\delta)$ are closed submodules of (G_A, A), $Ker(\pi) = G_A \cap \mathcal{E}_2 = \{\frac{a}{2^i}(1, 1) : a \in \mathbb{Z}\}$, while $Im(\delta) = G_A \cap \mathcal{E}_{-1} = \{b(1, -2) : b \in \mathbb{Z}\}$. Now $(0, 1) \in G_A$, so $(0, 1) = \frac{a}{2^i}(1, 1) + b(1, -2)$. Multiplying this equation by $(2, 1)^T$ yields $1 = \frac{3a}{2^i}$, which is impossible for $a \in \mathbb{Z}$.

On the other hand, there does exist a factor map (not constant-to-one) of degree 2^n, $\Sigma_A \to \Sigma_B$, namely a factor map $\Sigma_A \to \Sigma_B$ of degree 1 (which exists by [M, Theorem 6]), followed by an endomorphism of Σ_B of degree 2^n. (For example, such an endomorphism can be defined to be g^n, where $(g(x_i x_{i+1}))_i = x_i + x_{i+1}$ mod 2. See [He, Theorem 6.6].)

EXAMPLE 2.8. In this example, we show that A being shift equivalent to $B \bigoplus C$ does not imply the existence of a constant-to-one factor map $\Sigma_A \to \Sigma_B$ (of any degree).

$$A = \begin{bmatrix} 0 & 1 & 1 \\ 0 & 0 & 1 \\ 1 & 1 & 0 \end{bmatrix}$$

and

$$B = \begin{bmatrix} 0 & 1 \\ 1 & 1 \end{bmatrix}.$$

Note that Σ_A factors onto Σ_B by a right resolving map whose relation matrix is

$$R = \begin{bmatrix} 0 & 1 \\ 1 & 0 \\ 0 & 1 \end{bmatrix}.$$

Also, A is similar over $\mathbb{Z}$ to the matrix

$$D = \begin{bmatrix} 0 & 1 & 0 \\ 1 & 1 & 0 \\ 0 & 0 & -1 \end{bmatrix}$$

since $PAP^{-1} = D$, where

$$P = \begin{bmatrix} 1 & 0 & 0 \\ 0 & 1 & 1 \\ 1 & 0 & -1 \end{bmatrix}.$$

So A is shift equivalent over $\mathbb{Z}$ to D, and therefore to $B \bigoplus [-1]$. However, there cannot exist any constant-to-one factor map $f : \Sigma_A \to \Sigma_B$. For suppose there were. Note that $\lambda = (1 + \sqrt{5})/2$ is the Perron eigenvalue for A, $l = (1, \lambda, \lambda)$ is a left eigenvector for λ and $r = (\lambda, 1, \lambda)^T$ is a right eigenvector for λ. By [T2, Lemma 2.6], the degree of f would have to divide lr in $\mathbb{Z}[\frac{1}{\lambda}] = \mathbb{Z}[\lambda]$. (The last equality follows from the fact that λ is invertible in $\mathbb{Z}[\lambda]$, since $\lambda(\lambda - 1) = 1$.) Since $lr = 2\lambda + \lambda^2 = 3\lambda + 1$, this implies that the degree of f must be 1, which implies that Σ_A is conjugate to Σ_B, and this is clearly false, since A and B have different non-zero spectra.

This raises the obvious question of what additional hypotheses, besides A being shift equivalent to $B \bigoplus C$, are needed in order to guarantee that A eventually factors onto B by d-to-one everywhere maps. We do not know the answer.

We next state a special case of Theorem 2.4, under the assumption that χ_B^* and $\frac{\chi_A^*}{\chi_B^*}$ are relatively prime. We use the following notation: if $f(x) \in \mathbb{Q}[x]$, then $\mathcal{N}(f(A))$ denotes the null space of $f(A)$.

COROLLARY 2.9. *Let A and B be I.E.P. matrices and suppose that A eventually factors onto B by d-to-1 everywhere maps. Let $d = ab$ be the factorization of d over B, and assume that $\chi_B^* = f$ and $\frac{\chi_A^*}{\chi_B^*} = g$ are relatively prime. Then $aG_A \subseteq [\mathcal{N}(f(A)) \cap G_A] \bigoplus [\mathcal{N}(g(A)) \cap G_A]$.*

REMARK: If A eventually factors onto B, then it follows from [KMT, Corollary 4.13] that $\frac{\chi_A^*}{\chi_B^*} \in \mathbb{Z}[x]$, so the relatively prime condition makes sense.

PROOF: Let π and δ be the maps in Theorem 2.4. Since f and g are relatively prime, and $Ker(\pi)$ and $Im(\delta)$ are closed submodules, it follows by elementary linear algebra that $Im(\delta) = \mathcal{N}(f(A)) \cap G_A$ and $Ker(\pi) = \mathcal{N}(g(A)) \cap G_A$. Now apply Theorem 2.4.   ∎

Corollary 2.9 can be extended to the case in which there are eventual constant-to-one maps of different degrees.

COROLLARY 2.10. *Let A and B be I.E.P. matrices, and let $\{d_i : 1 \le i \le r\}$ be a set of positive integers such that for each i, $1 \le i \le r$, A eventually factors onto B by d_i-to-1 everywhere maps. Assume that $\chi_B^* = f$ and $\frac{\chi_A^*}{\chi_B^*} = g$ are relatively prime, and that $d_i = a_i b_i$ is the factorization of d_i over B, for $1 \le i \le r$. If a is the greatest divisor of $\{a_i : 1 \le i \le r\}$, then $aG_A \subseteq [\mathcal{N}(f(A)) \cap G_A] \bigoplus [\mathcal{N}(g(A)) \cap G_A]$. Furthermore, if $a = 1$, then A is shift equivalent over $\mathbb{Z}$ to $B \bigoplus C$ for a* *square integral matrix C.*

PROOF: Choose integers q_i such that $\sum_{i=1}^{r} q_i a_i = a$. Clearly, $aG_A \subseteq \sum_{i=1}^{r} q_i a_i G_A$. Since $q_i a_i G_A \subseteq a_i G_A \subseteq [\mathcal{N}(f(A)) \cap G_A] \bigoplus [\mathcal{N}(g(A)) \cap G_A]$, by Corollary 2.9, the first statement follows. The proof of the last statement is similar to that of Corollary 2.6.   ∎

I wish to thank Mike Boyle for helpful conversations concerning this paper.

REFERENCES

[AM] R. Adler and B. Marcus, *Topological entropy and equivalence of dynamical systems*, Memoirs A.M.S. 219, (1979).

[B]   M. Boyle, *Constraints on the degree of a sofic homomorphism and the induced multiplication of measures on unstable sets*, Israel J. Math 53, (1986), 52–68.

[BMT] M. Boyle, B. Marcus and P. Trow *Resolving maps and the dimension group for shifts of finite type*, Memoirs A.M.S. 377, (1987).

[CP] E. Coven and M. Paul *Endomorphisms of irreducible shifts of finite type*, Math. Syst. Theory 8, (1974), 167–175.

[He] G. A. Hedlund, *Endomorphisms of irreducible shifts of finite type*, Math. Syst. Theory 8, (1974), 320–375.

[Ki] B. Kitchens, *Ph.D. dissertation*, University of North Carolina, (1981).

[K]   W. Krieger, *On dimension functions and topological Markov chains*, Inventiones Math. 56, (1980), 239–250.

[KMT] B. Kitchens, B. Marcus and P. Trow, *Eventual factor maps and compositions of closing maps*, Ergodic Th. and Dynam. Syst. 11, (1991), 85–113.

[KR] K. H. Kim and F. Roush, *Some Results on Decidability of Shift Equivalence*, J. of Combinatorics, Info. and Sys. Sci. 4, (1979), 123–146.

[M]   B. Marcus, *Factors and extensions of full shifts*, Monats. fur Mathematik 88, (1979), 239–247.

[N] M. Nasu, *Constant-to-one and onto global maps of homomorphisms between strongly connected graphs*, Th. and Dynam. Sys. 3, (1983), 387–413.

[T1] P. Trow, *Degrees of constant-to-one factor maps*, Proc. A.M.S. 103, (1988), 184-188.

[T2] P. Trow, *Degrees of finite-to-one factor maps*, Israel J. Math. 71, (1990), 229–238.

[W] R. Williams, *Classification of shifts of finite type*, Annals of Math. 98, (1973), 120–153, Eratta. Annals of Math. 99 (1974), 380–381.

Contemporary Mathematics
Volume **135**, 1992

Faces of Markov chains and matrices of polynomials

SELIM TUNCEL

1. Introduction.

The main purpose of this paper is to refine an invariant of [MT1].

Let A be an irreducible matrix whose entries are polynomials with non-negative integral coefficients in the variables $x_1, \cdots, x_k$. As will be explained in section 2, such matrices may be regarded as defining matrices of Markov chains. We call a square matrix a *sub-matrix* of A if its indexing set is a subset of that of A and if each of its entries may be obtained from the corresponding entry of A by discarding some of the monomials appearing in the entry. [MT1] associated a polytope $W(A)$ to A and, for each face F of $W(A)$, defined a finite collection $A_{F,1}, \cdots, A_{F,\ell}$ of irreducible sub-matrices of A. These sub-matrices are disjoint, in the sense that they are indexed by mutually disjoint subsets of the indexing set of A. It was shown in [MT1] that the polytope and the submatrices $A_{F,1}, \cdots, A_{F,\ell}$ are respected by bounded-to-one block homomorphisms, yielding invariants for a number of classifications.

In the present paper we introduce "connecting paths" between some of $A_{F,1}, \cdots, A_{F,\ell}$. We do this by considering the beta function of β_A of A and a corresponding positive right eigenvector $r = r_A$: For $x_1, \cdots, x_k > 0$, the value $\beta_A(x_1, \cdots, x_k) > 0$ is defined to be the maximum eigenvalue of the non-negative real valued matrix $A(x_1, \cdots, x_k)$ and $r_A(x_1, \cdots, x_k)$ is a positive eigenvector satisfying $A\, r_A = \beta_A\, r_A$. For a face F of $W(A)$, we obtain a corresponding function β_F from $\beta = \beta_A$ by taking limits along certain curves. Equivalently, in terms of the irreducible matrices $A_{F,1}, \cdots, A_{F,\ell}$, we have $\beta_F = \max\{\beta_{A_{F,i}} : 1 \le i \le \ell\}$. This may be regarded as the continuous extension of β from the interior of $W(A)$ to the part of its boundary given by F. The problem we face is that the extended function need not be analytic: Let $\mathbb{R}^{++}$ denote the positive reals. The interior

1991 *Mathematics Subject Classification*. Primary 28D20. Secondary 15A48, 15A54, 60J10.
Partially supported by NSF Grant DMS-9004253
This paper is in final form and no version of it will be submitted for publication elsewhere.

of $W(A)$ is represented by $(\mathbb{R}^{++})^k$, and F can be represented by the hyperplane obtained by setting $x_k = 0$ and thus identified with $(\mathbb{R}^{++})^{k-1}$. Putting $x_k = z^c$ for a suitable positive integer c, we consider the set $D \subset (\mathbb{R}^{++})^{k-1}$ consisting of those $(a_1, \cdots, a_{k-1})$ such that $\tilde{\beta}(x_1, \cdots, x_{k-1}, z) = \beta(x_1, \cdots, x_{k-1}, z^c)$ is analytic at $(a_1, \cdots, a_{k-1}, 0)$. We show that the complement of D is contained in a proper algebraic variety. In particular D is an open dense subset of $(\mathbb{R}^{++})^{k-1}$. In addition, D has only finitely many connected components. Denote by $\mathcal{D} = \mathcal{D}(\beta, F)$ the finite collection consisting of the connected components of D. We work on a fixed domain $D_0 \in \mathcal{D}$. From the equation $A\,r = \beta\,r$, we obtain a sub-matrix $A_F = A_{F, D_0}$ of A and a vector r_F with the following properties. Both A_F and r_F have the same indexing set as A, the matrix A_F has no trivial rows, r_F is strictly positive on D_0 and $A_F\, r_F = \beta_F\, r_F$ on D_0. Moreover, $A_{F,1}, \cdots, A_{F,\ell}$ are precisely the irreducible components of A_F.

The definition of A_F is algebraic in nature, and A_F is invariant under relations that have algebraic formulations. Considering block isomorphism, we show that a strong shift equivalence of A, B induces a strong shift equivalence between $A_F = A_{F, D_0}$ and $B_F = B_{F, D_0}$. It follows that any block isomorphism of Σ_A and Σ_B must send the (usually reducible) shift of finite type Σ_{A_F} onto Σ_{B_F}. Similar situations obtain with shift equivalence and right closing maps. For the relations considered, the results given here both refine and give alternative (algebraic) proofs of results of [MT1]. The arguments of [MT1] were more dynamical in character; they did not capture the connecting paths between the components but applied to more general relations.

The statements made above for $A_F = A_{F, D_0}, B_F = B_{F, D_0}$ hold for every choice of $D_0 \in \mathcal{D}$. However, $A_F = A_{F, D_0}$ depends on both F and D_0. For fixed F, each A_{F, D_0} has $A_{F,1}, \cdots, A_{F,\ell}$ for irreducible components but, in general, the connections between these components vary with D_0, reflecting the changes in the roles played by the components. When $\beta = \beta_A$ is a polynomial such complications do not arise. In this case $D = (\mathbb{R}^{++})^{k-1}$, so that $D_0 = D = (\mathbb{R}^{++})^{k-1}$ is the only element of $\mathcal{D}$, and our definition of A_F is more transparent. For this reason, before taking up the general case in section 4, we consider in section 3 the case of polynomial beta functions. When $\beta = \beta_A$ is a polynomial, $W(A)$ coincides with the Newton polyhedron $W(\beta)$ of β; the first task of section 4 is to give an analogue of this for the general case. We associate a polytope $W(\beta)$ to an arbitrary β by considering its minimal polynomial $t^d + p_1 t^{d-1} + \cdots + p_d$ and using the Newton polyhedra $W(p_n)$ of the coefficient polynomials: $W(\beta)$ is defined as the convex hull of $\cup_{n=1}^d \frac{1}{n} W(p_n)$. We show $W(A) = W(\beta)$. Then we examine the analytic properties of β on a face F of $W(\beta)$. Once this is done, $A_F = A_{F, D_0}$ can be defined and studied in the general case; the material of section 3 generalizes by working on $D_0 \in \mathcal{D}$ and using power series instead of polynomials. Section 5 extends some results of [MT3] from the case of polynomial beta functions so that they hold on $D_0 \in \mathcal{D}(\beta, F)$ in the general case. The strict positivity of r_F on D_0 is among these. Section 6 consists of examples. Section

2 is devoted to background material. It includes a fairly detailed account of the relationship between Markov chains and matrices of polynomials, as such an account is not, to my knowledge, available elsewhere.

This paper was motivated by past and current work with Brian Marcus. Thanks, Brian, for such enjoyable collaborations, not to mention all those ideas you contributed to them.

Many people in the department suffered my questions and confusion as I grappled with some basic facts about algebraic functions in several variables. I thank, in particular, Valerio De Angelis, Doug Lind and Don Marshall for discussions and Garth Warner for referring me to Whitney's paper [W].

2. Markov chains and matrices of polynomials.

We establish some terminology and facts concerning non-negative matrices, Markov chains and matrices of polynomials.

Every matrix we encounter in this paper will be such that none of its rows is trivial, even when this is not explicitly stated. We will refer to the elements of the indexing set of a matrix as its *states*. First we recall some facts from the theory of non-negative matrices; standard references for this are [BP, S].

1. THEOREM. *Let M be a non-negative real valued matrix. After a permutation of states, M may be expressed as a block matrix $M = [M_{ij}]$ such that:*

(a) *the block $M_{ij} = 0$ whenever $i < j$,*

(b) *every diagonal block M_{ii} is either irreducible or a one by one zero matrix.*

The irreducible diagonal blocks of (1) will be referred to as the *irreducible components* of M. An irreducible component M_{ii} is called a *principal component* of M if we also have $M_{ij} = 0$ for all $i > j$; otherwise it is called *non-principal*.

If M is irreducible, the Perron-Frobenius theorem tells us that the spectral radius λ_M is a simple eigenvalue of M and provides us with a strictly positive vector r such that $Mr = \lambda_M r$. When M is reducible, its spectral radius $\lambda_M = \max_i\{\lambda_{M_{ii}}\}$ is still an eigenvalue of M and there exists a non-negative vector $r \neq 0$ with $Mr = \lambda_M r$. The reducible matrices M we encounter will have the special property that there exists a strictly positive r with $Mr = \lambda_M r$. The following characterization of such matrices is a (non-trivial) exercise in Perron-Frobenius theory.

2. THEOREM. *Let M be a non-negative real valued matrix, and let λ_M be its spectral radius (maximum eigenvalue). The following are equivalent.*

(i) *There exists a strictly positive vector r such that $Mr = \lambda_M r$.*

(ii) *The spectral radius of every principal component of M equals λ_M and the spectral radius of every non-principal component is strictly less than λ_M.*

Now specialize, and let M be a non-negative integral matrix. It is well-known that a directed graph $G(M)$ and a shift of finite type (Σ_M, σ_M) are associated with M: The vertices of $G(M)$ are the states of M and, for states I, J, there are $M(I, J)$ edges from I to J. The shift space Σ_M consists of all infinite walks $(e_n)_{n \in \mathbb{Z}}$ where the e_n are edges of $G(M)$ and, for each n, the terminal state of e_n equals the starting state of e_{n+1}. Our standing assumption that M has no trivial rows implies that Σ_M is non-empty. Σ_M is topologized in the usual way and supports the left shift transformation $\sigma_M : \Sigma_M \to \Sigma_M$. The pair (Σ_M, σ_M) is the *shift of finite type* defined by M. For further details on shifts of finite type and directed graphs, see [AM, MT1, PT3].

Before going on to a discussion of Markov chains, we introduce some notation. We will write $\mathbb{Z}^+$ and $\mathbb{Z}^{++}$ for the non-negative and positive integers, respectively. In several variables $x_1, \cdots, x_k$, we will consider the Laurent polynomial ring $\mathbb{Z}[x_1^{\pm}, \cdots, x_k^{\pm}]$ and its postive cone $\mathbb{Z}^+[x_1^{\pm}, \cdots, x_k^{\pm}]$. For a monomial $m = x_1^{w_1} x_2^{w_2} \cdots x_k^{w_k}$ in this ring, we will write $\log(m) = (w_1, w_2, \cdots, w_k)$. When A is a matrix over $\mathbb{Z}^+[x_1^{\pm}, \cdots, x_k^{\pm}]$, the matrix $A(1, \cdots, 1)$ obtained on setting $x_1 = x_2 = \cdots = x_k = 1$ is over $\mathbb{Z}^+$ and we put $G(A) = G(A(1, \cdots, 1))$, $\Sigma_A = \Sigma_{A(1, \cdots, 1)}$, $\sigma_A = \sigma_{A(1, \cdots, 1)}$. A non-zero entry $A(I, J)$ of A is a sum of monomials, the number of monomials in the sum equaling the number of edges from I to J. We bijectively assign the monomials of $A(I, J)$ to the edges from I to J and denote by $\mathrm{wt}_A(e)$ the monomial assigned to an edge e, calling it the *weight* of e. Periodic orbits of (Σ_A, σ_A) correspond to cycles of $G(A)$; we define the *weight* of a cycle $\gamma = e_1 \cdots e_n$ to be $\mathrm{wt}_A(\gamma) = \prod_{i=1}^n \mathrm{wt}_A(e_i)$ and the *weight-per-symbol* of γ to be $\mathrm{wps}_A(\gamma) = \frac{1}{n} \log(\mathrm{wt}_A(\gamma)) \in \mathbb{Q}^k$.

Let us now assume that the non-negative integral matrix M is irreducible and consider the weighted graph resulting from assigning a number $w(e) > 0$ to each edge e of $G(M)$. This gives us another non-negative matrix P indexed by the states of $G(M)$: An entry $P(I, J)$ of P is defined to be the sum of the weights $w(e)$ of the edges e running from I to J. Then P is irreducible, because M is. Hence, there is a row vector $\ell > 0$ and a column vector $r > 0$ with $\ell P = \lambda_P \ell$, $P r = \lambda_P r$; we make sure ℓ and r are normalized to also have $\ell r = 1$. A sequence $e_0 e_1 \cdots e_m$ of edges and $h \in \mathbb{Z}$ give us a cylinder set

$$[e_0, e_1, \cdots, e_m]_h = \{(x_n) \in \Sigma_M : x_h = e_0, x_{h+1} = e_1, \cdots, x_{h+m} = e_m\}.$$

For each such set we identify the starting state I of e_0 and the terminal state J of e_m and put

$$\mu([e_0, e_1, \cdots, e_m]_h) = \ell(I) \, w(e_0) \, w(e_1) \, \cdots \, w(e_m) \, r(J).$$

This defines a Markov measure μ on Σ_M. Thus, a *Markov chain* $(\Sigma_M, \sigma_M, \mu)$ is defined by the irreducible matrix M and the weight function $w : E(M) \to \mathbb{R}^{++}$ on the set $E(M)$ of edges of $G(M)$. Note that the support of μ equals Σ_M; the irreducibility of M is necessary to get a well-defined Markov measure with this property.

Let H be a finitely generated (multiplicative) subgroup of $\mathbb{R}^{++}$ such that H contains the set $\{w(e) : e \in E(M)\}$ defining $(\Sigma_M, \sigma_M, \mu)$. Choose a basis

$a_1, \cdots, a_k$ for H. Every $w(e)$ can be uniquely expressed as a product of integral powers of $a_1, \cdots, a_k$ and, by replacing each a_i by an indeterminate x_i, we obtain a monomial in $\mathbb{Z}[x_1^{\pm}, \cdots, x_k^{\pm}]$. For each pair (I, J) of states we define $A(I, J)$ to be the sum of the monomials corresponding to edges e running from I to J. This gives us an irreducible matrix A over $\mathbb{Z}^+[x_1^{\pm}, \cdots, x_k^{\pm}]$. Corresponding to A we have the graph $G(A)$ whose edges are weighted by the monomials $\mathrm{wt}_A(e)$. By evaluating these monomials at $(x_1, \cdots, x_k) = (a_1, \cdots, a_k)$ we, of course, recover the Markov chain $(\Sigma_M, \sigma_M, \mu)$. However, we may also evaluate at any other point of $(\mathbb{R}^{++})^k$. This gives us a family $(\Sigma_A, \sigma_A, \mu_{(x_1, \cdots, x_k)})$ of Markov chains indexed by $(x_1, \cdots, x_k) \in (\mathbb{R}^{++})^k$. This family may be regarded as a vector space of Markov measures, but we will not need this fact. The crucial observation is that a bounded-to-one block code sends one Markov measure to another if and only if it sets up a bijection between the corresponding families of measures. (The bijection turns out to be a vector space isomorphism.) To make this more precise, let $(\Sigma_N, \sigma_N, \nu)$ be another Markov chain, defined by an irreducible matrix N over $\mathbb{Z}^+$ and a positive weight function on $E(N)$. Making sure that H contains the image of this weight function, we obtain an irreducible matrix B over $\mathbb{Z}^+[x_1^{\pm}, \cdots, x_k^{\pm}]$ and a family $(\Sigma_B, \sigma_B, \nu_{(x_1, \cdots, x_k)})$ of Markov chains. Let $\phi: \Sigma_A \to \Sigma_B$ be a bounded-to-one continuous surjection such that $\phi \sigma_A = \sigma_B \phi$. We have $\mu \circ \phi^{-1} = \nu$ if and only if $\mu_{(x_1, \cdots, x_k)} \circ \phi^{-1} = \nu_{(x_1, \cdots, x_k)}$ for every $(x_1, \cdots, x_k) \in (\mathbb{R}^{++})^k$. These conditions are also equivalent to ϕ being weight-preserving in the sense that $\mathrm{wps}_B(\phi(\gamma)) = \mathrm{wps}_A(\gamma)$ for cycles γ of $G(A)$. We also remark that A and B define the same family of measures (i.e., $\mu_{(x_1, \cdots, x_k)} = \nu_{(x_1, \cdots, x_k)}$ for all $(x_1, \cdots, x_k)$) if and only if there exists a diagonal matrix D with monomials for its diagonal entries, and a monomial m, such that $B = m\Delta^{-1}A\Delta$.

The above viewpoint on Markov chains was developed by Parry and the author. It emerged gradually. All the essential ideas, though not the final steps, are given in [PT2]. The vector space structure on Markov measures, also developed by Parry and the author, is described in [BT, Ki].

Henceforth, we restrict our attention to matrices over $\mathbb{Z}^+[x_1^{\pm}, \cdots, x_k^{\pm}]$ and weight-preserving bounded-to-one block codes between their shifts of finite type. As explained above, in the case of irreducible matrices this is equivalent to the consideration of measure-preserving block codes between Markov chains. The point is, $\mathbb{Z}^+[x_1^{\pm}, \cdots, x_k^{\pm}]$ provides a natural way of taking into account the families of Markov measures defined above. Furthermore, weight-preserving maps make sense also for reducible matrices over $\mathbb{Z}^+[x_1^{\pm}, \cdots, x_k^{\pm}]$: In the reducible case there are serious problems with (unambiguously) defining Markov chains, let alone measure-preserving maps between them.

By passing directly to the setting of $\mathbb{Z}^+[x_1^{\pm}, \cdots, x_k^{\pm}]$, we are avoiding exponential functions; earlier papers, in particular [PT1, PT2, MT1, MT3], would opt for rings of exponential functions and pass to polynomial rings when necessary. With the increase in the need to do so, exponential functions become an intermediate step that is best avoided.

Put $R = \mathbb{Z}[x_1^{\pm}, \cdots, x_k^{\pm}]$, $R^+ = \mathbb{Z}^+[x_1^{\pm}, \cdots, x_k^{\pm}]$. Let A be a matrix over R^+. We will simply write $G(A)$ for the weighted directed graph defined by A, that is, the directed graph defined by A together with the monomials $\text{wt}_A(e)$ assigned to its edges. Similarly, we write Σ_A to mean the shift of finite type defined by A together with the weights assigned to its periodic orbits. We translate some definitions and facts from [T1, MT1]. For $x_1, \cdots, x_k > 0$, let $\beta_A(x_1, \cdots, x_k)$ be the maximum eigenvalue of the non-negative real valued matrix $A(x_1, \cdots, x_k)$. This defines a continuous function $\beta_A = (\mathbb{R}^{++})^k \to \mathbb{R}^{++}$ called the *beta function* of A. Let $W(A)$ denote the convex hull, in $\mathbb{Q}^k$, of $\{\text{wps}_A(\gamma) : \gamma \text{ is a cycle of } G(A)\}$. Since $G(A)$ has finitely many simple cycles and every cycle can be decomposed into simple ones, $W(A)$ is a polytope; it is called the *weight-per-symbol polytope* of A. Let F be a face of $W(A)$. Define a sub-graph of the weighted graph $G(A)$ as follows. Keep an edge e (and its weight $\text{wt}_A(e)$) if and only if there exists a cycle γ which contains e and has $\text{wt}_A(\gamma) \in F$. The resulting sub-graph consists only of principal components. Letting $\ell = \ell_F(A)$ be the number of these components and listing the R^+-matrices representing the components, we associate disjoint irreducible submatrices $A_{F,1}, \cdots, A_{F,\ell}$ of A to F. We define $\beta_{A,F} : (\mathbb{R}^{++})^k \to \mathbb{R}^{++}$ by $\beta_{A,F} = \max\{\beta_{A_{F,i}} : 1 \le i \le \ell\}$.

Let B also be matrix over R^+. A continuous, shift-commuting, weight-preserving surjection $\phi : \Sigma_A \to \Sigma_B$ is called a *block homomorphism*. A *block isomorphism* is a bijective block homomorphism. The basic relations are as follows.

3. THEOREM. *[MT1]. If $\beta_A = \beta_B$ then $W(A) = W(B)$ and, for each face F of $W(A) = W(B)$, we have $\beta_{A,F} = \beta_{B,F}$.*

4. THEOREM. *[MT1]. Let $\phi : \Sigma_A \to \Sigma_B$ be a bounded-to-one block homomorphism. Then:*

    (a) *$\beta_A = \beta_B$ and, therefore, $W(A) = W(B)$.*

    (b) *For each face F of $W(A) = W(B)$, there exists a surjection $\Pi_F : \{1, \cdots, \ell_F(A)\} \to \{1, \cdots, \ell_F(B)\}$ such that $\phi(\Sigma_{A_{F,i}}) \subset \Sigma_{B_{F,\Pi_F(i)}}$.*

    (c) *For each $j \in \{1, \cdots, \ell_F(B)\}$ there exists $i \in \Pi_F^{-1}(j)$ such that the restriction of ϕ to $\Sigma_{A_{F,i}}$ gives a (bounded-to-one) block homomorphism of $\Sigma_{A_{F,i}}$ onto $\Sigma_{B_{F,j}}$.*

    (d) *If ϕ is a block isomorphism then Π_F is a bijection and, for each $i \in \{1, \cdots, \ell_F(A)\}$, the restriction of ϕ to $\Sigma_{A_{F,i}}$ is a block isomorphism of $\Sigma_{A_{F,i}}$ onto $\Sigma_{B_{F,\Pi_F(i)}}$.*

Further details and proofs may be found in [MT1].

3. Faces of matrices with polynomial beta functions.

We consider, in this section, matrices with polynomial beta functions. As explained in the introduction, this case lacks some of the complications of the general case, making our definitions more transparent. In addition, this is a significant case. For instance, for any non-zero $\beta \in R^+$, we obtain many examples of matrices A with $\beta_A = \beta$ by simply requiring that the row (or column) sums of A equal β. Included among these are matrices corresponding to Bernoulli shifts. There are also many examples of A with $\beta_A \in R \setminus R^+$ (see [D]).

Let $p \in R = \mathbb{Z}[x_1^{\pm}, \cdots, x_k^{\pm}]$. Write

$$p = \sum_w a_w\, x_1^{w_1} x_2^{w_2} \cdots x_k^{w_k}\,,$$

with $w = (w_1, \cdots, w_k) \in \mathbb{Z}^k$, $a_w \in \mathbb{Z}$ and the a_w non-zero for only finitely many w. The *Newton polyhedron* of p, denoted $W(p)$, is the convex hull in $\mathbb{Q}^k$ of the finite set $\{w \in \mathbb{Z}^k : a_w \neq 0\}$. For a face F of $W(p)$, let

$$p_F = \sum_{w \in F} a_w\, x_1^{w_1} \cdots x_k^{w_k}\,.$$

We call p_F the *face of p corresponding to F*. These ideas were utilized by Handelman (in [H] and other papers) and, more recently, by Marcus and the author [MT3]. The polynomial p is said to be *Handelman* if $p_F > 0$ on $(\mathbb{R}^{++})^k$ for every face F of $W(p)$.

Let A be an irreducible matrix over $R^+ = \mathbb{Z}^+[x_1^{\pm}, \cdots, x_k^{\pm}]$ such that its beta function $\beta = \beta_A$ belongs to R. Let F be a (proper) face of the Newton polyhedron $W(\beta)$, and let β_F be the face of β corresponding to F. Since β is a polynomial, the weight-per-symbol polytope $W(A)$ coincides with the Newton polyhedron $W(\beta)$: Implicit in the proof of (5.1) of [MT1], this fact was first made explicit by De Angelis [D]. It is also a special case of *(12)*, which is proved in section 4. Let $A_{F,1}, \cdots, A_{F,\ell}$ be the disjoint irreducible sub-matrices of A associated to the face F of $W(A) = W(\beta)$. It will emerge below that $\beta_F = \beta_{A,F} = \max\{\beta_{A_{F,i}} : 1 \leq i \leq \ell\}$ on $(\mathbb{R}^{++})^k$. For now, however, β_F simply denotes the face of β as defined in the preceding paragraph. Let $r = r_A$ be a vector over R such that $Ar = \beta r$ and $r(x_1, \cdots, x_k) > 0$ for all $(x_1, \cdots, x_k) \in (\mathbb{R}^{++})^k$. For example, we could take r to be any column of the adjoint $\mathrm{Adj}(\beta I - A)$.

Corresponding to F, we will define a sub-matrix A_F of A and a subvector r_F of r. To this end, find $v = (v_1, \cdots, v_k) \in \mathbb{Z}^k$ such that, for some $b \in \mathbb{Z}$, the face $F \subset \{u \in \mathbb{Q}^k : u \cdot v = b\}$ while $W(\beta) \setminus F \subset \{u \in \mathbb{Q}^k : u \cdot v > b\}$. For $p \in R$, express p as a sum

$$p = \sum_{w \in S} a_w\, x_1^{w_1} \cdots x_k^{w_k}$$

with $a_w \neq 0$ for $w \in S \subset \mathbb{Z}^k$. Define $\delta(p) = \min\{w \cdot v : w \in S\}$. Note that $G = \{w \in W(p) : w \cdot v = \delta(p)\}$ is a face of $W(p)$. By a slight abuse of notation, we will write $p_F = p_G$. (Recall that F is a face of $W(\beta)$.) In other words,

$$p_F = \sum_{w \cdot v = \delta(p)} a_w\, x_1^{w_1} \cdots x_k^{w_k}\,.$$

If $p = 0$, we put $\delta(p) = \infty$ and $p_F = 0$. For a state I of A, we put $r_F(I) = r(I)_F$. This defines r_F, which is a sub-vector of r in the sense that each $r_F(I)$ is obtained from $r(I)$ by discarding some of the monomials appearing in $r(I)$. We define A_F by considering the weighted graph $G(A)$ and specifying which edges to keep. For an edge e, denote the initial state of e by $I(e)$ and the terminal state by $T(e)$. Recall that the weight (monomial) associated to e is denoted $\mathrm{wt}_A(e)$. For each state I, the equation $Ar = \beta r$ implies

$$(\dagger) \qquad\qquad \beta\, r(I) = \sum_{I(e)=I} \mathrm{wt}_A(e)\, r(T(e)).$$

Keep the edge e (and its weight $\mathrm{wt}_A(e)$) if and only if

$$(\ddagger) \qquad\qquad \delta\big(\mathrm{wt}_A(e)\, r(T(e))\big) = \delta\big(\beta\, r(I(e))\big).$$

This defines a sub-graph $G(A)_F$ of the weighted graph $G(A)$. Take A_F to be the corresponding sub-matrix of A, so that $G(A_F) = G(A)_F$. Think of A_F as the face of A corresponding to F. The basic properties of A_F, r_F are as follows.

5. THEOREM.

    (a) A_F and r_F have the same indexing set as A.

    (b) No row of A_F is trivial.

    (c) No entry of r_F is trivial and $r_F(x_1, \cdots, x_k) \geq 0$ for $x_1, \cdots, x_k > 0$.

    (d) $A_F\, r_F = \beta_F\, r_F$. In particular, $\beta_{A_F} = \beta_F$.

    (e) A cycle γ of $G(A)$ belongs to $G(A_F)$ if and only if $\mathrm{wps}_A(\gamma) \in F$. Consequently, $A_{F,1}, \cdots, A_{F,\ell_F(A)}$ are precisely the irreducible components of A_F and we have $\beta_F = \max \{\beta_{A_{F,i}} : 1 \leq i \leq \ell_F(A)\}$.

PROOF. Clearly r_F has the same indexing set as A. (c) is a consequence of the fact that $r > 0$ on $(\mathbb{R}^{++})^k$. Note also that β_F is non-trivial and $\beta_F \geq 0$ on $(\mathbb{R}^{++})^k$ because $\beta > 0$ on $(\mathbb{R}^{++})^k$. Let I be a state of A. We will show that

$$\beta_F\, r_F(I) = \sum_J A_F(I, J)\, r_F(J).$$

Since β_F and each $r_F(I)$ is non-trivial and non-negative on $(\mathbb{R}^{++})^k$, it will follow that the I-th row of A_F is non-trivial and we will have established (a)–(d). (For the second statement in (d), note that the first statement and (c) show that $\beta_{A_F} = \beta_F$ except where an entry of r_F vanishes and use the fact that both β_{A_F} and β_F are continuous.) Consider the equation

$$\beta\, r(I) = \sum_J A(I, J)\, r(J).$$

This equation, (c), and the fact that A is over R^+ imply that we have

$$\delta\big(A(I, J)\, r(J)\big) \geq \delta\big(\beta\, r(I)\big)$$

for all J. Hence

$$\beta_F\, r_F(I) = \big(\beta\, r(I)\big)_F = \sum\nolimits^* \big(A(I,J)\, r(J)\big)_F = \sum\nolimits^* A(I,J)_F\, r_F(J),$$

where the sums $\sum^*$ are over all J with $\delta\big(A(I,J)\,r(J)\big) = \delta\big(\beta\,r(I)\big)$. But the last sum is exactly equal to

$$(A_F\, r_F)(I) = \sum_J A_F(I,J)\, r_F(J).$$

This proves (a)–(d).

The second statement in (e) follows from the first statement and the definition of $A_{F,1}, \cdots, A_{F,\ell_F(A)}$. To prove the first statement of (e), let $\gamma = e_1 e_2 \cdots e_n$ be a cycle in $G(A)$ and put $I = I(e_1) = T(e_n)$. First suppose γ belongs to $G(A_F)$. It follows from the condition ($\ddagger$) in the definition of A_F that

$$\delta\big(\beta^n\, r(I)\big) = \delta\big(\mathrm{wt}_A(\gamma)\, r(I)\big).$$

Thus,

$$n\,\delta(\beta) + \delta\big(r(I)\big) = \delta\big(\mathrm{wt}_A(\gamma)\big) + \delta\big(r(I)\big)$$

and we have

$$b = \delta(\beta) = \frac{1}{n}\,\delta\big(\mathrm{wt}_A(\gamma)\big) = \frac{1}{n}\,\big[\log(\mathrm{wt}_A(\gamma))\cdot v\big] = \mathrm{wps}_A(\gamma)\cdot v,$$

which means that $\mathrm{wps}_A(\gamma) \in F$. Conversely suppose $\mathrm{wps}_A(\gamma) \in F$, so that $\mathrm{wps}_A(\gamma)\cdot v = b$. Reversing the argument above, we find

$$\delta\big(\beta^n\, r(I)\big) = \delta\big(\mathrm{wt}_A(\gamma)\, r(I)\big).$$

We claim this implies

$$\delta\big(\beta\, r(I(e_i))\big) = \delta\big(\mathrm{wt}_A(e_i)\, r(T(e_i))\big)$$

for each i. Since γ is a cycle, it suffices to prove this for $i = 1$. Suppose, for a contradiction, the equality

$$\delta\big(\beta\, r(I)\big) = \delta\big(\mathrm{wt}_A(e_1)\, r(T(e_1))\big)$$

fails. Then, using (†), we find

$$\delta\big(\beta\, r(I)\big) < \delta\big(\mathrm{wt}_A(e_1)\, r(T(e_1))\big).$$

Similarly, using

$$\beta^{n-1}\, r\big(T(e_1)\big) = \sum_J A^{n-1}\big(T(e_1),J\big)\, r(J),$$

we see that

$$\delta\big(\beta^{n-1}\, r(T(e_1))\big) \le \delta\big(\mathrm{wt}_A(e_2)\cdots \mathrm{wt}_A(e_n)\, r(I)\big).$$

Hence

$$\delta\big(\beta^n\, r(I)\big) < \delta\big(\mathrm{wt}_A(e_1)\,\beta^{n-1}\, r(T(e_1))\big) \le \delta\big(\mathrm{wt}_A(\gamma)\, r(I)\big),$$

which is a contradiction. $\qquad\square$

Our next observation concerns the uniqueness of A_F, r_F. Suppose $r' = r'_A$ is another right eigenvector over R. Then there exist $p, q \in R$ such that $p\, r = q\, r'$. As a result, $p_F\, r_F = q_F\, r'_F$. The equation $p\, r = q\, r'$ also implies that, for an edge e, we have $\delta\big(\beta\, r(I(e))\big) = \delta\big(\mathrm{wt}_A(e)\, r(T(e))\big)$ if and only if $\delta\big(\beta\, r'(I(e))\big) = \delta\big(\mathrm{wt}_A(e)\, r'(T(e))\big)$. This proves:

6. PROPOSITION. *A_F does not depend on the choice of the right eigenvector* $r = r_A$ *over R. If $r' = r'_A$ is another right eigenvector over R, then there exist* $p, q \in R$ *such that* $p\, r = q\, r'$ *and* $p_F\, r_F = q_F\, r'_F$.

Next, we note that a result of [MT3] leads to a strengthening of (5)(c).

7. THEOREM. *[MT3]. The entries of* $\mathrm{Adj}(\beta I - A)$ *are Handelman. In particular, there exists a right eigenvector of A whose entries are Handelman.*

8. PROPOSITION. *If the entries of r are Handelman, then* $r_F(x_1, \cdots, x_k) > 0$ *for all* $(x_1, \cdots, x_k) \in (\mathbb{R}^{++})^k$.

PROOF. By definition, $r_F(I) = r(I)_G$ for a suitable face G of $W\big(r(I)\big)$ and, when $r(I)$ is Handelman, $r(I)_G$ is strictly positive on $(\mathbb{R}^{++})^k$. $\qquad \square$

In particular, the following simple but significant constraints are imposed on the components of A_F by the beta function β of A.

9. COROLLARY. *Among the components* $A_{F,1}, \cdots, A_{F,\ell_F(A)}$ *of A_F there is at least one whose beta function equals β_F. For any component $A_{F,i}$ of A_F, either* $\beta_{A_{F,i}} = \beta_F$ *or* $\beta_{A_{F,i}} < \beta_F$ *on* $(\mathbb{R}^{++})^k$.

PROOF. The first assertion is a consequence of the fact that, since A_F has no trivial rows, it must have at least one principal component. For the second, suppose i and $\alpha = (\alpha_1, \cdots, \alpha_k) \in (\mathbb{R}^{++})^k$ are such that $\beta_0 = \beta_{A_{F,i}}$ has $\beta_0(\alpha) = \beta_F(\alpha)$. We will show $\beta_0 = \beta_F$. By (5), (6), (7) and (8), there is a vector r_F such that $A_F\, r_F = \beta_F\, r_F$ and $r_F > 0$ on $(\mathbb{R}^{++})^k$. Using this and the fact that $\beta_0(\alpha) = \beta_F(\alpha)$, we conclude from (2) that $A_{F,i}$ must be a principal component of A_F. Then, using (2) again, we see that $\beta_0 = \beta_F$ on $(\mathbb{R}^{++})^k$. $\qquad \square$

For example, if $\beta_A = \beta = 1 + x + x^2 + y$ and F is the face of $W(\beta)$ lying on the x-axis, A_F must have a component whose beta function equals $\beta_F = 1 + x + x^2$, but no component of A_F can have $3x$ or $2 + \sqrt{x}$ as its beta function.

We will make no further use of (7).

We will examine the invariance of faces of matrices under several relations. To define these relations, let A, B be square matrices over R^+. We do not insist, for the duration, that A, B be irreducible, or that they have polynomial beta functions. Recall, however, our standing assumption that the matrices we consider have no trivial rows. The matrices A, B are *strong shift equivalent* if there exist $L \in \mathbb{Z}^{++}$ and matrices $U_1, \cdots, U_L, V_1, \cdots, V_L$ over R^+ such that

$$A = U_1 V_1\,, \quad V_1 U_1 = U_2 V_2\,, \cdots, \quad V_{L-1} U_{L-1} = U_L V_L\,, \quad V_L U_L = B\,.$$

They are *shift equivalent* if there exist $L \in \mathbb{Z}^{++}$ and matrices U, V over R^+ such that

$$AU = UB \,, \; VU = B^L \,, \; VA = BV \,, \; UV = A^L \,.$$

These notions arise from dynamical considerations. The matrices A, B are strong shift equivalent if and only if Σ_A, Σ_B are block isomorphic. Shift equivalence of A, B is equivalent to isomorphism of their (future) dimension modules. It also characterizes eventual block isomorphism, that is, block isomorphism of $\Sigma_{A^n}, \Sigma_{B^n}$ for all large n. Strong shift equivalence implies shift equivalence. (See [Wi, PT2, T2, BMT, MT1].) Recently, Boyle [B] exhibited two irreducible matrices over R^+ which are shift equivalent but not strong shift equivalent.

A (weight-preserving) block homomorphism $\phi : \Sigma_A \to \Sigma_B$ is said to be *right closing* if, for each point $a = (a_n)_{n \in \mathbb{Z}}$ of Σ_A, its image $\phi(a)$ and its past $(a_{-n})_{n \in \mathbb{Z}^+}$ determine a. Right closing maps are necessarily bounded-to-one. Recall from section 2 that block homomorphisms are required to be weight-preserving: $\text{wps}_B(\phi(\gamma)) = \text{wps}_A(\gamma)$ for every cycle (periodic orbit) γ of A. We say that A *right closes* onto B if there exists a right closing block homomorphism $\phi : \Sigma_A \to \Sigma_B$. Let us say that A *algebraically right closes* onto B if there exist $L \in \mathbb{Z}^+$ and matrices U, V over R^+ such that $AU = UB, VU = B^L$. Similar to shift equivalence, A right closes onto B if and only if the (future) dimension module of A factors onto that of B. Right closing of A onto B implies algebraic right closing. In fact eventual right closing, that is, right closing of Σ_{A^n} onto Σ_{B^n} for all large n, is enough to guarantee algebraic right closing. (See [BMT, T2].) The converse of the last statement is false: There are examples of A, B such that A does not eventually right close onto B even though A algebraically right closes onto B. The first such example was given by Marcus. Subsequent (unpublished) work by Marcus and the author has revealed eventual right closing to involve finer dimension conditions.

Each of the above relations implies $\beta_A = \beta_B$. (See [T1, PT2].)

10. REMARK. Recall that a block homomorphism $\phi : \Sigma_A \to \Sigma_B$ is called a 1-block homomorphism if it is induced by a map from the edges $E(A)$ of $G(A)$ onto those of $G(B)$; that is, if there exists $\phi : E(A) \to E(B)$ such that for $a = (a_n) \in \Sigma_A$, we have $\phi(a) = ((\phi(a_n)))$. A 1-block homomorphism is *right resolving* if the associated map $\phi : E(A) \to E(B)$ has the following property: If $e_1 \in E(A)$ and $f_1, f_2 \in E(B)$ are such that $T(f_1) = I(f_2)$ and $\phi(e_1) = f_1$, then there exists a unique $e_2 \in E(A)$ such that $T(e_1) = I(e_2)$ and $\phi(e_2) = f_2$. A zero-one matrix S is called an *amalgamation matrix* if its columns are non-trivial and each row contains exactly one non-zero entry. According to a result of Kitchens (see [BKM]), a block homomorphism is right closing if and only if it is conjugate to a right resolving one. Combining this with [PT1], we find that we have a right closing map $\phi : \Sigma_A \to \Sigma_B$ if and only if there exist an amalgamation matrix S and matrices $\tilde{A}, \tilde{B}$ over R^+ such that $\Sigma_{\tilde{A}}, \Sigma_{\tilde{B}}$ are block isomorphic to Σ_A, Σ_B, respectively, and $\tilde{A}S = S\tilde{B}$. The equation $\tilde{A}S = S\tilde{B}$ gives a map $\psi : E(\tilde{A}) \to E(\tilde{B})$ with $\text{wt}_{\tilde{A}}(e) = \text{wt}_{\tilde{B}}(\psi(e))$: First define a map

on states by taking $\psi(I)$, for a state I of $\tilde{A}$, to be the unique state of $\tilde{B}$ with $S(I, \psi(I)) = 1$. Then for states I_0 of $\tilde{A}$ and J_1 of $\tilde{B}$ the equation $\tilde{A}S = S\tilde{B}$ implies

$$\sum_{\substack{I_1 \text{ with} \\ \psi(I_1) = J_1}} A(I_0, I_1) = B(\psi(I_0), J_1),$$

so that there is a weight-preserving bijection of $\{e \in E(A) : I(e) = I_0, \psi(T(e)) = J_1\}$ onto $\{f \in E(B) : I(f) = \psi(I_0), T(f) = J_1\}$. We define $\psi : E(\tilde{A}) \to E(\tilde{B})$ by choosing such a bijection for each pair (I_0, J_1), and ψ then gives a right resolving homomorphism $\psi : \Sigma_{\tilde{A}} \to \Sigma_{\tilde{B}}$ which is conjugate to ϕ. (See [PT1, BMT].) We will use this construction in the proof of the next theorem.

Now assume that the matrices A, B are irreducible. Since the beta function is invariant under all the relations we defined, assume $\beta = \beta_A = \beta_B$ and, further, that $\beta \in R$. Let $r = r_A$ be a vector such that $Ar = \beta r$ and $r > 0$ on $(\mathbb{R}^{++})^k$. Let F be a face of $W(\beta) = W(A) = W(B)$. A change of variables will enable us to view F, β_F, A_F, r_F very concretely: As before, find a vector $v \in \mathbb{Z}^k$ that exposes F by having, for some $b \in \mathbb{Z}$, the face $F \subset \{u \in \mathbb{Q}^k : u \cdot v = b\}$ while $W(\beta) \setminus F \subset \{u \in \mathbb{Q}^k : u \cdot v > b\}$. Make sure that the entries of v are relatively prime. Extend v to a basis of $\mathbb{Z}^k$ to find $M \in GL(k, \mathbb{Z})$ which has v as its k-th row. Then consider $P = M^{-1}$ and let $\bar{x}_i = x_1^{P(1,i)} x_2^{P(2,i)} \cdots x_k^{P(k,i)}$. After such a change of variables and dividing by $\bar{x}_k^b$, we assume that the variables $x_1, \cdots, x_{k-1}, y = x_k$ are such that $W(\beta) \subset \{(u_1, \cdots, u_k) \in \mathbb{Q}^k : u_k \geq 0\}$ and $F = \{(u_1, \cdots, u_k) \in \mathbb{Q}^k : u_k = 0\} \cap W(\beta)$. In other words, $v = (0, \cdots, 0, 1)$ and $b = 0$. For $p \in R$, the integer $\delta(p)$ now represents the lowest power of y appearing in p and p_F consists of the terms of p involving that power of y. More precisely, finding $n_0 \in \mathbb{Z}$ and non-trivial $p_0 \in \mathbb{Z}[x_1^{\pm}, \cdots, x_{k-1}^{\pm}]$ such that $p = p_0 y^{n_0} + q y^{n_0+1}$ for some $q \in \mathbb{Z}[x_1^{\pm}, \cdots, x_{k-1}^{\pm}, y]$, we have $\delta(p) = n_0$ and $p_F = p_0 y^{n_0}$. Note that $\delta(\beta) = 0$. Let D_A be the diagonal matrix whose I-th diagonal entry is $y^{\delta(r(I))}$. Define D_B similarly, using a right eigenvector r_B of B. Replace A, r, B, r_B by $D_A^{-1}AD_A, D_A^{-1}r, D_B^{-1}BD_B, D_B^{-1}r_B$. Then, in addition to $\delta(\beta) = 0$, we have $\delta(r(I)) = 0$ for every state I of A, so that β_F, r_F are obtained by putting $y = 0$:

$$\beta_F(x_1, \cdots, x_{k-1}, y) = \beta(x_1, \cdots, x_{k-1}, 0)$$

and

$$r_F(x_1, \cdots, x_{k-1}, y) = r(x_1, \cdots, x_{k-1}, 0).$$

We claim that, in addition, A is over $\mathbb{Z}^+[x_1^{\pm}, \cdots, x_{k-1}^{\pm}, y]$. To see this, suppose an entry $A(I, J)$ has $\delta(A(I, J)) < 0$. Consider the equation

$$\beta r(I) = \sum_K A(I, K) r(K).$$

We have $\delta(\beta r(I)) = 0$, while the RHS has $\delta(\sum_K A(I, K) r(K)) < 0$, which is absurd! So A is, indeed, over $\mathbb{Z}^+[x_1, \cdots, x_{k-1}^{\pm}, y]$. It follows that A_F is also obtained by setting $y = 0$:

$$A_F(x_1, \cdots, x_{k-1}, y) = A(x_1, \cdots, x_{k-1}, 0).$$

Similar statements are valid for r_B and B. In the first step of the proof of the next result, we will assume we are in this situation, so that β, A, r, B, r_B have the properties listed above. We will refer to this by saying they are F-adapted.

11. THEOREM. *Let A, B be irreducible matrices over $R^+ = \mathbb{Z}^+[x_1^{\pm}, \cdots, x_k^{\pm}]$ such that $\beta = \beta_A = \beta_B \in R = \mathbb{Z}[x_1^{\pm}, \cdots, x_k^{\pm}]$. Let F be a face of the polytope $W(\beta) = W(A) = W(B)$. Every strong shift equivalence, shift equivalence, right closing, algebraic right closing of A onto B induces, respectively, a strong shift equivalence, shift equivalence, right closing, algebraic right closing of the face A_F onto B_F.*

PROOF. As above, we assume that the variables $x_1, \cdots, x_{k-1}, y = x_{k-1}$ are such that $W(\beta) \subset \{(u_1, \cdots, u_k) \in \mathbb{Q}^k : u_k \geq 0\}$ and $F = \{(u_1, \cdots, u_k) \in \mathbb{Q}^k : u_k = 0\} \cap W(\beta)$. Consequently, $\delta(\beta) = 0$ and $\beta_F(x_1, \cdots, x_{k-1}, y) = \beta(x_1, \cdots, x_{k-1}, 0)$. It is easy to see that this assumption leads to no loss of generality.

Initially we also assume that A, $r = r_A$, B, r_B are F-adapted; we will later remove this assumption. If U is a matrix over R, put $\delta(U) = \min\{\delta(U(I, J))\}$, the minimum being taken over all entries $U(I, J)$. This defines $\delta(u)$, in particular, for a row vector u. Using $\delta(u)$, we define a row vector u_F of the same size as u: For each I, write $u(I) = \sum_{w \in \mathbb{Z}^k} a_w x_1^{w_1} \cdots x_k^{w_k}$ with $a_w \in \mathbb{Z}$ and non-zero for only finitely many w. Define

$$u_F(I) = \sum_{w \in \mathbb{Z}^k, w_k = \delta(u)} a_w x_1^{w_1} \cdots x_k^{w_k}.$$

For a matrix U, define U_F by rows: Letting e_I denote the I-th elementary basis vector, put $e_I U_F = (e_I U)_F$. Using the fact that A, r are F-adapted, we see that the definition just given is consistent with our earlier definition of A_F, r_F. (Remember that r is a column vector!) Similar statements apply to B, r_B.

Suppose that the matrices U, V over R^+ furnish an algebraic right closing of A onto B, so that $AU = UB, VU = B^L$ for some $L \in \mathbb{Z}^{++}$. We will show $A_F U_F = U_F B_F, V_F U_F = (B_F)^L$. The equation $AU = UB$ implies that $U r_B$ is a right eigenvector for A, and $U r_B \neq 0$ because $V(U r_B) = \beta^L r_B$. Put $r' = r'_A = U r_B$. Then $V r' = \beta^L r_B$. Recall that, since r_A, r_B are F-adapted, $\delta(r_A(I)) = 0$ for every state I of A and $\delta(r_B(J)) = 0$ for every state J of B. Letting $p, q \in R$ be such that $pr = qr'$ and putting $\delta = \delta(p) - \delta(q)$, we find that $\delta(r'(I)) = \delta$ for every state I of A. Considering the equation $r'(I) = \sum_J U(I, J) r_B(J)$, we then see that $\delta(e_I U) = \delta$ for every row $e_I U$ of U. Therefore, $\delta(U) = \delta$ and $\delta(e_I U) = \delta(U) = \delta$ for every I. It follows from this and the fact that A is F-adapted that $\delta(e_I AU) = \delta(U) = \delta$ for every state of I of A and $(AU)_F = A_F U_F$. Similarly $(UB)_F = U_F B_F$, and we have $A_F U_F = U_F B_F$. Next observe that the equation

$\beta^L r_B = V r'$ implies, in an entirely analogous manner, $\delta(e_J V) = \delta(V) = -\delta$ for every state J of B. Hence,

$$(B^L)_F = (VU)_F = V_F U_F \,.$$

It is easy to see that $(B^L)_F = (B_F)^L$. Thus, $AU = UB, VU = B^L$ imply $A_F U_F = U_F B_F$, $V_F U_F = (B_F)^L$. In the case of a shift equivalence we also have the equations $VA = BV$, $UV = A^L$ and these yield $V_F A_F = B_F V_F$, $U_F V_F = (A_F)^L$.

Now suppose U, V effect an algebraic right closing of A onto B and A, B are not F-adapted. Let D_A, D_B be diagonal matrices such that the diagonal entries of D_A, D_B are monomials and $\bar{A} = D_A^{-1} A D_A$, $\bar{B} = D_B^{-1} B D_B$ are F-adapted. Note that weights of edges that do not belong to any irreducible component have no bearing on the weights of cycles, so that conjugations by such diagonal matrices do not alter the weights of cycles. The matrices

$$\bar{U} = D_A^{-1} U D_B \,, \bar{V} = D_B^{-1} V D_A$$

effect an algebraic right closing of $\bar{A}$ onto $\bar{B}$. By the first part of our proof, $\bar{U}, \bar{V}$ induce matrices $\bar{U}_F, \bar{V}_F$ which effect an algebraic right closing of $\bar{A}_F$ onto $\bar{B}_F$. But

$$D_A \bar{A}_F D_A^{-1} \,, D_B \bar{B}_F D_B^{-1}$$

are precisely the faces A_F, B_F defined just before (5). Hence, $D_A \bar{U}_F D_B^{-1}$, $D_B \bar{V}_F D_A^{-1}$ give an algebraic right closing of A_F onto B_F. This argument also applies to shift equivalence, establishing the theorem for algebraic right closings and shift equivalence.

Consider what we have established for shift equivalence in the case $L = 1$: If $A = UV, VU = B$, then U, V induce matrices U_F, V_F such that $A_F = U_F V_F$, $V_F U_F = B_F$. Finitely many applications of this yield the desired result for strong shift equivalence.

To establish the required result for right closings, in view of (10) and what we have already proved for block isomorphism (strong shift equivalence), it suffices to assume that we have an amalgamation matrix S with $AS = SB$ and show that the 1-block map ψ defined in (10) sends Σ_{A_F} onto Σ_{B_F}. Let r_B be a vector over R with $B r_B = \beta r_B$, and put $r = r_A = S r_B$. Then $Ar = \beta r$. By the definition of ψ and the fact that $r = S r_B$, we have

$$\delta\big(\beta\, r(I(e))\big) = \delta\big(\mathrm{wt}_A(e)\, r(T(e))\big)$$

if and only if

$$\delta\big(\beta\, r_B(I(\psi(e)))\big) = \delta\big(\mathrm{wt}_B(\psi(e))\, r_B(T(\psi(e)))\big) \,,$$

which implies that ψ sends Σ_{A_F} onto Σ_{B_F}. $\square$

4. Faces of matrices: the general case.

We now turn to the general case.

Let A be an irreducible matrix over $R^+ = \mathbb{Z}^+[x_1^\pm, \cdots, x_k^\pm]$, and let $\beta = \beta_A$. The Perron-Frobenius theorem implies that β is analytic on $(\mathbb{R}^{++})^k$. (See [Hi,Ka].) Moreover, as β satisfies the characteristic polynomial of A, it is an algebraic function, an integer over $R = \mathbb{Z}[x_1^\pm, \cdots, x_k^\pm]$.

Let F be a face of the weight-per-symbol polytope $W(A)$, and let $A_{F,1}, \cdots,$ $A_{F,\ell}$ be the disjoint irreducible sub-matrices of A corresponding to F. According to (3), $W(A)$ and $\beta_F = \beta_{A,F} = \max\{\beta_{A_{F,i}} : 1 \le i \le \ell\}$ are determined by β and do not otherwise depend on A. We make these facts more explicit by showing how the polytope and β_F may be obtained directly from β.

Let $\chi(t) = t^d + p_1 t^{d-1} + \cdots + p_{d-1}t + p_d$ be the minimum polynomial of β over R. Since $p_1, \cdots, p_d \in R$, we have the Newton polyhedra $W(p_1), \cdots, W(p_d) \subset \mathbb{Z}^k$. Define $W(\beta)$ to be the convex hull of $\cup_{n=1}^d \frac{1}{n} W(p_n)$ in $\mathbb{Q}^k$. This definition generalizes the Newton polyhedron, since for $\beta \in R$ we have $\chi(t) = t - \beta$. We will prove:

12. THEOREM. *If A is an irreducible matrix over R^+ and $\beta = \beta_A$, then $W(A) = W(\beta)$.*

The proof of (12) requires that we deal with the companion matrix of β. For this purpose, we extend our tools to matrices over R. For $p \in R$, write $p = \sum_{w \in \mathbb{Z}^k} a_w x_1^{w_1} \cdots x_k^{w_k}$ and put $|p| = \sum_{w \in \mathbb{Z}^k} |a_w| x_1^{w_1} \cdots x_k^{w_k}$. Let C be a square matrix over R. Define $|C|$ by letting $|C|(I,J) = |C(I,J)|$, and take $W(C) = W(|C|)$. Note that C assigns a sign $s(e) = \pm 1$ to each edge e of the graph $G(|C|)$ according to the sign of the monomial $\text{wt}_{|C|}(e)$ in C. To each sub-matrix $|C|_0$ of $|C|$ there corresponds a sub matrix C_0 of C with

$$C_0(I,J) = \sum_{\substack{e \text{ edge of } G(|C|_0), \\ I(e)=I, T(e)=J}} s(e)\,\text{wt}_{|C|}(e).$$

When F is a face of $W(C) = W(|C|)$, we take $C_{F,1}, \cdots, C_{F,\ell_F(C)}$ to be the submatrices of C corresponding to the sub-matrices $|C|_{F,1}, \cdots, |C|_{F,\ell_F(|C|)}$ of $|C|$. (Alternatively, of course, we could associate a signed weighted graph $G(C)$ to C and define everything in terms of $G(C)$.) For $x_1, \cdots, x_k > 0$ we denote the spectral radius of $C(x_1, \cdots, x_k)$ by $\beta_C(x_1, \cdots, x_k)$. For a face F of $W(C)$ we also define $\beta_{C,F}$ on $(\mathbb{R}^{++})^k$ by $\beta_{C,F} = \max\{\beta_{C_{F,i}} : 1 \le i \le \ell_F(C)\}$.

Let $v = (v_1, \cdots, v_k) \in \mathbb{Q}^k$ and $b = \max\{W(C) \cdot v\}$. Consider the face

$$F = \{u \in W(C): u \cdot v = b\}$$

of $W(C)$. For $a_1, \cdots, a_k > 0$, consider the curve $f(t) = (a_1 e^{v_1 t}, \cdots, a_k e^{v_k t}), t \in \mathbb{R}$. Let γ be a cycle of $G(|C|)$ of length n and weight $\text{wt}_{|C|}(\gamma) = x_1^{u_1} \cdots x_k^{u_k}$. Writing $u = (u_1, \cdots, u_k)$ and evaluating $e^{-bnt}\text{wt}_{|C|}(\gamma)$ at $(x_1, \cdots, x_k) = f(t)$, we obtain

$$e^{(u \cdot v - bn)t} a_1^{u_1} \cdots a_k^{u_k}.$$

Letting $t \to \infty$, we get $a_1^{u_1} \cdots a_k^{u_k}$ or 0 according as $u \cdot v = bn$ or $u \cdot v < bn$, that is, according to whether $\text{wps}_{|C|}(\gamma) \in F$ or not. This implies that the spectral radius of $e^{-bt} C\big(f(t)\big)$ tends to $\beta_{C,F}(a_1, \cdots, a_k)$ as $t \to \infty$:

13. LEMMA. *Let C be a square matrix over R and $v = (v_1, \cdots, v_k) \in \mathbb{Q}^k$. Put $b = \max\{W(C) \cdot v\}$ and let $F = \{u \in W(C): u \cdot v = b\}$ be the corresponding face of $W(C)$. For $a_1, \cdots, a_k > 0$ we have*

$$\lim_{t \to \infty} e^{-bt} \beta_C(a_1 e^{v_1 t}, \cdots a_k e^{v_k t}) = \beta_{C,F}(a_1, \cdots, a_k).$$

PROOF OF (12). Let $\chi(t) = t^d + p_1 t^{d-1} + \cdots + p_d$ be the minimum polynomial of β, and let C be its companion matrix. By the definition of the companion matrix, the p_n determine the simple cycles of $G(|C|)$: For $1 \leq n \leq d-1$, if $p_n = \sum_{w \in \mathbf{Z}^k} a_w x_1^{w_1} \cdots x_k^{w_k}$ then the graph $G\big(|C|\big)$ contains exactly $|a_w|$ simple cycles of length n and weight-per-symbol $\frac{1}{n} w$, and this gives all of the simple cycles of $G(|C|)$. It follows from this and the definitions of $W(\beta), W(C)$ that $W(\beta) = W(C) = W(|C|)$. We will prove $W(\beta) = W(A)$ by showing

$$\max\{W(C) \cdot v\} = \max\{W(A) \cdot v\}$$

for every $v \in \mathbb{Q}^k$. To this end, fix $v \in \mathbb{Q}^k$. Set

$$w = \max\{W(C) \cdot v\}, \ w' = \max\{W(A) \cdot v\},$$

and let

$$F = \{u \in W(C): u \cdot v = w\}, \ F' = \{u \in W(A): u \cdot v = w'\}$$

be the corresponding faces of $W(C), W(A)$. Applying (13) to A, we have

$$\lim_{t \to \infty} e^{-w't} \beta(a_1 e^{v_1 t}, \cdots, a_k e^{v_k t}) = \beta_{F'}(a_1, \cdots, a_k) > 0$$

for all $a_1, \cdots, a_k > 0$. Similarly applying (13) to C and remembering $\beta_C = \beta$,

$$\lim_{t \to \infty} e^{-wt} \beta(a_1 e^{v_1 t}, \cdots, a_k e^{v_k t}) = \beta_{C,F}(a_1, \cdots, a_k) \geq 0$$

for all $a_1, \cdots, a_k > 0$. In order to conclude $w = w'$, we show that $\beta_{C,F}(a_1, \cdots, a_k) \neq 0$ for some $(a_1, \cdots, a_k) \in (\mathbb{R}^{++})^k$. Since $W(C) = W(\beta)$ is the convex hull of $\cup_{i=1}^d \frac{1}{n} W(p_n)$, we have $\max\{W(p_n) \cdot v\} \leq nw$ for every $1 \leq n \leq d$, and equality must hold for at least one value of n. Applying (13) to p_n, the polynomial p_n^o defined by

$$p_n^o(a_1, \cdots, a_k) = \lim_{t \to \infty} e^{-nwt} p_n(a_1 e^{v_1 t}, \cdots, a_k e^{v_k t})$$

is the zero polynomial if $\max\{W(p_n) \cdot v\} < nw$ and is a face of p_n if $\max\{W(p_n) \cdot v\} = nw$. Multiplying the identity

$$\chi(\beta(a_1 e^{v_1 t}, \cdots, a_k e^{v_k t})) = 0$$

by e^{-wdt} and letting $t \to \infty$, we find that $\beta_{C,F}$ is a root of

$$\chi_0(t) = t^d + p_1^o t^{d-1} + \cdots + p_d^o.$$

At least one of $p_1^o, \cdots, p_d^o \in R$ is not the zero polynomial. Moreover, it is not hard to see that (modulo a power of t) the polynomial $\chi_0(t)$ is the characteristic polynomial of the matrix C_0 consisting of the disjoint sub-matrices $C_{F,1}, \cdots, C_{F,\ell_F(C)}$ of C. Thus, $\beta_{C,F}$ is the maximal root of $\chi_0(t)$ and, for $a_1, \cdots, a_k > 0$, we have $\beta_{C,F}(a_1, \cdots, a_k) = 0$ if and only if $p_n^o(a_1, \cdots, a_k) = 0$ for every $1 \leq n \leq d$. It follows that $w = w'$. In addition $F = F'$ and, by continuity, $\beta_{C,F} = \beta_F$ on $(\mathbb{R}^{++})^k$, but we do not need these facts. $\square$

14. COROLLARY. *Let F be a face of $W(\beta) = W(A)$, and let $v \in \mathbb{Q}^k$ and $b = \max\{W(\beta) \cdot v\}$ be such that $F = \{u \in W(\beta): u \cdot v = b\}$. For $a_1, \cdots, a_k > 0$, we have $\beta_F(a_1, \cdots, a_k) = \lim_{t \to \infty} e^{-bt}\beta(a_1 e^{v_1 t}, \cdots, a_k e^{v_k t}) = \max\{\beta_{A_{F,i}}(a_1, \cdots, a_k) : 1 \leq i \leq \ell_F(A)\}.$*

15. REMARK. Note, for later use, a fact contained in the proof of (12). The value $\beta_F(a_1, \cdots, a_k) > 0$ is a real root of χ_0 and $\beta_F(a_1, \cdots, a_k)$ is greater than or equal to the absolute value of any other root of χ_0 at $(a_1, \cdots, a_k)$. We will improve on this later in the section.

[H, MT1, MT2, D] contain material related to the above. In particular, it is shown in [MT2] that the Legendre transform

$$(x_1, \cdots, x_k) \mapsto \frac{1}{\beta}\left(x_1\frac{\partial\beta}{\partial x_1}, \cdots, x_k\frac{\partial\beta}{\partial x_k}\right)$$

maps $(\mathbb{R}^{++})^k$ onto the (relative) interior of $W(\beta)$ taken in its affine hull over $\mathbb{R}$. Actually, it is shown that, once the variables are chosen to capture the dimension of $W(\beta)$, this map is a bijection. Thus, we may view β as being defined on the interior of $W(\beta)$. Our definition of β_F for each face F of $W(\beta)$ then extends β to a continuous function of $W(\beta)$. In that this extension need not be analytic, a fundamental difficulty presents itself.

16. EXAMPLE. Let $p(x), q(x) \in \mathbb{Z}^+[x]$ be polynomials with $p(1) = q(1)$. The function

$$\beta = \frac{1}{2}\left((p(x) + q(x)) + \sqrt{(p(x) - q(x))^2 + 4y}\right)$$

is a beta function. For example, it is the beta function of the matrix $\begin{bmatrix} p(x) & y \\ 1 & q(x) \end{bmatrix}$. This class of functions was considered in [MT1] to obtain counter-examples to the finite equivalence conjecture of [PT1]. The minimal polynomial of β is

$$\chi(t) = t^2 - (p(x) + q(x))t + (p(x)q(x) - y),$$

so that $W(\beta)$ is a triangle whose base lies on the x-axis. The face F we consider is the base of the triangle, obtained on putting $y = 0$. Let us first do this in the case $p(x) = 1, q(x) = x$. Then

$$\beta(x, 0) = \begin{cases} 1 & \text{if } 0 < x \leq 1, \\ x & \text{if } x \geq 1. \end{cases}$$

Clearly, $\beta(x, y)$ is not analytic at the point $(1, 0)$. It is analytic on the remainder of the positive x-axis.

17. EXAMPLE. Take $p(x) = q(x) = 1 + x$ in (16). Then

$$\beta(x, y) = 1 + x + \sqrt{y}$$

is not analytic at any point of the positive x-axis. However, if we put $y = z^2$, then we have the function

$$\tilde{\beta}(x, z) = \beta(x, z^2) = 1 + x + z,$$

which is analytic.

We are thus lead to examine the analytic properties of β on a face of $W(\beta)$. Fix a proper face F of $W(\beta)$. We will avoid the additional complication of having β_F defined via limits by choosing variables to suit the face under consideration: We assume, for the rest of the section, that the variables $x_1, \cdots, x_{k-1}, y = x_k$ are such that $W(\beta) \subset \{(u_1, \cdots, u_k) \in \mathbb{Q}^k : u_k \geq 0\}$ and

$$F = \{(u_1, \cdots, u_k) \in \mathbb{Q}_k : u_k = 0\} \cap W(\beta).$$

Then β_F depends only on $x_1, \cdots, x_{k-1}$ and

$$\beta_F(x_1, \cdots, x_{k-1}) = \beta(x_1, \cdots, x_{k-1}, 0).$$

As before, let $p_1, \cdots, p_d \in R$ be such that

$$\chi(t) = \chi(t, x_1, \cdots, x_{k-1}, y) = t^d + p_1 t^{d-1} + \cdots + p_d$$

is the minimal polynomial of β. Put

$$\chi_0(t) = \chi_0(t, x_1, \cdots, x_{k-1}) = \chi(t, x_1, \cdots, x_{k-1}, 0).$$

This is the polynomial (15) refers to. In particular,

$$\chi_0(\beta_F(x_1, \cdots, x_{k-1}), x_1, \cdots, x_{k-1}) = 0.$$

We will show that, though it may happen that β is not analytic at any point of the hyperplane $y = 0$, this can be overcome as in (17) and we then have a function which is analytic outside of a (proper) algebraic variety. Put $y = z^{d!}$ and, for $x_1, \cdots, x_{k-1} > 0, z \geq 0$, let

$$\tilde{\beta}(x_1, \cdots, x_{k-1}, z) = \beta(x_1, \cdots, x_{k-1}, z^{d!}).$$

For $(x_1, \cdots, x_k) \in C^k$, let $\alpha_1(x_1, \cdots, x_k), \cdots, \alpha_d(x_1, \cdots, x_k)$ be the zeros of $\chi(t, x_1, \cdots, x_k)$. The zeros may be ordered so that the functions $\alpha_1, \cdots, \alpha_d$ are continuous on C^k and analytic whenever $(x_1, \cdots, x_k)$ is such that $\chi(t, x_1, \cdots, x_k)$ has d distinct zeros. The discriminant of χ,

$$\Delta(\chi) = \prod_{i<j} (\alpha_i - \alpha_j)^2$$

is a symmetric polynomial in $\alpha_1, \cdots, \alpha_d$, so it is a polynomial in $p_1, \cdots, p_d$. In particular, using $W(\beta) \subset (\{(u_1, \cdots, u_k) \in \mathbb{Q}^k : u_k \geq 0\}$ and the definition of $W(\beta)$, we have $\Delta(\chi) \in \mathbb{Z}[x_1^{\pm}, \cdots, x_{k-1}^{\pm}, y]$. Let $N \in \mathbb{Z}^+$ and $Q \in \mathbb{Z}[x_1^{\pm}, \cdots, x_{k-1}^{\pm}, y]$ be such that

$$\Delta(\chi) = y^N Q$$

and y does not divide Q. Define $Q_0 \in \mathbb{Z}[x_1^{\pm}, \cdots, x_{k-1}^{\pm}]$ by

$$Q_0(x_1, \cdots, x_{k-1}) = Q(x_1, \cdots, x_{k-1}, 0).$$

Let $\chi_1, \cdots, \chi_n$ be the distinct irreducible factors of χ_0 in $\mathbb{Z}[x_1^{\pm}, \cdots, x_{k-1}^{\pm}]$ and set $\bar{\chi}_0 = \prod_{i=1}^{n} \chi_i$. Considering the discriminant $\Delta(\bar{\chi}_0)$, define $P = Q_0 \Delta(\bar{\chi}_0)$ and note that P is a non-trivial element of $\mathbb{Z}[x_1^{\pm}, \cdots, x_{k-1}^{\pm}]$. Put

$$E = \{(x_1, \cdots, x_{k-1}) \in (\mathbb{R}^{++})^{k-1} : P(x_1, \cdots, x_{k-1}) \neq 0\}$$

Multiplying P by a monomial so that $\bar{P} = x_1^{w_1} \cdots x_{k-1}^{w_{k-1}} P \in \mathbb{Z}[x_1, \cdots, x_{k-1}]$, we see that $(\mathbb{R}^{++})^{k-1} \setminus E$ is contained in the algebraic variety

$$\{(x_1, \cdots, x_{k-1}) \in \mathbb{R}^k : \bar{P}(x_1, \cdots, x_{k-1}) = 0\}.$$

In particular, E is dense in $(\mathbb{R}^{++})^{k-1}$.

18. LEMMA. *If $(a_1, \cdots, a_{k-1}) \in E$, then the function $\tilde{\beta}(x_1, \cdots, x_{k-1}, z)$ is analytic at $(a_1, \cdots, a_{k-1}, 0)$.*

PROOF. We shall make use of Osgood's lemma [GR], which states that a continuous function of several variables is analytic on an open subset of C^k if and only if it is analytic in each variable. Let $(a_1, \cdots, a_{k-1}) \in E$. Since $\Delta(\bar{\chi}_0)(a_1, \cdots, a_{k-1}) \neq 0$ and $Q_0(a_1, \cdots, a_{k-1}) \neq 0$ we can find $0 < \epsilon < \min\{a_1, \cdots, a_{k-1}\}$ such that

$$\Delta(\bar{\chi}_0)(x_1, \cdots, x_{k-1}) \neq 0 \text{ and } \Delta(\chi)(x_1, \cdots, x_{k-1}, y) \neq 0$$

whenever $|x_1 - a_1|, \cdots, |x_{k-1} - a_{k-1}| < \epsilon$ and $0 < |y| < \epsilon^{d!}$. Put

$$\mathcal{O}_0 = \{(x_1, \cdots, x_{k-1}) \in \mathbb{C}^{k-1} : |x_1 - a_1|, \cdots, |x_{k-1} - a_{k-1}| < \epsilon\}.$$

We shall extend $\tilde{\beta}(x_1, \cdots, x_{k-1}, z)$ so that it is analytic on

$$\mathcal{O} = \mathcal{O}_0 \times B(0, \epsilon) = \mathcal{O}_0 \times \{z \in \mathbb{C} : |z| < \epsilon\}.$$

Initially, $\beta, \tilde{\beta}$ are defined for $x_1, \cdots, x_{k-1} > 0$ and $y, z \geq 0$. For $0 < y < \epsilon^{d!}$, the discriminant $\Delta(\chi)$ is non-zero on $\mathcal{O}_0 \times \{y\}$, so that $\Delta(\chi)$ has d distinct roots at each point of $\mathcal{O}_0 \times \{y\}$ and β can be analytically extended on this set. Defining β by continuity at $y = 0$, we have extended β to a function on $\mathcal{O}_0 \times [0, \epsilon^{d!})$. Define $\tilde{\beta}$ on $\mathcal{O}_0 \times [0, \epsilon)$ by $\tilde{\beta}(x_1, \cdots, x_{k-1}, z) = \beta(x_1, \cdots, x_{k-1}, z^{d!})$. Fixing $(x_1, \cdots, x_{k-1}) \in \mathcal{O}_0$, the function

$$y \mapsto \beta(x_1, \cdots, x_{k-1}, y)$$

is analytic on $(0, \epsilon^{d!})$ and satisfies the equation $\chi(t, x_1, \cdots, x_{k-1}, y) = 0$. By the theory of algebraic functions of one complex variable [Hi], $\tilde{\beta}$ may be extended so that

$$z \mapsto \tilde{\beta}(x_1, \cdots, x_{k-1}, z)$$

is analytic for $|z| < \epsilon$. This defines $\tilde{\beta}$ on $\mathcal{O}$. It is not hard to check $\tilde{\beta}$ is continuous on $\mathcal{O}$. To check that $\tilde{\beta}$ is analytic, we use Osgood's lemma. For fixed $(x_1, \cdots, x_{k-1}) \in \mathcal{O}_0$,

$$z \mapsto \tilde{\beta}(x_1, \cdots, x_{k-1}, z)$$

is analytic for $|z| < \epsilon$ by the definition of $\tilde{\beta}$. For fixed non-zero z with $|z| < \epsilon$, the discriminant $\Delta(\chi)(x_1, \cdots, x_{k-1}, y) = \Delta(\chi)(x_1, \cdots, x_{k-1}, z^{d!})$ is non-zero on $\mathcal{O}_0$, so that the continuous function

$$(x_1, \cdots, x_{k-1}) \mapsto \tilde{\beta}(x_1, \cdots, x_{k-1}, z)$$

is also analytic. Finally, let $z = 0$. Then the continuous function

$$(\ddagger) \qquad (x_1, \cdots, x_{k-1}) \mapsto \tilde{\beta}(x_1, \cdots, x_{k-1}, 0) = \beta(x_1, \cdots, x_{k-1}, 0)$$

extends β_F and satisfies χ_0. Therefore, it satisfies $\bar{\chi}_0$. Since $\Delta(\bar{\chi}_0)$ is non-zero on $\mathcal{O}_0$, the function $(\ddagger)$ is analytic on $\mathcal{O}_0$. $\square$

Put

$$D = \{(x_1, \cdots, x_{k-1}) \in (\mathbb{R}^{++})^{k-1} : \tilde{\beta} \text{ is analytic at } (x_1, \cdots, x_{k-1}, 0)\}.$$

By (18), $E \subset D$. According to a result of Whitney [W], the complement of an algebraic variety in $\mathbb{R}^{k-1}$ has finitely many connected components. Applying this result to the variety

$$\{(x_1, \cdots, x_{k-1}) \in \mathbb{R}^{k-1} : x_1 \cdots x_{k-1} \bar{P}(x_1, \cdots, x_{k-1}) = 0\},$$

we see that $E \subset (\mathbb{R}^{++})^{k-1}$ has finitely many components. It follows that D is an open dense subset of $(\mathbb{R}^{++})^{k-1}$ with finitely many connected components. We denote by $\mathcal{D} = \mathcal{D}(\beta, F)$ the finite collection consisting of the components of D.

19. EXAMPLE. Take $p(x) = 4 + 4x^2, q(x) = 8x$ in (16). Then

$$\beta(x, y) = 2(x + 1)^2 + \sqrt{4(x - 1)^4 + y}.$$

Consider the face F of $W(\beta)$ obtained by putting $y = 0$. Since $\Delta(\chi) = 16(x - 1)^4 + 4y$, we have $Q_0(x) = 16(x - 1)^4$ and, also, $\Delta(\chi_0) = 16(x - 1)^4$. Thus, $E = \mathbb{R}^{++} \setminus \{1\}$. (As $\Delta(\chi)$ is not divisible by y, we do not need to put $y = z^2$; such a substitution would do no harm, but it would not change E.) It is easy to see that $\beta(x, y)$ is not analytic at $(1, 0)$, so $D = E = \mathbb{R}^{++} \setminus \{1\}$. By contrast, $\beta_F = 4 + 4x^2$ is analytic on $\mathbb{R}^{++}$. Since $\mathcal{D} = \{(0, 1), (1, \infty)\}$, we consider the intervals $(0, 1), (1, \infty)$ separately.

Note that we are taking $y = z^{d!}$ for specificity; in many cases a power less than $d!$ may work. In particular if, as in (19), $\Delta(\chi)$ is not divisible by y then $y = z$ already works.

Fix $D_0 \in \mathcal{D}$. The set $D_0 \subset (\mathbb{R}^{++})^{k-1}$ is then non-empty, open and connected. We shall also write D_0 for the set $D_0 \times \{0\} \subset (\mathbb{R}^+)^k$; it will always be clear from the context whether we are thinking of D_0 as a subset of $\mathbb{R}^{k-1}$ or as a subset of $\mathbb{R}^k$. Recall that $y = z^{d!}$, where d is the degree of β over R. Suppose we have a non-trivial function $f(x_1, \cdots, x_{k-1}, z)$ which is real-valued and analytic on D_0. For any $(a_1, \cdots, a_{k-1}) \in D_0$ we have a power series

$$\sum_{w \in (\mathbb{Z}^+)^k} a_w (x_1 - a_1)^{w_1} \cdots (x_{k-1} - a_{k-1})^{w_{k-1}} z^{w_k}$$

with $a_w \in \mathbb{R}$. The series is absolutely convergent in a neighborhood $\mathcal{O}$ of $(a_1, \cdots, a_{k-1}, 0)$ in $\mathbb{C}^k$ and converges to $f(x_1, \cdots, x_{k-1}, z)$ on $\mathcal{O} \cap (\mathbb{R}^+)^k$. Rearranging the terms and using analytic continuation, we can write

$$f(x_1, \cdots, x_{k-1}, z) = \sum_{n=n_0}^{\infty} f_n(x_1, \cdots, x_{k-1}) z^n,$$

where $n_0 \in \mathbb{Z}^+$, each f_n is analytic on D_0 and f_{n_0} is non-trivial. More generally, we allow f to be meromorphic in z so that we have an expansion

$$(*) \qquad f(x_1, \cdots, x_{k-1}, z) = \sum_{n=n_0}^{\infty} f_n(x_1, \cdots, x_{k-1}) z^n$$

with $n_0 \in \mathbb{Z}$, f_{n_0} non-trivial, each f_n analytic on D_0 and the (Laurent) series absolutely convergent on a neighborhood of D_0 in $\mathbb{C}^k$ (except, perhaps, on D_0). For such f, we define $\delta(f) = n_0$ and $f_F = f_{n_0} z^{n_0}$. For a polynomial $f \in \mathbb{Z}[x_1^{\pm}, \cdots, x_{k-1}^{\pm}, z^{\pm}]$, this notation is consistent with that of section 3, except that we are using z instead of $y = z^{d!}$. (See the discussion following (10).) As before, for the zero function $f = 0$ we take $\delta(f) = \infty$ and $f_F = 0$.

Writing $\tilde{R} = \mathbb{Z}[x_1^{\pm}, \cdots, x_{k-1}^{\pm}, z^{\pm}]$, note that $\tilde{\beta}$ and therefore elements of the ring $\tilde{R}[\tilde{\beta}]$ satisfy $(*)$. Moreover, using $y = z^{d!}$, the ring $R[\beta]$ may be embedded in $\tilde{R}[\tilde{\beta}]$. This gives meaning to f_F and $\delta(f)$ for $f \in R[\beta]$. When $f = \beta$, this is consistent with our earlier notation because $\beta(x_1, \cdots, x_{k-1}, 0) > 0$. Thus $\delta(\beta) = 0$,

$$\beta_F(x_1, \cdots, x_{k-1}) = \beta(x_1, \cdots, x_{k-1}, 0) = \tilde{\beta}(x_1, \cdots, x_{k-1}, 0) > 0$$

on D_0, and β_F is analytic on D_0. We pick a vector $r = r_A$ over $R[\beta]$ such that $Ar = \beta r$ and $r(x_1, \cdots, x_k) > 0$ for $(x_1, \cdots, x_k) \in (\mathbb{R}^{++})^k$. For example, we could take r to be any column of the adjoint $\mathrm{Adj}(\beta I - A)$. We define a vector r_F on D_0 by letting $r_F(I) = r(I)_F$ for each state I of A. Next we define a sub-matrix A_F of A in a way similar to the case of polynomial beta functions. We specify which edges of the weighted graph $G(A)$ to keep. For each state I we have

$$\beta \, r(I) = \sum_{I(e)=I} \mathrm{wt}_A(e) \, r(T(e)).$$

Keep the edge e (and its weight $\text{wt}_A(e)$) if and only if

$$\delta\big(\text{wt}_A(e)\, r(T(e))\big) = \delta\big(\beta\, r(I(e))\big)\,.$$

By definition, A_F is the R^+-matrix corresponding to the resulting weighted graph. Observe that $(A^n)_F = (A_F)^n$. The following are analogues of (5) – (8) of section 3. The proofs of (20) and (21) are just like those of (5) and (6); we just have to remember we are putting $x_k = y = z^{d!}$ and working with functions which satisfy (∗) on D_0.

20. THEOREM.

   (a) A_F and r_F have the same indexing set as A.

   (b) No row of A_F is trivial.

   (c) No entry of r_F is trivial and $r_F(x_1, \cdots, x_{k-1}) \geq 0$ for $(x_1, \cdots, x_{k-1}) \in D_0$.

   (d) $A_F\, r_F = \beta_F\, r_F$ on D_0. In particular, $\beta_{A_F} = \beta_F$ on D_0.

   (e) A cycle γ of $G(A)$ belongs to $G(A_F)$ if and only if $\text{wps}_A(\gamma) \in F$. Consequently, $A_{F,1}, \cdots, A_{F,\ell}$ are precisely the irreducible components of A_F.

21. PROPOSITION. A_F does not depend on the choice of the right eigenvector $r = r_A$ over $R[\beta]$. If $r' = r'_A$ is another right eigenvector over $R[\beta]$, then there exist $f, g \in R[\beta]$ such that $fr = gr'$ and $f_F\, r_F = g_F\, r'_F$.

22. THEOREM. Write $\text{Adj} = \text{Adj}(\beta I - A)$. Every entry $\text{Adj}(I, J)$ of the adjoint has the property that $\text{Adj}(I, J)_F > 0$ on D_0. Hence, any column of Adj presents a vector r such that $Ar = \beta\, r$, $r > 0$ on $(\mathbb{R}^{++})^k$, and $r_F > 0$ on D_0.

The proof of (22) will be described in section 5.

23. COROLLARY.

   (a) Among the components $A_{F,1}, \cdots, A_{F,\ell}$ of A_F, there is at least one whose beta function equals β_F on D_0.

   (b) For any component $A_{F,i}$ of A_F, either $\beta_{A_{F,i}} = \beta_F$ on D_0 or $\beta_{A_{F,i}} < \beta_F$ on D_0.

PROOF. Let $\chi_A(t) = \chi_A(t, x_1, \cdots, x_k)$ be the characteristic polynomial of A, and put

$$\chi_A^o(t) = \chi_A^o(t, x_1, \cdots, x_{k-1}) = \chi_A(t, x_1, \cdots, x_{k-1}, 0).$$

Letting $\chi_1, \cdots, \chi_n$ be the distinct irreducible factors of χ_A^o, consider $\bar{\chi}_A^o = \prod_{j=1}^n \chi_j$. Then the discriminant $\Delta = \Delta(\bar{\chi}_A^o)$ is a non-trivial element of $\mathbb{Z}[x_1^{\pm}, \cdots, x_{k-1}^{\pm}]$ and each $\beta_{A_{F,i}}$, $1 \leq i \leq \ell$, satisfies $\bar{\chi}_A^o$. Whenever $\Delta(x_1, \cdots, x_{k-1}) \neq 0$, distinct elements of $\{\beta_{A_{F,i}} : 1 \leq i \leq \ell\}$ take on distinct values at $(x_1, \cdots, x_{k-1})$. It follows that there exist $(x_1, \cdots, x_{k-1}) \in D_0$ and $1 \leq j \leq \ell$ such that $\beta_{A_{F,j}}$ is the unique element of $\{\beta_{A_{F,i}} : 1 \leq i \leq \ell\}$ which attains β_F at $(x_1, \cdots, x_{k-1})$. Then $\beta_F = \beta_{A_{F,j}}$ on a neighborhood of $(x_1, \cdots, x_{k-1})$ in D_0 and, since both $\beta_{A_{F,j}}$ and β_F are analytic on D_0, we have $\beta_F = \beta_{A_{F,j}}$ on D_0. This proves (a).

Part (b) is obtained from (20), (21), (22) and (2) in the same way that (9) was obtained from (5)–(8) and (2).   □

24. EXAMPLE. For an application of (23), consider again the beta function β of (19) and the face F lying on the x-axis. We have $\beta_F = 4 + 4x^2$ and $\mathcal{D} = \{(0,1),(1,\infty)\}$. So, if A is a matrix with $\beta_A = \beta$, then A_F must have a component whose beta function equals $4 + 4x^2$ and no component of A_F can have a beta function whose value equals (or exceeds) $4 + 4x^2$ for any $x \in \mathbb{R}^{++} \setminus \{1\}$. For example, $3 + 4x \leq 4 + 4x^2$ with equality when (and only when) $x = \frac{1}{2}$, and no component of A_F can have $3 + 4x$ for its beta function. On the other hand, there are examples of matrices A with $\beta_A = \beta$ such that A_F has components whose beta functions are $8x$ and $4\sqrt{2x(1+x^2)}$. (See (5.7) of [MT1].)

The reader may wish to take a look now at section 6, particularly at (32), for specific examples of faces A_F.

The following characterizes the components of $A_{F,j}$ with $\beta_F = \beta_{A_{F,j}}$ on D_0 for some $D_0 \in \mathcal{D}$.

25. PROPOSITION. *Consider the disjoint irreducible sub-matrices $A_{F,1}, \cdots,$ $A_{F,\ell}$ of A corresponding to F. For an element $A_{F,j}$ of this collection, the following are equivalent.*

(i) *There exists $D_0 \in \mathcal{D}$ such that $\beta_F = \beta_{A_{F,j}}$ on D_0.*

(ii) *There exists $(x_1, \cdots, x_{k-1}) \in (\mathbb{R}^{++})^{k-1}$ such that, for each $1 \leq i \leq \ell$, either $\beta_{A_{F,i}} = \beta_{A_{F,j}}$ as functions or $\beta_{A_{F,i}}(x_1, \cdots, x_{k-1}) < \beta_{A_{F,j}}(x_1, \cdots, x_{k-1})$. That is, $\beta_{A_{F,j}}$ is the unique element of the set $\{\beta_{A_{F,j}} : 1 \leq i \leq \ell\}$ which attains β_F at $(x_1, \cdots, x_{k-1})$.*

(iii) *$\beta_F = \beta_{A_{F,j}}$ on a non-empty open subset of $(\mathbb{R}^{++})^{k-1}$.*

PROOF. The proof of (23)(a) contains the fact that (i) implies (ii). That (ii) implies (iii) is clear. If (iii) holds then, since D is open and dense in $(\mathbb{R}^{++})^{k-1}$, we have $\beta_F = \beta_{A_{F,j}}$ on a non-empty open subset of a component D_0 of D. As $\beta_{A_{F,j}}$ and β_F are analytic on D_0, we conclude that $\beta_{A_{F,j}} = \beta_F$ on D_0. (We intentionally avoided the use of (22) and (23)(b).)   □

We will give a similar result for the conjugates of β restricted to the hyperplane $y = 0$. Let χ be the characteristic polynomial of β and let $\chi_0, \bar{\chi}_0$ be as defined between (17) and (18). Denote by n the degree of $\bar{\chi}_0(t) = \bar{\chi}_0(t, x_1, \cdots, x_{k-1})$ in t. For $(x_1, \cdots, x_{k-1}) \in \mathbb{C}^{k-1}$, let $\gamma_1(x_1, \cdots, x_{k-1}), \cdots, \gamma_n(x_1, \cdots, x_{k-1})$ be the zeros of $\bar{\chi}_0(t, x_1, \cdots, x_{k-1})$. By appropriately ordering the zeros, we assume that the functions of $\gamma_1, \cdots, \gamma_n$ are continuous on $\mathbb{C}^{k-1}$, and analytic whenever possible; in particular, they are analytic whenever we have $\Delta(\bar{\chi}_0)(x_1, \cdots, x_{k-1}) \neq 0$. We will only need those γ_i that are real-valued and analytic on $(\mathbb{R}^{++})^{k-1}$. Put $\mathcal{R} = \{\gamma_i : 1 \leq i \leq n \text{ and } \gamma_i \text{ is real analytic on } (\mathbb{R}^{++})^{k-1}\}$.

26. PROPOSITION. *$\mathcal{R}$ is non-empty and we have*

$$\beta_F = \max\{\gamma : \gamma \in \mathcal{R}\}$$

on $(\mathbb{R}^{++})^{k-1}$. For $\gamma \in \mathcal{R}$, the following are equivalent.

   (i)   *There exists $D_0 \in \mathcal{D}$ such that $\beta_F = \gamma$ on D_0.*
   (ii) *There exists $(x_1, \cdots, x_{k-1}) \in (\mathbb{R}^{++})^{k-1}$ such that $\bar{\gamma}(x_1, \cdots, x_{k-1}) <$ $\gamma(x_1, \cdots, x_{k-1})$ whenever $\bar{\gamma} \in \mathcal{R}$ and $\bar{\gamma} \neq \gamma$.*
   (iii) *$\beta_F = \gamma$ on a non-empty open subset of $(\mathbb{R}^{++})^{k-1}$.*

PROOF. As noted in (15), $\beta_F(x_1, \cdots, x_{k-1})$ satisfies χ_0. Therefore β_F satisfies $\bar{\chi}_0$ and, since β_F is analytic on each $D_0 \in \mathcal{D}$, there exists $1 \leq i \leq n$ such that $\beta_F = \gamma_i$ on D_0. Using (25), we also find $1 \leq j \leq \ell$ such that $\beta_F = \beta_{A_F, j} = \gamma_i$ on D_0. Since $\beta_{A_F, j}$ is real analytic on $(\mathbb{R}^{++})^{k-1}$, we conclude that $\gamma_i \in \mathcal{R}$ and $\beta_F = \max\{\gamma : \gamma \in \mathcal{R}\}$ on D_0. The fact that D is dense in $(\mathbb{R}^{++})^{k-1}$ and continuity then gives $\beta_F = \max\{\gamma : \gamma \in \mathcal{R}\}$ on $(\mathbb{R}^{++})^{k-1}$. The rest of the result is established by arguing as in the proofs of (23)(a) and (25), using $\bar{\chi}_0$ instead of $\bar{\chi}_A^o$.   □

Finally, we give the generalization of (11). Note that A may be F-adapted (actually, (F, D_0)-adapted) as in the remarks preceding (11), using a diagonal matrix whose diagonal is made up of the $z^{\delta(r(I))} = y^{\delta(r(I))/d!}$. (Similarly for B.) The proof of the general result then mimicks that of (11).

27. THEOREM. *Let A, B be irreducible matrices over $R^+ = \mathbb{Z}^+[x_1^\pm, \cdots, x_k^\pm]$ such that $\beta = \beta_A = \beta_B$. Let F be a face of the polytope $W(\beta)$, and let $D_0 \in \mathcal{D} = \mathcal{D}(\beta, F)$. Corresponding to F and D_0 we have the sub-matrices $A_F = A_{F, D_0}, B_F = B_{F, D_0}$ of A, B whose properties are described in (20). Every strong shift equivalence, shift equivalence, right closing, algebraic right closing of A onto B induces, respectively, a strong shift equivalence, shift equivalence, right closing, algebraic right closing of A_F onto B_F.*

As the statement of (27) makes clear, in the general case $A_F = A_{F, D_0}$ depends on both F and $D_0 \in \mathcal{D} = \mathcal{D}(\beta, F)$. For fixed F, each A_{F, D_0} has $A_{F, 1}, \cdots, A_{F, \ell}$ for irreducible components but, as the examples in section 6 will illustrate, the connections between these components vary with D_0, reflecting the changes in the roles played by the components. We also remark that each $D_0 \in \mathcal{D}(\beta, F)$ corresponds, via the Legendre transform, to an open connected subset of F.

The faces A_F are not, in general, respected by (bounded-to-one) block homomorphisms. In fact, it is easy to find algebraic left closings (that is, matrix equations $SA = BS, SR = B^L$ over R^+) that are not induced on the faces. This is a reflection of our use of a right eigenvector in the definition of A_F. An entirely similar construction can, of course, be carried out with a left eigenvector (or just consider the transpose A^T); this is then respected by (algebraic) left closings, shift equivalence and strong shift equivalence.

Most of the material in section 3, with the notable exception of (7)–(9), precedes [MT3]. In fact this material partly motivated [MT3], where the finite equivalence conjecture [PT1] is proved in the case of polynomial beta functions. The work given in the present section was done subsequently; Brian Marcus and I are in the process of investigating its implications for finite equivalence. In particular, it seems to be the case that β is a complete invariant of finite equivalence whenever its extension to $W(\beta)$ is analytic. We hope to discuss these developments in a future paper.

5. Positivity on domains of analyticity.

Let B be an irreducible matrix over R^+ and $\beta = \beta_B$. In this section, B will play the role of the matrix A of the preceding section. Other than this, we retain the notation of that section. In particular, F is a face of $W(\beta)$, and the variables $x_1, \cdots, x_{k-1}, y = x_k$ are such that F is obtained by putting $y = 0$. We fix $D_0 \in \mathcal{D} = \mathcal{D}(\beta, F)$. We will restrict our attention to functions f which, on D_0, satisfy (*) of section 4. We will consider the property

$$(\#) \qquad\qquad\qquad f_F > 0 \text{ on } D_0$$

and indicate a proof of (22). In fact, we sketch a proof of:

28. THEOREM. *Let B be an irreducible matrix over R^+ and, writing $\beta = \beta_B$, let F be a face of $W(\beta)$. Fix $D_0 \in \mathcal{D}(\beta, F)$. Suppose A is a sub-matrix of B.*

   (a) *If $A \neq B$, then $\chi_A(\beta) = \det(\beta I - A)$ satisfies (#).*
   (b) *If $A \neq 0$, every entry of the adjoint $\mathrm{Adj}(\beta I - A)$ satisfies (#).*

(28) generalizes theorem 1 of [MT3]; its proof mimics that of theorem 1 of [MT3], using power series instead of polynomials. We shall state the main steps in our present notation and, in each case, name in brackets the corresponding result in [MT3].

For a positive function f satisfying (*) and $(a_1, \cdots, a_{k-1}) \in (\mathbb{R}^{++})^{k-1}$, expand f_F as a power series in $(x_1 - a_1), \cdots, (x_{k-1} - a_{k-1})$. Let d be the lowest (total) degree of the terms which have a non-zero coefficient in the power series, and let f_{Fd} be the homogeneous polynomial consisting of all degree d terms of f_F (where the degree d terms retain their coefficients). Then:

29. LEMMA. *(Lemma 1). There exists $\epsilon > 0$ such that $f_{Fd}(x_1, \cdots, x_{k-1}) \geq 0$ whenever $0 \leq x_1 - a_1, \cdots, x_{k-1} - a_{k-1} \leq \epsilon$.*

30. THEOREM. *(Theorem 2). Let B be an irreducible $n \times n$ matrix over R^+ and put $\beta = \beta_B$. Let r be a vector over $R[\beta]$ such that $Br = \beta r$ and $r > 0$ on $(\mathbb{R}^{++})^k$. Suppose that for a state I of B, a non-empty subset $T \subset \{1, \cdots, n\}$, functions $U(J), J \in T$, which satisfy (*) and (#), and a function f satisfying (*), we have*

$$f\, r(I) = \sum_{J \in T} U(J)\, r(J).$$

Then f satisfies (#).

The crucial observation in the proof of (30) is that the entries of r_F have the same zeros and the same order at these zeros:

31. LEMMA. *(Lemma 2). Suppose that the state I_0 and $(a_1, \cdots, a_{k-1}) \in$ $(\mathbb{R}^{++})^{k-1}$ are such that $r_F(I_0)(a_1, \cdots, a_{k-1}) = 0$, and let $d(I_0)$ be the order of $(a_1, \cdots, a_{k-1})$ as a zero of $r(I_0)$. Then $r_F(J)(a_1, \cdots, a_{k-1}) = 0$ and the order $d(J) = d(I_0)$ for all $1 \le J \le n$.*

The proofs of (29)–(31) are just like those of the corresponding results of [MT3]. These appear in section 3 of that paper; the rest of the argument, given in section 4, is algebraic. Once we have (30), the rest of the proof lifts verbatim to the present setting. We need only observe that sums and products of functions satisfying ($*$) and (#) satisfy ($*$) and (#). The proof of (28) is then a combination of this fact, (30), and the inductive decomposition of the adjoint used in section 4 of [MT3].

Everything in this section is valid, as in [MT3], in the more general setting of matrices whose non-zero entries are Handelman polynomials.

6. Examples.

We will present matrices through labelled directed graphs. Each edge of the graph will be labelled by an element of R^+, specifying the corresponding entry of the matrix. We will sometimes label the states with the corresponding entries of a right eigenvector for the beta function of the matrix. The intent of the first example is to illustrate the construction of faces.

32. EXAMPLE. Taking $p(x) = 2$, $q(x) = 2x$ in (16), we have

$$\beta = 1 + x + \sqrt{(1-x)^2 + y}.$$

We consider the face F obtained by putting $y = 0$. In the notation of section 4, the minimal polynomial χ of β has $\Delta(\chi) = 4(1-x)^2 + 4y$, so that $Q_0 = \Delta(\chi_0) = 4(1-x)^2$. Therefore, by 18, $\beta(x, y)$ is analytic at the point $(x, 0)$ whenever $x \in \mathbb{R}^{++} \setminus \{1\}$. It is easy to see that $\beta(x, y)$ is not analytic at $(1, 0)$. Hence, putting $D_0 = (0, 1)$, $D_1 = (1, \infty)$, we have $\mathcal{D} = \{D_0, D_1\}$. We shall compute A_{F, D_0} and A_{F, D_1} for the matrix A corresponding to the following graph.

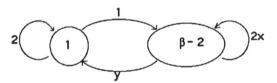

Order the states so that the right eigenvector is given by $(1, \beta - 2)^T$. First consider $D_0 = (0, 1)$. Then $\beta_F = 2$ and partial differentiation shows that the expansion ($*$) of β on D_0 is given by:

$$\beta = 2 + \frac{1}{2}\frac{1}{1-x}y - \frac{1}{8}\frac{1}{(1-x)^3}y^2 + \cdots .$$

Hence, $\delta(\beta-2)=1$ and $(\beta-2)_F=\frac{1}{2}\frac{1}{1-x}y$. It follows that $r_F=\left(1,\frac{1}{2}\frac{1}{1-x}y\right)^T$. At state 1, the equation $\beta r=Ar$ amounts to

$$\beta\cdot 1=2\cdot 1+1\cdot(\beta-2).$$

As $\delta(\beta r(1))=\delta(\beta)=0$ and $\delta(A(1,1)\,r(1))=\delta(2)=0$, $\delta(A(1,2)\,r(2))=\delta(\beta-2)=1$, in forming A_F we discard the edge from 1 to 2. At state 2, the equation $\beta r=Ar$ is

$$\beta\cdot(\beta-2)=2x\cdot(\beta-2)+2\cdot y.$$

Since $\delta(\beta(\beta-2))=\delta(2x(\beta-2))=\delta(2y)=1$, we retain all the edges coming out of 2. Therefore, A_{F,D_0} is given by:

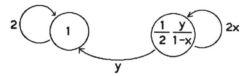

Multiplying r_F by $2(1-x)$, we obtain a simpler eigenvector of A_{F,D_0}:

If, in addition, we use the degrees $\delta(1)=\delta(2(1-x))=0$ and $\delta(\beta-2)=\delta(y)=1$ to F-adapt by conjugating with $\begin{bmatrix}1&0\\0&y\end{bmatrix}$, we obtain:

On $D_1=(1,\infty)$, we have $\beta_F=2x$ and the expansion $(*)$ takes the form

$$\beta=2x+\frac{1}{2}\frac{1}{x-1}y-\frac{1}{8}\frac{1}{(x-1)^3}y^2+\cdots.$$

Now $r_F=(1,2(x-1))^T$ and A is already (F,D_1)-adapted. Therefore, A_{F,D_1} is simply the matrix obtained from A by putting $y=0$:

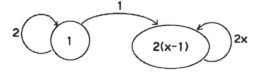

The difference between A_{F,D_0} and A_{F,D_1} reflects the switch in the roles of the two components given by the 1×1 matrices (2) and $(2\mathrm{x})$. When we go from

D_0 to D_1, the function β_F changes from 2 to $2x$ and, accordingly, the principal component becomes non-principal and vice versa.

33. EXAMPLE. With β, F, D_0, D_1 as in (32), we now consider the matrix A associated with the following graph.

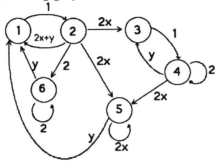

Then $\beta_A = \beta$ and, with the given ordering of states,

$$r = r_A = (1, \beta, 1, \beta, \beta - 2, \beta - 2x)^T$$

is a corresponding right eigenvector. We put $B = A^T$. A right eigenvector for B is

$$r_B = (y\beta, y, y, \beta - 2x, 2x(\beta - 1), 2\beta - 4x).$$

We will use F to dynamically distinguish A from its transpose B. Either of the domains D_0, D_1 can be used for this purpose; let us use $D_0 = (0, 1)$. Then, as in (32),

$$\beta = 2 + \frac{1}{2}\frac{1}{1-x}y - \frac{1}{8}\frac{1}{(1-x)^2}y^2 + \cdots,$$

and it is easy to check that $A_F = A_{F,D_0}$ and $B_F = B_{F,D_0}$ are given by the following graphs.

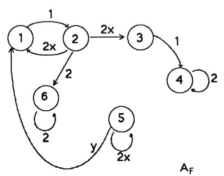

A_F

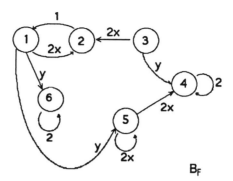

B_F

Recall that, in general, $(A_F)^n = (A^n)_F$. We claim that there is no $n \in \mathbb{Z}^{++}$ for which there exists a (weight-preserving) bounded-to-one block homomorphism from one of $\Sigma_{A_F^n}, \Sigma_{B_F^n}$ onto the other. To see this, note that, in each of $\Sigma_{A_F^n}$ and $\Sigma_{B_F^n}$, the periodic orbits of the form $(12)^\infty$ are the only ones of weight-per-symbol $x^{1/2}$. So, a block homomorphism must map the subshift

of one system onto the same subshift of the other. Similarly, the two-shift

in one system must map onto the same subshift in the other. But $\Sigma_{A_F^n}$ contains points that are left asymptotic to points of the form 5^∞ and right asymptotic to points of the form $(12)^\infty$ and $\Sigma_{B_F^n}$ has no such points. Hence, there is no bounded-to-one block homomorphism from $\Sigma_{A_F^n}$ to $\Sigma_{B_F^n}$. Likewise, $\Sigma_{B_F^n}$ has points that are left asymptotic to those of the form $(12)^\infty$ and right asymptotic to those of the form 5^∞ while $\Sigma_{A_F^n}$ has no such points, and there can be no bounded-to-one block homomorphisms from $\Sigma_{B_F^n}$ to $\Sigma_{A_F^n}$. In view of (27), it follows that neither A^n right closes onto B^n nor B^n onto A^n. In particular, A and B are not shift equivalent or block isomorphic.*

34. EXAMPLE. Now consider the matrices A, B given by the graphs below. These matrices were brought up by Mike Boyle and Jack Wagoner in relation to the finite order generation (FOG) conjecture for automorphism groups. Boyle and Wagoner asked if A, B are block isomorphic. The distinction of A, B motivated David Handelman's talk at the Yale symbolic dynamics conference for Roy Adler.

---

*(Added in proof.) It can be deduced from the argument given here that there is no non-trivial matrix S over $\mathbb{Z}^+[x^\pm, y^\pm]$ which satisfies the equation $AS = SB$. This will be discussed elsewhere.

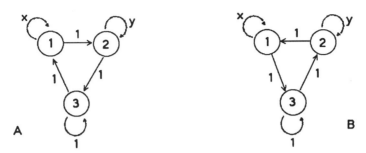

Clearly, $B = A^T$. We will distinguish A, B with the aid of faces, but without explicit calculation of their beta function $\beta = \beta_A = \beta_B$ or of their eigenvectors. The weight-per-symbol polytope $W(A)$ is easy to determine; $W(\beta) = W(A) = W(B)$ is the following triangle.

We will be interested in the slanted edge F with endpoints x, y. Corresponding to F, there are two components:

Using this and the density of the set D of analyticity of β on F, we see that D must have a component D_0 contained in $\{(x,y): x < y\}$. On D_0 the fixed point labelled y becomes the dominant component and that labelled x forms a non-dominant component. As a consequence of (20), each state of A must appear in $A_F = A_{F,D_0}$ and must have a path leading it to the state I, where the sole dominant component is. In addition, all cycles of A_F must have weight-per-symbol in F, and we are left with only one possible choice for A_F. A similar argument applies to $B_F = B_{F,D_0}$. The corresponding graphs are as follows.

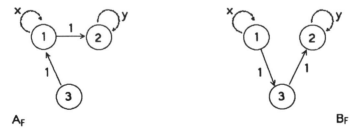

Since A, B have no multiple edges, there is a one-to-one correspondence between sequences of edges and sequences of states. For simplicity, we will use sequences of states to indicate points of Σ_A, Σ_B. Suppose that, for some $N \in \mathbb{Z}^{++}$, there is a right closing of A^N onto B^N. In view of the above graphs and (27), points of the form $1^\infty 2^\infty$ must be sent to points of the form $1^\infty 3\,2^\infty$. For large n, consider the periodic orbit of Σ_{A^N} given by the cycle $1^{nN}\,2^{nN-2}\,331$. This must be sent to a periodic orbit of Σ_B with the same weight-per-symbol. Hence, the corresponding cycle η must travel a total of four times along edges with weight 1 and use the central 3-cycle. This means γ visits each of the four edges of weight 1 precisely once. Since $1^\infty 2^\infty$ is sent to $1^\infty 3\,2^\infty$, the cycle η must contain a path of the form $1^{n_1}3\,2^{n_2}$. But this forces η to pass state 3 without visiting the self loop based there and, as the central 3-cycle is to be used only once, it cannot come back to state 3! This contradiction shows that A^N does not right close onto B^N. In particular, A, B are not shift equivalent or block isomorphic. A similar argument shows that B^N does not right close onto A^N.

Finally, we observe that a step of [B] may be viewed in our context. As a step in exhibiting two matrices A, B which, over R^+, are shift equivalent but not strong shift equivalent, Boyle [B] shows that the sub-matrices

$$(\dagger) \qquad \begin{bmatrix} aU & 0 \\ dB & eU \end{bmatrix}, \ \begin{bmatrix} aU & 0 \\ dI & eU \end{bmatrix}$$

must be respected by a block isomorphism. There are 9 variables, named $a, b, \cdots, i$. Consider the face F obtained by putting $i = 0$ and $D_0 \in \mathcal{D}(\beta, F)$ which intersects the subset of $(\mathbb{R}^{++})^8$ on which a exceeds the other variables. Then, as in (34), we see without computing β or right eigenvectors that aU is the only dominant component and the matrices $(\dagger)$ are sub-matrices of $A_F = A_{F,D_0}$ and $B_F = B_{F,D_0}$, so that (27) applies.

References

[AM] R. Adler and B. Marcus, *Topological entropy and equivalence of dynamical systems*, Mem. Amer. Math. Soc. **219** (1979).

[BP] A. Berman and R. Plemmons, *Nonnegative Matrices in the Mathematical Sciences*, Academic Press, New York, San Francisco, London, 1979.

[B] M. Boyle, *The stochastic shift equivalence conjecture is false*, preprint.

[BKM] M. Boyle, B. Kitchens and B. Marcus, *A note on minimal covers for sofic systems*, Proc. Amer. Math. Soc. **95** (1985), 403–411.

[BMT] M. Boyle, B. Marcus, and P. Trow, *Resolving maps and the dimension group for shifts of finite type*, Mem. Amer. Math. Soc. **377** (1987).

[BT] M. Boyle and S. Tuncel, *Infinite-to-one codes and Markov measures*, Trans. Amer. Math. Soc. **285** (1984), 657–684.

[D] V. De Angelis, Ph.D. thesis, University of Washington, in preparation.

[GR] R. Gunning and H. Rossi, *Analytic Functions of Several Complex Variables*, Prentice-Hall, Englewood Cliffs, New Jersey, 1965.

[H] D. Handelman, *Positive polynomials and product type actions of compact groups*, Mem. Amer. Math. Soc. **320** (1985).

[Hi] E. Hille, *Analytic Function Theory, Vol. 2*, Chelsea, New York, 1962.

[Ka] T. Kato, *Perturbation Theory for Linear Operators*, Springer-Verlag, Berlin, 1980.

[Ki] B. Kitchens, *Linear algebra and subshifts of finite type*, Contemp. Math. 26, Amer. Math. Soc., Providence, 1981, pp. 231–248.

[MT1] B. Marcus and S. Tuncel, *The weight-per-symbol polytope and scaffolds of invariants associated with Markov chains*, Ergod. Th. and Dynam. Sys. **11** (1991), 129–180.

[MT2] _____, *Entropy at a weight-per-symbol and embeddings of Markov chains*, Invent. Math. **102** (1990), 235–266.

[MT3] _____, *Matrices of polynomials, positivity, and finite equivalence of Markov chains*, preprint.

[PT1] W. Parry and S. Tuncel, *On the classification of Markov chains by finite equivalence*, Ergod. Th. and Dynam. Sys. **1** (1981), 303–335.

[PT2] _____, *On the stochastic and topological structure of Markov chains*, Bull. London Math. Soc. **14** (1982), 16–27.

[PT3] _____, *Classification Problems in Ergodic Theory*, LMS Lecture Notes 67, Cambridge Univ. Press, Cambridge, 1982.

[S] E. Seneta, *Non-negative Matrices and Markov Chains*, Springer-Verlag, Berlin, Heidelberg and New York, 1981.

[T1] S. Tuncel, *Conditional pressure and coding*, Israel J. Math. **39** (1981), 101–112.

[T2] _____, *A dimension, dimension modules, and Markov chains*, Proc. London Math. Soc. **46** (1983), 100–116.

[W] H. Whitney, *Elementary structure of real algebraic varieties*, Ann. Math. **66** (1957), 545–556.

[Wi] R. Williams, *Classification of shifts of finite type*, Ann. Math. **98**, 120–153;Errata, Ann. Math. **99** (1974), 380–381.

DEPARTMENT OF MATHEMATICS, UNIVERSITY OF WASHINGTON, SEATTLE, WA 98195

Contemporary Mathematics
Volume **135**, 1992

Classification of subshifts of finite type revisited

J.B. Wagoner

Dedicated to Roy Adler

Summary : This article describes developments over the last several years leading to the recent Kim-Roush example [KR4] showing that shift equivalence does *not* imply strong shift equivalence over the nonnegative integers for reducible subshifts of finite type with two primitive components. The problem is still open for primitive subshifts of finite type as of 15 January 1992 . M.Boyle has used the Kim-Roush example to show in [B] that eventual conjugacy does *not* imply conjugacy for primitive Markov chains where the measure is not the one of maximal entropy.

1 The classification problem

Subshifts of finite type (SFT's) appear in such fields as smooth dymnamics, ergodic theory, information theory , and statistical mechanics. See [DGS, Fr, PT]. In smooth dynamics, they are models for zero-dimensional basic sets of Axiom A diffeomorphisms. A topological conjugacy between the diffeomorphism restricted to one of its basic sets and a SFT (X_A, σ_A) constructed in the usual way [Fr] from a nonnegative integral matrix A can be obtained by using a Markov partition on the basic set. But there are many ways of choosing such Markov partitions, and this gives different matrices A such that the SFT's (X_A, σ_A) are all topologically conjugate. So

Partially supported by NSF Grant DMS 8801333

AMS(MOS) Subject Classification(1980) : 20B27 , 54H20 , 55D20

This paper is in final form. No version of it will be submitted for publication elsewhere.

1991 Mathematics Subject Classification. Primary 20B27, 54H20, 55D20.

this brings up the problem of trying to decide when two SFT's (X_A, σ_A) and (X_B, σ_B) are *conjugate*. That is, when does there exist a homeomorphism $H : X_A \longrightarrow X_B$ satisfying $\sigma_A H = H \sigma_B$? A weaker notion is *eventual conjugacy*. Namely, when does there exist a positive integer N so that for all $k \geq N$ the SFT's (X_A, σ_A^k) and (X_B, σ_B^k) are conjugate ?

In his very influential papers [Wi1, Wi2] , R.F.Williams introduced the concepts of strong shift equivalence and shift equivalence in order to give algebraic formulations of conjugacy and eventual conjugacy, respectively. An *elementary strong shift equivalence* $(R, S) : A \longrightarrow B$ between two square nonnegative integral matrices A and B is a pair of nonnegative integral matrices R and S satisfying

$$(1.1) \qquad\qquad A = RS \quad , \quad B = SR$$

Then A and B are said to be *strong shift equivalent* over the nonnegative integers Z^+ provided there is a chain of elementary strong shift equivalences over Z^+ from A to B. We will often write SSE for strong shift equivalence. Incidentally, the matrices appearing in (1.1) and , in fact , *all* the matrices appearing in this paper will be of finite size. They will also have the property that each row and each column has at least one non-zero entry.

Theorem 1.2 (Williams) *The shifts (X_A, σ_A) and (X_B, σ_B) are conjugate iff the matrices A and B are strong shift equivalent over the nonnegative integers.*

See [Fr, Wi2, W1] and the discussion below. The main difficulty is that SSE over Z^+ is very hard to determine.

A *shift equivalence* $R : A \longrightarrow B$ over the nonnegative integers Z^+ from A to B is a nonnegative integral matrix R with the property that there is a nonnegative integral matrix S and a positive integer k such that

$$(1.3) \qquad \begin{array}{rclcrcl} AR & = & RB & , & BS & = & SA \\ A^k & = & RS & , & B^k & = & SR \end{array}$$

The integer k is the *lag* and $k = 1$ corresponds to an elementary strong shift equivalence. We will often write SE for shift equivalence. The dynamical significance of shift equivalence over Z^+ is given by

Theorem 1.4 (Williams and Kim-Roush) *The shifts (X_A, σ_A) and (X_B, σ_B) are eventually conjugate iff the matrices A and B are shift equivalent over the nonnegative integers.*

See [KR1, Wi2] . Shift equivalence over Z^+ is much more tractable than strong shift equivalence over Z^+ . For example, Krieger proved [K, E] that A and B are SE over Z^+ iff their dimension groups are isomorphic ,and Kim-Roush proved in [KR1] that SE over Z^+ is decidable . It is easy to verify that strong shift equivalence over Z^+ implies shift equivalence over Z^+ . Williams' Problem of some 17 years standing asks

(SHIFT) *Does SE imply SSE over Z^+ ?*

Nonnegativity is important here , because Effros and Williams proved that SE does imply SSE over the integers Z . See [W4]. An affirmative partial result for certain 2x2 primitive matrices was obtained by Cuntz-Krieger in [CK]. Another result in the positive direction ,due to Kim-Roush in [KR3],is that SE implies SSE in the category of matrices over the nonnegative rationals Q^+ when A and B have only one nonzero eigenvalue .

Recently Kim-Roush gave the first counterexamples to SHIFT. See [KR4]. However, the examples are for reducible matrices with two mixing components , and as far as we know, SHIFT is still open for irreducible or even primitive matrices A and B . In the remainder of this section we will briefly discuss the primitive case and will present the Kim-Roush example below in (1.8). The main part of the paper will explain some of the developments which took place in the 1980's and which are the ingredients of the example. Then we give applications of these methods , including the short and ingenious proof of (1.8) due to Kim-Roush.

Possible Primitive Examples . For each $n \geq 2$, let

$$(1.5) \quad A_n = \begin{pmatrix} 1 & n(n-1) \\ 1 & 1 \end{pmatrix} , \quad B_n = \begin{pmatrix} 1 & n \\ n-1 & 1 \end{pmatrix}$$

and let

$$R_n = \begin{pmatrix} n-1 & n \\ 1 & 1 \end{pmatrix}$$

Then $A_n R_n = R_n B_n$ and A_n is SE over Z^+ to B_n . Note that $A_2 = B_2$. Also, Kirby Baker has shown A_3 is SSE over Z^+ to B_3 by exhibiting a chain of some 27 elementary SSE's between them, some of which involve 4×4 . At the present time it is not known whether A_n is SSE over Z^+ to B_n for $n \geq 4$. These examples are part of the folklore in the field and are not covered by the Cuntz-Krieger results in [CK].

The Kim-Roush Counterexamples [KR4]. Start by setting

(1.6) $$U = \begin{pmatrix} 0 & 0 & 1 & 1 \\ 1 & 0 & 0 & 0 \\ 0 & 1 & 0 & 0 \\ 0 & 0 & 1 & 0 \end{pmatrix}$$

and then let

(1.7) $$A = \begin{pmatrix} U & 0 \\ I & U \end{pmatrix} \quad \text{and} \quad B = \begin{pmatrix} U & 0 \\ U^n(U-I) & U \end{pmatrix}$$

Direct machine computation shows that B is nonnegative for $n \geq 23$.

Proposition 1.8 *If $n \geq 23$, then A and B are shift equivalent over Z^+ but are not strong shift equivalent over Z^+ .*

To check that A is shift equivalent to B over Z^+ , let

$$C = \begin{pmatrix} I & 0 \\ 0 & U^n(U-I) \end{pmatrix}$$

for $n \geq 23$. Then $B = CAC^{-1}$, and so A and B are strong shift equivalent over Z using $R = C^{-1}$ and $S = CA$. For each $k \geq n$, we have the following shift equivalence equations between A and B over Z using $R = C^{-1}B^k$ and $S = CA$:

(1.9) $$\begin{array}{rclcrcl} A(C^{-1}B^k) & = & (C^{-1}B^k)B & , & (CA)A & = & B(CA) \\ (C^{-1}B^k)CA & = & A^{k+1} & , & (CA)(C^{-1}B^k) & = & B^{k+1} \end{array}$$

The matrix CA is nonnegative for $n \geq 23$. Consider the matrix

$$C^{-1}B^k = \begin{pmatrix} U^k & 0 \\ kU^{k-1} & U^{k-n}(U-I)^{-1} \end{pmatrix}$$

The charateristic polynomial of U is $t^4 - t - 1$, and therefore

$$(U-I)^{-1} = U^3 + U^2 + U \geq 0$$

Hence $C^{-1}B^k$ is nonnegative for $k \geq n \geq 23$.

The four main ingredients which led up to the Kim-Roush example and which we will discuss below are

(A) Automorphisms and strong shift equivalence theory.
    See [W1, W2, W3, W4]
(B) The dimension group representation . See [BLR, K]
(C) Sign and gyration numbers . See [BK]
(D) Positivity methods and the sign-gyration-compatibility-
    condition homomorphism . See [H, KR2, KRW]

2 Automorphisms and strong shift equivalence theory

The automorphism group $Aut(\sigma_A)$ of a subshift of finite type (X_A, σ_A) consists of those homeomorphisms $H : X_A \to X_A$ satisfying $\sigma_A H = H\sigma_A$. The automorphism group $Aut(\sigma_2)$ of the Bernoulli 2-shift was studied in the 1960's by Hedlund and coworkers [He], and in the 1980's there was a renewal of interest in $Aut(\sigma_A)$, which usually is a very large countable group. For example, if A is primitive, then $Aut(\sigma_A)$ contains the direct sum of any countable collection of finite groups or of infinite cyclic groups [BLR] . More generally, Kim-Roush have shown it contains an isomorphic copy of any locally finite countable residually finite group . The automorphism group plays a key role in our present understanding of the classification problem. Just as Williams' theory of strong shift equivalence and shift equivalence allows one to formulate conjugacy and eventual conjugacy in "algebraic" terms, it is also possible give an algebraic description of the whole group $Aut(\sigma_A)$.

A square, nonnegative integral matrix $A = (A_{ij})$ determines a directed graph where there are A_{ij} edges from the vertex i to the vertex j. The *edge path presentation* of the SFT (X_A, σ_A) defines X_A to be the set of all bi-infinite sequences (e_n) of edges in the graph such that the terminal vertex of e_n is the initial vertex of e_{n+1} [Fr]. Thus any self-isomorphism of the directed graph induces an automorphism of (X_A, σ_A). A *simple automorphism* of (X_A, σ_A) is one of the form $\alpha\beta\alpha^{-1}$ where α is a conjugacy from (X_A, σ_A) to (X_B, σ_B) and β is an automorphism of (X_B, σ_B) coming from an isomorphism of the directed graph of B fixing the vertices. Let $Simp(\sigma_A)$ denote the subgroup of $Aut(\sigma_A)$ generated by the simple automorphisms. This subgroup was singled out by Nasu in [N]. It is , of course, a normal subgroup and is generated by elements of finite order . When A is primitive , it is infinite and very rich [BK, BLR, N] .

Now let $(R, S) : A \longrightarrow B$ be an elementary SSE over Z^+ . This determines an *elementary conjugacy*

$$(2.1) \qquad c(R, S) : (X_A, \sigma_A) \longrightarrow (X_B, \sigma_B)$$

by the well known method of bipartite coding . See [W3] for example. In general, c(R,S) is only defined modulo multiplication on the left and right respectively by elements of $Simp(\sigma_A)$ and $Simp(\sigma_B)$ .If A,B,R,and S are all zero-one matrices , then there is no indeterminacy in c(R,S) . A more precise statement of the main result in [Wi2] is

Theorem 2.2 (Williams) *Every conjugacy α from (X_A, σ_A) to (X_B, σ_B) is of the form*

$$\alpha = \prod_i c(R_i, S_i)^{\epsilon_i}$$

where the (R_i, S_i) give a chain of elementary strong shift equivalences over Z^+ from A to B and where $\epsilon_i = \pm 1$.

In general , the product expression in (2.2) is certainly not unique, but we can ask whether there are some natural and universal relations which are satisfied . Consider the diagram

(2.3)

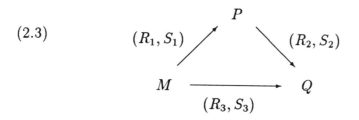

where *all* the matrices are zero-one. What are some conditions under which the composition of the two elementary conjugacies across the top is equal to the elementary conjugacy along the bottom ?

Proposition 2.4 *Assume that* $R_1 R_2 = R_3$. *Then*

$$c(R_3, S_3) = c(R_1, S_1)c(R_2, S_2)$$

if and only if

$$S_1 = R_2 S_3 \quad and \quad S_2 = S_3 R_1$$

See [W2] . So we write down the

Triangle Identities

(2.5)
$$\begin{aligned} R_3 &= R_1 R_2 \\ S_1 &= R_2 S_3 \\ S_2 &= S_3 R_1 \end{aligned}$$

Now let $\mathcal{E}$ denote the set of zero-one matrices and construct a locally finite CW-complex $RS(\mathcal{E})$ called the *space of strong shift equivalences* where

Vertices are square zero-one matrices A

Edges are elementary SSE's $(R, S) : A \longrightarrow B$ in $\mathcal{E}$

Triangles are diagrams like (2.3) for which the Triangle Identities hold

Etc. , ...

More precisely ,

Definition 2.6 $RS(\mathcal{E})$ is the realization of the simplicial set whose n-simplices are (n+1)-tuples $\langle A_0, A_1, \ldots, A_n \rangle$ where each A_i is in $\mathcal{E}$ together with SSE's $(R_{ij}, S_{ij}) : A_i \longrightarrow A_j$ in $\mathcal{E}$ for $i < j$ with the property that the Triangle Identities hold whenever $i < j < k$.

See [S] for background on simplicial sets .

Theorem 2.7 *(A) The set $\pi_0 RS(\mathcal{E})$ of path components of $RS(\mathcal{E})$ is the set of strong shift equivalence classes in $\mathcal{E}$.*
(B) There is an isomorphism $\pi_1 RS(\mathcal{E}) \cong Aut(\sigma_A)$.
(C) $\pi_i RS(\mathcal{E}) = 0$ for $i \geq 2$.

The fundamental group $\pi_1 RS(\mathcal{E})$ is taken with respect to a base point A. So it is really π_1 of the component of $RS(\mathcal{E})$ containing A . For simplicity, we have left this out of the notation and will do so throughout the remainder of the paper. Part(A) is merely a restatement of the definition of SSE in $\mathcal{E}$. Parts (B) and (C) are proved in [W2] and immediately imply that the path component of $RS(\mathcal{E})$ containing A is homotopy equivalent to the classifying space of the discrete group $Aut(\sigma_A)$. Such an independent description of the classifying space is often useful in studying Eilenberg-MacLane homology, and a step in this direction for $H_1 Aut(\sigma_A)$ is the discussed below in Section 5 where we construct the sign-gyration-compatibility-condition homomorphisms $SGCC_{A,m}$. The proof of (2.7) uses the action of the automorphism group on the contractible space of Markov partitions defined in [W1].

Next let $RS(Z^+)$ denote the *space of strong shift equivalences over the nonnegative integers* constructed just like $RS(\mathcal{E})$ but where all the matrices are allowed have nonnegative integral entries. In [W3] we proved

Theorem 2.8 *(A) The set $\pi_0 RS(Z^+)$ of path components of $RS(Z^+)$ is the set of strong shift equivalence classes over Z^+.*
(B) $\pi_1 RS(Z^+) \cong Aut(\sigma_A)$ modulo $Simp(\sigma_A)$

As above , Part(A) is a restatement of the definition of SSE over Z^+. It is part of the folkore and due to R.F.Willams that

(2.9) $$\pi_0 RS(\mathcal{E}) \longrightarrow \pi_0 RS(Z^+)$$

is a bijection of sets. See [W3] .

3   The dimension group representation

The dimension group is an invariant of eventual conjugacy which was introduced by Krieger in [K] using the unstable sets in the shift space X_A. Moreover, he used the action of the automorphisms on the unstable sets to obtain a representation of $Aut(\sigma_A)$ to the group of automorphisms of the dimension group. The purpose of this section is to give a different construction of this dimension group representation in terms of strong shift equivalence theory. Some useful references are [BLR, E, K, W1, W2, W3, W4].

Of the various ways to define the dimension group algebraically, here is the one which is the most concrete and direct for our purposes. See [BLR]. Let A be an $n \times n$ matrix over Z^+ and consider A as a linear transformation $A : Q^n \longrightarrow Q^n$. The *eventual range* R_A is the defined to be

$$R_A = A^p(Q^n)$$

where p is the least nonnegative integer such that A is an isomorphism from $A^p(Q^n)$ to $A^{p+1}(Q^n)$. Let the *dimension group* $G(A)$ be defined by the equation

$$G(A) = \{v \epsilon R_A | v A^k \epsilon Z^n \text{ for some } k \geq 0\}$$

and the set of *positive elements* be given by

$$G(A)_+ = \{v \epsilon R_A | v A^k \epsilon (Z^+)^n \text{ for some } k \geq 0\}$$

Let s_A denote the isomorphism of $G(A)$ induced by A. Then s_A is an automorphism of the ordered group $(G(A), G(A)_+)$.

Definition 3.1 The *automorphism group* $Aut(s_A)$ of the dimension group consists of those automorphisms of $G(A)$ which take $G(A)_+$ to itself and which commute with s_A .

Observe that a shift eqivalence $R : A \longrightarrow B$ over Z^+ induces an isomorphism

(3.2) $$g(R) : G(A) \longrightarrow G(B)$$

taking $G(A)_+$ to $G(B)_+$ and satisfying $s_A g(R) = g(R) s_B$. In particular, any shift equivalence $R : A \longrightarrow A$ over Z^+ gives rise to an element g(R) in $Aut(s_A)$.

The automorphism group $Aut(s_A)$ is often readily computed . For example, if A is invertible, then $G(A)$ is isomorphic to Z^n. Let λ be the Perron-Frobenius eigenvalue of A. Let w be a positive right eigenvector for A over the field $Q(\lambda)$. Suppose $det(tI - A)$ is irreducible over Q . Then the homomorphism θ taking v in $G(A)$ to the dot product vw imbedds $G(A)$ in the real numbers R with its usual ordering in such as way that the action of s_A corresponds to multiplication by λ . In many cases, θ induces isomorphisms

(3.3) $$\begin{aligned} G(A) &\cong Z[\lambda] \subset \mathbf{R} \\ Aut(s_A) &\cong \text{positive elements in the group} \\ &\quad \text{of units } Z[\lambda]^* \end{aligned}$$

See [BLR]. The simplest example of this is the one-by-one matrix $A = (n)$ where we get the full Bernoulli n-shift (X_n, σ_n) and have

$$\begin{aligned} G(A) &\cong Z[\tfrac{1}{n}] \subset \mathbf{R} \\ Aut(s_A) &\cong \text{positive elements in the group of units } Z[\tfrac{1}{n}]^* \\ &\cong \text{the free abelian multiplicative group} \\ &\quad \text{generated by the primes dividing n} \end{aligned}$$

Now we give a homotopy description of $Aut(s_A)$ as the fundamental group of the space $S(Z^+)$ of *shift equivalences over Z^+* where

Vertices are square matrices A over Z^+

Edges are shift equivalences $R : A \longrightarrow B$ over Z^+

Triangles are diagrams like

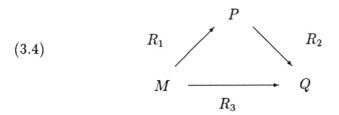

(3.4)

where $R_3 = R_1 R_2$.

Etc. , ...

More precisely ,

Definition 3.5 $S(Z^+)$ is the realization of the simplicial set whose n-simplices are (n+1)-tuples $\langle A_0, A_1, \ldots, A_n \rangle$ where each A_i is a square matrix over Z^+ together with shift equivalences $R_{ij} : A_i \longrightarrow A_j$ over Z^+ for $i < j$ with the property that whenever $i < j < k$ we have $R_{ik} = R_{ij} R_{jk}$.

Theorem 3.6 *(A) The set* $\pi_0 S(Z^+)$ *of path components of* $S(Z^+)$ *is the set of shift equivalence classes over* Z^+.
(B)There is an isomorphism $\pi_1 S(Z^+) \cong Aut(s_A)$ *obtained by taking* $R : P \longrightarrow Q$ *to* $g(R) : G(P) \longrightarrow G(Q)$

Part(A) is a restatement of the definition of SE over Z^+ . The proof of Part(B) is given in [W1] . But to indicate what is involved, we point out that the argument showing the map $\pi_1 S(Z^+) \longrightarrow Aut(s_A)$ is onto is essentially Krieger's proof in [E, K] showing that if the dimension group triples $(G(A), G(A)_+, s_A)$ and $(G(B), G(B)_+, s_B)$ are isomorphic, then A and B are shift equivalent over Z^+ .

The correspondence taking an edge $(R, S) : P \longrightarrow Q$ in $RS(Z^+)$ to the edge in $R : P \longrightarrow Q$ in $S(Z^+)$ yields a map of simplicial sets and gives rise to a natural map of spaces

$$RS(\mathcal{E}) \longrightarrow RS(Z^+) \longrightarrow S(Z^+)$$

We define the *dimension group representation*

(3.7) $$\delta_A : Aut(\sigma_A) \longrightarrow Aut(s_A)$$

to be the induced maps on fundamental groups

$$\delta_A : \pi_1 RS(\mathcal{E}) \longrightarrow \pi_1 RS(Z^+) \longrightarrow \pi_1 S(Z^+)$$

using the identifications $Aut(\sigma_A) \cong \pi_1 RS(Z^+)$ and $Aut(s_A) \cong \pi_1 S(Z^+)$ from (2.8) and (3.6) . This is compatible with Krieger's original definition in [K] .

If Λ is any ring , we can also form the spaces $RS(\Lambda)$ and $S(\Lambda)$ of strong shift equivalences and shift equivalences over Λ by allowing the matrices in (2.6) and (3.5) to have entries in Λ instead of just in $\mathcal{E}$ or in Z^+.The following is proved in [W4].

Theorem 3.8 *(A) If Λ is a principal ideal domain, then the natural map*

$$RS(\Lambda) \longrightarrow S(\Lambda)$$

is a homotopy equivalence . Moreover, $\pi_i RS(\Lambda) = 0$ for $i \geq 2$.
(B) $\pi_1 S(Z)$ is isomorphic to the group of automorphisms of the dimension group $G(A)$ which commute with s_A but which do not necessarily preserve $G(A)_+$.

If A is primitive, $\pi_1 S(Z) \cong \pi_1 S(Z^+) \oplus Z/2$.

Definition 3.9 The group of *inert automorphisms* $Aut_0(\sigma_A)$ is the kernel of δ_A .

$Aut(s_A)$ essentially a finite dimensional matrix group, whereas $Aut(\sigma_A)$ is generally huge . So most of the richness , size, and complexity of $Aut(\sigma_A)$ lies in $Aut_0(\sigma_A)$.

4   Sign and gyration numbers

We will briefly review the definitions in [BK] of the Boyle-Krieger sign and gyration number homomorphisms

$$OS_{A,m} : Aut(\sigma_A) \longrightarrow Z/2 \quad , \quad m \geq 1$$

and

$$GY_{A,m} : Aut(\sigma_A) \longrightarrow Z/m \quad , \quad m \geq 2.$$

Let $P_m^\bullet$ denote the set of periodic points of σ_A of period exactly m . Let $Aut(\sigma_A | P_m^\bullet)$ denote the group of permutations of $P_m^\bullet$ which commute with σ_A . Let σ denote σ_A restricted to $P_m^\bullet$. List the orbits of length m of σ_A and choose a base point x_i on the ith orbit . Now let α be an element of $Aut(\sigma_A | P_m^\bullet)$. Then $\alpha(x_i)$ will lie, say , on the jth orbit and we therefore have

$$\alpha(x_i) = \sigma^{k_i}(x_j)$$

for some integer k_i . Define

$$(4.1) \qquad gy(\alpha) = \sum_i k_i \qquad \mod m$$

Then $gy(\alpha)$ is does not depend on the various choices made and satisfies

$$gy(\alpha\beta) = gy(\alpha) + gy(\beta)$$

The *gyration number homomorphism* is defined to be the composition

$$(4.2) \qquad GY_{A,m} : Aut(\sigma_A) \longrightarrow Aut(\sigma_A | P_m^\bullet) \underset{gy}{\longrightarrow} Z/m$$

Example 4.3 Define α in $Aut(\sigma_A | P_m^\bullet)$ to be σ on a single orbit of length m and the identity on the other orbits of length m. Then $gy(\alpha) = 1 \mod m$.

Example 4.4 $GY_{A,m}(\sigma_A)$ is the number of orbits of length m .

If α is an element of $Aut(\sigma_A | P_m^\bullet)$, let

$$(4.5) \qquad os(\alpha) \quad = \quad \text{sign of } \alpha \text{ considered as a permutation} \atop \text{of the set of orbits of length m}$$

Then define the *orbit sign number homomorphism* to be the composition

$$(4.6) \qquad OSA_{,m} : Aut(\sigma_A) \longrightarrow Aut(\sigma_A | P_m^\bullet) \underset{os}{\longrightarrow} Z/2$$

Sign and gyration numbers have been used to study the action of $Aut(\sigma_A)$ on finite subsets of X_A .In particular, Boyle - Krieger [BK] discovered the "sign-gyration-compatibility-condition" , dubbed SGCC, which plays a key role in [F] and in the paper [BF] of Boyle-Fiebig characterizing the action of finite order inert automorphisms on periodic points of (X_A, σ_A) .

5   The sign-gyration-compatibility-condition homomorphism

The dimension group representation is essentially a finite dimensional matrix representation of $Aut(\sigma_A)$, and the sign and gyration number homomorphisms take $Aut(\sigma_A)$ into finite cyclic groups. The purpose of this section is to discuss how SGCC provides a link between these two constructions . In 1990 , Kim-Roush had the wonderful idea of turning the sign-gyration-compatibility- condition into the *sign-gyration-compatibility-condition homomorphism*

$$SGCC_{A,m} : Aut(\sigma_A) \longrightarrow Z/m$$

defined by the formula

(5.1) $$SGCC_{A,m} = GY_{A,m} + \sum_{i>0} OS_{A,m/2^i}$$

where we take $OS_{A,m/2^i} = 0$ whenever $m/2^i$ is not an integer and where,for m even , we consider $Z/2$ as the subgroup $\{0, m/2\}$ of Z/m . In particular, $SGCC_{A,m} = GY_{A,m}$ whenever m is odd , and

$$SGCC_{A,2} = GY_{A,2} + OS_{A,1}.$$

A basic fact proved by Nasu [N] is

(5.2) $SGCC_{A,m}$ vanishes identically on $Simp(\sigma_A)$

This was generalized by Fiebig , who showed that $SGCC_{A,m}$ vanishes on finite order inert automorphisms . Then in 1989 Kim-Roush [KR2] showed that $SGCC_{A,m}$ vanishes on all of $Aut_0(\sigma_A)$ when A is primitive . This implies that the value of $SGCC_{A,m}$ on any automorphism only depends on its image in $Aut(s_A)$. More generally, in [KRW] we prove

Theorem 5.3 (Kim-Roush-Wagoner) *Assume A is primitive.*
Then $SGCC_{A,m}$ factors through the dimension group representation.
That is, there is a homomorphism

$$sgcc_{A,m} : Aut(s_A) \longrightarrow Z/m$$

such that

$$SGCC_{A,m} = sgcc_{A,m} \circ \delta_A$$

The proof of this theorem uses "positivity methods" outlined in [KR2] , the first step of which is Handelman's theorem on eventually positive matrices [H]. Theorems (2.8) and (3.6) break the argument down into two stages

Step 1 . Show $SGCC_{A,m}$ factors through $\pi_1 RS(Z^+)$
Step 2 . Then show $SGCC_{A,m}$ factors through $\pi_1 RS(Z)$

Discussion of Step 1 .

Let $(R, S) : P \longrightarrow Q$ be an edge in $RS(Z^+)$. Choose an ordering of the vertices of the graphs P and Q . This determines a certain canonical lexicographical ordering of the periodic points of X_P and X_Q . So the base points on the orbits of X_P and X_Q can be chosen as the ones of least lexicographical order and the orbits can be ordered according to the order of their base points. This data allows us to then define the sign and gyration numbers of the elementary conjugacy $c(R, S) : X_P \longrightarrow X_Q$ similarly to the formulas (4.1) and (4.5). So it makes sense to define $SGCC_m(c(R, S))$ using the formula (5.1), and we let

$$(5.4) \qquad SGCC_m(R, S) = SGCC_m(c(R, S))$$

Once the orderings of the vertices for P and Q are chosen, the elementary conjugacy $c(R, S)$ is only defined modulo multiplication on the left and right respectively by elements of $Simp(\sigma_P)$ and $Simp(\sigma_Q)$. But using Nasu's result (5.2) , one shows that (5.4) does not change.

Now a closed loop α starting and ending at A in $RS(Z^+)$ gives a conjugacy

$$\alpha = \prod_i c(R_i, S_i)^{\epsilon_i}$$

as in (2.2), and we define

(5.5) $$SGCC_m(\alpha) = \sum_i \epsilon_i SGCC_m(R_i, S_i)$$

It follows from (2.4) and (5.2) that $SGCC_m(\alpha)$ is independent of the choices made in defining (5.4) and is independent of deformations of α using the Triangle Identities .

There is a polynomial formula for $SGCC_{A,m}$. For $m = 2$, it is

$$SGCC_2(R, S) =$$

(5.6)
$$\sum_{i<j,k>l} R_{ik} S_{ki} R_{jl} S_{lj} + \sum_{i<j,k \geq l} R_{ik} S_{kj} R_{jl} S_{li}$$
$$+ \sum_{i,k} \frac{R_{ik}(R_{ik} - 1)}{2} S_{ki}^2$$

taken modulo 2 .This implies that

$$SGCC_2(R_1 + mR_2, S_1 + nS_2) = SGCC_2(R_1, S_1)$$

whenever $m \equiv 0 \bmod 4$ and $n \equiv 0 \bmod 2$. See [KRW] .

Discussion of Step 2 .

To define $sgcc_m = sgcc_{A,m}$ on a closed loop α starting and ending at A in $RS(Z)$, the idea is to first convert α to a new closed loop $\alpha p(\alpha)$ in $RS(Z^+)$ corresponding to a certain nonnegative polynomial p(x) . Then we set

(5.7) $$sgcc_m(\alpha) = SGCC_m(\alpha p(\alpha))$$

and show that this definition does not depend upon p(x) nor upon deformation of α across triangles in $RS(Z)$ .This is where the "positivity methods" come into the picture.For simplicity , take the case where $m = 2$.One first uses methods of Handelman's theorem [H] to replace α by a new loop α in which, first of all, the edges (R, S) : $P \longrightarrow Q$ are *eventually positive* in the sense that there is some positive integer n so that the matrices P^n, Q^n, RQ^n,and SP^n are all strictly positive, and ,second of all , the matrices P,Q,R,S have rational entries with denominators congruent to 1 modulo 4 . Then one chooses a suitable polynomial

$$p(x) = (1 + 4d) \left(\frac{1 + 4a}{1 + 4b} + \frac{4c}{1 + 4b} x^n \right)$$

approximating $(1 + 4d)x^n$ sufficiently closely that $p(P)$, $p(Q)$, $Pp(P)$, $Qp(Q)$, $Rp(Q)$,and $Sp(P)$ are strictly positive and having the property that these matrices are also integral.Then the edges $(R, S) : P \longrightarrow Q$ are replaced by the edges

$$(Rp(Q), Sp(P)) : Pp(P)^2 \longrightarrow Qp(Q)^2$$

to form the new loop $\alpha p(\alpha)$. In particular, we have

$$sgcc_2(R, S) = SGCC_2(Rp(Q), Sp(P)).$$

Important Remark 5.8 The proof of (5.3) shows that if a closed loop in $RS(Z)$ has edges $(R, S) : P \longrightarrow Q$ which are already eventually positive, then $sgcc_2$ may be computed by first directly applying the formula (5.6) to compute the value $sgcc_2(R, S)$ for each edge and then adding up these contributions over all the edges in the loop.

6 Applications of $sgcc_{A,m}$

Example 6.1 **The dimension group representation is not always surjective . See [KRW] .**

Let U be the primitive 4×4 matrix given in (1.6). Let

$$R = U - I \text{ and } S = U^4 + U^3 + U^2$$

Since the charactristic polynomial of U is $t^4 - t - 1$,we have

$$U = RS = SR$$

So $(R, S) : U \longrightarrow U$ is a loop in $\pi_1 RS(Z)$. S is nonnegative and RU^n is nonnegative for $n \geq 23$. Hence R represents an element of $Aut(s_A)$. As observed in (5.8), we can apply the formula (5.6) to compute directly that $sgcc_{U,2}(R, S) = 1$. Hence, Theorem 5.3 shows R cannot be in the image of δ_A. This is because X_U has no fixed points nor orbits of length 2 , making $SGCC_{U,2}$ identically zero .

Example 6.2   Counterexamples to LIFT. See[KRW] .

One of the basic questions in studying the action of $Aut(\sigma_A)$ on periodic points of X_A has been whether the restriction homomorphism

$$Aut(\sigma_A) \longrightarrow Aut(\sigma_A | P_m)$$

is surjective where P_m denotes the fixed points of σ_A^m .It turns out the answer is no in general. The following example is closely related to one first proposed by U.Fiebig and confirmed by Kim-Roush in [KR2]. Let $\sigma_2 : X_2 \longrightarrow X_2$ be the Bernoulli 2-shift . Let α in $Aut(\sigma_2 | P_6)$ be constructed so that as in (4.3) it only shifts one orbit of period 6 and so that orbits of period less than 6 are pointwise fixed. If there were an element β in $Aut(\sigma_2)$ which restricted to α on P_6 , then from (4.3) we would have $SGCC_{2,6}(\beta) = 1 \bmod 6$. On the other hand, $Aut(s_A) \cong Z$ is generated by $s_2 = \delta_2(\sigma_2)$. From (4.4) and Theorem 5.3, we then have $sgcc_{2,6}(s_2) = SGCC_{2,6}(\sigma_2) = GY_{2,6}(\sigma_2) = 3 \bmod 6$. So α is not the restriction of any automorphism of the 2-shift .Incidentally, there is a misprint in (4.2) of [KRW] where this example is presented.The phrase "the period 6 points that shifts one orbit " should read " the points of P_6 that shifts one period 6 orbit " .

Here is another example from [KRW] .Let

$$A = \begin{pmatrix} 1 & 1 & 0 \\ 0 & 1 & 1 \\ 5 & 0 & 0 \end{pmatrix}$$

It is shown in [KRW] that $sgcc_{A,2}$ vanishes identically on $Aut(s_A)$. On the other hand , any automorphism α switching the two fixed points would have $SGCC_{A,2}(\alpha) = OS_{A,2}(\alpha) = 1$ because there are no period two orbits. Hence, Theorem 5.3 shows the fixed points cannot be interchanged by an element of $Aut(\sigma_A)$. It is also shown in [KRW] that it is not possible to switch the two fixed points of any SFT which is eventually conjugate to (X_A, σ_A) !

Example 6.3   The Kim-Roush counterexample to SE imples SSE for reducible matrices. See [KR4] .

Here is the Kim-Roush argument showing that the matrices A and B defined in (1.7) cannot be SSE over Z^+ . If there is a conjugacy between (X_A, σ_A) and (X_B, σ_B), it must induce an automorphism of dimension groups represented by an 8×8 matrix over Z of the form

$$\begin{pmatrix} R_{11} & 0 \\ R_{21} & R_{22} \end{pmatrix}$$

where the 4×4 matrices R_{ii} are invertible and represent elements of $Aut(s_U)$ which are induced by automorphisms of (X_U, σ_U) . We will reach a contradiction by showing that either $sgcc_{U,2}(R_{11}) = 1$ or $sgcc_{U,2}(R_{22}) = 1$,which cannot happen in view of Theorem 5.3 because (X_U, σ_U) has no fixed points nor points of period two.

The shift equivalence equation $AR = RB$ implies that

$$R_{11} + UR_{21} = R_{21}U + R_{22}B_{21}$$

or, alternatively,

$$UR_{21} - R_{21}U = R_{22}B_{21} - R_{11}.$$

In this case one computes directly that (3.3) holds .So $G(U) \cong Z[\lambda]$ in such a way that the action of U on $G(U)$ corresponds to multiplication by λ on $Z[\lambda]$. For any X in the number field $Q(\lambda)$, we have $XU = UX$. Hence,

$$\begin{aligned} 0 &= \mathrm{Trace}(XUR_{21}) - \mathrm{Trace}(UXR_{21}) \\ &= \mathrm{Trace}(XUR_{21}) - \mathrm{Trace}(R_{21}UX) \\ &= \mathrm{Trace}(X(UR_{21} - R_{21}U)X) \\ &= \mathrm{Trace}((UR_{21} - R_{21}U)X^2) \\ &= \mathrm{Trace}((R_{22}B_{21} - R_{11})X^2) \end{aligned}$$

Since $Q(\lambda)$ has a real imbedding, the values $\pm X^2$ are dense in $Q(\lambda)$. We therefore conclude that

$$R_{22}B_{21} - R_{11} = 0$$

by the non-degeneracy of Trace on number fields. Hence

$$B_{21} = R_{11}R_{22}^{-1}.$$

In Example 6.1 we observed that $sgcc_{U,2}(U-1) = 1$ .We also have $sgcc_{U,2}(U^n) = SGCC_{U,2}(\sigma_A^n) = 0$, because (X_U, σ_U) has no fixed points nor points of period two. So

$$
\begin{aligned}
1 &= sgcc_{U,2}(U-1) = sgcc_{U,2}(U^n(U-1)) \\
&= sgcc_{U,2}(B_{21}) \\
&= sgcc_{U,2}(R_{11}R_{22}^{-1}) \\
&= sgcc_{U,2}(R_{11}) - sgcc_{U,2}(R_{22})
\end{aligned}
$$

Therefore either $sgcc_{U,2}(R_{11}) = 1$ or $sgcc_{U,2}(R_{22}) = 1$, which is impossible as explained above .

References

[B]   M.Boyle ,*The stochastic shift equivalence conjecture is false,* these proceedings

[BF]   M.Boyle and U.Fiebig ,*The action of inert finite order automorphisms on finite subsystems,* ETDS **11** (1991) , 413-425

[BK]   M.Boyle and W.Krieger , *Periodic points and automorphisms of the shift* , TAMS **302** (1987) , 125-149

[BLR]   M.Boyle , D.Lind , and D. Rudolph , *The automorphism group of a shift of finite type,*TAMS **306** (1988) , 71 -114

[CK]   J.Cuntz and W Krieger , *Topological Markov chains and dicyclic dimension groups,* J. Reine Angew. Math. **320** (1980),44-51

[DGS]   M.Denker,C.Grillenberger, and K.Sigmund ,*Ergodic theory on compact spaces* , Springer-Verlag LNM 527 (1976)

[E]   E.Effros ,*Dimensions and C*-algebras* , CBMS, No.46, AMS, Providence , RI , 1981

[F]   ————,*Gyration numbers for involutions of subshifts of finite type*, Math. Forum (to appear)

[Fr]   J.Franks , *Homology and dynamical systems* , CBMS, No.49, AMS, Providence , RI , 1982

[H]   D. Handelman , *Positive matrices and dimension groups*, Jour. Operator Theory **6** (1981) ,55-74

[He]   G. Hedlund , *Endomorphisms and automorphisms of shift dynamical systems* , Math. Systems Theory **3** (1969) , 320-375

[K]   W. Krieger , *On dimension functions and topological Markov chains*, Invent, Math. **56** (1980) , 239-250

[KR1]   K.H.Kim and F.W.Roush , *Some results on decidability of shift equivalence*, J. Comb. Info. Sys. Sci.**4** (1979) , 123-146

[KR2]   ———————————— , *On the structure of inert automorphisms of subshifts* , PU.M.A. Ser.B bf 2 (1991) , 3-22

[KR3]   ———————————— , *Full shifts over R^+ and invariant tetrahedra* , PU.M.A. Ser.B bf 4 (1990) , 251-256

[KR4]   ———————————— , *Williams' conjecture is false for reducible matrices*, JAMS (to appear,Jan. 1992)

[KRW]   K.H.Kim ,F.W.Roush, and J.B.Wagoner , *Automorphisms of the dimension group and gyration numbers* , JAMS (to appear,Jan.1992)

[N]   M.Nasu , *Topological conjugacy for sofic systems and extensions of automorphisms of finite subsystems of topological Markov chains*, Springer-Verlag LNM 1342 (1988), 564-607

[PT]   W.Parry and S. Tuncel , *Classification problems in ergodic theory*, LMS Lecture Notes **67** , Cambridge Univ. Press, 1982

[S]   G.Segal , *Classifying spaces and spectral sequences*,Pub.
           Math. IHES, No.34(1968), 105-112

[W1]   J.B. Wagoner , *Markov partitions and K_2* , Pub. Math.
           IHES, No.65 (1987) , 91-129

[W2]   ——————— , *Triangle identities and symmetries of a sub-
           shift of finite type*, Pac.Jour. Math. **144** (1990),
           181-205

[W3]   ——————— , *Eventual finite order generation for the ker-
           nel of the dimension group representation*,TAMS **317** (1990),
           331-350

[W4]   ——————— , *Higher dimensional shift equivalence is the
           same as strong shift equivalence over the integers* , PAMS
           109 (1990) , 527-536

[Wi1]   R.F.Williams ,*Classification of one dimensional attractors*,
           Proc. Symp. Pure Math. vol. 14 , AMS, Providence,
           RI,1970,314-361

[Wi2]   ——————- ,*Classification of subshifts of finite type*, Ann.
           of Math.(2) **98** (1973),120-153; Errata ibid.**99** (1974),
           380-381

Department of Mathematics,
University of California ,
Berkeley , CA 94720
USA
wagoner@math.berkeley.edu

Contemporary Mathematics
Volume **135**, 1992

Strong shift equivalence of matrices in $GL(2, \mathbb{Z})$

R. F. Williams

Shift equivalence and strong shift equivalence were introduced by the author [W1], [W2] to classify certain limits (either direct or indirect, or, in the case of sub-shifts of finite type, both direct and indirect). One starts with 2 self-maps, $f : X \to X$ and $g : Y \to Y$. Then form the limit spaces (inverse in our illustration):

$$\widehat{X} : \quad X \xleftarrow{f} X \xleftarrow{f} X \xleftarrow{f} \cdots .$$

$$\widehat{Y} : \quad Y \xleftarrow{g} Y \xleftarrow{g} Y \xleftarrow{g} \cdots .$$

and the induced maps $\hat{f} : \widehat{X} \to \widehat{X}$, $\hat{g} : \widehat{Y} \to \widehat{Y}$ defined, e.g., by $f(\hat{x}) = \hat{f}((x_1, x_2, \ldots)) = (f(x_1)f(x_2) \ldots .)$. ($\hat{f}$ is at times called the natural extension of f.)

Problem. Under what conditions upon the original maps $f : X \to X$, $g : Y \to Y$, will there be a topological conjugacy $h : \widehat{X} \to \widehat{Y}$? I.e., a homeomorphism h such that the following diagram commutes?

$$
\begin{array}{ccc}
\widehat{X} & \xrightarrow{\hat{f}} & \widehat{X} \\
h \downarrow & & \downarrow h \\
\widehat{Y} & \xrightarrow{\hat{g}} & \widehat{Y}
\end{array}
$$

The answer, valid in fairly great generality: if and only if [W1,

Supported in part by National Science Foundation Grant MCS 8002177. This note was written as a preprint in the early 80's and only a few lines have been updated (final version).

Subject Classification: 58F20, 47A35, 54H20

1991 Mathematics Subject Classification. Primary 58F20, 47A35, 54H20.

Theorem 3.3, p.348] there exist maps $r : X \to Y$, $s : Y \to X$ and an integer in such that

$$
\begin{array}{ccc}
X & \xrightarrow{\ f\ } & X \\
r\downarrow & & \downarrow r \\
Y & \xrightarrow[\ g\ ]{} & Y
\end{array}
\qquad\qquad
\begin{array}{ccc}
X & \xrightarrow{\ f\ } & X \\
s\uparrow & & \uparrow s \\
Y & \xrightarrow[\ g\ ]{} & Y
\end{array}
$$

commute where $s \circ r = f^m$ and $r \circ s = g^m$. Under these conditions, say f is (*weak*) *shift equivalent to* g, $f \sim_s g$, and m is called the *lag* of the shift equivalence. Now consider the big "triangle" in which f occurs m times.

$$
X \xrightarrow{\ f\ } X \xrightarrow{\ f\ } \cdots \xrightarrow{\ f\ } X
$$

$$
r \searrow \qquad\qquad \nearrow s
$$

$$
Y
$$

In most categories (e.g., groups and endomorphisms, compact spaces and maps, branched manifolds and immersions) one can, in case $f \sim_s g$, subdivide this into little triangles. We illustrate for the case $m = 3$.

$$
\begin{array}{ccccccc}
X & \xrightarrow{f} & X & \xrightarrow{f} & X & \xrightarrow{f} & X \\
& r_1 \searrow \nearrow s_1 & & r_1 \searrow \nearrow s_1 & & r_1 \searrow \nearrow s_1 & \\
& W & \xrightarrow[f_1]{} & W & \xrightarrow[f_1]{} & W & \\
& & r_2 \searrow \nearrow s_2 & & r_2 \searrow \nearrow s_2 & & \\
& & Z & \xrightarrow[f_2]{} & Z & & \\
& & & r_3 \searrow \nearrow s_3 & & & \\
& & & Y & & &
\end{array}
$$

Here the spaces and maps are quotient spaces and induced maps; thus in the same category. Thus one has an "elementary equivalence" from f_1 to f_2, i.e., a shift equivalence of lag 1. Thus the terminology f is *strong shift equivalent* to g. $f \approx_s g$ was introduced in [W2] to indicate that there is a finite string of elementary equivalences from f to g: $f = f_1 = r_1 s_1, s_1 r_1 = f_2 = r_2 s_2, s_2, r_2 = f_3, \ldots, f_m = g$. (A problem here is that the dimensions of the domains of $f_2, f_3, \ldots$ might be bigger than those of f and g, making this relation highly non computable.) The proof (valid in many categories, but *not* for matrices over $\mathbf{Z}^+$) that the weak version implies the stronger is so elementary that strong shift equivalence was not even defined in the first paper [W1]. The proof given there sufficed for our purposes at the time with no logical gap. But to show specifically that these two equivalence relations are the same takes an additional step as several authors have noticed, e.g., Effros [E].

Remark. *Given $f : X \to X$, $g : Y \to Y$ which are shift equivalent of lag m in any of the usual categories, there is a strong shift equivalence of at most $2m$ steps from f to g in this category.*

Proof. In [W1, lemma 4.6, p.350] making use only of quotient spaces and maps induced on quotient spaces, we found a sequence $f = f_0, f_1, \ldots, f_m = g \,|\, \mathrm{im}(r)$ of m steps constituting a strong shift equivalence from f to $g \,|\,$ (image of r). Here $r : X \to Y$, $s : Y \to X$ is a (weak) shift equivalence of lag m from f to g.

Now let $Y_i = \mathrm{im}(r) + g^i(Y)$ and $g_i = g \,|\, Y_i$, in which the $+$ is interpreted simply as $\cup$ in case we are in a purely topological category. Then, following Effros, [E, Section 2], the elementary equivalences

$$
\begin{array}{ccc}
Y_i & \xrightarrow{\;g_i\;} & Y_i \\
{\scriptstyle g_i}\downarrow & \nearrow & \downarrow \\
Y_{i+1} & \xrightarrow[\;g_{i+1}\;]{} & Y_{i+1}
\end{array}
$$

show that $g = g_0 \approx g_m = g \,|\, \mathrm{im}(r)$; here the diagonal map is the inclusion.

The principal problem remains: does weak shift equivalence imply strong shift equivalence in the category of *primitive* (added 1992) non-negative integral matrices? There have been several papers by Krieger [K] and others in which they look more closely at the limits (mostly direct) of the sort

$$G \xrightarrow{A} G \xrightarrow{A} G \xrightarrow{A} \cdots$$

in which G is free abelian and A is a matrix over $\mathbf{Z}^+$, or the non negative rationals, Q^+. Though this machinery has not resolved the principal problem, there are many important results. In particular, Cunz and Krieger [CK] proved the proposition of the current note, except for the equivalence of part (4). See Boyle [Bo] for a survey on subshifts and a large bibliography.

Let me close this introduction by mentioning a very interesting new result:

Theorem (Baker [Ba]). *For A, B 2×2, positive with positive determinant the assumption $A = R^{-1}BR$ where $R \in SL(2, \mathbf{Z})$ implies A is strong shift equivalent to B. Dimension 3 is required but nothing higher.*

The matrix $\binom{1\,2}{3\,1}$ is strong shift equivalent to $\binom{1\,6}{1\,1}$ as shown by Baker who proved this and found that dimension 3 does *not* suffice, using computers. (8 hours on an Apple; dimension 4 seems too big for main frame machines!) (This last is a very dated remark.) Kim and Roush have shown that weak shift equivalence is "decidable."

Proposition. *Given A, B 2×2 over $\mathbf{Z}^+$ with $\det A = \pm 1$. Then the following are equivalent:*

1) *$A = RBR^{-1}$, for some $R \in GL(2, \mathbf{Z})$*

2) *$A \sim_s B$*

3) *$A \approx_s B$*

4) *there exists non-negative integral matrices R, S such that $A = RS$ and $B = SR$.*

Proof. That $4 \Rightarrow 3$ is a tautology. $3 \Rightarrow 2$ and $2 \Rightarrow 1$ are easy and shown in previous papers [PW]. In addition, we may assume some power of A (and B) has all positive terms, as the other case reduces to $A = \begin{pmatrix} 1 & x \\ 0 & 1 \end{pmatrix}$ which is easy, even trivial. So assume 1) is true and let $\mathbf{v}A = \lambda\mathbf{v}$ be the Frobenius-Perron eigenvector and eigenvalue. In particular $\lambda > 0$ and $\mathbf{v}$ has positive entries. Now $\mathbf{v}AR = \mathbf{v}RB = \lambda\mathbf{v}R$ so that $\mathbf{v}R$ is a Perron-Frobenius eigenvector of B and hence is either a positive vector or a negative one. We can assume it is positive by replacing R with $-R$ if necessary. So that R satisfies the equation $AX = XB$. It follows that A^iR also satisfies this equation for $i \in \mathbf{Z}$. For sufficiently big i, A^iR is positive, which is definitely not true for very negative i. Then choose the smallest n such that $A^{n+1}R \geq 0$.

Case I. Both basis vectors are taken out of the first quadrant by A^nR. As the cone of these vectors contains $\mathbf{v}$ it follows that $(A^nR)^{-1}$ takes both basis vectors inside the first quadrant, and hence is positive. Then

$$A = (A^{n+1}R)(A^nR)^{-1} \quad \text{and} \quad (A^nR)^{-1}(A^{n+1}R) = B$$

which proves 4).

Case II. A^nR knocks only one basis vector outside the first quadrant. Then A^nR has one and only one negative entry since $A^{n+1}R > 0$. We may assume it is in the first row. Let $\bar{R} = A^nR$.

Case IIa. $\bar{R} = \begin{pmatrix} -a & b \\ c & d \end{pmatrix}$, $a, b, c, d \geq 0$. Then $\det \bar{R} = -ad - bc = \pm 1$.

Case IIa1. $ad = 1$, $bc = 0$. Cannot have $\bar{R} = \begin{pmatrix} -1 & b \\ 0 & 1 \end{pmatrix}$ as then

$A\bar{R}$ is not ≥ 0. Thus $\bar{R} = \begin{pmatrix} -1 & 0 \\ c & 1 \end{pmatrix}$ and $A^n RB = \bar{R}B$ has a negative term. But this is $A^{n+1}R$ which is positive.

Case IIa2. $ad = 0$, $bc = 1$. Then $\bar{R} = \begin{pmatrix} -a & 1 \\ 1 & 0 \end{pmatrix}$ and $\bar{R}^{-1} = \begin{pmatrix} 0 & 1 \\ 1 & a \end{pmatrix} \geq 0$. Then $(A^{n+1}R), (A^n R)^{-1} = \bar{R}^{-1}$ is the pair required for 4).

Case IIb1. $ad = 1$, $bc = 0$, so that $\bar{R} = \begin{pmatrix} 1 & -b \\ 0 & 1 \end{pmatrix}$ and $\bar{R}^{-1} = \begin{pmatrix} 1 & b \\ 0 & 1 \end{pmatrix}$. Then $(A^{n+1}R), \bar{R}^{-1}$ is the good pair. This leaves only

Case IIb2. $ad = 0$, $bc = 1$. Then either $a = 0$ or $d = 0$ and these cases are seen impossible as follows: If $\bar{R} = \begin{pmatrix} a & -1 \\ 1 & 0 \end{pmatrix}$, $\bar{R}^{-1} = \begin{pmatrix} 0 & 1 \\ -1 & a \end{pmatrix}$ so that $(A^{n+1}R)\bar{R}^{-1}$ has a negative term which is impossible. If $\bar{R} = \begin{pmatrix} 0 & -1 \\ 1 & d \end{pmatrix}$ then $A^{n+1}R = A^n RB = \bar{R}B$ has a negative term, contradicting our assumptions again. This completes the proof of the proposition.

Bibliography

[Ba] Baker, K.A., *Strong shift equivalence of 2 × 2 matrices of non-negative integers*, Ergodic Theory and Dynamical Systems **3** (1983) 541–558.

[Bo] Boyle, M., *Symbolic dynamics and matrices*, to appear in the proceedings of a conference held at The Institute for Mathematics and its Applications held in November, 1991.

[CK] Cunz, J. and Krieger, W., *Topological Markov chains with dicyclic dimension groups*, J.f.d. reine und ang. Math. **320** (1980), 44–51.

[E] Effros, E.G., *On Williams' problem for positive matrices*, preprint, UCLA, to appear in these proceedings (?).

[K] Krieger, W., *On dimension functions and topological Markov chains*, Inventiones Math. **56** (1980), 239–250.

[KR] Kim, K.H. and Roush, F.W., *Some results on the decidability of shift equivalence*, J. Combinatorics 4 (1979), 123–146.

[PW] Parry, W. and Williams, R., *Block coding and a zeta function for finite Markov chains*, Proc. London Math. Soc. **25** (1977), 483–495.

[W1] Williams, R.F., *Classifications of 1-dimensional attractors*, Proc. Sym. Pure Math. vol. 14, Amer. Math. Soc., Providence, RI 1970, 341–361.

[W2] Williams, R.F., *Classification of sub shifts of finite type*, Annals of Math. **98** (1973), 120–153 and Errata, Annals of Math. **99** (1974), 380–381.

Robert F. Williams
Department of Mathematics
The University of Texas at Austin
Austin, TX 78712-1082

Recent Titles in This Series

(Continued from the front of this publication)

(See the AMS catalog for earlier titles)